Host Guest Complex Chemistry Macrocycles

Synthesis, Structures, Applications

Editors: F. Vögtle, E. Weber

With 174 Figures and 46 Tables

Springer-Verlag
Berlin Heidelberg New York Tokyo
1985

Professor Dr. Fritz Vögtle
Dr. Edwin Weber

Institut für Organische Chemie und Biochemie
Universität Bonn
Gerhard-Domagk-Straße 1
5300 Bonn 1

ISBN-13: 978-3-540-13950-8   e-ISBN-13: 978-3-642-70108-5
DOI: 10.1007/978-3-642-70108-5

Library of Congress Cataloging in Publication Data. Main entry under title:
Host guest complex chemistry/macrocycles.
Bibliography: p. Includes index.
1. Electron donor-acceptor complexes--Adresses, essays, lectures. I. Vögtle, F. (Fritz), 1939 –
II. Weber, E.
QD474.H67   1985   541.2'8   84-23569

2152/3020-543210

Table of Contens

Crown-Type Compounds — An Introductory Overview

Dr. Edwin Weber, Prof. Dr. Fritz Vögtle

Institut für Organische Chemie und Biochemie der Universität Bonn, Gerhard-Domagk-Straße 1, D-5300 Bonn 1, FRG

Table of Contents

1 Introduction

The scientific and practical interest in coronands (crown ethers), cryptands, podands as complexing agents for cations as well as for anions and neutral low molecular species is undeniable [1, 2]. The chemistry of crown compounds is steadily increasing. About 250 original papers dealing with crown chemistry appeared only in 1980. New molecules with crown ether properties are constantly synthesized and new applications discovered.

Owing to lack of space, only a small number of the original publications is mentioned here. Thus, in the literature compilation only some, but relevant works are selected for each chapter. Whenever possible, reference is made to reviews or review-like articles alone by means of which original works can be consulted. The reviews given under ref. [1] are considered to be the most relevant. The formulae presented in the figures should be understood as representative structures outlining a specific field.

2 Classification of Oligo-/Multidentate Neutral Ligands and of their Complexes

Today, a distinction is made between the classical ring oligoethers *(crown ethers)* and monocyclic *coronands*, oligocyclic spherical *cryptands* and the acyclic *podands* with respect to topological aspects [3]. This classification and the topology are illustrated in Fig. 1, each figure representing the minimum number of donor atoms and chain segments characteristic of each class of compounds. Multidentate monocyclic ligands with any type of donor atoms are called coronands ("crown compounds"), while the term crown ether should be reserved for cyclic oligoethers exclusively containing oxygen as donor atom. Moreover, a subdivision of each of the three respectively four classes of ligands according to the number of arms or bridges is possible.

Because of their ability to take up ions and to transfer them across a lipophilic medium, these types of ligands also are often called *ionophores*, comparable to the structurally related polyether antibiotics, the ionophoric behaviour of which had been discovered first [4].

In order to differentiate the crown ether ligands from their metal ion complexes, the terms *coronand* and *coronate* were suggested for the uncomplexed and complexed species, respectively [3]. Analogously used are the terms *cryptand/cryptate* and *podand/podate* (cf. Sects. 3.3. and 3.4.).

3 Crown-Type Ligands

3.1 Historical Crown Ethers, Nomenclature

The first synthetic ionophores described by Pedersen in 1967 [5], were the cyclic hexaethers *1* and *2*, which have been simply called [18]crown-6 and dibenzo[18]-crown-6 [2f], in contrast to their cumbersome and less illustrative IUPAC nomenclature [6] (Fig. 2).

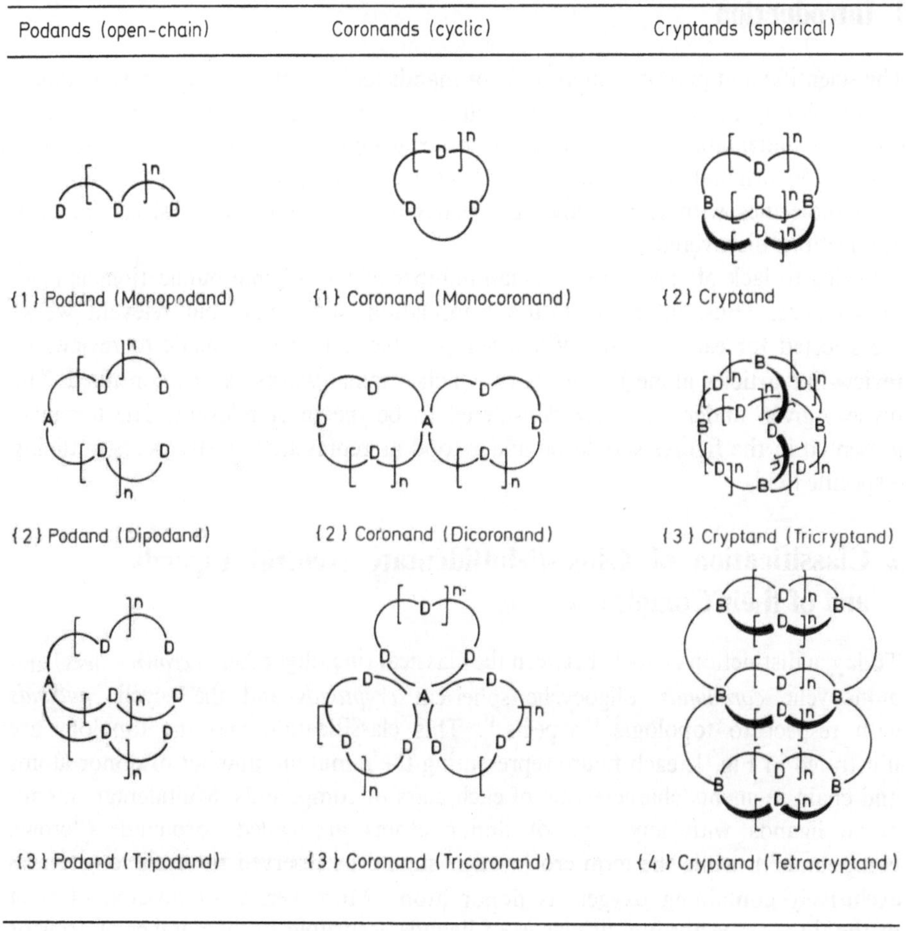

Podands (open-chain)	Coronands (cyclic)	Cryptands (spherical)

{1} Podand (Monopodand) | {1} Coronand (Monocoronand) | {2} Cryptand

{2} Podand (Dipodand) | {2} Coronand (Dicoronand) | {3} Cryptand (Tricryptand)

{3} Podand (Tripodand) | {3} Coronand (Tricoronand) | {4} Cryptand (Tetracryptand)

Fig. 1. Topology and classification of organic neutral ligands [3] (D = donor atom, A = anchoring group, ∩ = chain segment without donor atom, B = bridgehead atom)

"*Dibenzo*" stands for both of the benzene nuclei annexed to the ring while "[18]" in square brackets means the number of ring atoms. The class specification "*crown*" is followed by the number of heteroatoms in the ring; in this case "6".

The above notation is now commonly accepted and generally serves to give a rough characterisation of medium- to many-membered cyclic polyethers in which the oxygen atoms are mostly connected via ethano bridges and in which annexed benzene and cyclohexane rings may also be present [7] (Fig. 2).

As Pedersen's crown notation is not unequivocally defined with regard to the location of the donor atoms, the benzene nuclei, the cyclohexane units or other ring components, even in simple cases, a more systematic and widely applicable nomenclature covering all kinds of cyclic and noncyclic ligands and also their complexes has been proposed recently (cf. Figs. 1 and 2) [3]. In principle, the same symbolisms are retained but in addition it is specified as follows: The number preceding the angular brackets "⟨ ⟩" indicates the ring size. In the presence of aromatic and heteroaromatic units in the ring, the shortest way to the next donor atom is considered. The angular brackets contain in the order given: 1) donor

Structure:	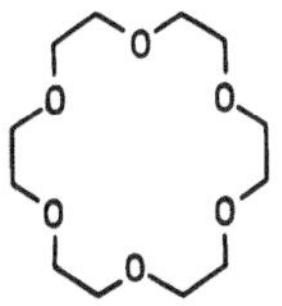	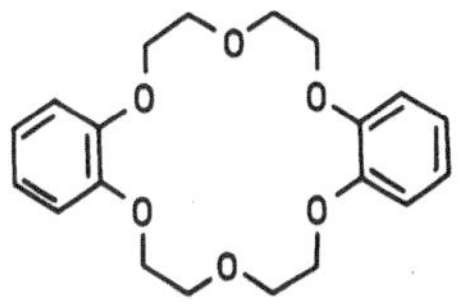	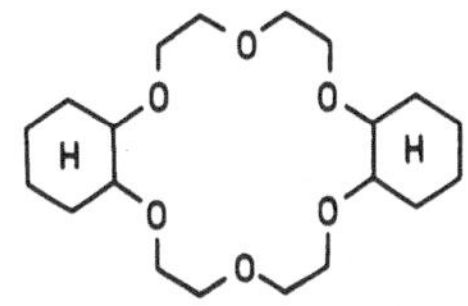
Nr.	*1*	*2*	*3*
IUPAC-designation	1,4,7,10,13,16-hexaoxacyclooctadecane	2,5,8,15,18,21-hexaoxatricyclo[20.4.0.0^{9,14}] hexacosa-1(22),8,11,13,23,24-hexaene[a]	2,5,8,15,18,21-hexaoxatricyclo [20.4.0.0^{9,14}]-hexacosane
Short name (Pedersen's crown nomenclature)	[18]crown-6	dibenzo[18]crown-6	dicyclohexano[18]crown-6
Notation	[18]C-6	DB[18]C-6	DCH[18]C-6
New nomenclature system	18⟨O$_6$coronand-6⟩	18⟨O$_6$(1,2)benzeno.2$_2$.-(1,2)benzeno.2$_2$coronand-6⟩	18⟨O$_6$(1,2)cyclohexano.2$_2$.(1,2) cyclohexano.2$_2$coronand-6⟩

[a] Phane nomenclature: 1,4,7,14,17,20-hexaoxa[7.7] (1,2)benzenophane

Fig. 2. Nomenclature of crown ethers

heteroatoms expressed by elemental symbols; 2) bridges, i.e. C—C chains between the donor atoms, denoted by numbers which correspond to the bridging C-atoms, bridge units like aromatic nuclei or more complex groups (position marked in round brackets). The designation "2" for ethano, the most common bridge, is omitted if only this kind of bridge unit is present or if such a procedure does not curtail the clarity of the structure (cf. 18⟨O$_6$-coronand-6⟩); 3) the class name (e.g. coronand), and 4) the total number of donor heteroatoms.

In the case of mixed O/N/S macrocycles where ether oxygens are successively replaced by other heteroatoms (combinations s.b.), the sequence of donor sites in the ligand skeleton is given by heteroelemental symbols arranged in the order of priority laid down by the IUPAC rules. Heterocycles with donor sites (s.b.) are treated as single atoms. The sequence for the chain segments without donor atoms correspond to that of the heteroatoms, beginning with the donor atom of highest priority (cf. structures *2* and *3* in Fig. 2).

Substituents and functional groups in the basic skeleton (class name) are denoted by prefixes and suffixes. The numbering is principally carried out according to the IUPAC rules (this is also valid for the cryptands but does not strictly apply to podands; s.b.).

Altogether, the new proposal of nomenclature extended in Sections 3.3. and 3.4. allows an essentially easier recognition of the ligand type and other characteristics such as topology and donor centers.

3.2 Variation Possibilities with Coronands (Fig. 3–11)

Since the discovery of crown ethers [8] efforts have not diminished to synthesize crown species with other *distributions*, *numbers* and *types* of donor heteroatoms like nitrogen or sulfur [1, 2]:

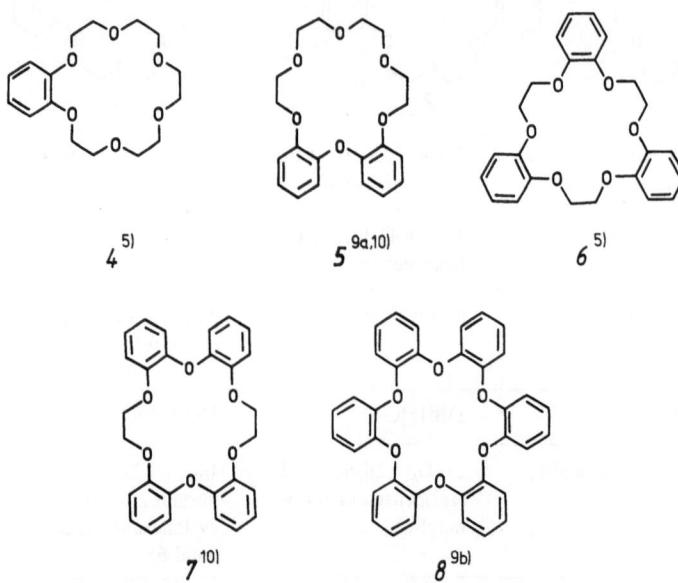

Fig. 3. Stiffening of the [18]crown-6 skeleton by benzo condensation

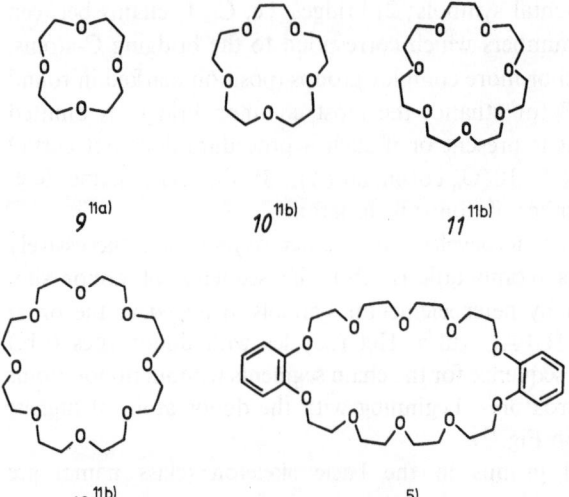

9 [11a)] 10 [11b)] 11 [11b)]

12 [11b)] 13 [5)]

Fig. 4. Crown ethers of different ring size

a) *Ring-stiffened aryl ether ligands* (Fig. 3) [5,9–10)]: Basicity and donor ability of the oxygen atoms are reduced (*4–8*).

b) *Varied ring size* (Fig. 4) [5,11)]: Every number of ring members and oxygen donors is possible (*9–13*).

c) *Geometric arrangement of the donor atoms in the ring* (Fig. 5) [5,12)]: The oxygen atoms can be separated (*14, 15*) or brought together (*16–18*).

d) *Sulfur as alternative donor site* (Fig. 6) [13,14)]: All O,S-sequence combinations (*19–22*) and even pure thiacrowns like *23* are possible.

e) *Nitrogen as donor site* (Fig. 7) [15)]: Incorporation of any number of N-atoms into the ring and at any position is possible (*24–28*).

6

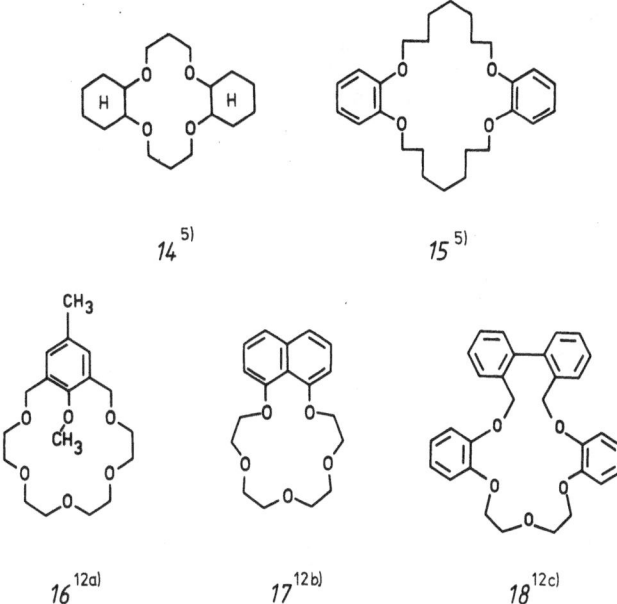

Fig. 5. Different arrangement of the oxygen atoms in the crown ether ring

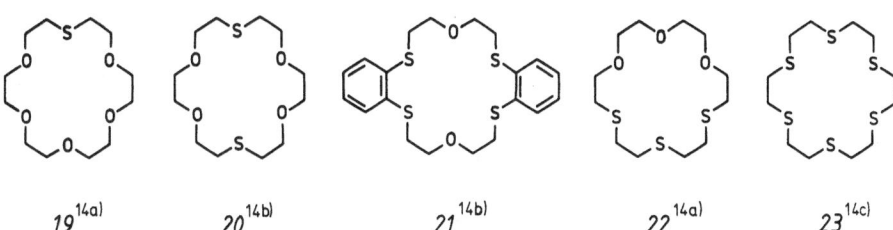

Fig. 6. Coronand sulfides

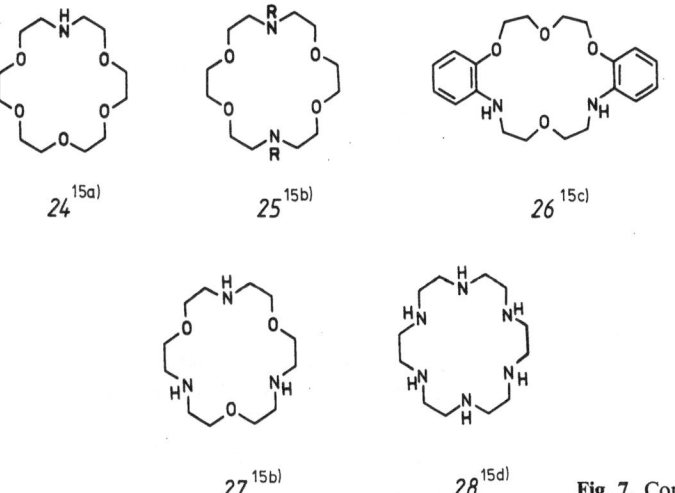

Fig. 7. Coronand amines

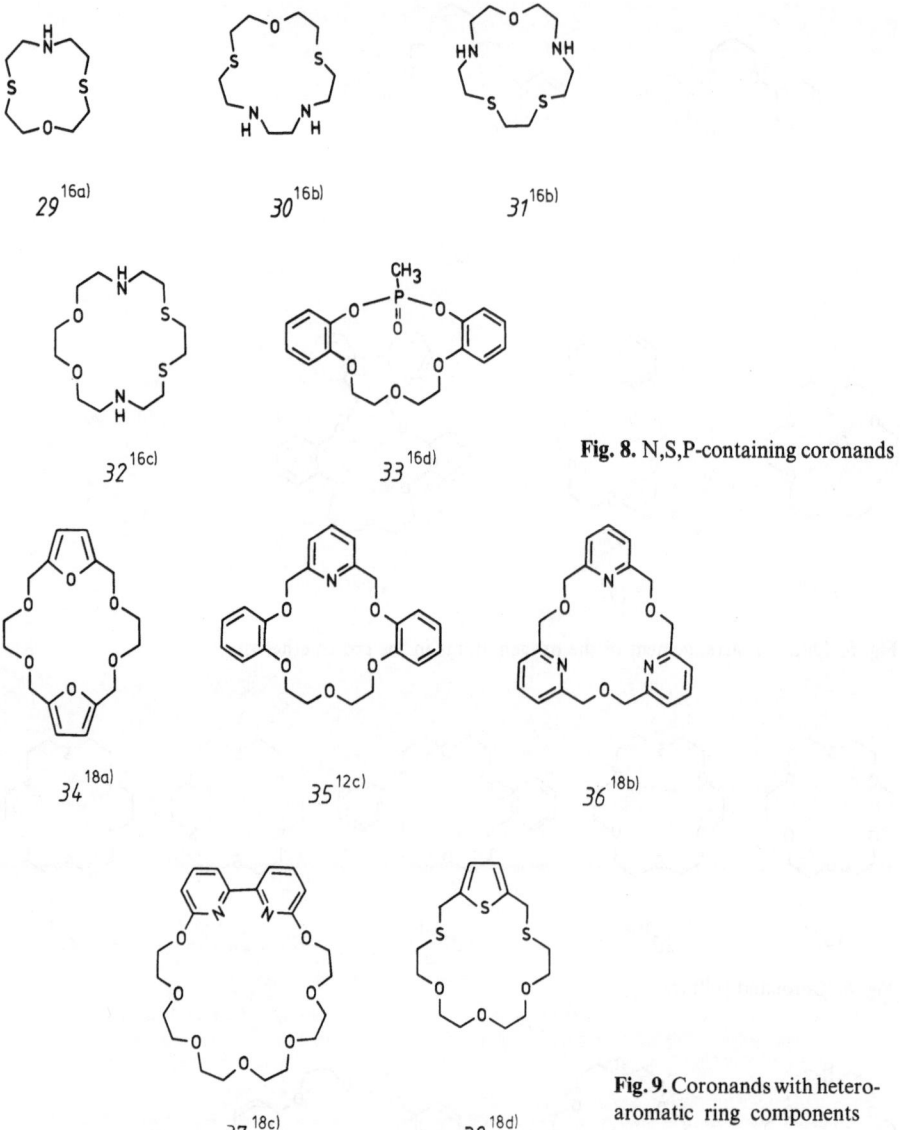

Fig. 8. N,S,P-containing coronands

Fig. 9. Coronands with hetero-aromatic ring components

f) *Mixed O,N,S,P-coronands 29–33* (Fig. 8) [16)]: Ligands with special properties are obtained when nitrogen as well as sulfur are introduced at different sites into the parent substance (*29–32*). Phosphor atoms have also been incorporated into the cyclic framework (*33*).

g) *Heteroaromatic coronands 34–38* (Fig. 9) [17, 18)]: A new crown concept, stimulated by cyclophane chemistry, led to the incorporation of heteroaromatic nuclei into the crown ether ring [18d)] whereby the donor atom of the heterocycle (furane, pyridine, thiophene) is located in the ethano position which yields the optimum donor stereochemistry.

h) *Donor sites incorporated into functional groups* (Fig. 10) [12b, 19, 20)]: This concept combines ligand stiffening (e.g. amide bonding) with highly polarizable and

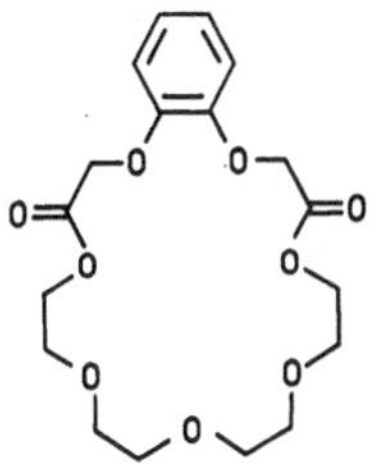

Fig. 10. Coronands with functional groups as building blocks

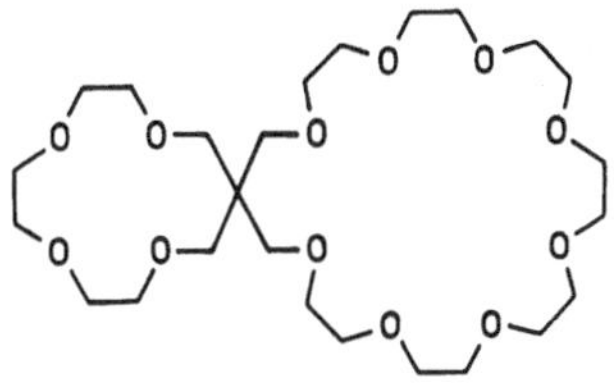

39 [20a)] 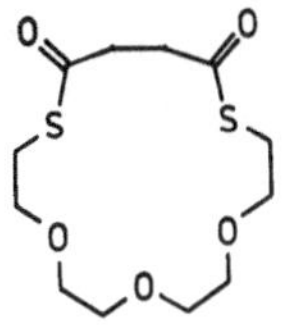 40 [20b)] 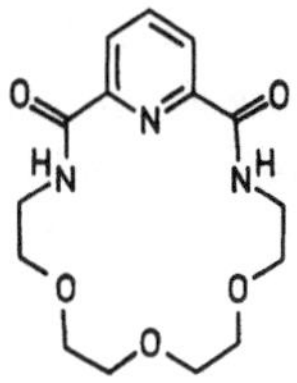 41 [12b)] 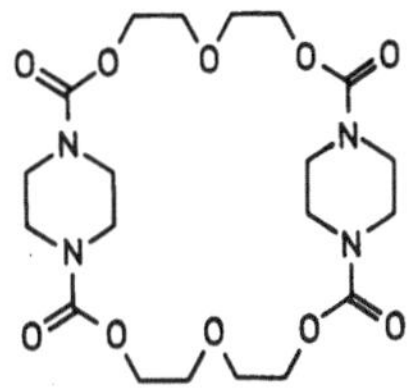 42 [20c] 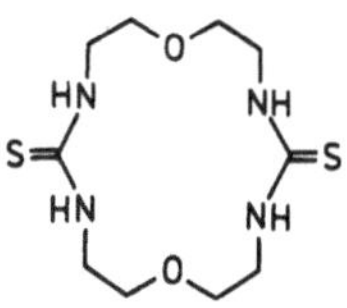 43 [20d)]

44 [21a)] 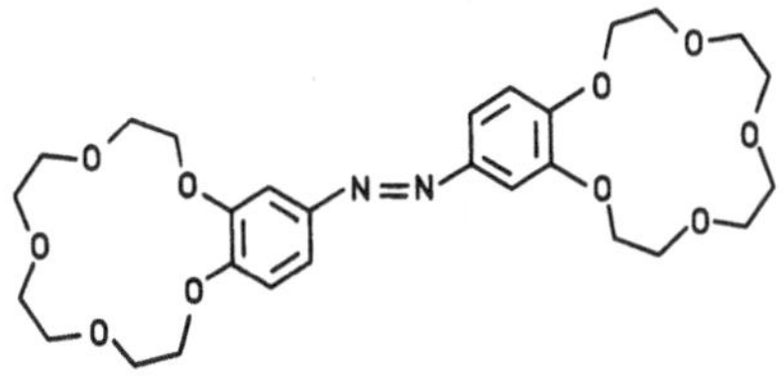 46 [21b)] 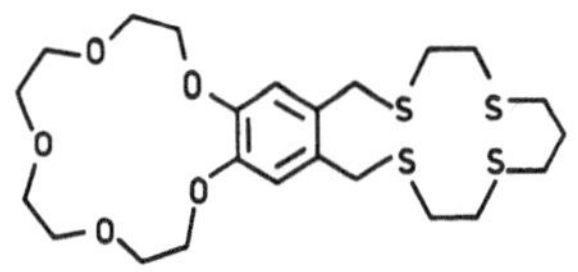 45 [10)] 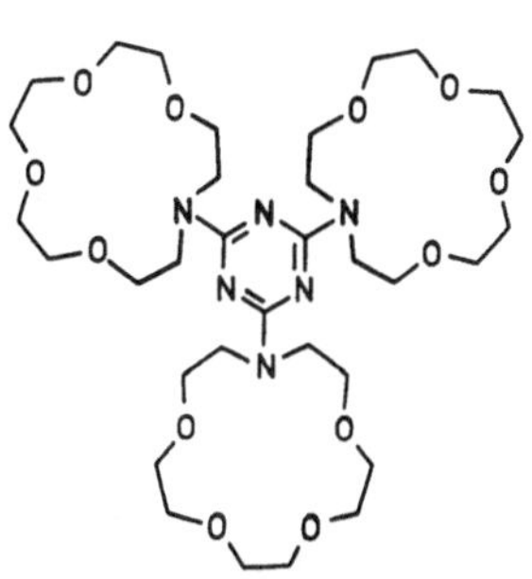 47 [10)] 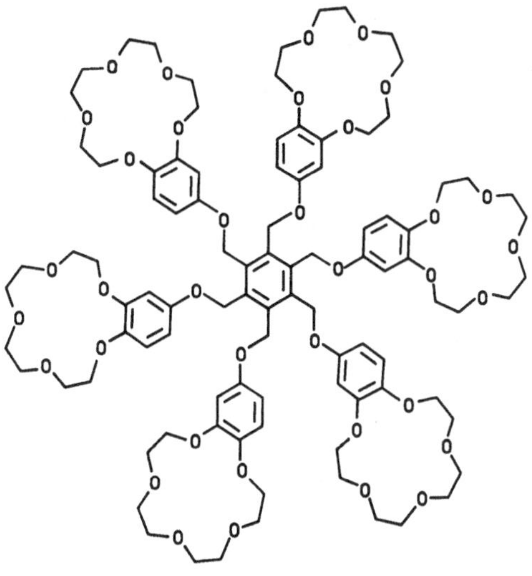 48 [10)]

Fig. 11. Multisite crown compounds

9

selective coordination areas provided by ester- (39), thioester- (40), amide- (41), urethane- (42) and thiourea units (43).

i) *Multisite crown compounds ({n}coronands)* (Fig. 11) [10, 21]: Assemblies of linearly (44, 45), angularly (46), radially (47) or spherically (48) bound crown ether and coronand rings ("multiloops") which can act cooperatively. Combinations of crown compounds differing in size (44) and/or donor characteristics (45) are possible.

For chiral crown-/coronand skeletons and such substituted by lipophilic alkyl chains (phase-transfer catalysts) or functional groups (lateral discrimination) see Sections 5.7, 5.8, 6.2, as well as the contribution from Montanari et al.

3.3 Cryptands (Fig. 12)

Bridging of a classical monocyclic crown ether with an additional oligoether chain led to a novel *bicyclic type of ligand* [22]. Lehn, their discoverer, called them "cryptands" (derived from the Greek word cryptos = cave) in order to express their special topological shape [23]. The corresponding complexes are termed cryptates [2b, 2d].

Conventional cryptands (49) [24a, 24b] possess two bridgehead nitrogen atoms as characteristic units which are joined by three oligooxa chains of different length and number of donor atoms (Fig. 12). However, some cryptands lacking nitrogen, e.g. 53 have been described recently [24e]. Also, sulfur-analogous cryptands like 51 [24c] as well as those with additional N-atoms (52) [24d] have been synthesized and are, in part, commercially available.

A further variation is revealed by replacement of donor sites by heteroaromatic rings (54 [24f] and 55) [24g]. Additional bridging allows the approach of sophisticated tricyclic ligands (56, 57) [24h, 24i]; the latter is generally known under the nick-name "soccer ball molecule" on account of its spherical three-dimensional skeleton. However, tetracyclic and higher ring systems have been infrequently reported because of their multi-step synthesis [24h].

A simple notation is used for bicyclic, nitrogen-bridgehead containing precurcors and — incorrectly — also for the monocyclic precursors [2d]. It is derived from the number of *bridges* and the *oxygen donor atoms* in each bridge. Thus, the term "[1.1.1]cryptand" for 49a follows from the three bridges equipped with one oxygen, respectively. The [2.2.2]cryptand (49d), the most common member of the cryptand family, accordingly possesses three bridges containing two oxygens each. Hetero-cryptands 51 and 52 designated as [2.2$_{SS}$.2$_{SS}$] and [3$_{NN}$.3$_{NN}$.3$_{NN}$] barely comply with this notation because here it is ambiguous with regard to the position of the nitrogen atoms in the case of 52.

The above described new notation/nomenclature system [3], which has successfully been used in the coronand series (cf. Sect. 3.1.), can also be employed for any cryptand applying some extended rules: In angular brackets is named first the *bridgehead* atom of highest priority and in square brackets follows the sequence of the individual *bridges*, starting from the bridgehead of highest priority. The bridge containing the highest number of donor atoms enjoys highest priority. With equal numbers of donor atoms in the bridges, the priority is determined by that of the heteroatoms and secondly by that of substitution. Thereupon and

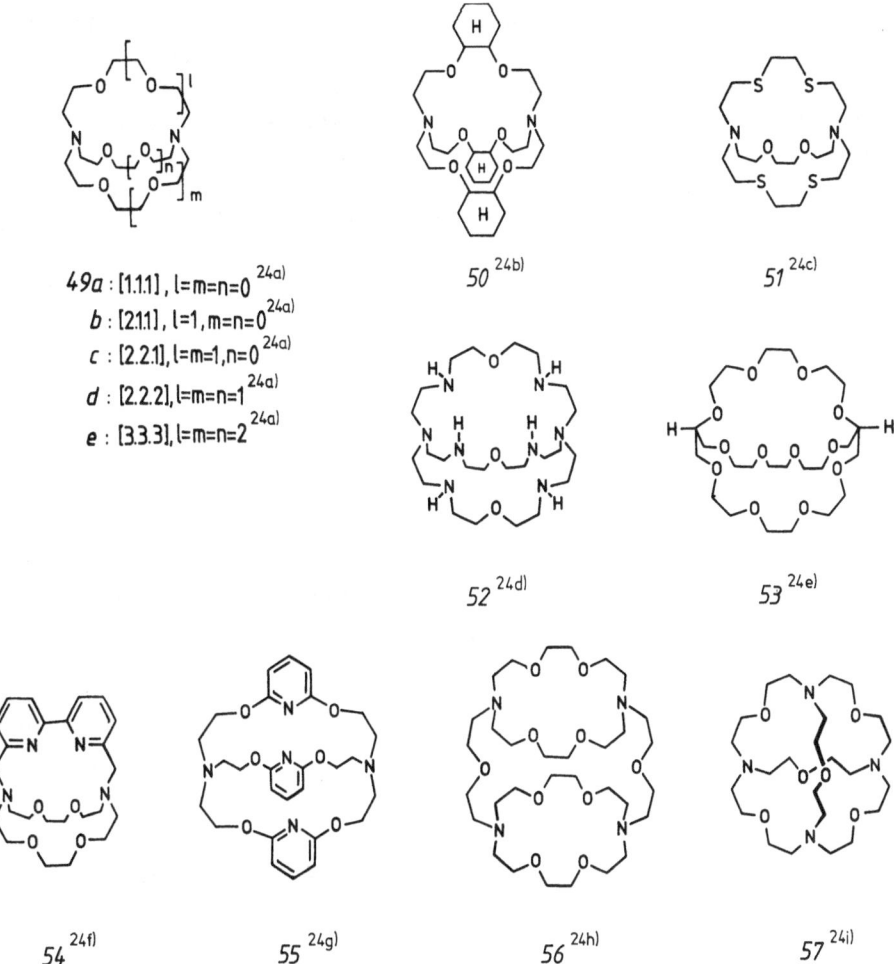

Fig. 12. Cryptands of spherical (*49, 50, 57*), ellipsoidal (*52, 53*) and cylindrical dimension (*56*), and with different donor characteristics (*51, 54, 55*)

in the given order are specified: the elemental symbol of the second bridgehead atom, the class name *cryptand* and finally the total *number of donor atoms* assembled in the ligand skeleton. Additional bridges in tri- to oligocyclic systems are expressed before the class name by separate square brackets with superscripts indicating their location. It can be helpful in the latter cases to make use of the topological information by the braces "{ }". Application of the rules represented above results in the notations $\langle N[O_2]_3N$-cryptand-8$\rangle$ for *49d* and $\{3\}\langle N[ONO]_2 [O] N\langle[O]^{7,19}$-cryptand-10$\rangle$ for *57*.

3.4 Podands (Open-Chain Crown Compounds and Cryptands, Octopus Molecules; Figs. 13–15)

The trend nowadays — supported by the interest of industry — is towards producing simple and cheap *open-chain crown-type compounds* [25]. While the solvating and complexing effect of oligoethylene glycol dimethyl ethers — the so-called *glymes* [e.g. pentaglyme or glyme-6 (58)] — on alkali/alkaline earth metal ions was already known and exploited much earlier [26], definite crystalline complexes of these simple crown-analogues had not been obtained [5,7].

The first "*open-chain crown compound*" with glyme-like structure, which — in analogy to their cyclic counterparts · easily allowed the isolation of stoichiometric complexes with alkali and alkaline earth metal ions, was the bis(quinoline)oligoether 59 (n = 2) [27a] (commercially available as "Kryptofix-5"). As a characteristic feature it contains two quinoline residues bound at both ends of a glyme polyether moiety. This construction principle, namely stiffening of a polyether chain by more or less rigid donor terminal groups, has proved very useful ("*end-group concept*") [27b]. Besides 8-quinolinole, 2-methoxyphenol, 2-nitrophenol and salicyclic ester are particularly suitable as donor end groups [28] (60). The ether moiety along the backbone can also be stiffened by annexation of aromatic or heteroaromatic units, as in cyclic crown ethers (61) [27c]. A new type of podand built up exclusively by rigid cyclooxalkane chain segments was found very recently [27d]. Another group of modified oligoethylene glycols possesses (besides terminal alkyl chains) a diamide structure as a main characteristic (62) [29].

After the complexing abilities of open-chain crown compounds had been discovered, the donor end-group concept was combined with the cryptand idea, resulting in three- and four-armed "*open-chain cryptands*" [25,28] (cf. Fig. 14).

Apart from 8-quinolinole (63) [30a] and 2-methoxyphenol (64) [30b], the tropolone unit (e.g. in 66) [30c] has proved to be an efficient donor end group. A stiffened and

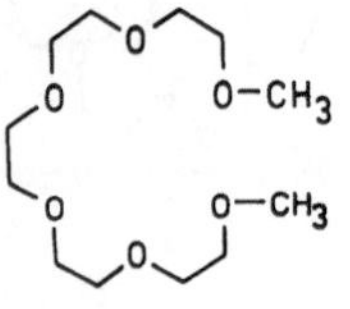

Pentaglyme
(Glyme-6)

58 [26]

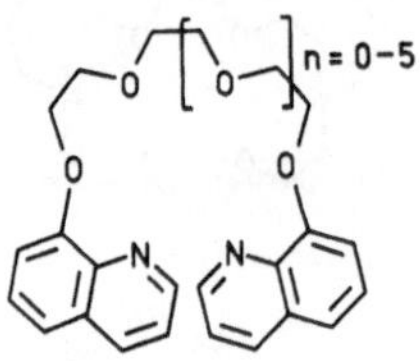

n=2: Kryptofix-5 [27a]

59 [27b]

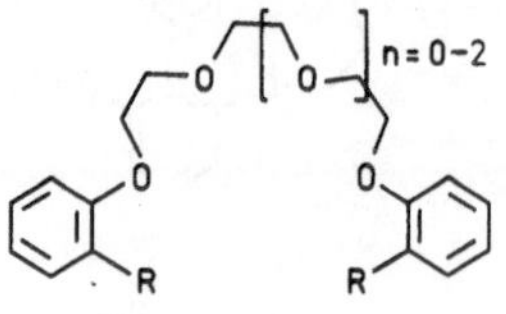

R=H, OCH$_3$, NO$_2$,
COOH, COOEt,
CONHR, NHCOR

60 [27b,28]

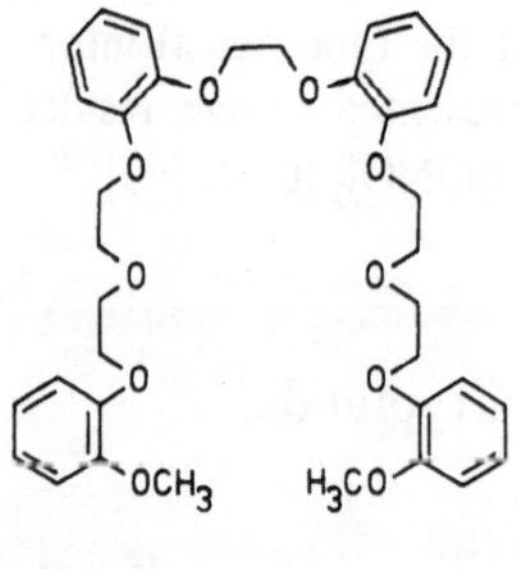

61 [27c]

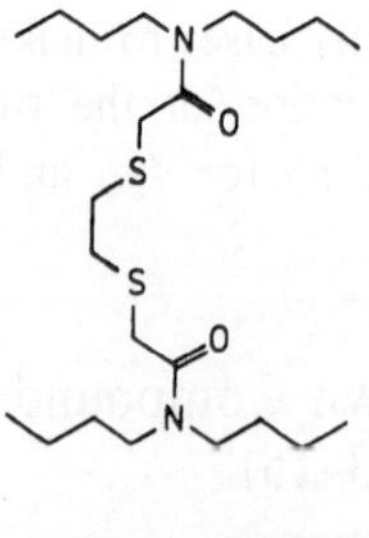

62 [29b]

Fig. 13. Podands (noncyclic crown-type compounds)

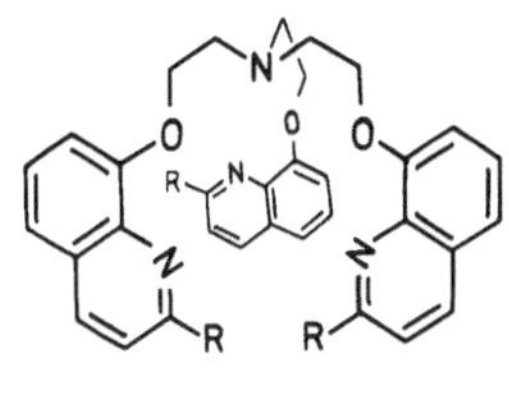

R = H, CH3

63 [30a)]

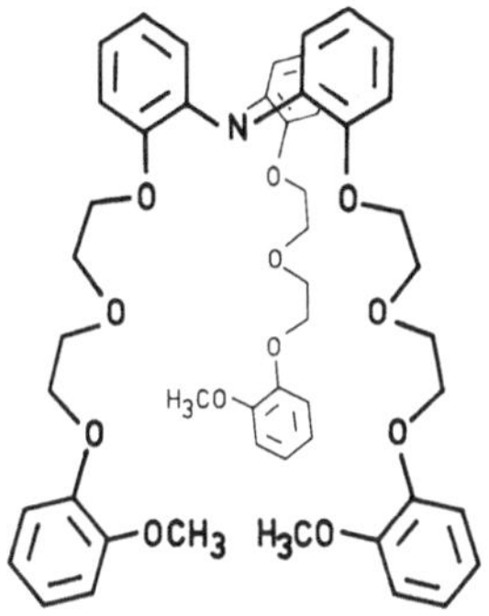

65 [30c)]

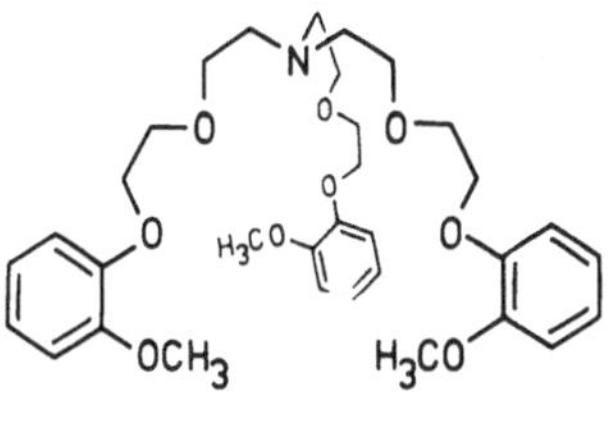

64 [30b)]

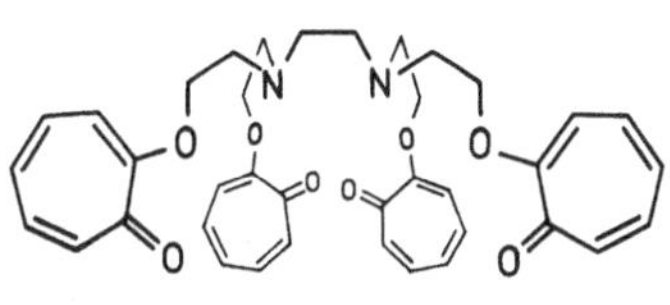

66 [30d)]

Fig. 14. Open-chain cryptands

geometrically more favourable environment at the nitrogen junction is obtained by the introduction of a triphenylamine unit (65) [30b)]. Nitrogen as a typical anchoring site for attaching three pendant groups was also replaced by a carbon atom [31].

The number of complexation arms can be further increased when multifunctional units such as benzene or triazine [31] nuclei are present as anchoring groups [25] (Fig. 15). On this basis, it is possible to synthesize ligands with six or more "tentacles", if necessary (67–69) [31,32]. The phenomenal structure resembling that of a cuttle-fish has imparted the nick-name "octopus molecules" to this type of ligand.

Generally, podands include all ligands which possess the characteristics of an open-chain oligoether or which consist of chains bearing heteroatoms in a particular array.

Using the generalized notation system proposed recently [3] (cf. Sect. 3.1.), the number of donor-active arms is given by the topological expression within the braces "{ }", e.g. {3}podand for 63–65, {4}podand for 66 and {6}podand for 67–69. The sequence and numbering begins with the donor atoms of highest priority at one end of the chain (selection principle see coronands, Sect. 3.1.). Oligopodands are analyzed by first selecting a main chain, containing the highest number of donor atoms, which is named first using the common symbolism followed by naming of the side arms. The latter are denoted in decreasing order of priority, where the sequence number of donor atoms, type of donor atom, priority of end groups (according to IUPAC rule C-14.1) is applied. Superscripts indicate their location on the main chain. As examples, the names for 58, 59 (n = 2), 63 (R = H) or 66 have been

13

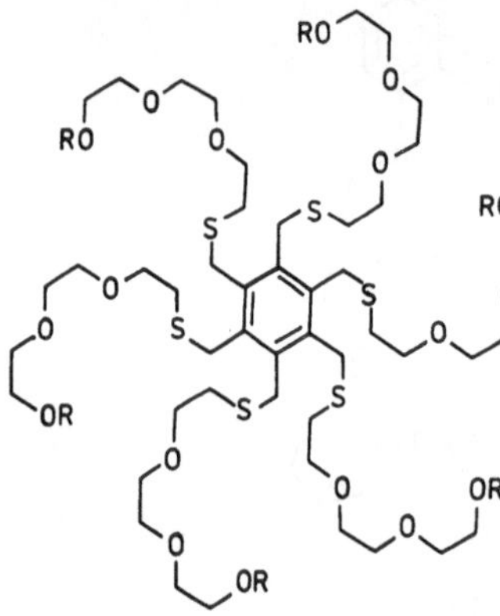

67 [32a)]

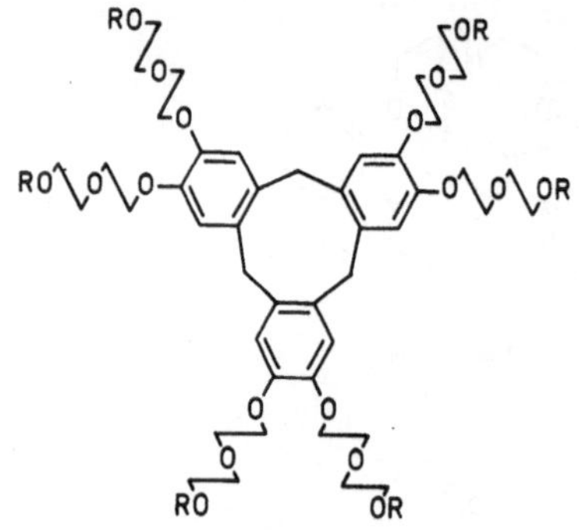

68 [32b)]

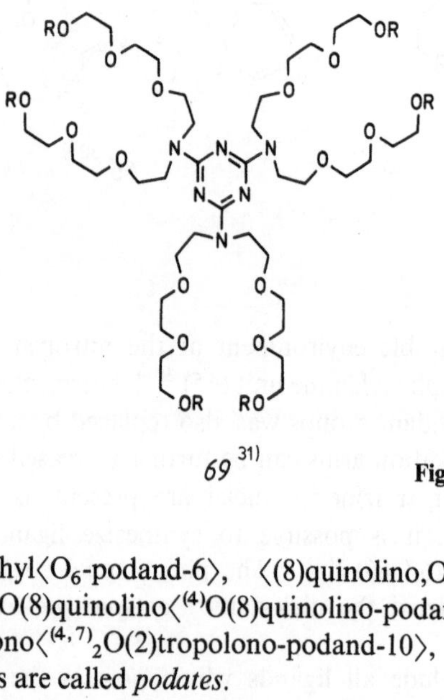

69 [31)]

Fig. 15. Octopus molecules

derived: 1,16-dimethyl⟨O₆-podand-6⟩, ⟨(8)quinolino,O₅(8)quinolino-podand-7⟩, {3}⟨(8)quinolino,ONO(8)quinolino⟨⁽⁴⁾O(8)quinolino-podand-7⟩ and {4}⟨(2)tropo-lono,ON₂O(2)tropolono⟨⁽⁴,⁷⁾₂O(2)tropolono-podand-10⟩, respectively. The corresponding complexes are called *podates*.

3.5 Podandocoronands (Fig. 16)

This relatively new type of ligand topologically lies just at the borderline between coronands and cryptands, and includes structural features of all three classes of ligands described before [3)]. The underlying idea is to combine some of their individual advantages [10, 33a)] (s.b.). Obviously, the character of the donor-active side arms dictated by the end groups (*70, 71*) is responsible as to whether a spherical cryptand-like or a two-dimensional coronand-like arrangement is favoured (*70–73*) [10, 33a)]. In other cases, highly structured complexing sites are displayed by latterally attached pendant groups, for example of tubular dimension (*74* [33c)]), cf. also Fig. 23c).

14

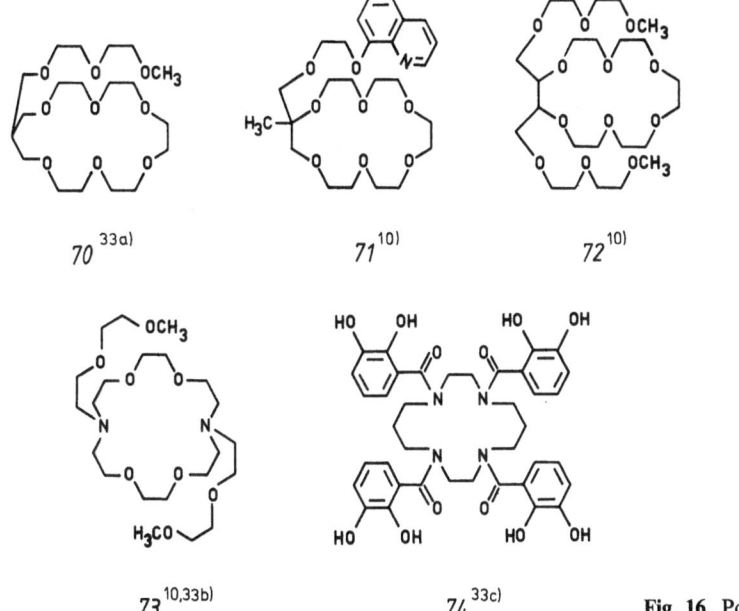

$70^{\,33a)}$ $71^{\,10)}$ $72^{\,10)}$

$73^{\,10,33b)}$ $74^{\,33c)}$ **Fig. 16.** Podando-coronands

3.6 Macrocyclic Oligoketones (cf. Fig. 17)

The principal structure of the cyclic oligoketones ("oxocrowns", 75 [34a)], 76 [34b)]) is derived from the concept of donor site-incorporation into functional groups (cf. Sect. 3.2.h). At present, only a few examples of this special type of organic neutral ligands are known [34)]. However, their general complexation tendency seems to be established so far. Going from the monoketone 75 to the hexaketone 76,

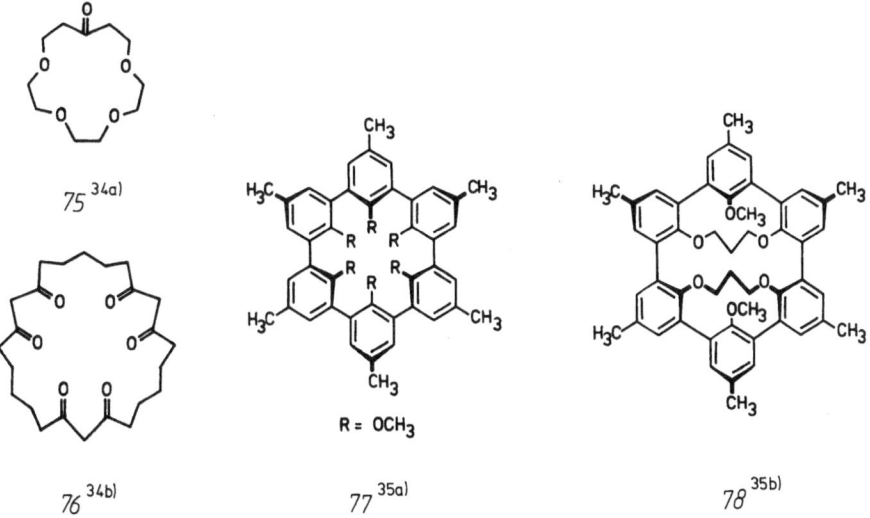

$75^{\,34a)}$

$76^{\,34b)}$ $77^{\,35a)}$ $78^{\,35b)}$

R = OCH$_3$

Fig. 17. Macrocyclic oligoketones and spherands

crown ether/coronand properties are retained, despite the absence of any ether oxygen atoms.

3.7 Spherands (cf. Fig. 17)

A further new type of ligand is revealed by the recently described ligand systems *77, 78* called spherands which, by means of intraanular crowding of donor groups attached to a rigid framework, end up with a spherical complex arrangement [35]. Actually, they represent the most rigid arrangements of donor sites known. Details are given in the contribution from D. J. Cram et al.

4 Ligand Synthesis [36]

4.1 Coronands

The synthesis of coronands (path A, Scheme 1) often proceeds in a surprisingly smooth way to give high yields of products even without applying the high dilution principle [6].

A cation catalysis, as depicted in Scheme 1 (path B), is considered to be the driving force behind this synthesis. This kind of cyclisation assistance is called *template effect* [37]. It might be effective in all crown syntheses starting from several components when cations participate. The extent to which the template effect is noticeable depends on the type of cation present [11a] (cf. Sect. 5.2.1.); in a few cases, Cs^+ has proved very suitable [38]. Details on the latter subject are given in the contribution by R. M. Kellogg et al.

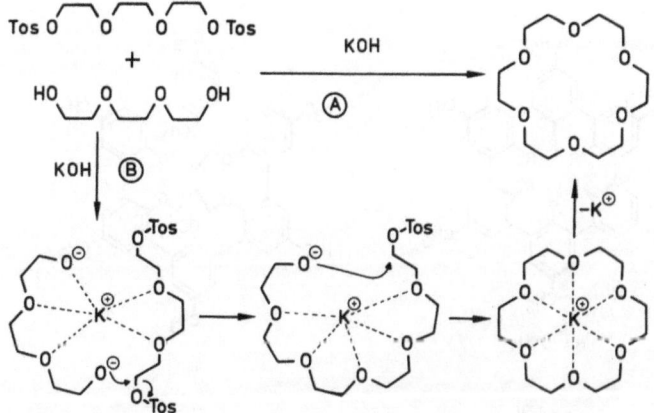

Scheme 1. Synthesis of coroand [18]crown-6 (*1*); the template effect

4.2 Cryptands

The synthesis of cryptands requires more steps and is thus more laborious [1a, 1b] (Scheme 2). Moreover, due to unfavourable or non-existing template effects, high dilution conditions are usually necessary [24]. However, a good-yield one-step synthesis of *49d* using simple starting materials has been developed recently [39].

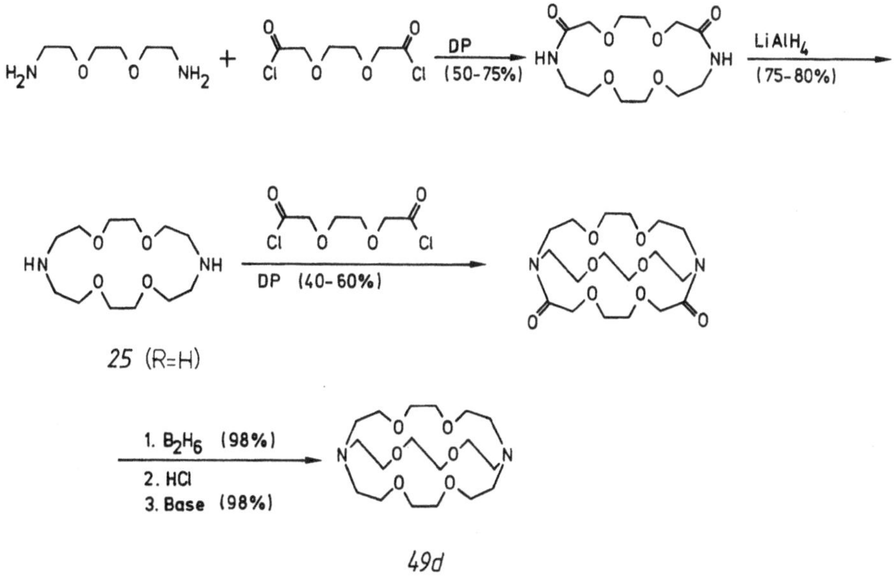

Scheme 2. Synthesis of [2.2.2]cryptand *(49d)* [24a]

4.3 Podands

Podand syntheses are much simpler in principle [25] (Scheme 3). Since no ring cyclisations are required, reagents can be reacted in excess, and both high dilution conditions and template catalysis are irrelevant; they play at the most only a minor role.

The high yields obtainable via simple and cheap reaction steps — an important

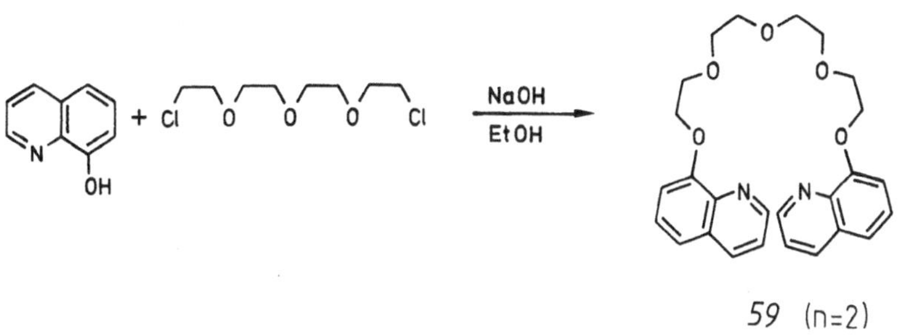

Scheme 3. Synthesis of podand *59* (n = 2) [27a]

aspect of the chemistry of open-chain ligands — generally make podands technically attractive and economically promising for the future [1m,40].

5 Properties of Crown Compounds [41]

5.1 Hydrophilicity/Lipophilicity Balance

Crown ethers and other coronands, cryptands and podands, consist of a series of lipophilic (methylene groups, alkyl chains) and hydrophilic structural elements (ether oxygen, nitrogen atoms or functional groups, cf. Sect. 3.) [23,25,42]. Their behaviour in hydrophilic media can thus be likened to that of a (molecular) fat droplet in water. In lipophilic media, the polarisation can be reversed, assuming a satisfactory degree of flexibility of the molecular backbone; the crown compound behaves in a certain way as a water droplet in oil. In the first case, an *endolipophilic* cavity is formed, in the latter case, an *endohydrophilic* one [28].

The lipophilic interior in the fat-droplet model is likely to be too small for the uptake of a lipophilic guest molecule, i.e. for the ligand to function as a lipophilic receptor (cf. Sect. 5.10.). On the other hand, the hydrophilic, electronegative cavity is ideally suited for alkali metal and alkaline earth metal cations according to their size [23].

5.2 Crown Ethers as Cation Receptors

5.2.1. Optimal Spatial Fit Concept and Circular Recognition

A comparison of the hydrophilic cavity sizes of various crown ethers and the diameters of unsolvated ("naked") ions of a few alkali and alkaline earth metal ions (Table 1) shows that [12]crown-4 (9) and Li^+, [15]crown-5 (10) and Na^+, [18]crown-6 (1) and K^+ are well matched [2h] (cf. "template effect", Sect. 4.1.).

The theoretically predicted spatial relationships with interior-directed oxygen atoms are quite in keeping with an ion-ball model of the [18]crown-6-K^+ complex [2e]. The measured complexation constants (s.b.) confirm the excellent "fit" of K^+ into the [18]crown-6 ring *(circular recognition)* [43].

5.2.2 Structure of Crystalline Complexes [41,44] (Fig. 18)

The fine structure of a classical crown ether complex is typified by the symmetrical coplanar array of the oxygen atoms. These atoms contact the cation located in the

Table 1. Comparison of cation- and cavity diameters

cation	cation diameter [Å]	crown ether	cavity diameter [Å]
Li^+	1.36	[12]crown-4 (9)	1.2–1.5
Na^+	1.90	[15]crown-5 (10)	1.7–2.2
K^+	2.66	[18]crown-6 (1)	2.6–3.2
Cs^+	3.38	[21]crown-7 (11)	3.4–4.3

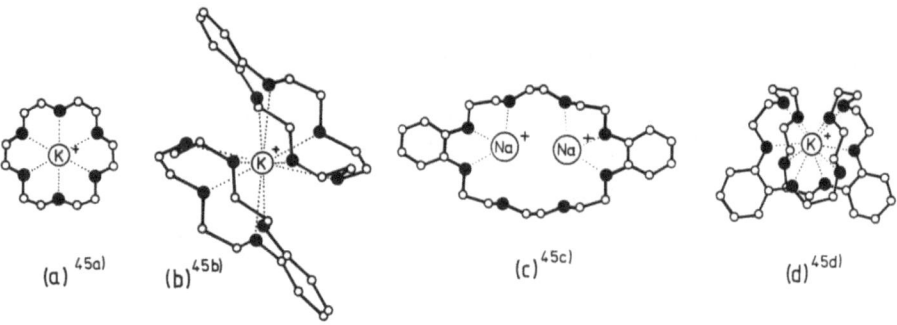

Fig. 18. Illustration of typical cation/ligand arrangements in crystalline crown ether complexes

center at an arithmethically calculated interatomic distance as realized in the K^+-[18]crown-6 complex (Fig. 18a) [45a]. Spatially less fitting cations are tolerated within certain limits, e.g. by deformation of the ligand skeleton or by movement of the cation out of the ring plane. In case of drastical discrepancy, for instance cation diameter $\gg$ cavity diameter and vice versa *sandwich complexes* with 2:1-stoichiometry (Fig. 18b) [45b], 1:2-stoichiometrical *two-nuclei* complexes (Fig. 18c) [45c] or *coiled* structures having still 1:1-stoichiometry (Fig. 18d) [45d] respectively, are formed.

However, conclusions on the real tendency of formation and the stability of a crown ether complex as well as the ligand conformation of the complex in solution can only be drawn to a limited extent via the crystal structure.

5.2.3 Complex Stability in Solution [41,43] (Fig. 19)

The degree of stability of crown ether complexes is represented by the *stability constant* K_s [s from stability; sometimes also denoted by K_f (f from formation)] which is defined by the law of mass equilibrium of the complexation reaction and is based on a specific solvent [2j]. From Fig. 19, in which the K_s-values for several crown ether/metal ion combinations are plotted in a logarithmic way, a clear relationship between cation diameter/cavity size and the stability constant of the appropriate complexes can be defined, e.g. high stability of *2-* or *80*-K^+-, *79*-Na^+- and *82*-Cs^+-complexes.

For a more detailed discussion [41,43] also the cation charge, the flexibility of the ligand and numerous other factors summarized in Table 2 have to be taken into consideration.

5.2.4 Selectivity of Crown Ether Complexation [41]

In order to distinguish different cations from one another, ligands must display differences in their complexation stabilities [23].

The discrimination ability (complexation selectivity) of a crown ether towards two different cations is expressed numerically by the quotient of the corresponding log K_s values whereby the ion considered appears in the numerator and the one to be differentiated in the denominator. Numerically high factors correspond to high

19

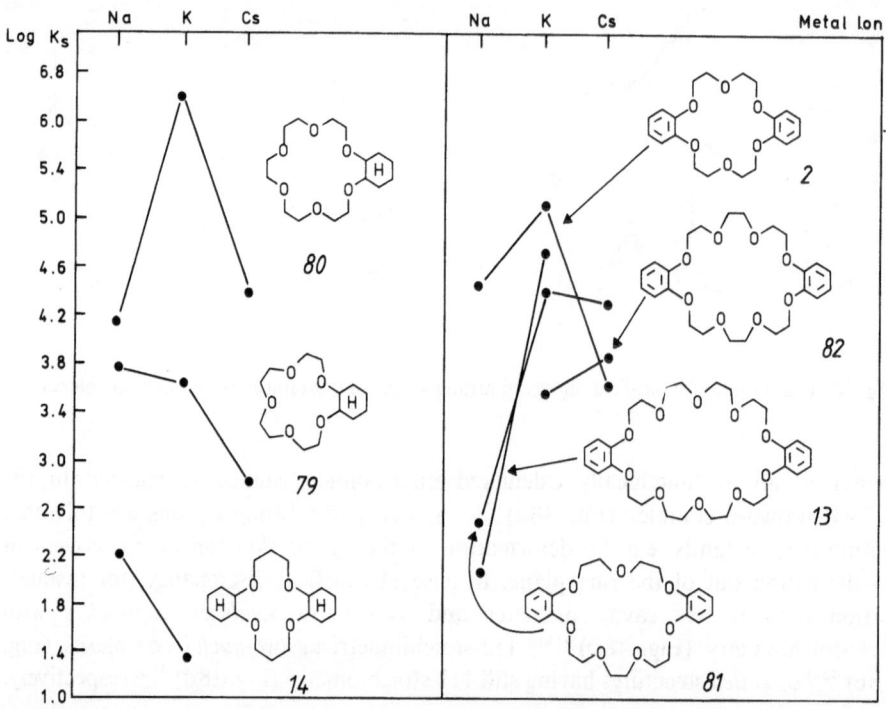

Fig. 19. Chart of K_s values of several crown ether/cation complexes (in methanol) [43]

Table 2. Information code and information storage of crown compounds

---

1) *Ligand topology*: dimension (open-chain, cyclic, bicyclic . . . polycyclic)
   bonding state (molecular skeleton)
   binding of atoms
   shape (spherical, elliptical etc.)
   size
   conformation
   chirality
   ligand dynamics, flexibility
2) *Bonding type, donor site*:
   chemical nature, electronic properties (charge, polarity, polarizability, van der Waals effects)
   number, form, size, arrangement
3) *Ligand shell*: thickness of ligand shell
   lipophilicity or hydrophilicity on the whole polarity (endolipophile-exolipophile or endopolaro-phile-exolipophile)
4) *Environment ("reaction medium")*:
   general and specific solvation effects in the complexed and uncomplexed states
5) *Counterion*: cation/anion interactions (ion pairing), influence of medium on cation/anion inter-actions

---

selectivity, low factors mean low discrimination. If high selectivity is supplied by a crown ether for several ions (e.g. of similar size or charge) but for another group of ions (e.g. of smaller ionic radius), the selectivity is much lower in general, one speaks of "*plateau selectivity*" [23] (cf. Fig. 19, *79* exhibiting low discrimination of Na[+]/K[+] and high discrimination of K[+]/Cs[+] and *81* vice versa).

In other cases, where the crown ether clearly distinguishes all foreign ions, irrespective of their size and charge, one speaks of *"peak selectivity"* [23] (cf. *2, 80* in Fig. 19; *80* shows distinct discrimination towards K^+ either between Na^+ or Cs^+). In between lies the domain of the weakly discriminating more or less unselective ligands.

To a first approximation, it is to be expected that peak selectivity is preferentially revealed by spatially rigid crown ethers possessing a well-defined cavity as well as by cryptands (cf. Sect. 5.4.).

5.3 Coronates — The Concept of Donor-Site Variation

Apart from the cavity geometry (stereochemistry of ligands), the nature of donor sites plays the most important role in complexation selectivity [23,41,43]. For example, the incorporation of (soft) *sulfur* atoms, as shown in Fig. 6, enhances the complexation of (soft) transition metal ions, e.g. of Ag^+, and hampers simultaneously that of (hard) alkali metals (cf. K_s values in Table 3) [46a,46b]. *Nitrogen* as a donor site (Fig. 7), on the other hand, plays the role of a mediator by promoting the complexation of transition metal ions (Ag^+) without drastically diminishing that of alkali metals as well (Table 3) [46a].

In addition, nitrogen in the form of a pyridine unit (cf. Fig. 9) seems to favour the complexation of Na^+ over K^+ [18d,46c].

Carbonyl groups of esters and amides (Fig. 10) reveal affinity to alkaline earth metal ions, e.g. to Ca^{2+}, Sr^{2+}, which is in agreement with higher electrostatic attraction [4b,46d].

The 24-membered cyclic hexaketone *76* (Fig. 17) behaves as a specific complexing agent for uranyl cations [34b].

Further possibilities for new binding site-controlled cation selectivities may be

Table 3. Comparison of log K_s values of K^+- and Ag^+-complexes of [18]crown-6 and of some sulfur and nitrogen-analogous coronands

LIGAND	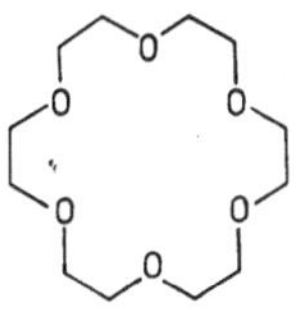	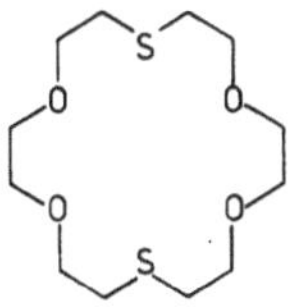	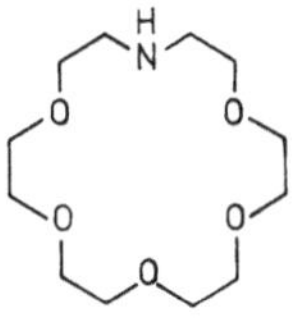	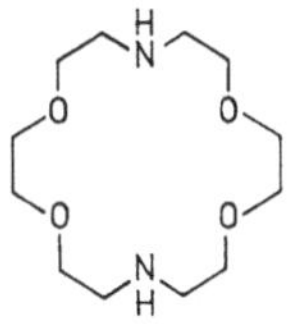
No.	*1*	*20*	*24*	*25*
$K^{\oplus}$ [a]	6,10	1,15	3,90	2,04
$Ag^{\oplus}$ [b]	1,60	4,34	3,30	7,80

[a] in Methanol ; [b] in water

21

opened by incorporation of phosphors or arsen donor atoms into the ring which has been realized till now only in a few cases [16d, 16e]

5.4 Complexation of Cryptands, Cryptate Effect and Spherical Recognition (for Formulae see Fig. 12)

The spheroidal hydrophilic inside of macrobi-/tricyclic ligands (cf. *49, 50, 57,* Fig. 12), is in shape particularly well adapted to a "ball-like" cation [1e, 23]. As a result, cryptands of this type reveal more pronounced receptor properties for spherical cations than crown ethers or coronands (*spherical recognition*) [47]. Variation of the length of the ethyleneoxy bridges in bicyclic systems gives rise to a gradual change in the cavity dimension (cf. *49 a–e*) which, with regard to an optimum spatial fit, greatly affects the complex stabilities with cations of different size (*size selectivity*) [47]. Thus, the cryptands [2.1.1] (*49b*), [2.2.1] (*49c*) and [2.2.2] (*49d*) complex preferentially Li$^+$, Na$^+$, K$^+$, respectively.

The most ideal topology for complexing spherical cations is displayed by the macrotricyclic architecture of the soccer-ball molecule *57*, having a cavity of about 3.4 Å diameter, lined with ten binding sites [24i]. The four nitrogens are located at the corners of a tetrahedron and the six oxygen atoms at the corners of an octahedron. This ligand offers optimum requirements for embedding of a Cs$^+$-ion (diameter 3.38 Å). As a consequence, the Cs$^+$-complex formed by *57* appears to be the most stable known to date (log K$_s$ = 3.4, in water at 25 °C).

The ions, which favourably fit into the center of the three-dimensional cavities, are shielded from their environment by the ligand skin; they behave as if hidden inside the cavity of the host molecule — there upon the name cryptate [2d] (cf. X-ray structure of Rb$^+$-[2.2.2], Fig. 20) [48].

The three-dimensional encapsulation strongly increases the complex stabilities, e.g. the K$^+$-[2.2.2] cryptate is more stable than the K$^+$-complexes of its corresponding monocyclic coronand (cf. *25*, R = CH$_3$, Fig. 7) by a factor 10^5 and even about 10^4 times more stable than the K$^+$-complex of the natural ionophor valinomycin [41]. This may illustrate what is called "*(macrobicyclic) cryptate effect*", which is found to be of higher efficiency (3–5 orders of magnitude) than the related "macrocyclic effect" operating in coronands [47a].

As the bicyclic and even more the tricyclic topology resists contraction or expansion of the intramolecular cavity more than a less rigid molecule, high peak selectivities in complexation are observed for a suitable cation [47]. Only for the larger dimensioned and therefore more flexible bicyclic cavities (e.g. *49e*) are plateau selectivities formed. Nevertheless, for almost any pair of alkali/alkaline earth ions

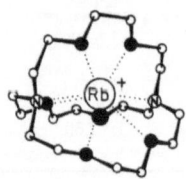

Fig. 20. Structure of the RbSCN-[2.2.2]cryptate showing the rubidium ion completely enclosed by the cryptand [48]

there is a cryptand used which provides higher selectivities than those known before.

5.5 Podates

As open-chain neutral ligands (cf. Figs. 13–15), podands usually do not have a preformed intramolecular cavity inside which a cation can nest [25,28]. However, they are able to build up appropriate cavities during complexation in various ways [49a], covering complex arrangements similar to crowns (cf. Fig. 21 a) [49b] as well as to cryptands (Fig. 21 c) [49d]. A unique wrapping of cations is revealed by the end group-containing podands (e.g. *59–61*) having a backbone of convenient chain length and donor-site capacity (Fig. 21 b) [49c].

For a more detailed representation of this complexation architecture with non-cyclic ligands see the contribution from Saenger et al. The variety of how podands can be coiled or bent around a given cation without being geometrically supported as in the case of cyclic coronands or oligocyclic cryptands is surprising [25,28,49a]. The whole ligand skeleton is held in the podate conformation only by cation-donor atom interactions. This causes a loss of entropy which is reflected in the decrease of the complex stabilities [41]. Podates are hence ordinarily less stable than coronates and much less than cryptates, which are favoured by the "macrocyclic" and the

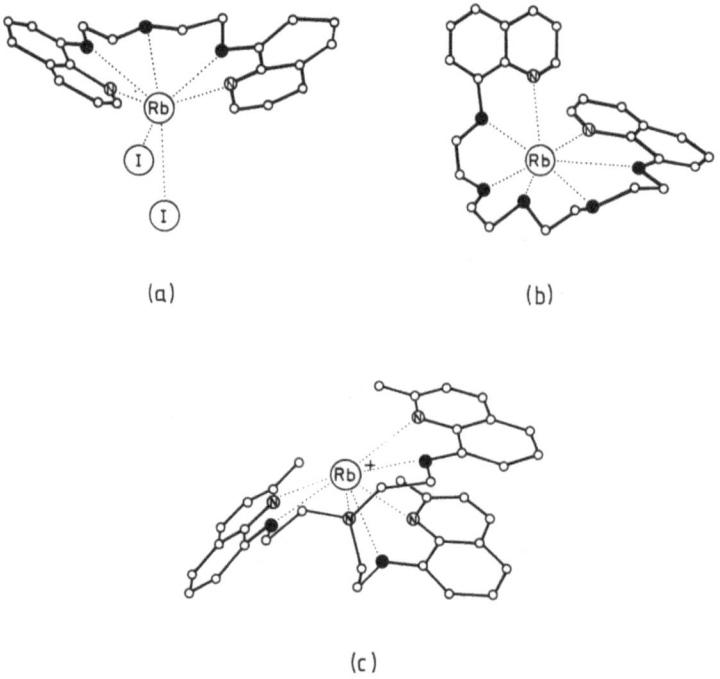

(a)

(b)

(c)

Fig. 21. Representative geometries of some podand-RbI complexes **a)** *59* (n = 0) [49b], **b)** *59* (n = 2) [49c], **c)** *63* (R = CH$_3$) [49d] as ligands

23

"cryptate effect", respectively. Typical orders of magnitude for the K_s values (in methanol) are 10^2–10^4 for podates, 10^4–10^5 for coronates and 10^6–10^8 for cryptates.

In accordance with the guideline that high complex stabilities do not imply high complexation selectivities at the same time, podands can display considerable discrimination abilities between different cations which clearly depend on specific ligand characteristics like the nature and the number of the donor atoms, the rigidity of the ligand chain and especially on the type of the end groups [46c, 50]. An illustration of the influence due to the end group effect is given in Fig. 22. As the donor-inactive phenyl residues in *83* are successively replaced by powerful 8-quinolyl end groups (*84, 59*), the complex stabilities drastically rise; however, the complexation selectivities of potassium ion either to larger or smaller cations drop.

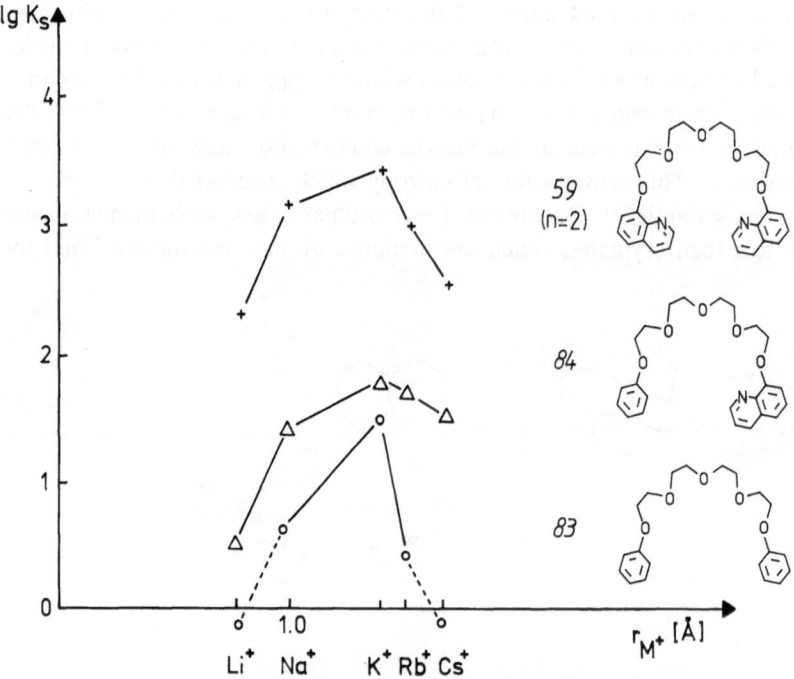

Fig. 22. Stabilities (K_s) of podand cation complexes as a function of the ionic radius [50]. Stepwise substitution of uneffective (phenyl) by powerful (8-quinolyl) end groups

These findings also suggest that the cavity which is formed by an open-chain ligand cannot be contracted or expanded to any dimension, without affecting the complex stability within a wide range. In analogy to the original cryptands, the cavity size effect is more pronounced for the three-armed podands of the pseudo cryptand-type [30c, 30d] (cf. *63–65*). A practical application of podand selectivities is realized in ion-selective membrane electrodes [1h]. Preferably lipophilic analogues like *62* are in use here [29]. On the other hand, podands are marked by their rapid complexation kinetics which make them particularly interesting for biological investigations [46c, 50].

5.6 Complexation Kinetics of Neutral Ligands

Stability and selectivity of crown complexes cannot be understood without considering the kinetics of complexation: "dynamic stability" of complexes [41,51a].

Molecular kinetics, i.e. the dynamic behaviour — complexation/decomplexation — of a system composed of ligand, cation and solvent give information about the lifetime of a complex [4g]:

$$M^+_{solvated} + Ligand_{solvated} \overset{\vec{k}}{\underset{\overset{\leftarrow}{k}}{\rightleftharpoons}} [M^+Ligand]_{solvated}$$

The ratio of the rate constant of complexation $(\vec{k})$ to that of decomplexation $(\overset{\leftarrow}{k})$ is thus directly connected with the stability (K_s) of a crown complex:

$$K_s = \frac{\vec{k}}{\overset{\leftarrow}{k}}$$

Metal complexation in solution is generally a rapid reaction. However, complex formation does not occur instantaneously and it is not a simple one-step reaction between ligand and cation [4g]. It often includes a series of one or several solvent molecules from the inner coordination shell of the metal ion and/or conformational rearrangements of the ligand, in particular when the ligand is a multidentate one.

Fast exchange rates are favoured by low cation solvation energy, ligand flexibility and not too high complex stability [41]. Thus, the most stable cryptates behave as *cation receptors* releasing the cation only slowly [51b], whereas the flexible podands with rapid exchange can be described as *cation carriers* [46c,50]. Coronands may serve either as receptors or carriers, depending on their molecular structure [51c].

5.7 Recognition and Complexation of Molecular Cations

Molecular recognition requires the ability of a ligand (molecular receptor) to select and to bind a specific substrate out of a medley of particles [11,41,52]. The binding makes use of all kinds of intermolecular interactions, electrostatic attractions being the most important in the fixation of a cationic species (cf. metal-ion recognition) but also hydrogen bonding, van der Waals forces, short-range repulsion, etc. [42]. Successful receptor-substrate binding leads to an assembly of two or more molecules which can be termed a "*supermolecule*" [52a]. The special design of the receptor is decisive of the binding characteristic and substrate selectivity. In order to get a high selectivity, receptor-substrate complementarity in topology and in binding features are desirable [23].

For instance, the simplest molecular cation, which is the ammonium ion, cannot be discriminated from K^+ by size very effectively. However, there is a marked difference in the charge distribution being spherical in the latter and tetrahedral in the former case. The tetrahedral arrangement of the nitrogen-binding sites makes cryptand *57* (Fig. 12) the topologically optimal receptor for the ammonium

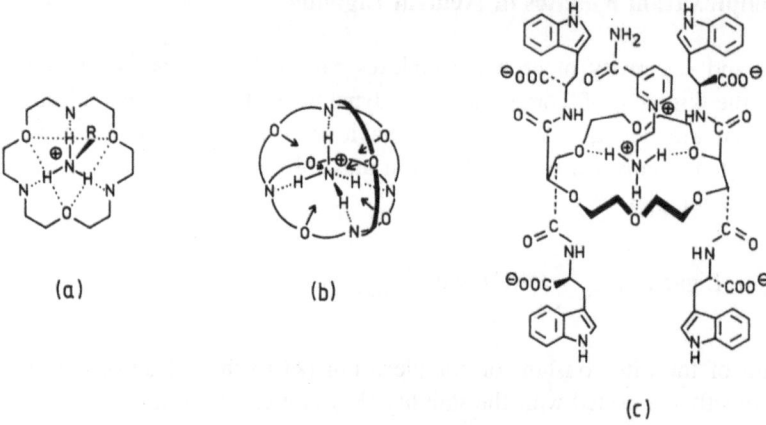

(a)　　　　　　　　(b)

(c)

Fig. 23. Receptor-substrate interaction of various organo-ammonium ions with host ligands. **a)** trigonal [15b], **b)** tetragonal recognition [47,52], **c)** laterally supported fixation of a dication [52a]

ion (*tetrahedral recognition*, Fig. 23b) [47,52]. The NH_4^+ is held inside the cavity by a tetrahedral array of H_3^+N—H ... N hydrogen bonds. Additional stabilization results from electrostatic and hydrogen bonding O—NH_4^+ interactions. The structure has been confirmed by X-ray analysis [53]. Further special features of this cryptate are high stability, low exchange rate and a shift of the effective pK_a value of the complexed ammonium ion to about 14 [52a].

A primary ammonium ion cannot longer nest into *57* for sterical reasons. Applied to the same lock-key concept, a trigonal symmetrical arrangement of NH-binding sites is deduced as the main receptor characteristic for primary ammonium ions. The coronand *27* (Fig. 7) meets this characteristic (*trigonal recognition*, Fig. 23a) [15b]. While the receptor *27* can select by *central discrimination* between primary and higher substituted ammonium ions, *lateral discrimination*, which is revealed in the podando-coronands *85* (Fig. 24), enables them to distinguish between primary ammonium ions, differing in the nature of the organic residues (Fig. 23c) [52a], due to electrostatic, hydrophobic and charge-transfer interactions.

Flattened guanidinium and imidazolium ions necessitate a larger ring size (*circular recognition*) [54a]. Another interesting concept [54b] deals with the selective binding of ammonium ions, due to their distances between the charge locations, using as receptors multi-site coronands (cf. Fig. 11) or dinuclear cryptand systems, e.g. of type *56* (cf. Fig. 12 and also Sect. 5.9.).

5.8 Chiral Crown Compounds and Chiral Recognition
   (Figs. 24, 25)

In order to enable the recognition of a chiral guest, a host ligand must satisfy at least two conditions: 1) as a crown compound it has to make available a suitable cavity for complexation to which 2) a fine structure can be imparted by incorporation of chirality barriers — a *diastereomeric* host/guest relationship must be able to develop so that out of two enantiomeric guest molecules only one may undergo the particularly tight energetically favourable interactions with the host (*chiral discrimination*) [11,41,42,55].

Apart from asymmetric centers, e.g. in sugars [55e] (*85* [52,56a,56b], *86* [56c]) or amino acids (*87*) [56d], the so-called "binaphthyl hinge" (*89* [56e], *90* [56f]) has proved particularly useful [42,55a−55c] (Fig. 24). With optically pure ligand *89*, a thorough enantiomeric separation of ammonium salt racemates (e.g. of protonated amino acid ester *91*, Fig. 25) is possible [57a]. The separation is based on the different stabilities of the complexes as pictured in Fig. 25 for the combinations (S,S)- *89* with guest ion (R)- or (S)-*91* [57b]. Enantiomer (R)-*91* is held inside the cavity by a strong "4-point" interaction whereas (S)-*91* does not spatially fit so well and also a weaker

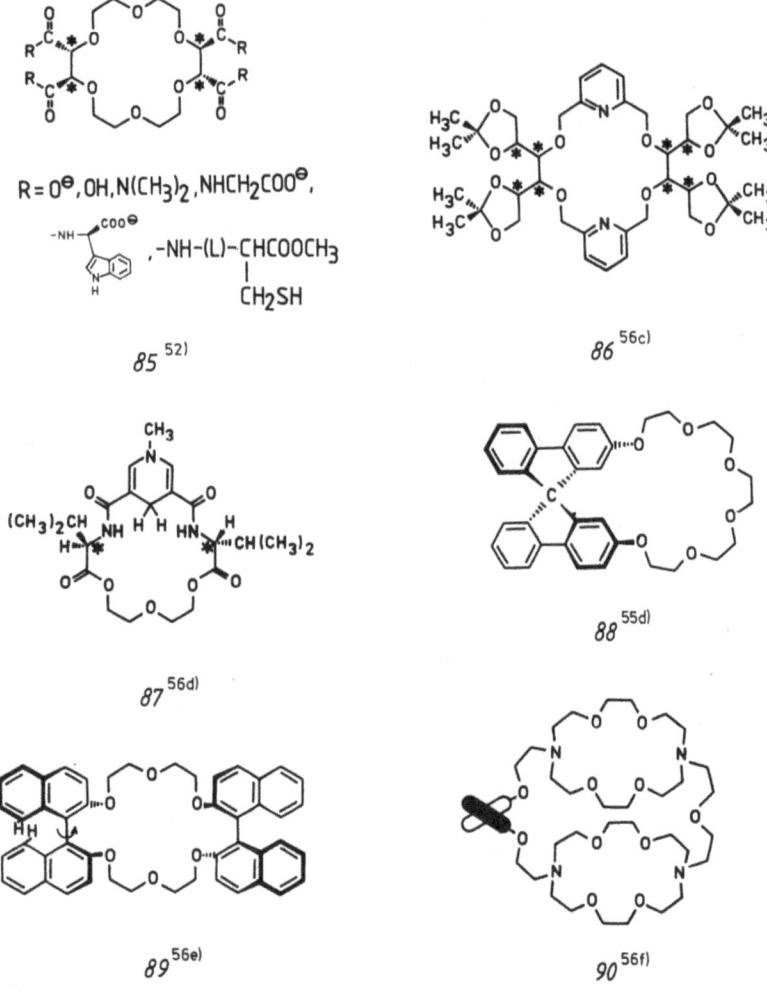

Fig. 24. Representative structures of chiroselective coronand/cryptand ligands

27

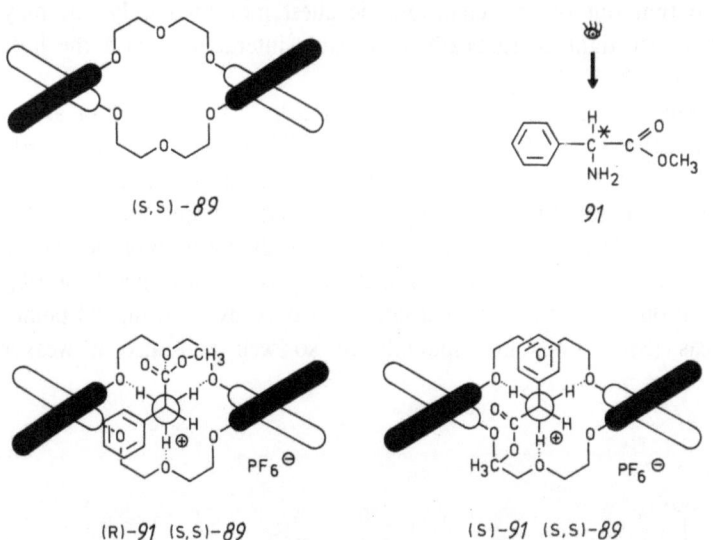

(S, S) - *89*

91

(R)-*91* (S, S)-*89*   (S)-*91* (S, S)-*89*

Fig. 25. Illustration of a chiral recognition by host-guest interaction [52b]. Host (*S, S*)-*89*, guest (*R*)-*91* or (*S*)-*91*

"3-point" interaction is present. More detailed information on this subject can be found in the contribution from Cram et al.

With the help of *87*, it has been possible to achieve a stereodifferentiating reduction of carbonyl compounds in high enantiomeric yields [56d]. The histidinyl-substituted *85*, displaying both central and lateral chirality, was employed in enantioselective thioester hydrolysis [56b]. The spirobisfluorene crown ether *88* performs a chiroselective phase transfer of optically active ammonium salts across a lipophilic membrane [55d].

On this basis, an essential step towards the realization of enzyme models [55] has been made (see contribution from Kellogg).

5.9 Multisite Receptors, Cascade Complexation and Ion-Pair Complexes

Macrocyclic and also open-chain ligands incorporating two or more separate receptor sites for metal ions are preconditioned to form bi- or polynuclear inclusion complexes [21,28,52a,58]. As outlined in the design, symmetric or dissymmetric (containing equivalent or nonequivalent binding subunits), they may prefer identical or different substrates (*homonuclear* or *heteronuclear* complexes). Among the topologically different structures known so far (cf. Figs. 11, 12, 15), cryptands of type *52* [24d] and *56* [24h] have most extensively been studied. The former is composed of two coaxially arranged tripode subunits linked by three bridges and builds up an ellipsoidal cavity. The latter is characterized by face-to-face linkage of two macrocycles enclosing a cylindrical interior. Both systems readily take up two identical metal ions (alkali metal or transition metal) being located on the poles of the ellipsoid or on top and bottom of the cylinder, respectively [59].

In the case of *56* also a heterometallic cryptate (Ag^+/Pb^{2+}) has been observed [59a]. The distances between the complexed cations clearly depend on the lengths of the bridges linking the subunits. This provides a novel approach to study *cation-cation interactions* at short interatomic distances, such as magnetic coupling, electron transfer, modification of redox potentials, etc. [58]. At larger interatomic separation, a central cavity gains in significance and enables the inclusion of additional substrates by binding across the two cations. This kind of double selection process, which implies selection of the cation by the ligand receptor sites and subsequent control of the substrate by the nature and arrangement of the complexed cations, is a main feature of *cascade complexation* (Fig. 26) [52b, 58].

Evidences for such a process are obtained from the changes in electronic spectra occurring on addition of H_2O, CN^-, N_3^- to the dinuclear complexes [24d]. Oxidation of the bis-Co^{2+}-complex of *52* yields a bridged μ-peroxo-μ-hydroxo species [58]. The imidazolate group, an especially attractive unit, however, is too large in size to intercalate into a bicyclic system, but forms similar cascade-type complexes with appropriate monocycles [59c]. The cascade complexation concept is also applicable to the chiroselective extraction of mandelate ion using the cryptand *90* as an optically active ion-pair carrier [59d].

Polytope receptors thus provide an entry into higher forms of molecular behaviour which can be termed as *cooperativity*, *allostery* and *regulation*.

Another interesting aspect is the internal cation exchange between the two macrocycles in singly occupied *56*-type cryptates, resembling the elementary jump process which might occur in biological cation channels [59e].

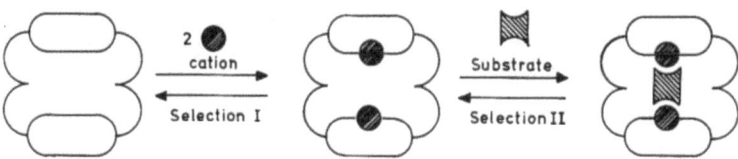

Fig. 26. Sequential formation of a cascade complex with a *56*-type tricyclic cylindrical cryptand [58] (schematic)

5.10 Anion Receptors and Neutral Molecular Complexes (Fig. 27)

Protonation of suitable catapinands [60a, 60b] and cryptands [24i, 60c, 60d] yields cavities suited for *anions* (Fig. 27). Anion-selective cryptand systems have been described for halide, azide and similar anions [28, 52]; for details see contribution from Vögtle et al.

Recently, it has became evident that not only cations but also *uncharged organic molecules* can be complexed by crown compounds [11, 41, 61]. Guest molecules, stoichiometrically bound, are especially NH-, OH- and CH-acidic compounds like urea [62a], hydrazines [62b], amides [61, 62c], malonic acid derivatives [62d], acetonitrile [61, 62e], nitromethane [62d], etc. [62f]. Crowns apparently may bind neutral guest molecules other than by hydrogen bonding in the crystal lattice as clathrates [63].

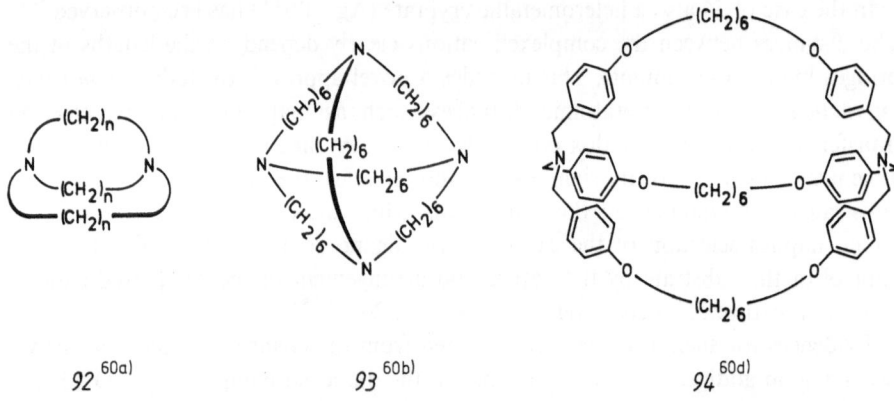

92 60a)   93 60b)   94 60d)

Fig. 27. Spherical and ellipsoidal cavities for anion inclusion (after protonation)

A survey of the hitherto obtained results is given in the contribution from Vögtle et al. (s.b.).

6 Consequences of the Properties of Crown Compounds

6.1 Lipophilisation and Phase Transfer of Ions [1f, 1m, 64]

Complex formation with crown compounds transforms a relatively small alkali metal/alkaline earth metal ion into a voluminous organic cation species which, in principle, is much more soluble in organic media [65] (Fig. 27, cf. also Sect. 5.1.). As a result, solubilization of inorganic salts in organic solvents of low polarity occurs (solid-liquid phase transfer, for this subject see contribution from Montanari et al.).

In the same way, using a water/organic solvent two-phase system, the distribution of a salt in the organic layer is enhanced (liquid-liquid phase transfer, ion-pair extraction, see contribution from Montanari et al.). Reasonably, the solubility of a crown cation complex in an organic medium is facilitated by a high lipophilicity of the ligand skeleton, apart from the complex stability. Typical phase-transfer catalysts of the crown type therefore contain cyclohexane rings or alkyl side chains as promoting groups [1m].

A second effect results from the lipophilic nature of the counter-anion [65]. With

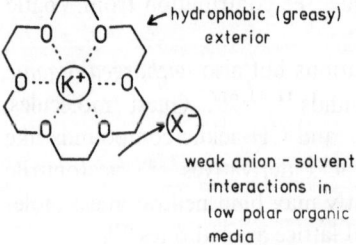

hydrophobic (greasy) exterior

weak anion - solvent interactions in low polar organic media

Fig. 28. Endopolarophilicity/exolipophilicity and anion activation illustrated by [18]crown-6 complexation of a potassium salt

anions of highly localized negative charge, solubilization may become difficult. Thus, salts containing such less polarizable (hard) anions as fluoride or hydroxide can hardly be dissolved in low-polar organic solvents like benzene, even by highly lipophilic and strongly complexing ligands such as cryptand *50* (Fig. 12).

6.2 Ion-Pairing and Ion-Aggregation Effects

Ionic structures can occur in every imaginable form ranging from separate-solvated ions (*I*) via solvent-separated (*II*), loose and tight ion-pairs (*III*) to contact ion pairs (*IV*) or complicated aggregates of three, four or more contact ion pairs [26,66] (Scheme 4).

Separately solvated ions predominate in good solvating solvents. In less polar media, contact ion pairs and associated aggregates become more significant. Upon addition of complexing agents like crown ethers or more particularly of cryptands these species are converted more or less into *ligand-separated* ion pairs (*V, VI*) [1f]. In the case of crown complexes a metal cation-anion contact always can occur from the open faces of the ring plane (cf. *V*). Direct cation-anion pairing is expected to be weaker or absent if more shielding cryptands are used ("ligand-solvated cation", cf. *VI*). However, completely unsolvated so-called *"naked anions"* [67] (cf. *VII*) are never generated with crown ethers or coronands and even with cryptated cations ion-pairing still happens, forming complexed ion pairs. Totally isolated naked anions in the real sense will exist only in the absence of any species suited for interaction, e.g. under gas-phase conditions [65b].

Facts about the structures of ion pairs that are formed and the type of their solvation can be gathered; nevertheless, of more importance are the resulting effects concerning their chemical reactivity as well as their influence on the mechanism and the stereochemistry of ion involved reactions.

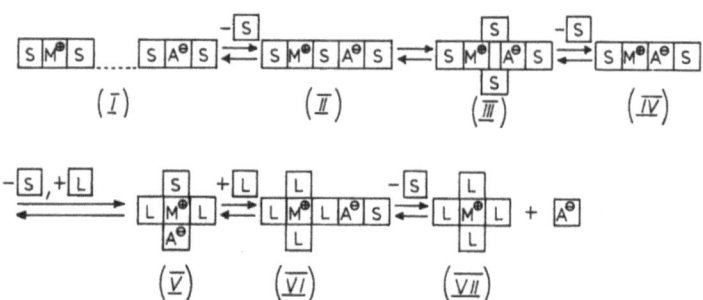

Scheme 4. Ion-pair solvation and complexation effects, schematically (S = solvent, M$^+$ = cation, A$^+$ = anion, L = ligand)

6.3 Modifications in Chemical Reactivity [1f, 1g, 1m, 65, 68]

6.3.1 Anion Activation

Weakly solvated, almost "naked" anions formed by complexation in solvents of low polarity differ essentially in two respects from their solvated ionic species [65,68]:

a) on account of their smaller size they possess a higher effective *charge* and b) they do not have a (tight) *solvation shell* that must be removed or broken in the course of a reaction. The anion is already in a state prepared for a reaction, so to say: it is activated (*anion activation by crown compounds* [1f,68c,68d], cf. Fig. 28). In order to cause also a large cation-anion separation (weak ion-pairing), bulky ligands especially of the cryptand type are required [65b]. This may lead to a drastic increase in the rate constant for a given reaction, and has been demonstrated in practice by numerous examples, ranging from simple nucleophilic substitutions on saturated or unsaturated carbon atoms to C—C-bond formations, eliminations, additions, oxidations, reductions, gas extrusions, rearrangements, polymerisations and some more [1g,1m,64b,65a,68]. Even a few of these reactions are only possible in the presence of complexed species.

6.3.2 Nucleophilicity/Basicity Balance

According to the anion properties affected by complexation, the balance between nucleophilicity and basicity (cf. hard/soft principle) [69], can be strongly influenced, depending on the ion type [1f].

As an example, the *fluoride ion* is generally known for its difficulties to undergo a nucleophilic substitution at an aliphatic carbon atom [70]. Owing to its high charge density it is surrounded by a thick and tight solvation shell in solvents in which inorganic fluorides readily dissolve. On the other hand, in less polar solvents, where the anion would be less solvated, the high lattice energy prevents the solubilization.

Poorly solvated fluoride ("naked fluoride"), generated by complexation, however, reacts as a strong (hard) *base* and as a more effective *nucleophile* compared to its higher solvated analogue. These features are reflected in substitution (A) *and* elimination reactions (B) (cf. Scheme 5) [71a].

On account of their larger size "naked" *acetate ions* display less basic character than "naked" fluoride, but high nucleophilic potential (preferentially acetate formation) [71b].

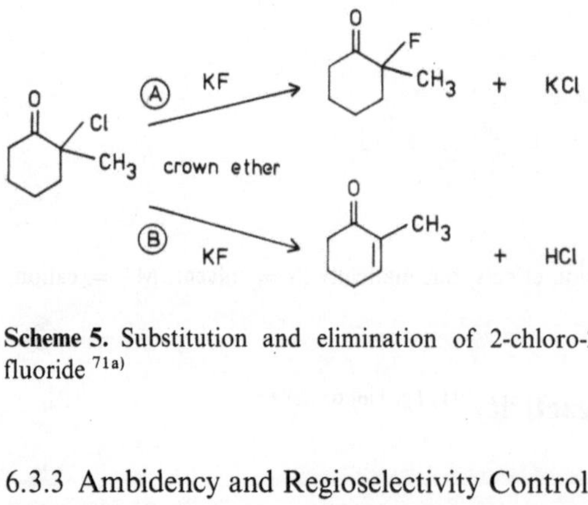

Scheme 5. Substitution and elimination of 2-chloro-2-methyl-cyclohexane induced by "naked" fluoride [71a]

6.3.3 Ambidency and Regioselectivity Control

Molecular systems containing both a basic and a nucleophilic center, are especially influenced by complexing agents [65]. The potassium enolate of ethyl acetoacetate which exhibits configurational mobility, can in principle adopt several arrangements (Scheme 6) [72a].

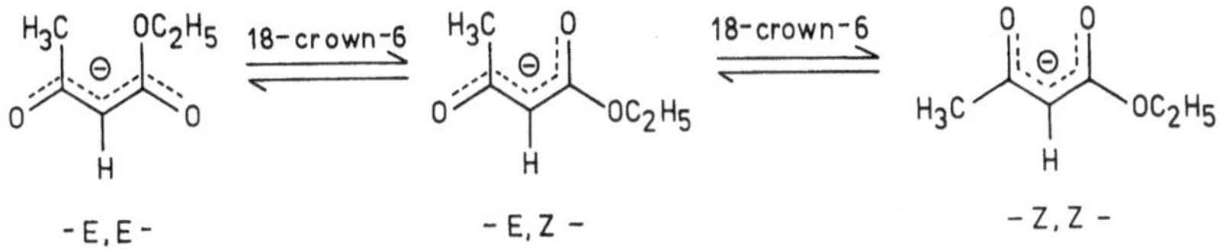

Scheme 6. Crown ether affected configurational flexibility of the delocalized ethyl acetoacetate anion [72a]

In the absence of a complexone, the Z-configuration with chelate-supported ion-pair structure dominates and C-alkylation occurs with ethyl iodide. However, if [2.2.2]cryptand (*49d*) or [18]crown-6 (*1*) is present, this arrangement is destabilized and depressed at the expense of E, Z which proceeds in a rate of O-alkylation of 21% for *49d* and 3% for *1*. The reaction is accelerated by a factor of 4000 and 100, respectively.

Thus, anion activation does not necessarily mean reactivity increase in a particular direction [72b]. Rather, in each special case — taking into consideration ionic structures, basicity, nucleophilicity and hardness or softness of ions — a sum of estimations, which may be influenced by crown compounds, must be made. By appropriate choice of ligand system and solvent, the desired activity of an anion can be widely varied and made to suit optimum reaction conditions.

6.3.4 Effects of Cation Capture

In contrast to anion activation, cation-assisted reactions are handicapped by complexation [65b]. Such effects have been found particularly in reductions of *carbonyl groups* by metal hydrides [73a] and on the addition of organic lithium compounds [73b], indicating that the coordination of the cation at the carbonyl group is a fundamental step of the reactions. Thus, cryptation of $LiAlH_4$ or of $NaBH_4$ by [2.1.1] (*49b*) or [2.2.1] (*49c*) causes a marked decrease of the reaction rates or even inhibition [73a]. α,β-Unsaturated carbonyl compounds preferably undergo reduction at the carbonyl group with $LiAlH_4$ (1,2-addition). However, in the presence of [2.1.1]cryptand, the reaction rates are slowed down and 1,4-addition is predominantly observed (regioselectivity) [73c]. As a conclusion, complex formation by crown compounds has to be considered as a balance between *anion activation* and *cation restriction*. This is impressively illustrated by the Knoevenagel-type condensation of acetonitrile anion with benzaldehyde [65b]: If the countercation is lithium, addition of [2.1.1]-cryptand is shown to cause a drastic decrease of the reaction rate (cation participation). For K^+ and [2.2.2]cryptand, an increase of the reaction rate is observed (anion activation), and in the case of Na^+ and [2.2.1]cryptand, the two effects seem to compensate (no corresponding effect is found). Taken together by coronate and cryptate formation possibilities are also supplied for mechanistic studies of organic reactions.

33

6.4 Modification of Reaction Mechanisms and of their Stereochemistry

Among the numerous reactions where an ion-association effect can be analyzed, base-promoted *β-eliminations* are most suitable [1h].

In the arrangements depicted in Fig. 29 (transition states of the reaction), X or OTos resp. represent the leaving group, B^- the base and M^+ the counterion. The cyclic transition state *VIII* involves an *associated* base particle. The base association should alleviate the transition state of the *syn*-elimination by allowing a simultaneous coordination of the counterion with the base and the leaving group to occur.

Investigations of the base system potassium tert-butoxide/tert-butanol, a system with typical ion-aggregate structure, confirmed this postulate [74]: In the presence of equimolar amounts of [18]crown-6 ligand-separated (dissociated) tert-butoxide ions are formed as active bases (transition state *IX*). As a result, the rate of *syn*-elimination product is drastically reduced at the expense of an increase in that of the *anti*-product. Consequently, in the presence of equimolar amounts of [12]crown-4 (*9*, Fig. 4), whose cavity is too small for K^+ ions (cf. Sect. 5.2.1. and Table 1), a corresponding "crown-ether effect" is unobserved.

The transition states *X* and *XI* included in Fig. 29 illustrate in a defined way (Newman projection) the significance of steric factors in a specific example (potassium tert-butoxide elimination of 2-butyl tosylate). Principally, it gains more importance with bulky bases (associated potassium tert-butoxide) and diminishes for separated ions. In accordance with this estimation, one can observe in the first case a clear preference of (Z)-butene over (E)-butene. On addition of a suitable crown ether, the relative amount of the (E)-product (transition state *XI*) increases.

The findings mentioned above, allow an extensive explanation of the old controversies concerning the direct relationship between base strength, base size and orientation in elimation reactions [75a].

Similar geometrical factors also account for the fact that the addition of crown ethers can prevent the formation of the specific transition state in the *Cannizzaro reaction* so that the reaction gradually slows down or does not occur at all [75b].

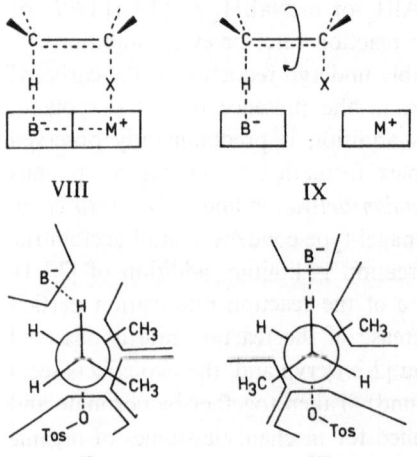

Fig. 29. Stereochemistry of base-promoted β-eliminations showing the influence of ion pairing

The *Wittig olefin synthesis* is also sensitive towards changes of the ionic structure in the transition state. In the presence of crown ethers principally more E-olefin is isolated [75c].

As a conclusion of these results, the complexation with crown compounds can also serve as a useful tool in stereoselective synthesis (cf. contribution from Kellogg in this volume).

7 Application Possibilities

As outlined above, there are many application possibilities of neutral ligands in *chemical syntheses* [1f, 1g, 1m, 65, 68] which, however, have not yet been fully utilized.

In principle, any reaction involving ions in whatever form may be considered to be (selectively) varied and modified by crown compounds and their analogues; some reactions are only possible in their presence.

For the introduction into technical fields an economically feasible crown-compound catalysis (phase transfer) is of importance [1f, 40, 64] (see contribution from Montanari et al.).

On the other hand, neutral ligands gain increasingly more attention in the *analytical* sector, e.g. in extractive enrichment procedures using ion selection and isotopic separations [1h, 1m, 76]. Remarkable successes have also been achieved in ion separations and ion determinations by means of crown ether resins, in salt inversions and crystal water determinations [77]. These subjects are discussed in detail in the contribution from Blasius et al.

Another approach in analytical chemistry, making use of the ion discriminating properties of crown compounds, is based on ion-selective membrane electrodes [29a, 78]. Multi-detecting systems for the continous ion-selective and electrochemical-enzymatic direct measurement on human beings have already become reality [79].

Crown-type compounds are also of general relevance as synthetic model substances for the study of *physiological* ion-transport processes in biological membranes, for the investigation of receptor and enzyme interactions as well as for the salt balance and for metabolic processes of the living organism [1i, 1j, 4]. In the latter case, a *pharmacological* application in the form of pharmaceutical crown-type compounds may be taken into consideration [1j].

There are some hints that the chemistry of macrocyclic ligands may also solve some of the future problems of energy storage [80].

8 Acknowledgement

The authors thank Prof. R. George Newkome and Prof. George W. Gokel for stimulating discussions and Miss Barbara Jendrny for her assistance.

9 References

1. Comprehensive presentations:
    a) Struct. Bonding *16*, Springer-Verlag, Berlin, Heidelberg, New York, 1973;
    b) Izatt, R. M., Christensen, J. J. (Eds.): Synthetic Multidentate Macrocyclic Compounds, Academic Press, New York, San Francisco, London, 1978;

   c) Izatt, R. M., Christensen, J. J. (Eds.): Progress in Macrocyclic Chemistry, Vol. 1, John Wiley and Sons, New York, 1979;

   d) Melson, G. A. (Ed.): Coordination Chemistry of Macrocyclic Compounds, Plenum Press, New York, London, 1979;

   e) Vögtle, F., Weber, E.: Kontakte (Merck) *1977* (1), 11;

   f) Vögtle, F., Weber, E.: ibid. *1977* (2), 16;

   g) Weber, E., Vögtle, F.: ibid. *1977* (3), 36;

   h) Weber, E., Vögtle, F.: ibid. *1978* (2), 16;

   i) Vögtle, F., Weber, E., Elben, U.: ibid. *1978* (3), 32;

   j) Vögtle, F., Weber, E., Elben, U.: ibid. *1979* (1), 3;

   k) Vögtle, F., Weber, E., Elben, U.: ibid. *1980* (2), 36;

   l) Weber, E., Vögtle, F.: ibid. *1981* (1), 24

  m) Weber, E.: ibid. *1981* (3), in press

2. Fundamentals[a−e] and general reviews[f−l]:

   a) Christensen, J. J., Hill, J. O., Izatt, R. M.: Science *174*, 459 (1971);

   b) Truter, M. R., Pedersen, C. J.: Endeavour *1971*, 142;

   c) Klamberg, H.: Chem. Lab. Betrieb *29*, 97 (1971);

   d) Dietrich, B., Lehn, J. M., Sauvage, J. P.: Chem. uns. Zeit *7*, 120 (1973);

   e) Weber, E., Vögtle, F.: Chem. Exp. Didakt. *2*, 115 (1976);

   f) Pedersen, C. J., Frensdorff, H. K.: Angew. Chem. *84*, 16 (1972); Angew. Chem., Int. Ed. Engl. *11*, 16 (1972);

   g) Black, D. St. C., Hartshorn, A. J.: Coord. Chem. Rev. *9*, 219 (1972);

   h) Vögtle, F., Neumann, P.: Chemiker-Ztg. *97*, 600 (1973);

   i) Kappenstein, C.: Bull. Soc. Chim. Fr. *1974*, 89;

   j) Christensen, J. J., Eatough, D. J., Izatt, R. M.: Chem. Rev. *74*, 351 (1974);

   k) Bradshaw, J. S., Stott, P. E.: Tetrahedron *36*, 461 (1980);

   l) Hubberstey, P.: Coord. Chem. Rev. *34*, 1 (1981)

3. Weber, E., Vögtle, F.: Inorg. Chim. Acta *45*, L65 (1980)

4. *Reviews*:

   a) Chock, P. B., Titus, E. O.: Progr. Inorg. Chem. *18*, 287 (1973);

   b) Simon, W., Morf, W. E., Meier, P. Ch. in Lit. [1a]

   c) Ovchinnikov, Yu. A., Ivanov, V. I., Shkrob, A. M.: Membraneactive Complexones, B. B. A. Library Vol. 12, Elsevier, Amsterdam, Oxford, New York, 1974;

   d) Pressman, B. C.: Ann. Rev. Biochem. *45*, 501 (1976);

   e) Fenton, D. E.: Chem. Soc. Rev. *6*, 325 (1977);

   f) Thoma, A. P., Simon, W.: Metal-Ligand Interactions in Organic Chemistry and Biochemistry (Eds. Pullman, B., Goldblum, N.), part 2, D. Reidel Publishing Company, Dordrecht, Holland, 1977;

   g) Burgermeister, W., Winkler-Oswatitsch, R.: Inorganic Biochemistry II, Top. Curr. Chem. *69*, Springer-Verlag, Berlin, Heidelberg, New York, 1977

5. Pedersen, C. J.: J. Am. Chem. Soc. *89*, 2495, 7017 (1967)

6. Cf. a) Pedersen, C. J.: Org. Synth. *52*, 66 (1972);

   b) Gokel, G. W. et al.: Org. Synth. *57*, 30 (1977)

7. Pedersen, C. J. in lit. [1b], p. 1

8. For the history of the discovery of crown ethers see Aldrichimica Acta *4*, 1 (1971)

9. a) Torizuka, A., Sato, T.: Org. Magn. Res. *12*, 190 (1979);

   b) Kime, D. E., Norymberski, J. K.: J. Chem. Soc. Perkin I *1977*, 1048

10. Weber, E.: unpublished

11. a) Cook, F. L., Caruso, T. C., Byrne, M. P., Bowers, C. W., Speck, D. H., Liotta, C. L.: Tetrahedron Lett. *1974*, 4029; Gokel, G. W. et al.: J. Org. Chem. *39*, 2445 (1974);

   b) Dale, J., Kristiansen, P. O.: Acta Chem. Scand. *26*, 1471 (1972);

   c) cf. also Wingfield, J. N.: Inorg. Chim. Acta *45*, L157 (1980)

12. a) Newcomb, M., Cram, D. J.: J. Am. Chem. Soc. *97*, 1257 (1975); Koenig, K. E., Helgeson, R. C., Cram, D. J.: ibid. *98*, 4018 (1976);

   b) Pedersen, C. J.: J. Am. Chem. Soc. *92*, 391 (1970);

   c) Weber, E., Vögtle, F.: Chem. Ber. *109*, 1803 (1976);

   d) cf. also Newcomb, M., Moore, S., Cram, D. J.: J. Am. Chem. Soc. *99*, 6405 (1977)

13. *Review*: Bradshaw, J. S., Hui, J. Y. K.: J. Heterocycl. Chem. *11*, 649 (1974)
14. a) Bradshaw, J. S. et al.: J. Heterocycl. Chem. *11*, 45 (1974);
    b) Pedersen, C. J.: J. Org. Chem. *36*, 254 (1971);
    c) Black, D. St. C., McLean, I. A.: Tetrahedron Lett. *1969*, 3961; see also Meadow, J. R., Reid, E. E.: J. Am. Chem. Soc. *56*, 2177 (1934)
15. a) Gokel, G. W., Garcia, B. J.: Tetrahedron Lett. *1977*, 317;
    b) Lehn, J. M., Vierling, P.: ibid. *21*, 1323 (1980);
    c) Lockhart, J. C., Thompson, M. E.: J. Chem. Soc. Perkin I *1977*, 202;
    d) Richman, J. E., Atkins, T. J.: J. Am. Chem. Soc. *96*, 2268 (1974)
16. a) Black, D. St. C., McLean, I. A.: J. Chem. Soc. Chem. Commun. *1968*, 1004;
    b) Pellisard, D., Louis R.: Tetrahedron Lett. *1972*, 4589;
    c) Dietrich, B., Lehn, J. M., Sauvage, J. P.: J. Chem. Soc. Chem. Commun. *1970*, 1055;
    d) Kudrya, T. N., Shtepanek, A. S., Kirsanov, A. V.: Zh. Obhsh. Khim. *48* (4), 927 (1978) [Chem. Abstr. *89*, 197645g (1978)]; Kirsanov, A. V. et al.: Dokl. Akad. Nauk SSSR *247*, 613 (1979); cf. also Ciampolini, M. et al.: Inorg. Chim. Acta *45*, L239 (1980); Dutasta, J. P., Martin, J., Robert, J. B.: Heterocycles *14*, 1631 (1980);
    e) Arsen-analoga see: Kyba, E. P., Chou, S. S. P.: J. Am. Chem. Soc. *102*, 7012 (1980); Ennen, J., Kauffmann, Th.: Angew. Chem. *93*, 117 (1981).
17. Newkome, G. R. et al.: Chem. Rev. *77*, 513 (1977)
18. a) Timko, J. M. et al.: J. Am. Chem. Soc. *96*, 7097 (1974); Timko, J. M., Cram, D. J.: ibid. *96*, 7159 (1974);
    b) Newcomb, M. et al.: ibid. *99*, 6392 (1977);
    c) Newkome, G. R., Hager, D. C.: ibid. *100*, 5567 (1978);
    d) Vögtle, F., Weber, E.: Angew. Chem. *86*, 126 (1974); Angew. Chem., Int. Ed. Engl. *13*, 149 (1974); cf. also Weber, E., Vögtle, F.: Liebigs Ann. Chem. *1976*, 891
19. *Review* (macrocyclic esters): Bradshaw, J. S. et al.: Chem. Rev. *79*, 37 (1979)
20. a) Frensch, K., Oepen, G., Vögtle, F.: Liebigs Ann. Chem. *1979*, 858;
    b) Bradshaw, J. S. et al.: J. Chem. Soc. Perkin Trans. I, *1976*, 2505;
    c) Buhleier, E., Frensch, K., Luppertz, F., Vögtle, F.: Liebigs Ann. Chem. *1978*, 1586;
    d) Bogatsky, A. V., Luk'yanenko, N. G., Kirichenko, T. I.: Tetrahedron Lett. *21*, 313 (1980); Bogatskii, A. V., Luk'yanenko, N. G., Kirichenko, T. I.: J. Org. Chem. USSR *16*, 1124 (1980)
21. a) Weber, E.: Angew. Chem. *91*, 230 (1979); Angew. Chem., Int. Ed. Engl. *18*, 219 (1979); Cf. Czugler, M., Weber, E.: J. Chem. Soz. Chem. Commun. *1981*, 472
    b) Shinkai, S. et al.: Chem. Lett. *1980*, 283;
    c) Cf. also: Kimura, K. et al.: Chem. Lett. *1979*, 611; Rebek, J. et al.: J. Am. Chem. Soc. *102*, 7398 (1980)
22. Dietrich, B., Lehn, J. M., Sauvage, J. P.: Tetrahedron Lett. *1969*, 2885
23. *Review*: Lehn, J. M. in lit. [1a], p. 1
24. a) Dietrich, B. et al.: Tetrahedron *29*, 1629 (1973);
    b) Cheney, J., Kintzinger, J. P., Lehn, J. M.: Nouv. J. Chim. *2*, 411 (1978); Landini, D., Montanari, F., Rolla, F.: Synthesis *1978*, 223;
    c) Dietrich, B., Lehn, J. M., Sauvage, J. P.: J. Chem. Soc. Chem. Commun. *1970*, 1055;
    d) Lehn, J. M. et al.: J. Am. Chem. Soc. *99*, 6766 (1977);
    e) Haines, A. H., Karntiang, P.: J. Chem. Soc. Perkin Trans. I, *1979*, 2577; cf. also Coxon, A. C., Stoddart, J. F.: ibid. *1977*, 767; Hanson, I. R., Parsons, D. G., Truter, M. R.: J. Chem. Soc. Chem. Commun. *1979*, 486;
    f) Buhleier, E., Wehner, W., Vögtle, F.: Chem. Ber. *111*, 200 (1978); cf. also Wehner, W., Vögtle, F.: Tetrahedron Lett. *1976*, 2603;
    g) Newkome, G. R. et al.: J. Am. Chem. Soc. *101*, 1047 (1979);
    h) Lehn, J. M., Simon, J., Wagner, J.: Angew. Chem. *85*, 621 (1973); Angew. Chem., Int. Ed. Engl. *12*, 578 (1973); Lehn, J. M., Simon, J., Wagner, J.: Nouv. J. Chim. *1*, 77 (1977); see also lit. [24b];
    i) Graf, E., Lehn, J. M.: J. Am. Chem. Soc. *97*, 5022 (1975)
25. *Review*: Vögtle, F., Weber, E.: Angew. Chem. *91*, 813 (1979); Angew. Chem., Int. Ed. Engl. *18*, 753 (1979)
26. *Review*: Smid, J.: Angew. Chem. *84*, 127 (1972); Angew. Chem., Int. Ed. Engl. *11*, 112 (1972);

cf. also Szwarc, M.: Ions and Ion Parsi in: Org. React. Vol. 1 and 2, Wiley, New York, 1972 and 1974

27. a) Weber, E., Vögtle, F.: Tetrahedron Lett. *1975*, 2415;
    b) Vögtle, F., Sieger, H.: Angew. Chem. *89*, 410 (1977); Angew. Chem., Int. Ed. Engl. *16*, 396 (1977);
    c) Vögtle, F., Heimann, U.: Chem. Ber. *111*, 2757 (1978);
    d) Schultz, W. J. et al.: J. Am. Chem. Soc. *102*, 7981 (1980)

28. *Review*: Vögtle, F.: Chimia *33*, 239 (1979)

29. a) *Review*: Morf, W. E. et al. in lit. [1c], p. 1.; see also Pretsch et al.: Helv. Chim. Acta *63*, 191 (1980);
    b) Schneider, J. K. et al.: Helv. Chim. Acta *63*, 217 (1980)

30. a) Vögtle, F. et al.: Angew. Chem. *89*, 564 (1977); Angew. Chem., Int. Ed. Engl. *16*, 548 (1977);
    b) Heimann, U., Herzhoff, M., Vögtle, F.: Chem. Ber. *112*, 1392 (1979);
    c) Vögtle, F. et al.: Chem. Ber. *112*, 899 (1979)

31. a) Fornasier, R. et al.: Tetrahedron Lett. *1976*, 1381
    b) Haines, A. H., Karntiang, P.: Carbohydr. Res. *78*, 205 (1980)

32. a) Vögtle, F., Weber, E.: Angew. Chem. *86*, 896 (1974); Angew. Chem., Int. Ed. Engl. *13*, 814 (1974);
    b) Hyatt, J. A.: J. Org. Chem. *43*, 1808 (1978)

33. a) Gokel, G. W., Dishong, D. M., Diamond, C. J.: J. Chem. Soc. Chem. Commun. *1980*, 1053;
    b) cf. also Lehn, J. M., Sauvage, J. P.: J. Am. Chem. Soc. *97*, 6700 (1975);
    c) Weitl, F. L., Raymond, K. N., Smith, W. L., Howard, T. R.: ibid. *100*, 1170 (1978)

34. a) Beresford, G. D., Stoddart, J. F.: Tetrahedron Lett. *21*, 867 (1980);
    b) Tabushi, I., Kobuke, Y., Nishiya, T.: ibid. *1979*, 3515;
    c) Alberts, A. H., Cram, D. J.: J. Am. Chem. Soc. *101*, 3545 (1979)

35. a) Cram, D. J. et al.: J. Am. Chem. Soc. *101*, 6753 (1979); cf. also Koenig, K. E. et al.: ibid. *101*, 3553 (1979);
    b) Cram, D. J. et al.: J. Chem. Soc. Chem. Commun. *1979*, 948

36. *Reviews*: a) Bradshaw, J. S. in lit. [1b], p. 53;
    b) Stoddart, J. F. in: S. Patai (Ed.), The Chemistry of the Ether Linkage Supplement E, part 1, Wiley, London, 1981

37. *Review*: DeSousa Healy, M., Rest, A. J.: Adv. Inorg. Chem. Radiochem. *21*, 1 (1978)

38. Piepers, O., Kellogg, R. M.: J. Chem. Soc. Chem. Commun. *1978*, 383; van Keulen, B. J., Kellogg, R. M., Piepers, O.: J. Chem. Soc. Chem. Commun. *1979*, 285

39. Kulstad, S., Malmsten, L. A.: Tetrahedron Lett. *21*, 643 (1980)

40. Cf. *review*: Schwind, R. A., Gilligan, T. J., Cussler, E. L. in lit. [1b], p. 289

41. *Review*: Vögtle, F., Weber, E. in: Patai, S. (Ed.), The Chemistry of the Ether Linkage, Supplement E, part 1, p. 59, Wiley, London 1981

42. *Reviews*: a) Cram, D. J. et al.: Pure Appl. Chem. *43*, 327 (1975);
    b) Cram, D. J. in: Application of Biochemical Systems in Organic Chemistry, part II (Eds. Jones, J. B., Sih, C. J., Perlmann, D.), Techniques of Chemistry, Vol. X, Wiley, New York, 1976

43. *Review*: Izatt, R. M., Eatough, D. J., Christensen, J. J. in lit. [1a], p. 161

44. *Reviews*: a) Truter, M. R. in lit. [1a], p. 71;
    b) Kent Dalley, N. in lit. [1b], p. 207

45. a) Seiler, P., Dobler, M., Dunitz, J. D.: Acta Crystallogr. *B30*, 2744 (1974);
    b) Mallinson, P. R., Truter, M. R.: J. Chem. Soc. Perkin Trans. 2, *1972*, 1818;
    c) Hughes, D. L.: J. Chem. Soc. Dalton Trans. *1975*, 2374;
    d) Bush, M. A., Truter, M. R.: J. Chem. Soc. Perkin Trans. 2, *1972*, 345

46. a) Frensdorff, H. K.: J. Am. Chem. Soc. *93*, 600 (1971)
    b) Izatt, R. M. et al.: J. Am. Chem. Soc. *99*, 6134 (1977); Izatt, R. M. et al.: Inorg. Chim. Acta *30*, 1 (1978);
    c) Tümmler, B. et al.: J. Am. Chem. Soc. *99*, 4683 (1977);
    d) Ammann, D., Pretsch, E., Simon, W.: Helv. Chim. Acta *56*, 1780 (1973); Bradshaw, J. S. et al.: J. Chem. Soc. Chem. Commun. *1975*, 874; Petranek, J., Ryba, O.: Tetrahedron Lett. *1977*, 4249

47. *Reviews*: a) Lehn, J. M.: Acc. Chem. Res. *11*, 49 (1978);
    b) Lehn, J. M.: Pure Appl. Chem. *49*, 857 (1977)
48. Metz, B., Moras, D., Weiss, R.: J. Chem. Soc. Chem. Commun. *1970*, 217; Moras, D., Metz, B., Weiss, R.: Acta Crystallogr. *B29*, 388 (1973)
49. a) *Review*: Saenger, W., Suh, I. H., Weber, G.: Isr. J. Chem. *18*, 253 (1979);
    b) Saenger, W., Reddy, B. S.: Acta Crystallogr. *B35*, 56 (1979);
    c) Saenger, W., Brand, H., Vögtle, F., Weber, E. in (Pullman, B., Goldblum, N., Eds.): Metal-Ligand Interactions in Organic Chemistry and Biochemistry, Part 1, p. 363, D. Reidel Publishing Company, Dordrecht-Holland, 1977; Saenger, W., Brand, H.: Acta Crystallogr. *B35*, 838 (1979);
    d) Weber, G., Sheldrick, G. M.: Inorg. Chim. Acta *45*, L35 (1980)
50. Tümmler, B. et al.: J. Am. Chem. Soc. *101*, 2588 (1979)
51. a) *Review*: Liesegang, G. W., Eyring, E. M. in lit. [1b], p. 245;
    b) E.g. Lehn, J. M., Sauvage, J. P., Dietrich, B.: J. Am. Chem. Soc. *92*, 2916 (1970); Ceraso, J. M., Dye, J. L.: ibid. *95*, 4432 (1973); Henco, K., Tümmler, B., Maaß, G.: Angew. Chem. *89*, 567 (1977); Angew. Chem., Int. Ed. Engl. *16*, 538 (1977); Cox, B. G., Schneider, H., Stroka, J.: J. Am. Chem. Soc. *100*, 4746 (1978);
    c) E.g. Shchori, E. et al.: J. Am. Chem. Soc. *93*, 7133 (1971); Chock, P. B.: Proc. Nat. Acad. Sci. USA *69*, 1939 (1972); Shchori, E., Jagur. Grodzinski, J., Shporer, M.: J. Am. Chem. Soc. *95*, 3842 (1973); Laidler, D., Stoddart, J. F.: J. Chem. Soc. Chem. Commun. *1976*, 979; DeJong, F., Reinhoudt, D. N., Huis, R.: Tetrahedron Lett. *1977*, 3985
52. *Reviews*: a) Lehn, J. M.: Pure Appl. Chem. *50*, 871 (1978);
    b) Lehn, Pure Appl. Chem. *51*, 979 (1979)
53. Metz, B., Rozalky, J. M., Weiss, R.: J. Chem. Soc. Chem. Commun. *1976*, 533
54. a) Lehn, J. M., Vierling, P., Hayward, R. C.: J. Chem. Soc. Chem. Commun. *1979*, 296;
    b) Johnson, M. R., Sutherland, I. O.: J. Chem. Soc. Chem. Commun. *1979*, 306, 309; Magswaran, R., Magswaran, S., Sutherland, I. O.: J. Chem. Soc. Chem. Commun. *1979*, 722; Johnson, M. R., Sutherland, I. O.: J. Chem. Soc. Perkin Trans. 1, *1980*, 586
55. *Reviews*: a) Cram, D. J., Cram, J. M.: Science *183*, 803 (1974);
    b) Hayward, R. C.: Nachr. Chem. Techn. Lab. *25*, 15 (1977);
    c) Cram, D. J., Cram, J. M.: Acc. Chem. Res. *11*, 8 (1978);
    d) Prelog, V.: Pure Appl. Chem. *50*, 893 (1978);
    e) Stoddart, J. F.: Chem. Soc. Rev. *8*, 85 (1979)
56. a) Girodeau, J. M., Lehn, J. M., Sauvage, J. P.: Angew. Chem. *85*, 813 (1975); Angew. Chem., Int. Ed. Engl. *14*, 764 (1975); Helv. Chim. Acta *63*, 2096 (1980);
    b) Lehn, J. M., Sirlin, C.: J. Chem. Soc. Chem. Commun. *1978*, 849;
    c) Laidler, D. A., Stoddart, J. F.: ibid. *1976*, 979;
    d) De Vries, J. G., Kellogg, R. M.: J. Am. Chem. Soc. *101*, 2759 (1979); see also Zinic, M., Bosnic-Kasnar, B., Kolbah, D.: Tetrahedron Lett. *21*, 1365 (1980);
    e) Kyba, E. P. et al.: J. Am. Chem. Soc. *95*, 2691, 2692 (1973);
    f) Dietrich, B., Lehn, J. M., Simon, J.: Angew. Chem. *86*, 443 (1974); Angew. Chem., Int. Ed. Engl. *13*, 406 (1974)
57. a) Sousa, L. R. et al.: J. Am. Chem. Soc. *96*, 7100 (1974); Gokel, G. W., Timko, J. M., Cram, D. J.: J. Chem. Soc. Chem. Commun. *1975*, 394; Sousa, L. R., Sogah, G. D. Y., Hoffmann, D. H., Cram, D. J.: J. Am. Chem. Soc. *100*, 4569 (1978);
    b) Helgeson, R. C. et al.: J. Am. Chem. Soc. *96*, 6762 (1974)
58. *Review*: Lehn, J. M.: Pure Appl. Chem. *52*, 2441 (1980); see also Nelson, S. M.: ibid. *52*, 2461 (1980)
59. a) Lehn, J. M., Simon, J.: Helv. Chim. Acta *60*, 141 (1971);
    b) Alberts, A. H., Annunziata, R., Lehn, J. M.: J. Am. Chem. Soc. *99*, 8502 (1977);
    c) Coughlin, P. K. et al.: J. Am. Chem. Soc. *101*, 265 (1979); Drew, M. G. B., McCann, M., Nelson, S. M.: J. Chem. Soc. Chem. Commun. *1979*, 481; Drew, M. G. B. et al.: ibid. *1980*, 1122;
    d) Lehn, J. M., Simon, J., Moradpour, A.: Helv. Chim. Acta *61*, 2407 (1978);
    e) Lehn, J. M., Stubbs, M. E.: J. Am. Chem. Soc. *96*, 4011 (1974)
60. a) Simmons, H. E., Park, C. H.: J. Am. Chem. Soc. *90*, 2428 (1968); Park, C. H., Simmons, H. E.: ibid. *90*, 2431 (1968);

b) Schmidtchen, F. P.: Angew. Chem. *89*, 751 (1977); Angew. Chem., Int. Ed. Engl. *16*, 720 (1977); Chem. Ber. *113*, 864 (1980)

c) Graf, E., Lehn, J. M.: J. Am. Chem. Soc. *98*, 6403 (1976); Lehn, J. M., Souveaux, E., Coillard, A. K.: ibid. *100*, 4914 (1978);

d) Wester, N., Vögtle, F.: J. Chem. Res. (S) *1978*, 400; Wester, N., Vögtle, F.: Chem. Ber. *112*, 3723 (1979)

61. Short overview: Vögtle, F., Müller, W. M., Weber, E.: Chem. Ber. *113*, 1130 (1980)

62. a) Pedersen, C. J.: J. Org. Chem. *36*, 1690 (1971); Raßhofer, W., Vögtle, F.: Tetrahedron Lett. *1978*, 309; Vögtle, F., Oepen, G., Raßhofer, W.: Liebigs Ann. Chem. *1979*, 1577; Weber, E., Müller, W. M., Vögtle, F.: Tetrahedron Lett. *1979*, 2335;

b) Vögtle, F., Müller, W. M.: Chem. Ber. *113*, 2081 (1980);

c) Knöchel, A. et al.: J. Chem. Soc. Chem. Commun. *1978*, 595;

d) El Basyony, A. et al.: Z. Naturforsch. *31b*, 1192 (1976); Kaufmann, R. et al.: Chem. Ber. *110*, 2249 (1977);

e) Gokel, G. W. et al.: J. Org. Chem. *39*, 2445 (1974);

f) Goldberg, I.: Acta Crystallogr. *B31*, 754 (1975); Oepen, G., Vögtle, F.: Liebigs Ann. Chem. *1979*, 2114; Vögtle, F., Müller, W. M.: Naturwissenschaften *67*, 255 (1980)

63. Weber, E., Vögtle, F.: Angew. Chem. *92*, 1067 (1980); Angew. Chem., Int. Ed. Engl. *19*, 1030 (1980)

64. *Reviews*: a) Dehmlow, E. V.: Angew. Chem. *86*, 187 (1974); Angew. Chem., Int. Ed. Engl. *13*, 170 (1974); Angew. Chem. *89*, 521 (1977); Angew. Chem., Int. Ed. Engl. *16*, 493 (1977);

b) Weber, W. P., Gokel, G. W.: Phase-transfer Catalysis in Organic Synthesis, Reactivity and Structure Concept in Organic Chemistry Vol. 4, Springer, Berlin, 1977

65. *Reviews*: a) Liotta, C. L. in lit. [1b], p. 111;

b) Lehn, J. M.: Pure Appl. Chem. *52*, 2303 (1980)

66. *Review*: Gutmann, V.: Chimia *31*, 1 (1977)

67. Cf. Chronik, Chem. uns. Zeit. *8*, 126 (1974)

68. *Reviews*: a) Gokel, G. W., Durst, H. D.: Aldrichimica Acta *9*, 3 (1976);

b) Gokel, G. W., Durst, H. D.: Synthesis *1976*, 168;

c) Knipe, A. C.: J. Chem. Educ. *53*, 618 (1976);

d) Gokel, G. W., Weber, G. W.: ibid. *55*, 350, 429 (1978)

69. *Review*: Ho, T. L.: Hard and Soft Acids and Bases Principle in Organic Chemistry, Academic Press, New York, 1977

70. Cf. *review*: Clark, J. H.: Chem. Rev. *86*, 429 (1980)

71. a) Liotta, C. L., Harris, H. P.: J. Am. Chem. Soc. *96*, 2250 (1974);

b) Liotta, C. L. et al.: Tetrahedron Lett. *1974*, 2417

72. a) Cambillau, C., Sarthou, P., Brahm, B.: Tetrahedron Lett. *1976*, 281; Cambillau, C. et al.: Tetrahedron *34*, 2675 (1978); cf. also Noe, E. A., Raban, M.: J. Am. Chem. Soc. *96*, 6184 (1974);

b) See Akabori, S., Tuji, H.: Bull. Chem. Soc. Jpn. *51*, 1197 (1978); Smith, S. G., Hanson, M. P.: J. Org. Chem. *36*, 1931 (1971); Kurts, A. J. et al.: Zh. Org. Khim. *9*, 1313 (1973); Zangg, H. E. et al.: J. Org. Chem. *37*, 2249 (1972); Pierre, J. L. et al.: Tetrahedron Lett. *1978*, 3259; Whitney, R. R., Jaeger, D. A.: Tetrahedron *36*, 769 (1980); Sakakibara, T., Haraguchi, K.: Bull. Chem. Soc. Jpn. *53*, 279 (1980)

73. a) Handel, H., Pierre, J. L.: Tetrahedron Lett. *1976*, 741 and former papers of this series; Loupy, A., Seyden-Penne, Tchoubar, B.: ibid. *1976*, 1677;

b) Pierre, J. L., Handel, H., Perraud, R.: Tetrahedron Lett. *1977*, 2013;

c) Handel, H., Pierre, J. L.: Tetrahedron *31*, 2799 (1975); Loupy, A., Seyden-Penne, J.: Tetrahedron Lett. *1978*, 2571

74. *Review*: Bartsch, R. A.: Acc. Chem. Res. *8*, 239 (1975)

75. a) Cf. also Alunni, S., Perucci, P., Ruzziconi, R.: Gazz. Chim. Ital. *110*, 261 (1980);

b) Gokel, G. W., Gerdes, H. M., Rebert, N. W.: Tetrahedron Lett. *1976*, 653;

c) Boden, R.: Synthesis *1975*, 784; Mikolajczyk, M. et al.: Synthesis *1975*, 278; Delmas, M., Le Bigot, Y., Gaset, A.: Tetrahedron Lett. *21*, 4831 (1980)

76. *Review*: Irving, H. M. N.: Pure Appl. Chem. *50*, 1129 (1978); Kolthoff, I. M.: Anal. Chem. *51*, 182 (1979)

77. *Review*: Blasius, E. et al.: Z. Anal. Chem. *284*, 337 (1977)

78. *Reviews*: Simon, W. et al.: Pure Appl. Chem. *44*, 613 (1975); Ammann, D. et al. in: Ion and Enzyme Electrodes in Biology and Medicine (Kessler, M., Ed.), p. 22, Urban and Schwarzenberg, München, Berlin, Wien, 1976; Pretsch, E. et al. in: Analytical Chemistry, Essays in Memory of Anders Ringbohm (Wänninen, E., Ed.), p. 231, Pergamon Press, Oxford, New York, 1977

79. a) *Reviews*: Schindler, J. G., Dennhardt, R., Simon, W.: Chimia *31*, 404 (1977); Schindler, J. G.: Biomed. Techn. *22*, 235 (1977);
    b) Schindler, J. G., v. Gülich, M.: J. Clin. Chem. Clin. Biochem. *19*, 49 (1981)

80. *Review*: Turro, N. J., Grätzel, M., Braun, A. M.; Angew. Chem. *92*, 712 (1980); Angew. Chem., Int. Ed. Engl. *19*, 675 (1980)

Structural Chemistry of Natural and Synthetic Ionophores and their Complexes with Cations

Rolf Hilgenfeld and Wolfram Saenger

Abteilung Chemie, Max-Planck-Institut für experimentelle Medizin,
Hermann-Rein-Straße 3, D-3400 Göttingen, FRG

Table of Contents

1 Introduction: Ionophores as Host Molecules for Cationic Guests

Although the antibiotics nigericin and lasalocid as first representatives of the naturally occurring ionophores were isolated as early as 1951 from *Streptomyces* cultures [1], it was only in the late sixties that the function of these membranes affecting compounds as complexing and transporting agents for alkali metal ions was established [2-4]. Once their outstanding properties as selective alkali cation carriers had been understood, however, they attracted the attention of numerous investigators, and many details have been learnt about their structures, complex formation and physiological activity. Furthermore, a considerable number of model compounds were synthesized, above all the so-called polyether ligands which did not only help to understand the mechanism of action of the naturally occurring ionophores but also led to the rapid development of phase-transfer techniques in organic chemistry, to mention only one application.

Ionophores can be characterized as receptors which form stable, lipophilic complexes with charged hydrophilic species such as Na^+, K^+, Ca^{2+} etc., and thus are able to transport them into lipophilic phases, for example across natural or artificial membranes. Very often, the processes of complexation and transport are highly specific: Many of the ionophores display the ability of discrimination between alkali metal ions of different size. Thus, the antibiotic valinomycin has a 10^4 times greater affinity to potassium than to sodium ions [5]. Both the specifity of complexation and the transport of highly polar entities into non-polar media requires certain structural features of the ligands, and in this review we shall endeavour to point out what these features are.

The elucidation of the spatial structures of ionophores and their ion complexes is essential for understanding the detailed mechanisms of their biological action. From this point of view, the nature of the conformational rearrangements accompanying complex formation is of special interest, because these very conformational changes contribute considerably to ion binding selectivity. For that reason, this review will focus on comparisons between ligand conformations in both the complexed and the uncomplexed state whenever data on both are available.

Since X-ray diffraction today is still the unique method of providing information of high accuracy on molecular conformations, we shall consider mainly results obtained from these studies. However, although in general representing a form of low energy, ligand conformations in crystals may differ from those in solution. Therefore, we shall try to compare both the solid state and the solution conformations if investigations of the latter have been carried out.

In view of the limited space available, this review cannot be encyclopaedic. Rather, we will select certain structures which we consider to be of interest. However, after each section we will give a compilation of published crystal structures to stimulate further studies.

Now, what are the essential requirements for ligand to be an effective ionophore?

Metal ion complexes of ionophores can be considered as host-guest complexes in which the guest entity is of spherical shape and entrapped in a cavity-like structure formed by the cyclic or open-chain host molecule. This cavity site can either be preformed to accept the metal ion without major conformational changes or it can adopt its final shape upon complexation of the cation, associated with structural rearrangements. In all cases, a mutual geometrical, topological fit between host and guest molecules is essential for adduct stabilization, the adduct being in general a 1:1 complex for the ionophores. For an alkali metal ion as a spherical guest, the optimum complementary structural feature is a cavity of corresponding size, lined with polar groups in order to provide maximum interaction through ion-dipole forces. The polar ligand groups, which usually contain electronegative atoms such as oxygen, nitrogen or, more rarely, sulfur, should be situated in such a way that they can step by step replace the solvation shell of the cation during complex formation. The exterior of the ligand molecules, however, should be lipophilic to provide an appropriate surface for the non-polar medium into which the metal ion is being transferred.

These structural features are maintained more or less by all the ionophores in their complexed forms. However, mention should also be made of exceptions to these general structural principles. Thus, in the case of some ionophores, cation complexes of other than 1:1 stoichiometry have also been described in which the formation of "sandwich"-type arrangements is favored [6].

Generally, two different modes of transmembraneous transport have been established: the "carrier" and the "channel" mechanism. The ionophores considered here act by the carrier mechanism. They form discrete antibiotic cation complexes at one interface of the membrane which then migrate across the membrane to the other interface where the metal ion is released. This kind of transport is displayed by the depsipeptide-type antibiotics which form positively charged complexes with metal ions. This is also true for the macrotetrolide nactins whereas the open-chain poly-ether antibiotics of the nigericin family mainly lead to electrically neutral metal ion complexes by dissociation of their carboxyl group. For the latter type of carriers, the ion transport of metal ions is coupled with a transfer of protons in the opposite direction.

Some linear peptides such as the gramicidins A, B, and C, alamethicin, suzukacillin, and trichotoxin A-40 do not act as carriers but they form transmembrane channels across which alkali metal ions can migrate. Just as the carrier cavities, these channels display a hydrophilic interior and a lipophilic exterior, but in contrast to the former they exhibit poor ion selectivity. Since no complete X-ray studies of any of these "channel" forming agents are available [7], only few facts are known about their conformations. Therefore, they will not be treated in this review.

A main section of this review will be devoted to the stereochemistry of synthetic ionophores and their ion complexes. The first analogs of the ionophorous antibiotics were cyclic *crown ethers* [12] some of which were shown to display similar selective complexation and transfer as the naturally occurring ligands [8–11]. Later, macrobi-cyclic oligoethers providing three-dimensional receptor cavities were designed [13]. Although they cannot be considered as model compounds for natural ionophores, these so-called *cryptands* are still very useful because, just as the crown ethers, they have found considerable application in organic chemistry as reagents in-fluencing the rate and stereochemistry of reactions.

Finally, a number of open-chain polyether ligands — frequently called *podands* — has been synthesized [14]. From the viewpoint of a structural chemist, the latter are perhaps the most interesting synthetic ligands because of their high flexibility which often allows for an optimum adoption to the shape of the guest entity. Though some synthetic polyethers have been shown to transport alkali metal ions across lipid bilayers, the membrane-affecting properties of many of these ligands have not yet been investigated. This is why Ovchinnikov and coworkers prefer the name "complexones" rather than "ionophores" for these compounds in order to emphasize their alkali cation-chelating properties [15]. However, in accordance with the more general definition of an ionophore given above we shall use the latter notation for all ligands that are capable of transferring group IA and IIA metal ions to lipophilic phases (which do not necessarily have to be membranes).

Apart from metal ions, many of the ionophores are also capable of complexing neutral or charged organic guest molecules. In these complexes, interaction between host and guest is mainly achieved by hydrogen bonding and dipole-induced forces. In contrast to group IA and IIA metal ions, which show spherically symmetric electron-acceptor properties, binding sites are of a more directional character in the case of molecular guests.

Since adducts between organic molecules are discussed in other reviews [345, 346] they will not be treated here, i.e. the discussion is restricted to complexes between ionophores and metal ions.

2 Depsipeptides: The "Prime" Ionophores

2.1 Valinomycin

Valinomycin is a macrocyclic dodecadepsipeptide with 12 subunits — amino- and hydroxycarboxylic acids — which are connected by alternate peptide and ester bonds. It consists of three identical fragments D-HyIv-D-Val-L-Lac-L-Val[1] (Fig. 1). Although the valinomycin structure has been discussed in a recent review on complex formation of monovalent cations with biofunctional ligands [16], it is described here in some detail because the exceptionally high K^+/Na^+ discrimination displayed by the antibiotic is unequalled by any other ionophore which led to its designation as a "prime ionophore" [17]. Yet another reason is the fact that the conformations of valinomycin in various solvents and in lipid bilayer membranes are still the object of intensive research so that periodic reviewing is justified.

Preliminary crystallographic results on the potassium complex of valinomycin were first reported by Steinrauf et al. in 1969 [18], and in 1975, a more detailed study by Neupert-Laves and Dobler was published [19]. Both revealed close to threefold symmetry of the complex, the potassium ion being located at the center of the 36-membered ring (Fig. 2). The latter is folded into six β-turns which are stabilized by intramolecular N—H ... O=C hydrogen bonds of the common $4 \rightarrow 1$ type [20], i.e. reverse hydrogen bonding between the C=O of residue i and the N—H of residue i + 3. As can be seen from Fig. 5a, hydrogen bonds of this kind lead to the formation of 10-membered rings.

---

1 D-HyIv = D-hydroxyisovalerate

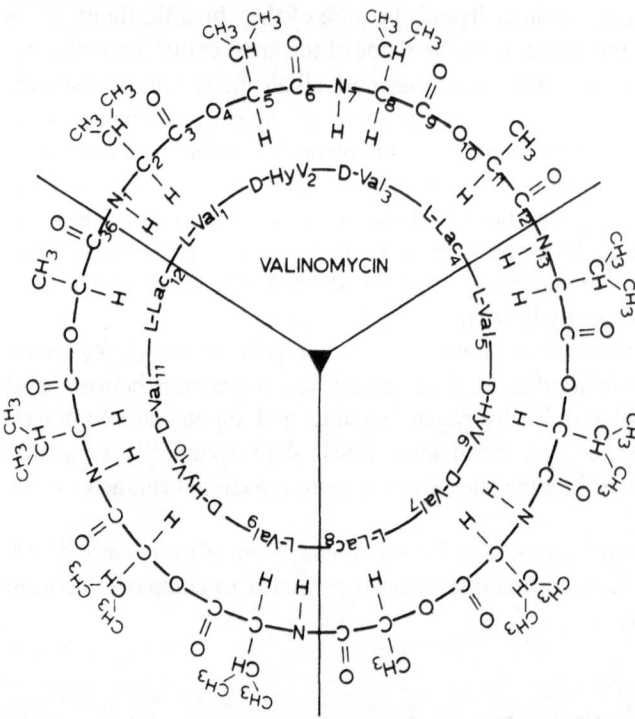

Fig. 1. Primary structure of valinomycin. The potential center of threefold symmetry is indicated

All the carbonyl oxygens involved in hydrogen bonding belong to amide groups; the ester CO groups are coordinated to the metal ion in an almost perfect octahedral arrangement, thus providing effective screening of the central cation from solvent interaction. All the lipophilic side chains are oriented toward the periphery of the molecule. Space filling models of the complex molecule are shown in Fig. 3.

The six hydrogen bonds are responsible for the limited flexibility of the depsipeptide ligand in the complex. Thus, valinomycin is not able to adjust itself to cations of different radius which explains its high selectivity. Whereas K^+ is of optimum size to fill the ligand cavity, Na^+ is too small to fully interact with all the ester oxygens and cannot be complexed without an energetically unfavorable breaking of intramolecular hydrogen bonds. On the other hand, Cs^+ as guest entity is too large to fit into the host binding site (for ionic radii see Table 1) and thus valinomycin complexes cesium ions less readily than K^+. Presumably, association with Cs^+ results in steric strain in the ligand backbone and/or considerable weakening of hydrogen bonding.

Only rubidium whose ionic radius is close to that of potassium is bound at least as effectively as the latter. An X-ray structural analysis of the valinomycin-RbAuCl₄ complex revealed essentially the same structure as in the K^+ complex, the octahedral cage of the oxygen atoms being by only 0.04 Å wider due to minor rotations of the ester bonds [21].

Numerous spectroscopic studies [22-28] and a conformational energy calculation [29] of the solution conformation of the valinomycin-K^+ complex are in good agreement

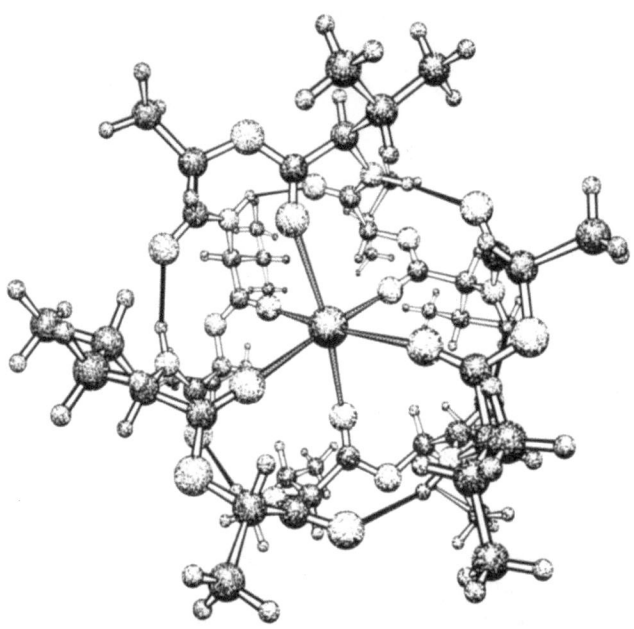

Fig. 2. Structure of the K^+ complex of valinomycin viewed from a 10 Å distance. Coordinative bonds are shaded, hydrogen bonding is indicated by narrow solid lines. Atomic radii are of arbitrary size and decrease in the order O > N > C > H. Similar graphical details are also followed in subsequent diagrams of molecular structures

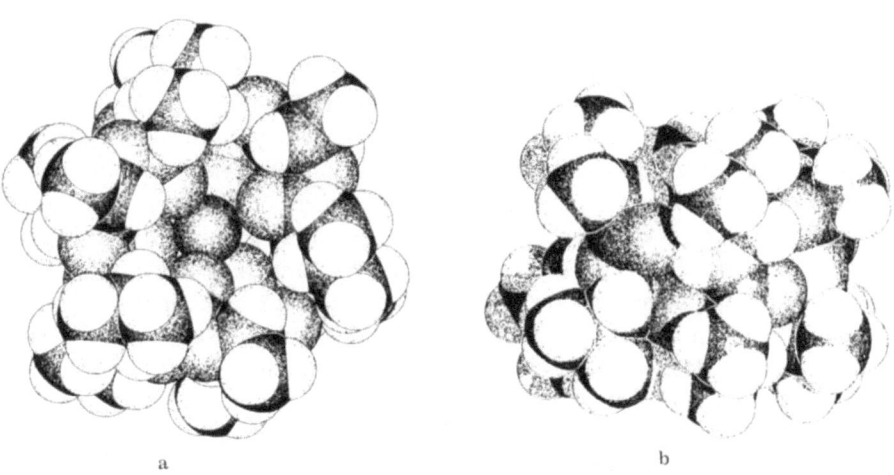

a   b

Fig. 3. Space-filling models of the valinomycin-K^+ complex: **a** view taken from the "top" of the molecule to demonstrate the optimum fit between host and guest, **b** view taken from one side of the complex showing the lipophilic periphery. Metal cation is completely covered and therefore not seen

Tab. 1. Ionic radii and charge densities of alkali and alkaline earth cations

Ion	r [Å]	Charge density $\times 10^{20}$ [Coulomb/Å^3]
Li$^+$	0.68	12.1
Na$^+$	0.95	4.46
K$^+$	1.33	1.62
Rb$^+$	1.48	1.18
Cs$^+$	1.69	0.79
Be^{2+}	0.35	178
Mg^{2+}	0.66	26.8
Ca^{2+}	0.99	7.90
Sr^{2+}	1.12	5.46
Ba^{2+}	1.34	3.17

with the results provided by crystallographic methods. This is not true, however, for the uncomplexed form of the antibiotic.

As indicated in Fig. 1, there is a potential center of threefold symmetry in the primary structure of valinomycin. However, X-ray diffraction studies on the uncomplexed molecule [30–33] have shown that this symmetry is not maintained in the crystalline state. Instead, the macrocycle adopts a somewhat ellipsoidal shape with a pseudo-center of symmetry (see Fig. 4), which is stabilized by six intramolecular N—H ... O=C hydrogen bonds. Four of these belong to the above mentioned 4→1 type and involve amide NH and CO groups, the corresponding N ... O distances being in the range of 2.81–3.13 Å (mean 2.95 Å) [32]. The remaining two hydrogen bonds are of the rare 5→1 type, which was actually first discovered in valinomycin. Ester rather than amide carbonyl oxygens act as acceptors of these bonds which are considerably weaker than those of the 4→1 type. Their mean N ... O distance is 3.05 Å, and the N—H ... O angles deviate largely from linearity. In contrast to 4→1 type hydrogen bonds, those of the 5→1 type build up 13-membered rings in the structure, as can be seen from Fig. 5b.

These 5→1 bonds are largely responsible for the oval shape of uncomplexed valinomycin. Moreover, they direct two of the ester carbonyl oxygens toward the surface of the molecule (see Fig. 4). These might serve as initiators of complex formation by interaction with the metal ion prior to the placement of the latter in the ligand cavity. Based on this assumption, Smith and Duax developed a simplified model for the complexation process [33,34]. They proposed that the formation of the initial loose complex with the potassium ion is followed by cleavage of both the 5→1 type hydrogen bonds in order to enable all the other ester carbonyl oxygens to interact with the cation and to replace the molecules of its solvation shell one after another.

The three X-ray diffraction studies of uncomplexed valinomycin actually revealed structures of five independent molecules. They all exhibit largely the same conformation, although the crystals were grown from solvents of different polarity and showed different modes of molecular packing and solvent contents. This finding strongly suggests that crystal packing forces do not markedly affect the conformation of valinomycin molecules in the solid state and justifies the assumption that this

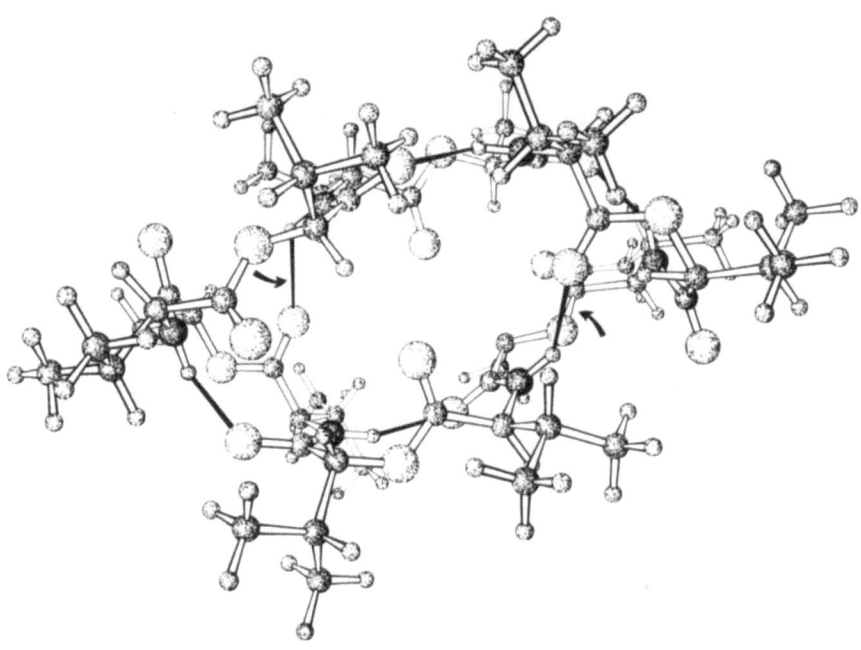

Fig. 4. Structure of uncomplexed valinomycin in the crystalline state. The two $5 \to 1$ type hydrogen bonds are indicated by arrows

low energy form also occurs in solution, at least in non-polar solutions where the solvent does not compete for hydrogen bonding [34].

There have been considerable efforts to elucidate the valinomycin conformation in various solvents by application of spectroscopic methods such as NMR, CD, ORD, IR, and Raman [35–45] as well as by conformational energy calculations [46]. However, little evidence has been provided for a solution conformation involving $5 \to 1$ type hydrogen bonds as observed in the solid state. Ovchinnikov and coworkers found that the conformation of uncomplexed valinomycin is highly dependent on solvent polarity. They proposed an equilibrium of three main forms which are usually referred to as A, B, and C (Fig. 6) [45]. Form A is believed to occur mainly in non-polar solvents such as CCl_4, octane etc. In the "bracelet"-like structure, there are six intramolecular NH ... O=C hydrogen bonds all of which are of the $4 \to 1$ type. Thus, in non-polar solvents the threefold symmetry of the molecule suggested by the primary structure (Fig. 1) is apparently maintained. However, Rothschild et al. reported evidence from Raman studies for hydrogen-bonded ester C=O groups in valinomycin dissolved in non-polar solvents [42] which indicates that the conformer found in the crystal is among those present in solution. Davies and Abu Khaled interpreted ^{1}H-NMR long-range proton coupling constants of valinomycin in $CDCl_3$ solution as consistent with a time-averaged conformation containing both $4 \to 1$ and $5 \to 1$ type H-bonds which are rapidly interconverting [35]. From conformational energy calculations, Maigret and Pullman concluded that the asymmetric conformer found in crystals is one of three possible structures existing in solution [46].

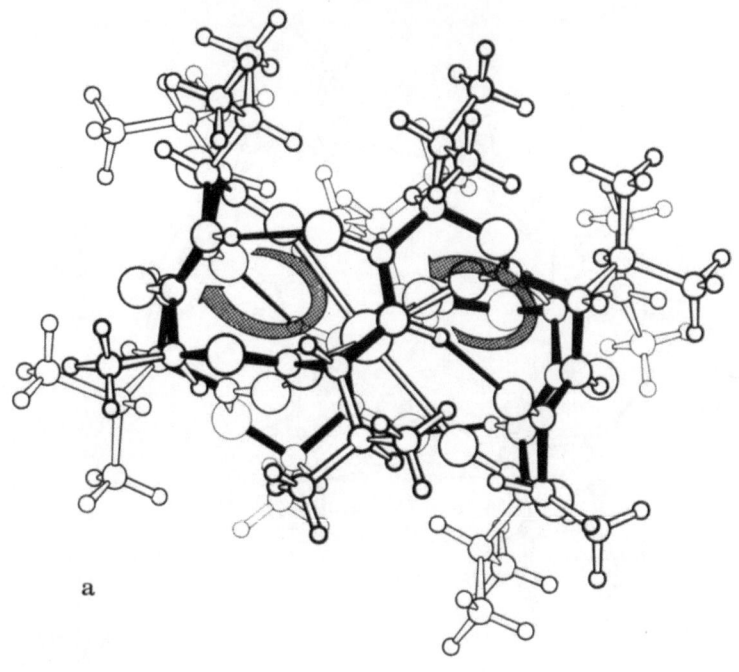

a

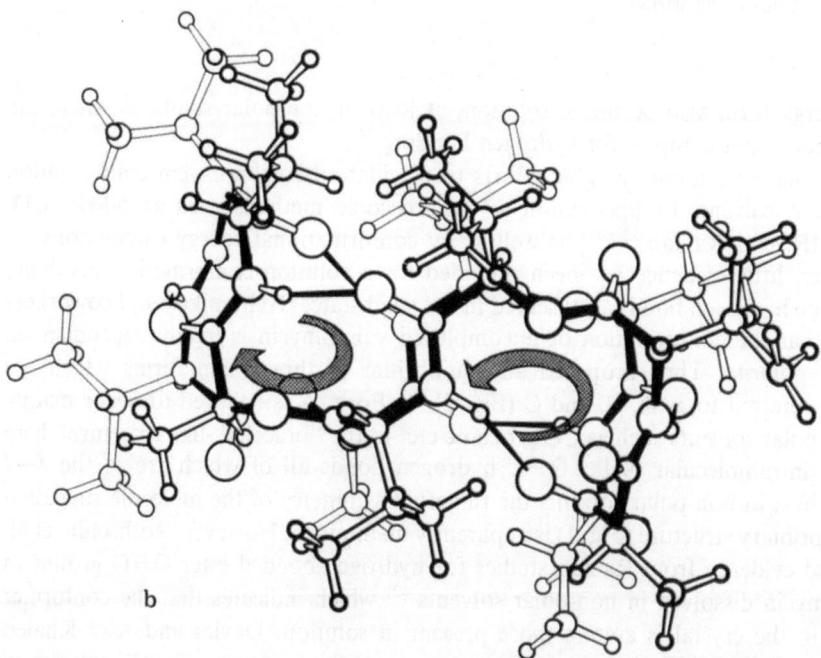

b

Fig. 5a and b. Hydrogen bonding in valinomycin-K$^+$ and in the free antibiotic: **a** Side view of the valinomycin potassium complex structure showing the ten-membered rings made up by $4{\rightarrow}1$ type hydrogen bonding. The antibiotic backbone is indicated by full bonds. **b** Side view of the uncomplexed valinomycin molecule. A 13-membered and a 10-membered ring, made up by $5{\rightarrow}1$ and $4{\rightarrow}1$ type hydrogen bonds, respectively, are shown

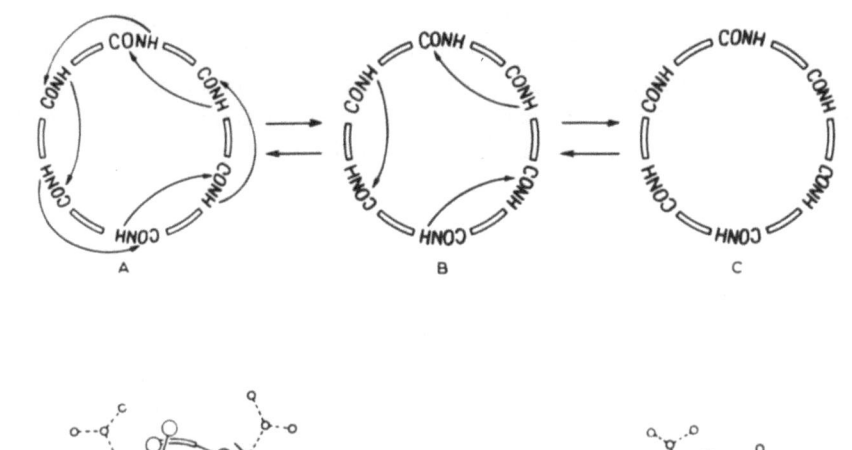

a

b

∘C ◯O ⓃN ══H-bond

Fig. 6a. Equilibrium between the three basic solution forms of valinomycin (from Ref. [15]). b Form A (left) is believed to adopt a bracelet-like conformation, form B resembles a propeller (from Ref. [15])

Recently, ^{1}H-NMR spectroscopic results for uncomplexed valinomycin in a phospholipid bilayer revealed largely the same conformation as for the molecule in non-polar solvents. This suggests a preferential location of the free carrier in the apolar interior of the bilayer [43].

The "propeller"-like form B of the Ovchinnikov model predominates in solvents of medium polarity. There are only three intramolecular hydrogen bonds. Form C eventually was proposed for polar solutions where all the NH groups of the molecule are apparently hydrogen bonded to solvent molecules.

Besides these, more conformers exhibiting various degrees of hydrogen bonding have been detected by ultrasonic absorption experiments [44]; they are believed to represent intermediates between the described basic forms [23].

Concerning the question of the occurrence of 5→1 type hydrogen bonds, crystal structures of synthetic valinomycin analogs are of interest. Surprisingly enough, there are no hydrogen bonds of this type in uncomplexed meso-valinomycin where L-lactate is replaced by L-hydroxyisovalerate [47]. All of the six H-bonds are of the 4→1 type, and the molecule adopts a bracelet-like conformation. Only one 5→1 H-bond is reported for isoleucinomycin, in which the valinomycin Val groups are substituted by Ile [48]. Apparently, the latter model compound is an intermediate between the extreme conformations of valinomycin and meso-valinomycin in their crystalline states. This variety of conformations of highly similar compounds suggests that the energetic barriers between the different forms are in fact quite low. Evidence is added to this viewpoint by the observation of a second form of valinomycin crystals provided by Raman spectroscopic methods. This crystal form grown from o-dichlorobenzene shows no hydrogen bonding to ester carbonyl oxygens [49,50].

2.2 Enniatins and Beauvericin

Until recently, valinomycin has been regarded as a classic monocarrier which only forms complexes of 1:1 stoichiometry with alkali metal ions. However, in 1975 Ivanov reported evidence for the formation of adducts with a 2:1 valinomycin:cation ratio, and proposed a sandwich-type structure for the latter [6].

Somewhat better than for valinomycin, these sandwich complexes have been characterized in the case of the enniatin antibiotics the pecularities of which will be discussed in this section. In fact, though 1:1 enniatin:metal ion complexes have been shown to exist as well, it has been suggested that their membrane-affecting activity is due to the formation of sandwich aggregates [17].

The enniatins are cyclic hexadepsipeptides, i.e. 18-membered macrocycles, which exhibit considerable lower cation selectivity than does valinomycin. Thus, they are efficient ionophores for sodium as well as for potassium and caesium ions [51].

Four different species with closely related primary structures have been described: enniatins A, B, C and beauvericin. The latter is produced by the fungus *Beauveria bassiana* [52] whereas the former can be isolated from various *Fusarium* cultures [53]. Their general formula is cyclo-(L-MeX-D-HyIv)$_3$, where X is Ile for enniatin A, Val for enniatin B, Leu for enniatin C, and Phe for beauvericin (see Fig. 7). One important structural difference to valinomycin is the methylation of the amide

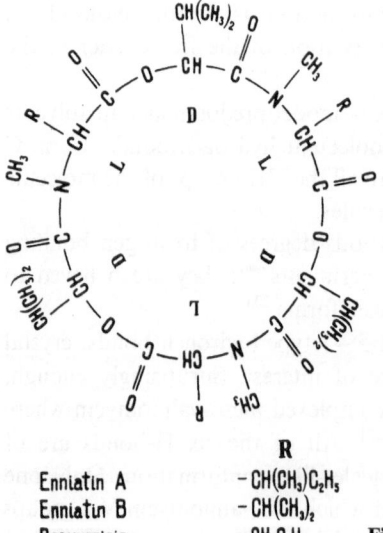

	R
Enniatin A	$-CH(CH_3)C_2H_5$
Enniatin B	$-CH(CH_3)_2$
Beauvericin	$-CH_2C_6H_5$

Fig. 7. Primary structure of enniatins and beauvericin

nitrogens. Thus, no formation of intramolecular hydrogen bonds is possible which makes the molecular backbone much more flexible.

According to Ovchinnikov and coworkers, it is this relative flexibility that accounts for the diminished selectivity of the enniatins toward alkali metal ions. Based on spectral data these authors postulated a structure of the 1:1 complexes with monovalent cations in which the metal ion occupies the center of the macrocyclic cavity and is coordinated by all of the six carbonyl oxygens in a octahedral arrangement [54,55]. Following this model, the ligand is able to easily adapt itself to the size of the metal ion by varying the orientation of the carbonyl groups. However, as pointed out by Steinrauf and Sabesan [21], this arrangement would lead to M^+ ... C (carbonyl) distances shorter than any observed earlier, especially in the case of the larger cations, but even for sodium. Moreover, the $C=O$ ... M^+ angles would all be in the unfavorable range of 90–100°, as compared to 152–162° found for valinomycin-K^+.

The structure of the enniatin B complex with potassium iodide has been studied by X-ray crystallography [56]. Unfortunately, from this investigation it could not be concluded with certainty whether the metal ion is entrapped in the central cavity or, instead, occupies a site between two adjacent ligand molecules. An arrangement of the latter type has been observed in the crystal structure of the 1:1 complex between RbNCS and the synthetic LDLLDL isomer of enniatin B [57]. In this case, Rb^+ ions are coordinated by five carbonyl oxygens (three of the upper and two of the lower depsipeptide molecules) and the nitrogen atom of the isocyanate anion, thus forming infinite sandwiches.

Evidence is added to the assumption of a binding of the metal cation outside the macrocyclic cavity also in the 1:1 complexes by the fact that these, though being more stable than sandwich-type aggregates, are apparently ineffective in membrane transport [17]. This suggests that the metal ion is not sufficiently shielded from solvent or counter ion interactions which would be in agreement with a metal ion position outside the macrocyclic cavity.

In comparison to the 1:1 complexes, the enniatin sandwiches display a higher ion selectivity, their actual stability constants decreasing in the order $K^+ > Cs^+ > Na^+$. Besides adducts of 2:1 stoichiometry, a 3:2 "club sandwich" has been proposed for the Cs^+ complex [17].

Among the enniatin antibiotics, beauvericin is the one characterized in greater detail. This carrier is most interesting with respect to an *anion*-dependence of its transport properties [58]. Moreover, in contrast to valinomycin, it is capable of complexing alkaline earth as well as alkali metal ions [59]. A study of the effects of beauvericin on the conductivity of artificial lipid membranes in the presence of both mono- and divalent cations revealed a second-order relationship between conductance and antibiotic concentration [60,61]. Finally, Prince, Crofts, and Steinrauf detected an apparent charge of *plus one* for calcium in the beauvericin-mediated transport across bacterial chromatophore membranes [62].

These most unusual findings could be well explained after elucidation of the crystal structure of the beauvericin complex with barium picrate [61,63]. This adduct turned out to be a 2:2 dimer structure of the form $(Bv \cdot Ba \cdot Pic_3 \cdot Ba \cdot Bv)^+ Pic^-$ (Fig. 8) which is very unique inasmuch as three of the four picrate anions are incorporated into the space between the antibiotic molecules. Both the Ba^{2+} ions

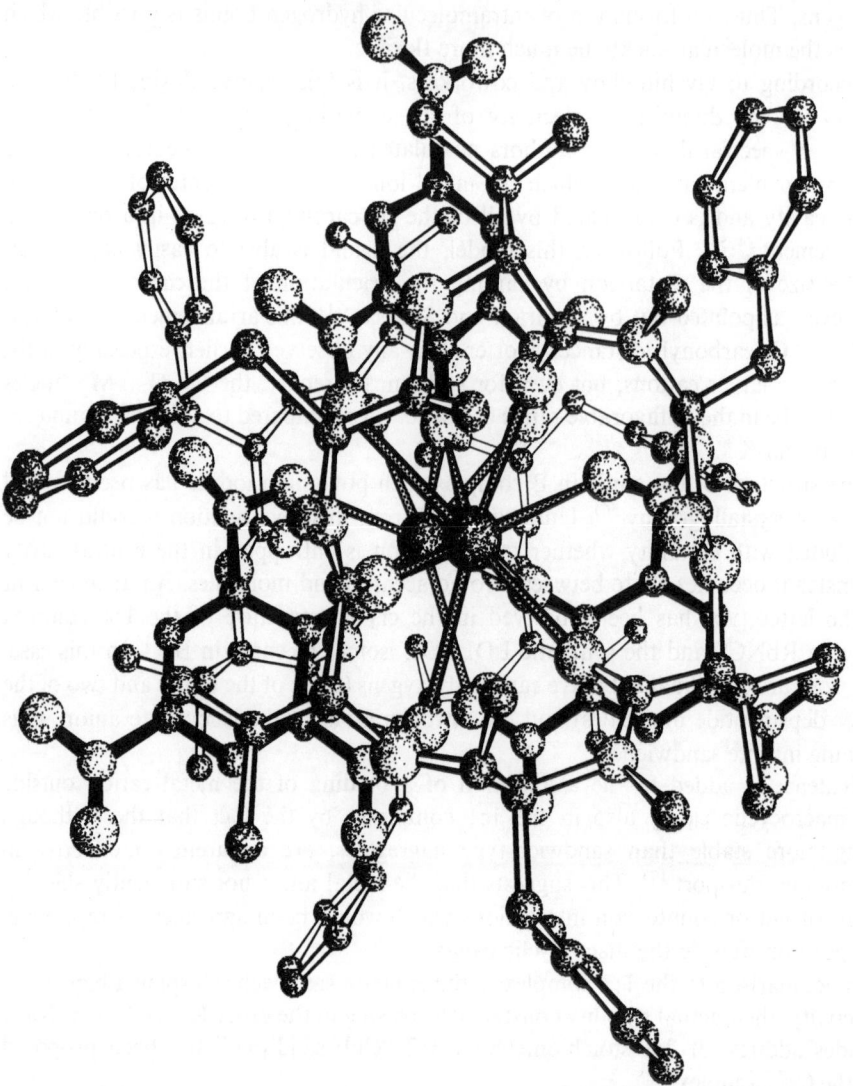

Fig. 8. Sandwich structure of the 2:2 complex between beauvericin and barium picrate shown from a 30 Å distance. From "top" to "bottom": 1) the "upper" beauvericin molecule (strong open lines), 2) one Ba^{2+} ion (dark sphere), 3) three picrate anions (full bonds), 4) the other Ba^{2+} ion, 5) the "lower" beauvericin molecule (weak open lines). For clarity, hydrogen atoms have been omitted

are *not* situated in the macrocyclic cavities but are displaced toward the center of the dimer, being connected to each other at a distance as short as 4.13 Å by the bridging picrate anions. Each of the metal ions is ninefold coordinated by the three amide oxygens of the ionophore (Ba ... O distances 2.64–2.77 Å), by three phenolate oxygens (one from each picrate anion, Ba ... O distances 2.72–2.78 Å), and by three nitro group oxygen atoms, again one contributed by each picrate (Ba ... O distances 2.96–3.07 Å).

The outer surface of the beauvericin-barium picrate complex is highly hydrophobic due to the orientation of the isopropyl, N-methyl, and phenyl groups toward the exterior of the molecule. The bulky phenyl residues provide an effective screening of the enclosed picrate ions.

A comparable 2:2 sandwich complex has also been found in a second crystal modification which showed the same overall structure only distinguished by a different orientation of the phenylalanyl residues [64]. This suggests that the described aggregation is indeed a very stable one. Moreover, crystals of the corresponding complexes with potassium [61] and rubidium [65] picrate apparently contain similar cluster units.

It is very likely that this beauvericin metal ion complex as revealed by X-ray structure analysis is the membrane-active species as well, because it is actually the only possible structure that is able to rationalize the aforementioned pecularities of ion transport induced by beauvericin. Above all, this is obvious for the residual net charge of plus one for the ionophore complex. *In vivo*, the picrate ions might be replaced by carboxylates which exhibit a similar chelating ability.

The structure of uncomplexed beauvericin as well has been investigated by single-crystal X-ray diffraction [65]. Apart from minor rotations of the side chains, its conformation is almost identical to that of the complexed antibiotic. Similar conclusions have been arrived at from a spectroscopic study of the free ligand in polar solvents [54] and from conformational energy calculations [65, 66].

From the structural studies of beauvericin and its picrate complexes two things can be learnt:

1) The intriguing power of X-ray crystallography as a method of explaining the structure-activity relationships of biomolecules: It was only the X-ray structure of the barium picrate complex of beauvericin that provided an understanding of the anion-dependent activity of the antibiotic. All spectroscopic approaches failed to do so.

2) The different modes of binding displayed by the ionophores: Whereas valinomycin undergoes conformational rearrangements including the breaking of hydrogen bonds to replace the solvation shell of the metal ion and to finally enclose the latter in the macrocyclic cavity, beauvericin does not adjust its conformation to fit the ion but lipophilizes the polar guest entities by encapsulating the cation-anion complex as a whole. Since the host cavity does not maintain its function as selective binding site, this process is less cation specific. In view of the involved anion specifity, however, it still represents a high stage of molecular organization.

3 Macrotetrolides: Cubic Cages for Alkali Ions

In the cases of valinomycin and enniatin depsipeptides described in Section 2. the explanation for the structural origins of the more or less pronounced ion selectivities exhibited by these antibiotics was, though being plausible, somewhat tentative, as completed X-ray analyses were available only for complexes with a single ion species, i.e. with K^+ for valinomycin and Ba^{2+} for beauvericin. However, a detailed discussion of the structural features that lead to metal ion selectivities should be based on a whole set of comparable data on complex structures with various metal ions of

Table 2. Published X-ray structures of depsipeptide complexes: Number and lengths of coordinative bonds. Bond distances in this and subsequent tables were either taken from the original publication or calculated employing the Cambridge Crystallographic Data Package [343]. In cases where no numerical values are given, both these sources did not provide crystallographic coordinates. A dash indicates that there is no coordination to the heteroatoms in question. All distances are given in Å units

Compound	Coord. no.	Bonding distances			Comments, if any	Ref.
		M — O	M — anion	M — solvent		
[valinomycin · K]+I3-/I5-	6	2.69–2.83	—	—		19)
[valinomycin · K]+AuCl4-	6		—	—		18)
[valinomycin · Rb]+AuCl4-	6		—	—		21)
[prolinomycin · Rb]+ pic-	6		—	—	peptide analog of valinomycin	67)
[enniatin B · K]+I-	6	2.6–2.8	—	—		56)
[LDLDL-enniatin B · RbNCS]	6	2.88–2.98	3.06	—	infinite "sandwich" spiral	57)
enniatin B · 6.25 H2O · Na+Ni2+(NO3-)3	6	—	—	—	both Na+ and Ni+ are not situated in the central cavity	68)
[(beauvericin)2 · Ba2 · (pic)3]+ pic-	9	2.64–2.77	2.72–3.07	—	2:2 "sandwich" structure	61,63)
[antamanide · Li · (MeCN)]+ Br-	5	2.04–2.23	—	2.07		69,70)
[Phe4, Val6-antamanide · Na · (EtOH)]+ Br-	5	2.25–2.36	—	2.28		69)

pic = picrate

Table 3. Uncomplexed depsipeptides investigated by crystallographic methods

Compound	Constitution	Comments	Ref.
valinomycin	cyclo-(D-HyIv-D-Val-L-Lac-L-Val)$_3$	contains two 5→1 type hydrogen bonds	30–33)
isoleucinomycin	cyclo-(D-HyIv-D-Ile-L-Lac-L-Ile)$_3$	contains one 5→1 type hydrogen bond	48)
meso-valinomycin	cyclo-(D-HyIv-D-Val-L-HyIv-L-Val)$_3$	contains only 4→1 type hydrogen bonds	47)
(Me · Ala^2)valinomycin		conformation close to native valinomycin	71,73)
(Me · Ala^{2,6})octavalinomycin	cyclo-(D-HyIv-L-Ala-L-Lac-L-Val)$_2$	contains 5→1 and 3→1 type hydrogen bonds	72,73)
enniatin B	cyclo-(L-Me-Val-D-HyIv)$_3$	molecular cavity occupied by two water molecules	57,73–75)
DLLLLL-enniatin B			76)
beauvericin	cyclo-(L-N-MePhe-D-HyIv)$_3$	conformation very similar to that in Ba^{2+} complex	65)
antamanide	cyclo-(Val-Pro-Pro-Ala-Phe-Phe-Pro-Pro-Phe-Phe)	four water molecules serve as intramolecular bridges	77)
(Phe^4, Val^6)antamanide	cyclo-(Val-Pro-Pro-Phe-Phe)$_2$	three water molecules serve as intramolecular bridges	78–80)

HyIv = hydroxyisovalerate

59

different size. Fortunately, information of this kind has been provided for the macrotetrolide antibiotics by a large number of X-ray crystallographic studies of some of the free ionophores [81–83] as well as of their complexes with all the alkali metal ions (except lithium) [84–89] and with NH_4^+ [90–92].

The macrotetrolide antibiotics are 32-membered cyclic tetralactons which can be isolated from various *actinomyces* species. Five homologs of the general formula indicated in Fig. 9 are known, which are referred to as nonactin, monactin etc. depending on the number of methyl replaced by ethyl groups. The compounds are built up by four ω-hydroxycarboxylic acid subunits of alternating enantiomerism condensated to each other by esterification.

These ionophores which are frequently also called *nactins*, exhibit high selectivity in complex formation with alkali metal ions [93–94] as well as in ion transport through biological and artificial membranes [95,96]. In acetone, the stabilities of their K^+ complexes are increased by as much as 100 times compared to those of the corresponding Na^+ complexes. It is of interest that this selectivity is even enhanced by additional ethyl groups attached to the ligand backbone [96]. The general selectivity sequence for alkali metal ions displayed by the macrotetrolides is $K^+ > Rb^+ > Cs^+ \approx Na^+ \gg Li^+$, i.e. identical with that for valinomycin. However, even more stable than the K^+ complexes are those with NH_4^+ as guest ion which is not true for valinomycin.

The most extensive structural data available on macrotetrolides refer to *nonactin* and *tetranactin*. The structures of both the antibiotics in their uncomplexed states have been analyzed by X-ray crystallography [81,82]. Surprisingly enough, their conformations differ considerably from each other. Whereas nonactin displays a spatial form with S_4 symmetry as indicated in Fig. 10, the tetranactin molecule is of somewhat elongated shape (Fig. 11). On the basis of force-field calculations the latter conformer was found to be more stable by about 25 kJ · mol^{-1} than the former [87]. A recent X-ray study of uncomplexed *dinactin* revealed an asymmetric structure which can approximately be described as an intermediate between the nonactin- and tetranactin-type conformations. Actually, the dinactin conformer has the lowest conformational energy among the three [83].

Spectroscopic investigations of the corresponding solution conformations [97–99] showed that free macrotetrolides display appreciable rotational freedom of the macrocyclic skeleton. Most probably, all types of conformers found in the crystalline state are among those present in solution.

Upon complex formation, the ionophores undergo considerable conformational rearrangements, mainly resulting in a turning of the ester carbonyl oxygens toward the interior of the molecule. Just as in the case of valinomycin, but in contrast to beauvericin, the central ligand cavity serves as binding site in all the ion complexes

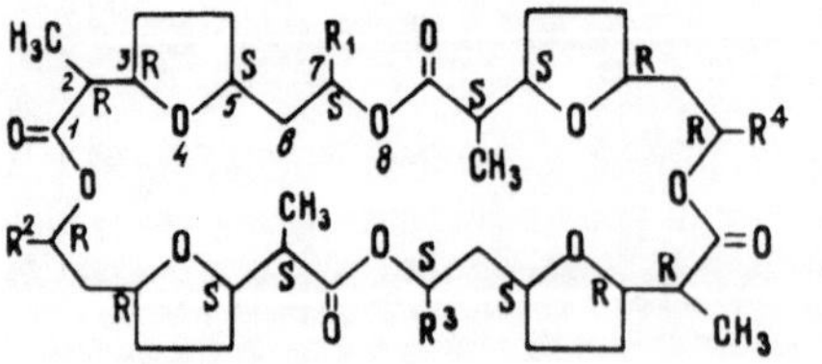

Nonactin: $R^1 = R^2 = R^3 = R^4 = CH_3$
Monactin: $R^1 = R^2 = R^3 = CH_3$, $R^4 = C_2H_5$
Dinactin: $R^1 = R^3 = CH_3$, $R^2 = R^4 = C_2H_5$
Trinactin: $R^1 = CH_3$, $R^2 = R^3 = R^4 = C_2H_5$
Tetranactin: $R^1 = R^2 = R^3 = R^4 = C_2H_5$

Fig. 9. Primary structure of macrotetrolide nactins

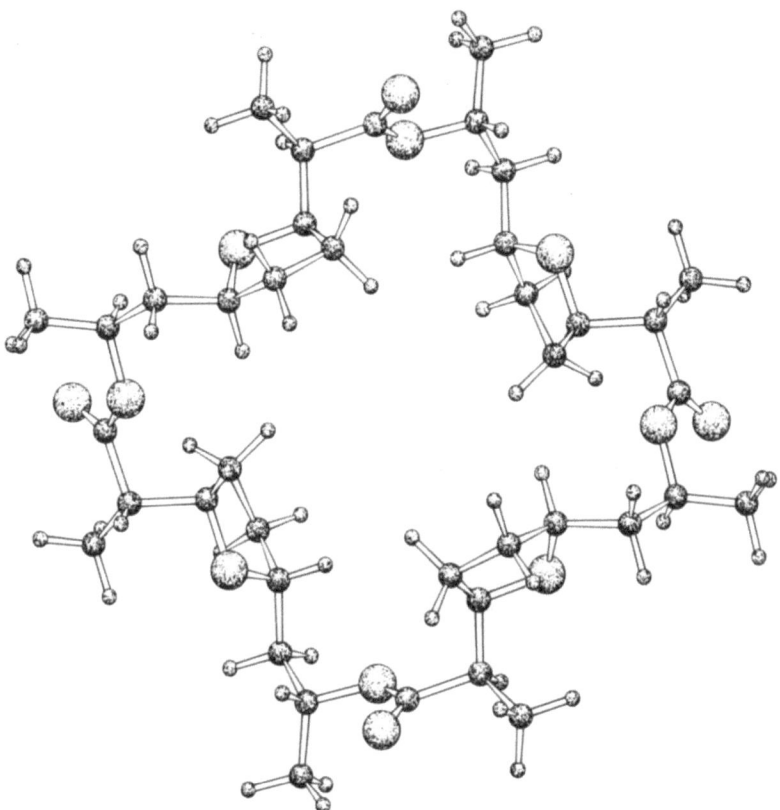

Fig. 10. Spatial structure of free nonactin

of the macrotetrolides. Though being distinct in the free state, nonactin and tetra-nactin exhibit very much the same conformation in their alkali cation complexes. Moreover, the overall structures of the complexes are retained upon replacement of one alkali ion by another. As an example, the spatial structure of the nonactin-Na$^+$ complex is shown in Fig. 12. The guest ions are eightfold coordinated in a more or less distorted cubic arrangement by the four ester carbonyl oxygens and by the four ether oxygens from the tetrahydrofuran moieties. The structures display all the features required for an effective alkali metal ion transport. The periphery of the complex molecules is highly hydrophobic. This holds above all for the tetranactin complexes where the ester oxygens are shielded from the exterior by ethyl groups [87] whereas nonactin lacking the latter presents a somewhat less lipophilic surface in its complexes which may account for the above mentioned diminished stability of nonactin complexes as compared to those of tetranactin.

Now, if the main spatial structures of all the complexes are similar, how can their different stabilities be rationalized? It is in fact the structural niceties that are responsible for the cation selectivity. In the K$^+$ complexes [84, 85, 87], the distances between the potassium ion and the ether oxygens are slightly larger than those to the carbonyl oxygens, the latter being close to the sum of the ionic radius of K$^+$

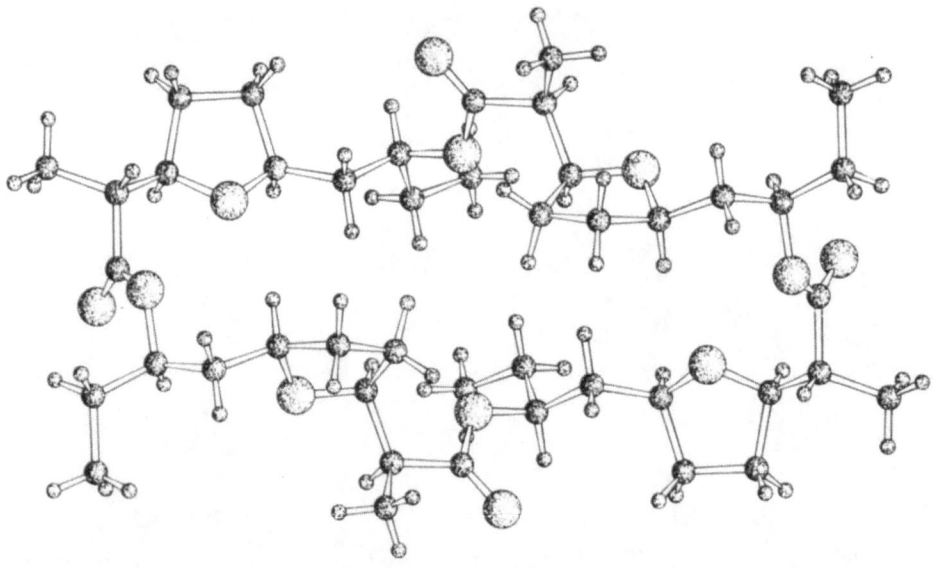

Fig. 11. Molecular structure of tetranactin in the crystalline state

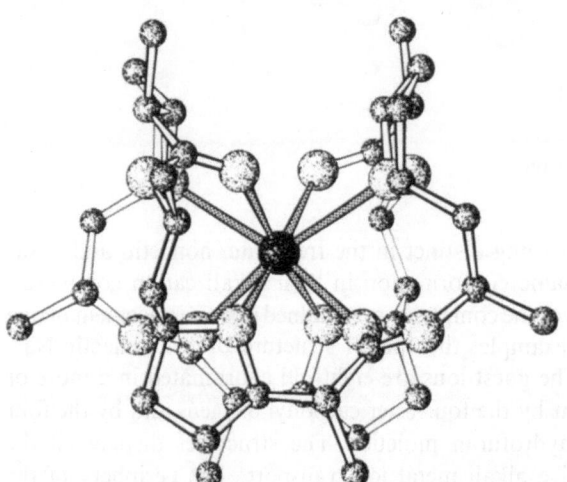

Fig. 12. Structure of the nonactin complex with Na^+. Hydrogen atoms are not shown

(1.33 Å) and the Van der Waals radius of oxygen (1.40 Å). When potassium is replaced by sodium, the carbonyl oxygen atoms approach the metal ion (Na^+ ... O distances 2.40–2.45 Å), but the ether oxygens are obviously restricted in doing so due to intramolecular steric hindrace [86, 87]. The Na^+ ... O (ether) distances are as long as 2.70–2.94 Å, i.e. very much larger than 2.35 Å, the sum of the Van der Waals radius of oxygen and ionic radius of sodium, thus providing considerably less electrostatic interaction and leading to distortion of the cubic coordination arrangement. The ligand cavity is only partially able to adjust to the smaller size of the guest ion which accounts for the reduced stability of the Na$^+$ complexes. Binding of lithium ions by nactins has not been observed [89]. Clearly, the ionic radius of Li$^+$ (0.60 Å) is too small to provide favorable interaction with all the ligand oxygen atoms. Furthermore, the high hydration energy of Li$^+$ ($\Delta G_H^\circ = -503 \text{ kJ} \cdot \text{mol}^{-1}$) may prevent complex formation.

The distances between alkali metal ion and carbonyl as well as ether type oxygen atoms as found in various X-ray structures of nonactin and tetranactin are plotted against the ionic radii of the guests in Fig. 13a. Incorporation of larger ions such as Rb$^+$ and Cs$^+$ into the macrocyclic cavity leads to its expansion which involves essentially a slight outward displacement of the carbonyl oxygens. This is reflected in increasing rotation angles about the ester linkages, as indicated in Fig. 13b for nonactin complexes. The observation of Cs$^+$ ... O (ether) distances shorter than the sum of the ionic radius of Cs$^+$ and the Van der Waals radius of oxygen in the cesium-tetranactin complex [89] suggests the presence of steric strain in the ligand molecule which would explain the diminished stability of the complex.

The solution conformations of the nactin complexes with alkali cations have been investigated by NMR [97,98,100,101], IR [102], and Raman [103,104] spectroscopic methods. All these studies revealed structures almost identical with those found in the crystalline state. Whereas in the alkali cation complexes of the macrotetrolides the distances between metal ion and ether oxygens are generally longer than the corresponding distances to carbonyl oxygens (see Fig. 13), the opposite is true for the NH$_4^+$ complexes [90-92]. In the latter, the guest moiety is complexed by four strong N—H ... O hydrogen bonds to the *ether oxygens* (N ... O distances ≈ 2.86 Å), while it

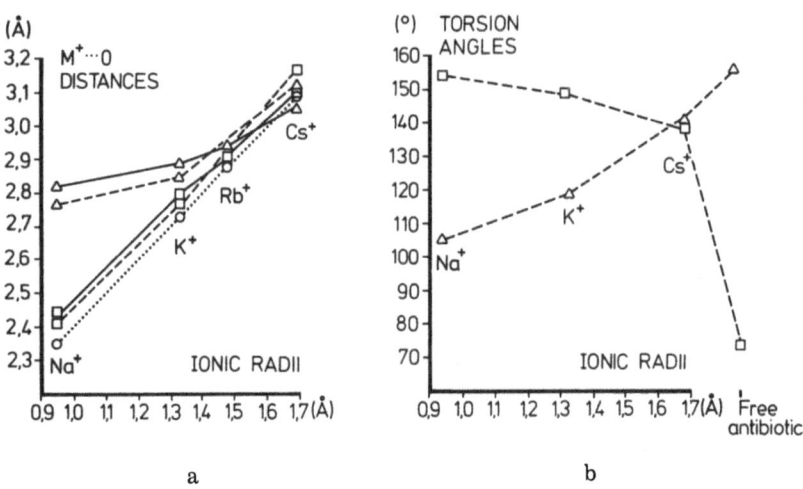

a   b

Fig. 13. **a** Distances between alkali metal ion and ligand donor atoms in macrotetrolide complexes as observed by X-ray crystallographic studies. Values for the nonactin (broken lines) and the tetranactin (solid lines) complexes are plotted against the radii of the cations. △, Mean distance between the cation and ether oxygen atoms. □, Mean distance between the cation and carbonyl oxygen atoms. ○, Sum of the ionic radius of the central cation and the oxygen Van der Waals radius (theoretical M$^+$ ··· O distance) (after ref. [89]). **b** Torsion angles about the ester linkages in nonactin complexes. Values for torsion angles about the C(7)–O(8) (△) and the C(1)–C(2) (□) bonds are plotted against the ionic radii of the cations. The atom numbering refers to Fig. 9

Table 4. Published crystal structures of nactin and of boromycin complexes

Compound	Coord. no.	Bonding distances		Comments, if any	Ref.
		M—O (carbonyl)	M—O (ether)		
[nonactin · Na]⁺ NCS⁻	8	2.40–2.44	2.74–2.79		86)
[nonactin · K]⁺ NCS⁻	8	2.73–2.81	2.81–2.88		84,85)
[nonactin · Cs]⁺ NCS⁻	8	3.13–3.18	3.07–3.16		89)
[tetranactin · Na]⁺ NCS⁻	8	2.43–2.45	2.70–2.94		87,88)
[tetranactin · K]⁺ NSC⁻	8	2.75–2.81	2.85–2.92		87,88)
[tetranactin · Rb]⁺ NCS⁻	8	2.88–2.93	2.90–2.98		87,88)
[tetranactin · Cs]⁺ NCS⁻	8	3.06–3.16	3.03–3.10		89)
[des-valino-boromycin · Rb]	8	—	2.80–3.17ᵃ	*anionic* ligand is a Böeseken complex of boric acid with a macrodiolide	105,106)

ᵃ coordination distances to hydroxy and ester oxygens are included here

Table 5. Uncomplexed macrotetrolide antibiotics and boromycin

Compound	Comments	Ref.
nonactin	molecule shows S_4 symmetry	[81]
dinactin	molecule adopts a twisted asymmetric conformation	[83]
tetranactin	shape of the molecule fairly elongated	[82]
des-valino-des-boron-boromycin	ligand backbone shows largely the same conformation as in Rb^+ complex of des-valino-boromycin	[106]

interacts only weakly with the more distant carbonyl oxygens by $N^+ \ldots O^{\delta-}$ electrostatic forces (mean $N \ldots O$ distances 3.00 Å–3.07 Å). Most probably, it is this additional host-guest hydrogen bonding interaction that accounts for the high stability of the NH_4^+-nactin complexes.

4 Polyether Antibiotics: Pseudo-Cavities

4.1 Classification

The ionophores we shall discuss in this section are referred to as *nigericin antibiotics* because nigericin was the first compound of this family to be discovered [1]. Other common designations are *carboxylic acid ionophores* and *polyether antibiotics*, describing essential structural features of these biomolecules which are unique among the ionophores described so far because they are linear and contain
1) a terminal carboxy group
2) one or two hydroxy groups at the other end of the molecule
3) several ether oxygen atoms provided by tetrahydrofuran and tetrahydropyran rings which may or may not be connected to each other by spiro-type junctions
   In contrast to the above described neutral carriers which form positively charged complexes with cations, polyether antibiotics form neutral salts L^-M^+ with monobasic metal ions because their carboxy group is dissociated at physiological pH. Recently, however, it has been shown that they can also act as neutral ionophores to produce charged complexes such as $LHM^{+\ [107,108]}$.
   Some of the carboxylic acid ionophores are capable of complexing divalent cations. Whether or not they have this ability can be used to classify the various members of the nigericin family. According to Westley [109,110], those ionophores *not* being able to transport divalent cations are referred to as *monovalent polyethers* some of which are shown in Fig. 14. These may be further divided into two subgroups, the distinction between which rests on whether the ionophore contains a hexapyranose moiety attached to the polycyclic ligand backbone or not. The former type antibiotics such as dianemycin, lenoremycin and A204A are called *monovalent monoglycoside polyether antibiotics*. The sugar unit is bound as an α-glycoside in antibiotic A204A, and as a β-glycoside in the other members of this class.

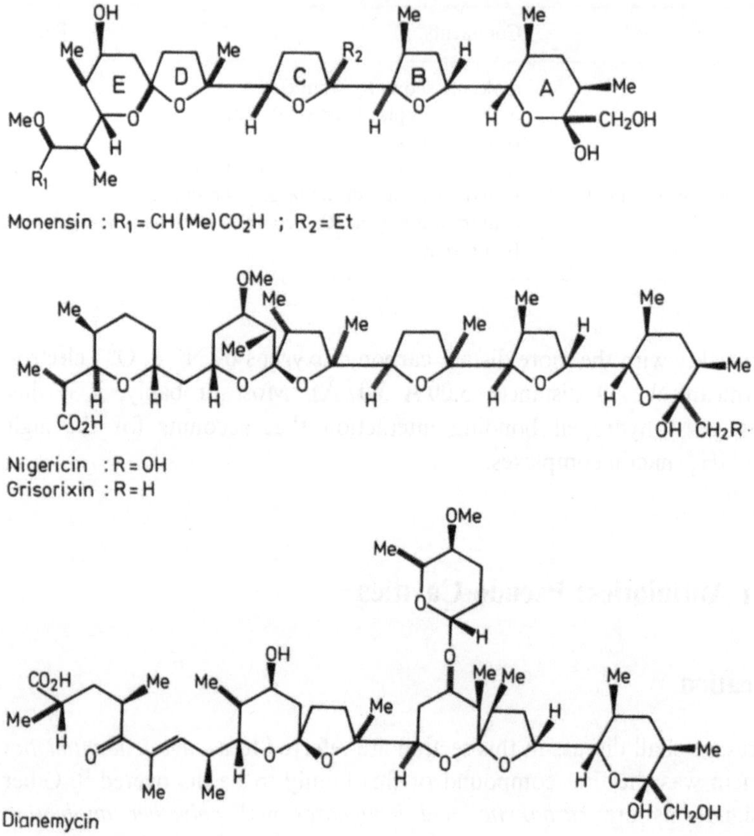

Monensin : R₁ = CH(Me)CO₂H ; R₂ = Et

Nigericin : R = OH
Grisorixin : R = H

Dianemycin

Fig. 14. Some monovalent polyether antibiotics mentioned in the text

The *divalent polyether antibiotics* are much fewer in number than those of the first group. The most popular representatives are lasalocid, formerly known as antibiotic X-537A, and A23187. More recently discovered examples are lysocellin [111], ionomycin [112,113], and antibiotic X-14547 A [114] (see Fig. 15). Actually, the Ca^{2+}-specific ionophores A23187 and X-14547 A should be classified as pyrrole ethers rather than polyethers, because they possess only two and one ether functions, respectively. In fact, they are the only monocarboxylic bioionophores reported thus far to contain nitrogen.

The ion selectivities displayed by the polyether antibiotics are given in Table 6. They are somewhat lower than those of the nactins, but in some cases still considerable. Thus, monensin exhibits specifity for Na^+ ions whereas nigericin prefers K^+ [115,116].

The discovery of the calcium-transporting properties of lasalocid and A23187 has prompted numerous studies concerning their physiological activities [117] and they were shown to be potential cardiovascular agents. Generally, polyether antibiotics are effective against gram-positive bacteria and fungi, and some of them have been claimed to be powerful insecticides or pesticides [109]. Although clinical applications

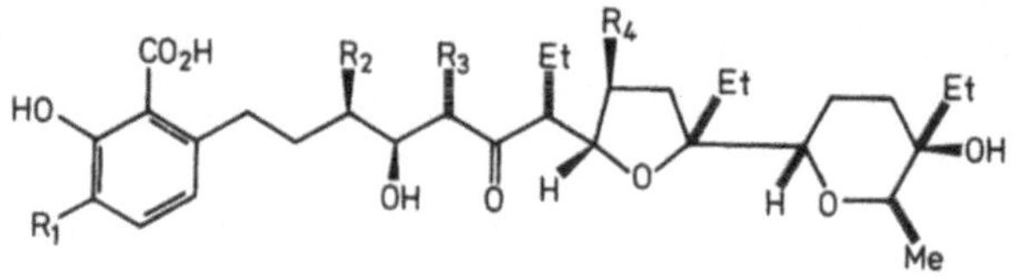

Lasalocid A : $R_1 = R_2 = R_3 = R_4 = Me$

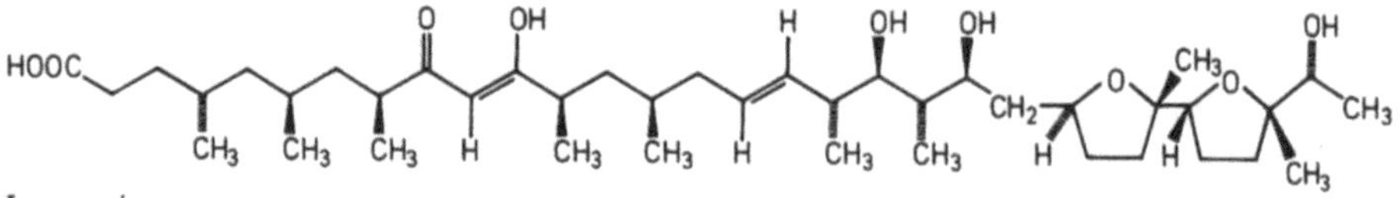

Ionomycin

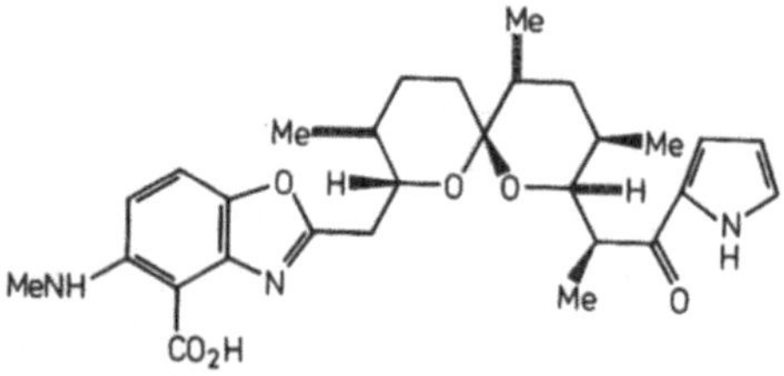

Antibiotic A 23187

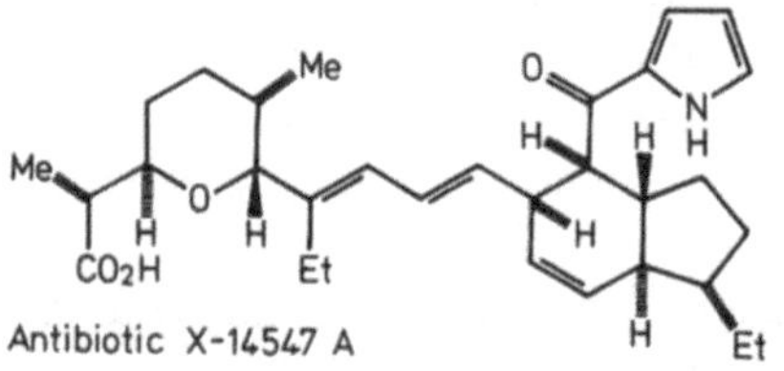

Antibiotic X-14547 A

Fig. 15. Divalent polyether and pyrrole ether antibiotics mentioned in the text

have been hampered by their parental toxicity, almost all of these antiotics have become particularly important as coccidiostatica for poultry industries.

It is probably these commercial applications that caused extensive search for new members of the nigericin family. Up to now, more than 45 distinct polyether antibiotics have been isolated from various *Streptomycetes*, and still a few ones are added every year. Fortunately, the structures of many of these have been elucidated by X-ray crystallography so that quite a large body of data on ligand conformation is available.

Table 6. Ion selectivity patterns found for polyether antibiotics

monensin	$Na^+ \gg K^+ > Rb^+ > Li^+ > Cs^+$
nigericin	$K^+ > Rb^+ > Na^+ > Cs^+ > Li^+$
dianemycin	$Na^+ \approx K^+ > Rb^+ \approx Cs^+ > Li^+$
lasalocid A	$Ba^{2+} \gg Cs^+ > Rb^+ \approx K^+ > Na^+ \approx Ca^{2+} \approx Mg^{2+} > Li^+$

4.2 Monovalent Polyether Antibiotics

Monensin was the first polyether antibiotic to have its structure solved by crystallographic methods. In the silver salt, the monensin anion is wrapped around the cation and held in this conformation by two strong *head-to-tail hydrogen bonds* between the carboxylate group and the two hydroxy functions of the terminal tetrahydropyran moiety as indicated in Fig. 16 [118,119]. Two water molecules are also involved in hydrogen bonding to the ligand. The silver ion is coordinated in an irregular arrangement by six oxygen atoms, four of which are of the ether and the remaining two of the hydroxy type. The carboxylate oxygens do not interact with the cation but are involved in hydrogen bonding with hydroxy groups at the other end of the molecule. The metal ion is thus entrapped in a hydrophilic cavity (to be more exact, in a pseudocavity owing to the linearity of the ligand), while the exterior of the molecule is highly lipophilic thus fulfilling the requirements for an effective transmembrane transport.

Recently, crystal structures of three different monensin complexes with sodium have been published [108,120]. One of these is a NaBr complex in which the carboxy group is not deprotonated, the antibiotic thus acting as a neutral ligand rather than as anion [108]. Apart from minor differences in the hydrogen bonding scheme, the molecular conformation turned out to be very close to that in the Ag$^+$ salt. This does also hold for an anhydrous and a hydrated form of the monensin sodium salt [120]. From these results, it appears difficult to deduce to which conformational properties the marked specifity for sodium is due. However, Steinrauf and Sabesan have carried out computer simulations based on the available crystallographic data [21]. They found that larger cations would lead to a rotation of the primary alcohol group of ring A (see Fig. 14 for notation) away from the guest ion, thus weakening the hydrogen bond to the carboxylate and exposing the hydrogen atom to the solvent. It seems conceivable that this could account for a diminished complex stability.

Evidence is added to this viewpoint by inspection of the structures of the *grisorixin* complexes with Ag$^+$ and Tl$^+$ (ionic radii: 1.26 and 1.47 Å, resp.) as revealed by X-ray analysis [121,122]. The cavity containing the cation is slightly dilated in the

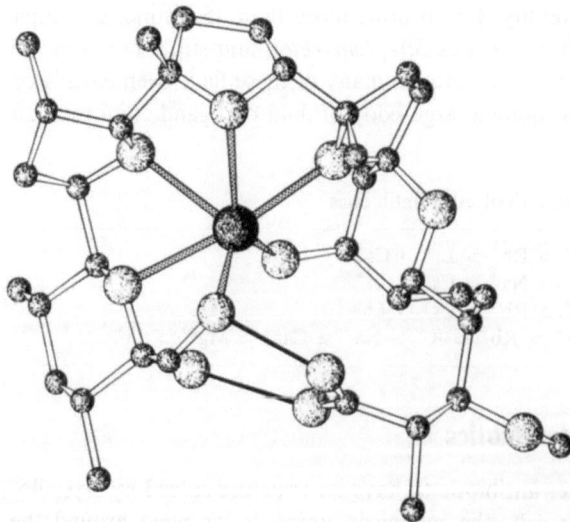

Fig. 16. Spatial structure of the Ag$^+$ complex of monensin. Note the two head-to-tail hydrogen bonds (full narrow lines) in the lower part of the figure. Two water molecules that are hydrogen-bonded to the terminal hydroxy groups have been omitted in the figure for the sake of clarity, as have been all hydrogen atoms

thallium salt, resulting in an orientation of one of the carboxylic oxygens toward the exterior of the molecule where it accepts a hydrogen bond of a water molecule not present in the Ag^+ salt. This indicates a somewhat less effective screening of the polar ligand groups from solvent interactions in the Tl^+ complex. Moreover, the length of the head-to-tail hydrogen bond is increased to 2.73 Å, as compared to 2.64 Å in the silver salt.

The free acid of monensin has also been subjected to an X-ray crystallographic study [123]. It revealed that the pseudo-cyclic conformation secured by head-to-tail hydrogen bonding is also present in the uncomplexed ionophore, the hydrogen bonding scheme, however, being somewhat different (see Fig. 17). One important feature of this structure is the presence of a water molecule located within the ligand cavity and hydrogen bonded to three oxygens of the antibiotic.

The conformation of the free ionophore is comparable to that found in the metal ion complexes. This is not surprising because of the relative rigidity of the ligand owing to the numerous methyl substituents attached to the ether rings. Moreover, the spiro junction between the D and E rings prohibits rotations. Therefore, this part of the molecule is in fact invariant in all the published crystal structures of monensin [120] whereas the rest of the ligand shows limited flexibility. Thus, even though changes of torsion angles are small, they add up to shift three oxygen atoms by as much as 2 Å in going from the uncomplexed to the complexed form.

Among the five cyclic ethers, ring C is the most flexible one as is indicated by the conformational changes it suffers upon complex formation. Based on this observation, Duax et al. proposed that the ion-entrapping process might be initiated by interaction of the ring C ether oxygen with the metal ion, followed by replacement of additional water molecules out of the hydration sphere by adjacent ether oxygens [120].

Spectroscopic studies including ^{1}H-, ^{13}C-, ^{23}Na-, ^{7}Li-, and ^{205}Tl-NMR [107,124-127] have provided conformational assignments for monensin and its ion complexes in

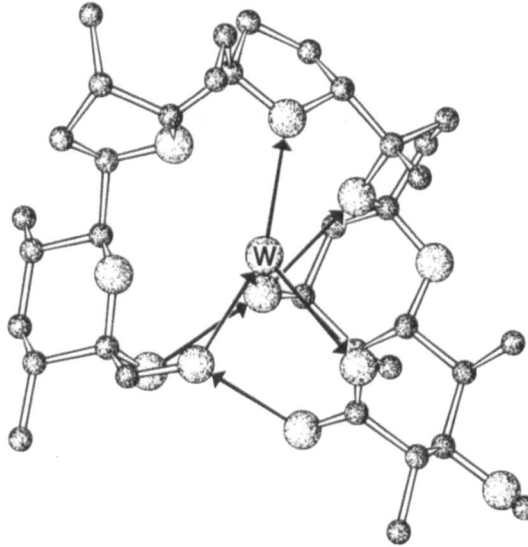

Fig. 17. Uncomplexed monensin shown in a similar orientation as for the Ag^+ salt in Fig. 16. A water molecule (indicated by "W") occupies the central cavity. Arrows mark the direction of hydrogen bonds from the donor to the acceptor atom. Hydrogen atoms are not shown

solution that are largely consistent with the results of X-ray work. Obviously, the free ionophore retains a water molecule also in chloroform solution, whereas no evidence was obtained for water being involved in the hydrogen bonding of the metal salts.

The above described structural properties of monensin and its metal complexes have essentially been found in all the monovalent polyether antibiotics. In some cases, e.g. nigericin and lonomycin, the carboxylate group is also involved in the coordination of the cation, thus providing additional ion-pair contribution to the binding energy. The head-to-tail hydrogen bonding which may include one or two hydrogen bonds is a common structural feature displayed by all the polyether antibiotics. Details of the coordination and hydrogen bonding patterns are given in Table 7a.

Monensin appears to be a good example to demonstrate how a biological host molecule utilizes a number of minor conformational changes in order to rearrange its binding sites for optimum complex formation with a guest entity.

This cooperative mechanism which may be compared with the induced-fit model of enzyme-substrate interaction [128] seems to be in some contrast to the nactins where only two torsion angles are subject to significant alteration (see Sect. 3).

4.3 Divalent Polyether Antibiotics

The term "divalent polyether" does *not* imply that these antibiotics are generally not able to complex monovalent cations, as can be seen from the ion selectivities given in Table 6. The most versatile antibiotics in the "divalent group" are the lasalocids which have been shown to effectively bind and transport Ba^{2+}, Ca^{2+}, Cs^+, Na^+ and Fe^{2+} [129] ions, and even catecholamines [130]. Interestingly, the barium ion, although the most strongly complexed, is not the most rapidly transported. This finding illustrates that in the ion-carrier mechanism tight metal ion binding of the ionophore ligand does not necessarily correlate with good carrier qualities. Rather it is of importance that not too strong binding allows for a sufficiently rapid release of the metal ion after having passed through the membrane — a subtle interplay of several components.

The lasalocids have several unique features such as a salicyclic acid moiety and a carbonyl group. Furthermore, their ligand backbones are considerably shorter than those of the other polyether antibiotics (see Fig. 15). This accounts for the inability of lasalocid ionophores to fully shield the complexed metal ion from the solvent by folding around it. The resulting complex has two distinct surfaces one of which is much more polar than the other, as demonstrated in Fig. 18 for lasalocid A, the most abundant species. In non-polar media, this unfavorable situation is overcome by the formation of dimeric complexes of stoichiometries $(M^+L^-)_2$ and $M^{2+}(L^-)_2$ for mono- and divalent metal ions, respectively. In fact, these aggregates present a highly lipophilic surface to the solvent.

An arrangement of this type was first shown by X-ray crystallography to exist in the barium complex of lasalocid A [131, 132]. In this complex, two antibiotic anions coordinate to the Ba^{2+} ion in different ways. The metal ion occupies a polarophilic "pocket" mainly provided by one of the ligands and is coordinated by two ether, two hydroxyl and one of the carboxylate oxygens. The ninefold coordination is

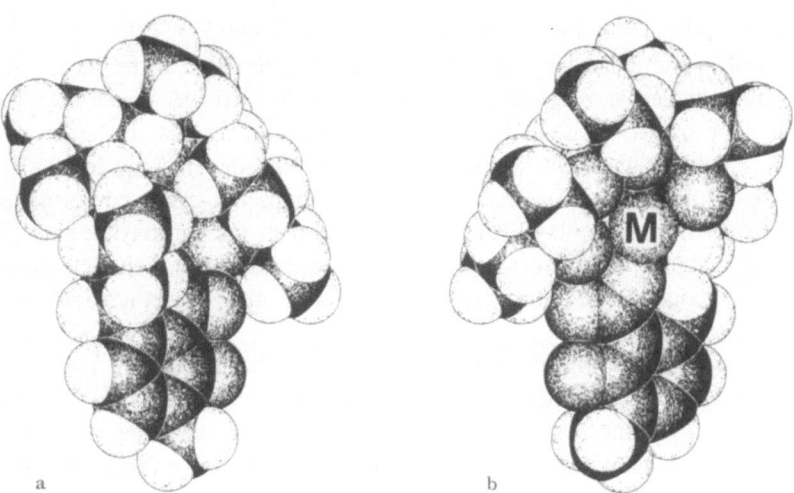

Fig. 18a and b. Space filling model of a 1:1 lasalocid complex with a metal ion (M): **a** from the non-polar side (metal ion completely covered) **b** from the polar side

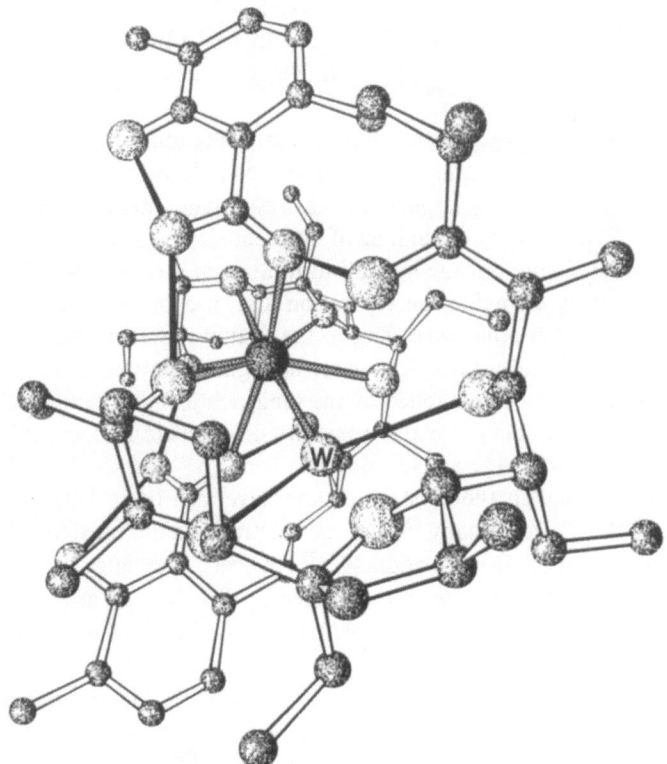

Fig. 19. Spatial structure of the Ba^{2+} complex of lasalocid A viewed from a 10 Å distance. The Ba^{2+} ion is mainly coordinated by the "far" ligand molecule to yield a 1:1 complex of the type shown in Fig. 18. This is covered by a second lasalocid molecule (the "near" one). The water molecule is indicated by a "W". Hydrogen atoms have been omitted for clarity

completed by two oxygens contributed by the second lasalocid anion, and a hydrogen-bonded water molecule occupying the cavity of the latter (see Fig. 19).

If the salicylate and the tetrahydropyran moieties are referred to as the "head" and the "tail" of the molecule, respectively, the two anions are arranged as a *head-to-tail dimer*. There are apparently no hydrogen bonds between the ligands, the latter being held together by their guest entities, i.e. the Ba^{2+} ion and the H_2O molecule. Each of the lasalocid molecules, however, is folded into a pseudo-cyclic conformation stabilized by an intramolecular hydrogen bond between the carboxylate and the terminal hydroxy group. This interaction is similar to the aforementioned "head-to-tail" hydrogen bond found in monensin and all the other polyether antibiotics belonging to this family.

Discrete dimers of the "head-to-tail" type have also been found in the crystal structures of the 2:2 lasalocid complexes with Ag^+ [133)] and Na^+ [134)] when crystallized from the nonpolar carbon tetrachloride. In both complexes, each metal ion is fivefold coordinated in the cavity of one anion, and there is an additional interaction with the second one, i.e. with its aromatic ring in the case of Ag^+ or with one of its carboxylate oxygens if the cation is Na^+.

However, when the sodium complex was crystallized from a solvent of medium polarity such as acetone, a *head-to-head dimer* structure was obtained [134)]. This is also true for the free acid of 5-bromolasalocid [135)] crystallized from C_6H_{14}/CH_2Cl_2 in which the antibiotic molecules are connected to each other by a hydrogen-bonded water molecule and by an additional hydrogen bond from the carboxy group of one ligand to the carbonyl group of the other.

It should be emphasized that in eighteen crystallographically independent molecular structures of lasalocid itself and of its cation complexes including the p-bromophenethylamine salt [130)], the ionophore displays essentially the same conformation. Complexation of lasalocid by metal ions of different sizes is associated only with minor conformational changes of the ligand but to a much greater extent with configurational adjustment by changing the separation and orientation of the two ligand molecules relative to each other according to the spatial requirements of the guest cation. This simple mechanism provides proper host-guest adaptation similar, in a sense, to the complexation mode exhibited by the sandwich-forming beauvericin described in Section 2.2, and accounts for the versatility of lasalocid in complexing cations as different in size and charge as Na^+ and Ba^{2+}, or Cs^+ and Ca^{2+}.

The dimeric structures found in the crystalline state have also been detected in non-polar solutions by NMR spectroscopy [136)] but the spectral data obtained in polar solvents where not consistent with such adducts [137)]. Instead, they suggested the presence of simple 1:1 complexes. To assess the influence of solvent polarity, Paul and coworkers examined the crystal structures of free lasalocid and of its sodium complex crystallized from methanol, and, indeed, found these to be *monomers* [138, 139)]. Both structures display the familiar head-to-tail hydrogen bonding. In the sodium complex, the Na^+ is coordinated to the same five ligand oxygens as in the dimeric structure, but it is capped by the oxygen of a methanol molecule.

Based on these structural data, it is reasonable to assume that monomeric forms are involved in cation uptake and release in polar media, e.g. at the exterior of the membrane, whereas the actual transport process is achieved by the lipophilic dimer [138)]. A model for monomer — dimer transition has been provided by a recent

X-ray structural analysis of a 2:2:2 adduct between lasalocid, Na^+, and water obtained from 95% ethanol solution [140]. In this complex, each of the sodium ions is mainly associated with a single lasalocid anion in a manner reminiscent of the aforementioned monomeric Na^+ complex. Coordination of the metal ions is completed by two water molecules which are accommodated between the antibiotic molecules. The dimer is held together by hydrogen bonds *via* these water molecules. This suggests that the adduct might be a result of the initial association of two Na^+-lasalocid monomers, thus representing an intermediate in the complexation process.

The detection of both monomeric and dimeric structures in the solid state, depending on the polarity of the solvent from which the crystals were grown, and the consistency of these results with the solution structures as provided by spectroscopic methods nicely demonstrates that crystallographic data can be sensitive even to solvent influences. Moreover, the crystal structure of the "intermediate complex" shows the importance of solid-state conformational analysis for the elucidation of complex formation mechanisms or, more generally speaking, for molecular dynamics.

In contrast to lasalocid, antibiotic A23187 (see Fig. 15 for formula) does undergo considerable conformational changes upon transition from the free to the complexed state which is expressed in drastic alterations of four torsion angles [34]. The X-ray crystallographic analysis of the free acid [141] revealed a monomeric structure with the familiar head-to-tail hydrogen bond between the pyrrole nitrogen and the carboxy group.

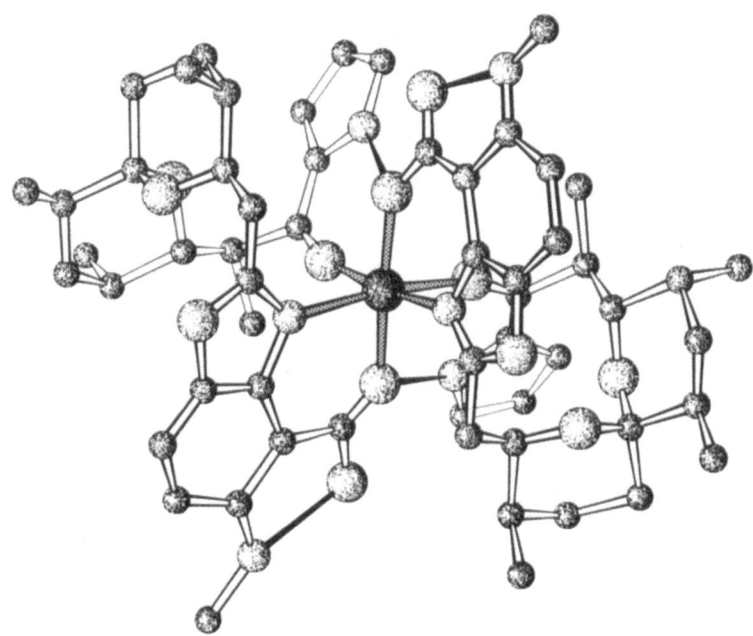

Fig. 20. The 2:1 complex between antibiotic A23187 and Ca^{2+}. Hydrogens are not shown

Table 7a. Coordination and hydrogen bonding in metal ion complexes of monovalent polyether antibiotics

Antibiotic	Cation	Total coord. number	Liganding oxygen atoms				Number and lengths of "head-to-tail" H-bonds	Comments, if any	Ref.
			ether	hydroxy	carboxylate	Solvent			
monensin	Na^+	6	4 / 2.36–2.54	2 / 2.34, 2.45	—	—	2 / 2.51, 2.64	refers to a hydrated sodium salt	120)
	Na^+	6	4 / 2.41–2.53	2 / 2.35, 2.38	—	—	2 / 2.58, 2.62	refers to an anhydrous form of Na^+ salt	120)
	Na^+	6	4 / 2.35–2.50	2 / 2.37, 2.42	—	—	2 / 2.73, 2.76	refers to NaBr complex	108)
	Ag^+	6	4 / 2.40–2.69	2 / 2.43, 2.45	—	—	2 / 2.51, 2.65	structure contains two H_2O molecules	118,119)
nigericin	Na^+	5	4 / 2.38–2.52	—	1 / 2.25	—	—		146)
	Ag^+	5	4 / 2.47–2.66	—	1 / 2.26	—	1 / 2.59		147–149)
	K^+	7	5 / 2.67–3.09	1 / 3.06	1 / 2.58	—	2 / 2.63, 2.73	two of the oxygens interact only weakly with K^+	150)
grisorixin	Ag^+	5	4 / 2.4–2.7	—	1 / 2.20	—	1 / 2.64		121)
	Tl^+	5	4 / 2.6–3.0	—	1 / a)	—	1 / 2.73	a water molecule is hydrogen-bonded to carboxy group	122)
Ionomycin (emericid, DE-3936)	Na^+	6	4 / 2.40–2.51	—	2 / 2.38, 2.45	—	2 / 2.66		151)
	Ag^+	6	4 / 2.50–2.77	—	2 / 2.41, 2.65	—	1 / 2.73		151,152)
	Tl^+	6	4	—	2	—	1		153,154)
X206	Ag^+	6	2	1	—	—	1 / 2.69		155,156)

Ionophore	Cation													Ref.
alborixin	K^+	8	3	2.76–3.07	4	2.69–2.98	—	2.89	—	—	1	2.64		157)
dianemycin	Na^+	7	4		2		—		—		1			158)
	K^+	7	4		2		—		—		1			158)
	Tl^+	7	4		2		—		1 (H_2O)	—	—	—	A water molecule is inserted into the "head-to-tail" hydrogen bonding	158)
lenoremycin (Ro21-6150, A130A)	Ag^+	8	6	2.46–2.88	2	2.38, 3.01	—		—	—	2	2.60, 2.67	pyranose oxygen is coordinated to the cation	159,160)
A204A	Na^+	6	4	2.72–2.85	—		2	2.71, 2.97	—	—	1	2.69	sugar ring well removed from the central cavity	161)
carriomycin	Tl^+	6	4	2.82–3.00	—		2	2.73, 3.00	—	—	1	2.76	sugar ring does not interact with Tl^+	162,163)
K-41	Na^+	6	4		—		2		—		1		the p-iodo- and p-bromobenzoate derivatives have been employed in the X-ray analysis	164)
6016	Tl^+	6	4		—		2		—		1		glycoside moiety does not interact with the cation	165)

Table 7b. Divalent polyether antibiotics: Bonding distances in their complexes with metal ions. In the case of 2:1 complexes, the ionophore molecules are referred to as "A" and "B".

Antibiotic	Cation	Total coord. number	Liganding atoms and bonding distances (Å)						Number and lengths of intra-ligand "head-to-tail" H-bonds	Comments	Ref.
			ether	hydroxy	car-boxy-late	keto	benzoxazole nitrogen	Solvent			
lasalocid	Na$^+$	6	ligand A: 2	2	—	1	a)	—	1	refers to Na$^+$ salt crystallized from non-polar solvents; Structure is a 2:2 "head-to-tail" dimer	134)
			ligand B: —	—	1	—	a)	—	1		
	Na$^+$	6	ligand A: 2	2	—	1	a)	—	1	refers to Na$^+$ salt crystallized from solvents of medium polarity. Structure is a 2:2 "head-to-head" dimer	134)
			ligand B: —	—	—	1	a)	—	1		
	Na$^+$	6	2	2	—	1	a)	1 (MeOH)	1	refers to Na$^+$ salt crystallized from polar solvents. The structure is monomeric	138)
	Na$^+$ (1)	7	ligand A: 2 2.42, 2.47	2 2.56, 2.67	—	1 2.67	a)	1 (H$_2$O) 2.45	?	2:2:2 Na$^+$-lasalo-cid-water complex, cryst. from 95% ethanol. The two Na$^+$ ions show different coordi-nation	140)
			ligand B: —	—	—	1 2.41	a)	—	—		
	Na$^+$ (2)	6	ligand A: —	—	—	—	a)	1 (H$_2$O) 2.37	?		

Ionophore	Cation	CN	Ligand						H₂O		Remarks	Ref
	Ag^+	5	ligand B: 2	2.40, 2.44	2 2.47, 2.72	—	—	a)	1 (H_2O) 2.40	—	2:2 "head-to-tail" dimer	133)
			ligand A: 2		2		1	a)		1 2.60		
	Ba^{2+}	9	ligand A: 2	2.86, 2.98	2 2.71, 3.08	1 2.81	1 2.80	a)	—	1 2.64	2:1 (L:M) "head-to-tail" structure	131, 132)
			ligand B: —									
lysocellin	Ag^+	6		1 2.84	1 2.64	1 2.99	a)	a)	1 (H_2O) 2.74	1 2.72	1:1 complex	166, 167)
ionomycin	Ca^{2+}	6	2	2.38, 2.46	2 2.45, 2.55	1 2.55	1 2.99	a)	—	1 2.62	intermolecular hadrogen bonding leads to 2:2 dimers	145)
			1	2.45	2 2.44, 2.44	1 2.28	2 2.26, 2.28	a)	—	1 2.66		
	Cd^{2+}	6	1	2.38	2 2.38, 2.40	1 2.30	2 2.25, 2.25	a)	—	1 2.69		145)
	Ca^{2+}	6	ligand A: —		a)	1 2.01	1 2.10	1 2.21	—	—	2:1 (L:M) "head-to-tail" dimer	142)
			ligand B: —		a)	1 1.92	1 2.02	1 2.22	—	—		
A23 187	Ca^{2+}	7	ligand A: —		a)	1 2.27	1 2.37	1 2.69	1 (H_2O) 2.38	—	2:1 (L:M) "head-to-tail" dimer	143)
			ligand B: —		a)	1 2.28	1 2.38	1 2.58	—	—		

a) does not apply   L = ligand, M = metal ion; Crystallographic coordinates are not available in cases where bonding distances are not given

Two different structures of an A23187 complex with calcium have been reported [142, 143]. In both of these, the Ca^{2+} ion is coordinated by two antibiotic anions
each contributing a carboxy oxygen, a carbonyl oxygen, and the nitrogen of the
benzoxazole ring system. The dimer is held together by coordination to the metal ion
and by hydrogen bonds between the pyrrole nitrogen and the carboxy oxygen of the
other A23187 ligand (see Fig. 20). The two dimeric forms, however, differ in the
coordination geometry. In one complex form, the Ca^{2+} ion is sixfold coordinated [142]
but in the second case it has sevenfold coordination [143]. The additional coordination
site is occupied by a water molecule which is placed at the exterior of the complex.
The fact that one complex form involves a water molecule at the surface while the
other does not, suggests that the former is an intermediate in the complexation
process, as pointed out by Smith and Duax [34].

Recently, a new divalent carboxylic ionophore called ionomycin has been isolated [112]. It was shown to be an effective calcium complexone [113] and to cause
catecholamine release from rat pheochromocytoma cells [144]. Ionomycin is unique
among the polyether antibiotics in chelating Ca^{2+} as a *dibasic* acid whereas all the
other members of the nigericin family are *monobasic*. The dianion results from the
ionization of the carbonyl and the enolized β-diketon moieties (see Fig. 15 for
formula). A crystallographic study of several very similar forms of the Ca^{2+} and
Cd^{2+} salts [145] revealed on octahedral coordination of the metal ion provided by one
of the carboxylate oxygen atoms, both oxygens of the β-diketonate group, two
hydroxy oxygens, and an ether oxygen of one of the tetrahydrofuran rings (Fig. 21).
In the crystalline state, the complexes are joined in patris by two hydrogen bonds.
The resulting dimers exhibit primarily lipophilic surfaces.

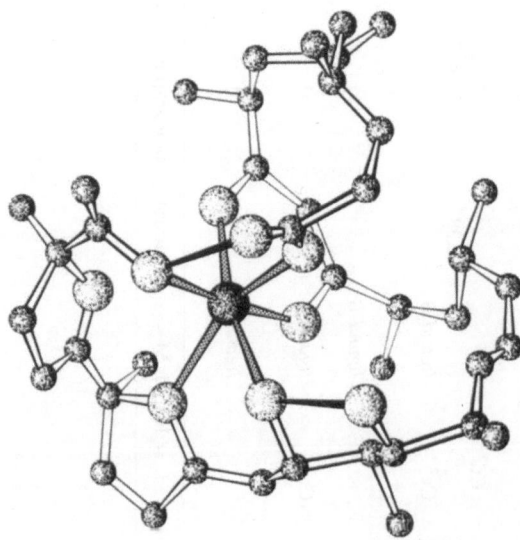

Fig. 21. Ca^{2+} complex of ionomycin.
Hydrogen atoms have been omitted

Table 8. Uncomplexed polyether antibiotics and their adducts with chiral ammonium salts

Antibiotic	Number and lengths of internal "head-to-tail" hydrogen bonds	Comments	Ref.
monensin	1 2.66	The ligand shows several subtle changes of conformation as compared to the M^+ complexes. A water molecule occupies a site in the central cavity	123)
grisorixin	1	The ligand displays the same conformation as in the Ag^+ complex. A water molecule takes the place of the metal ion	168)
salinomycin	—	The p-iodophenacyl ester derivative was employed for X-ray analysis. The molecule adopts a helical structure lacking head-to-tail hydrogen bonding	169, 170)
X-206	1	Conformation similar to that in the Ag^+ complex. A water molecule occupies the central cavity	156)
septamycin	—	The p-bromophenacyl ester derivative was employed for X-ray analysis	171)
A204A	1 2.99	The central cavity is occupied by a water molecule	172)
lasalocid A	1	Refers to the 5-bromo derivative crystallized from non-polar solvents. The structure is a "head-to-tail" dimer enclosing a water molecule	135)
lasalocid A	1 2.53	Refers to the free acid crystallized from methanol. Monomeric structure. A hydrogen-bonded methanol molecule is enclosed in the central cavity	138, 139)
isolasalocid A	—	The compound does not exist in the characteristic cyclic conformation	173)
A23 187	1	The hydrogen bond is of the $N-H \cdots O$-type	140)
lasalocid A 4-bromophenethylamine	1 2.75	1:1 Host-guest complex. The guest molecule is hydrogen-bonded to the antibiotic	130)
X-14 547A 4-bromophenethylamine	—	2:1 Host-guest complex. Only one of the antibiotic molecules is ionized	174, 175)

5 Crown Ethers: Synthetic Macrocyclic Multidentates

5.1 Basic Stereochemistry of the 1,4-Dioxa Group in Polyether Complexes

As we have seen, X-ray studies of the ionophorous antibiotics and their cation complexes were able to explain many of the steric factors that determine the selectivity patterns shown by these ligands. However, more systematic investigations on the relationship between host-cavity size and guest-ion radius could only be carried out using simpler synthetic ligands as models. In 1967, Pedersen reported the synthesis and complexing properties of a new class of compounds named crown ethers [256] which are able to mimic effectively their natural counterpieces.

The most common crown ethers are shown in Fig. 22. Since application of IUPAC rules to polyethers leads to somewhat cumbersome designations, we will follow the simple "crown" nomenclature proposed by Pedersen [256].

Polyethers are built up from 1,4-dioxa units, $O-CH_2-CH_2-O$. The minimum energy conformation of these units is staggered with torsion angles about C-C bonds being synclinal (60°) and about C-O bonds being antiperiplanar (≈ 180 °C) (for definitions see Fig. 23). These preferences, however, do not preclude deviations if required by ring formation or cation complexation.

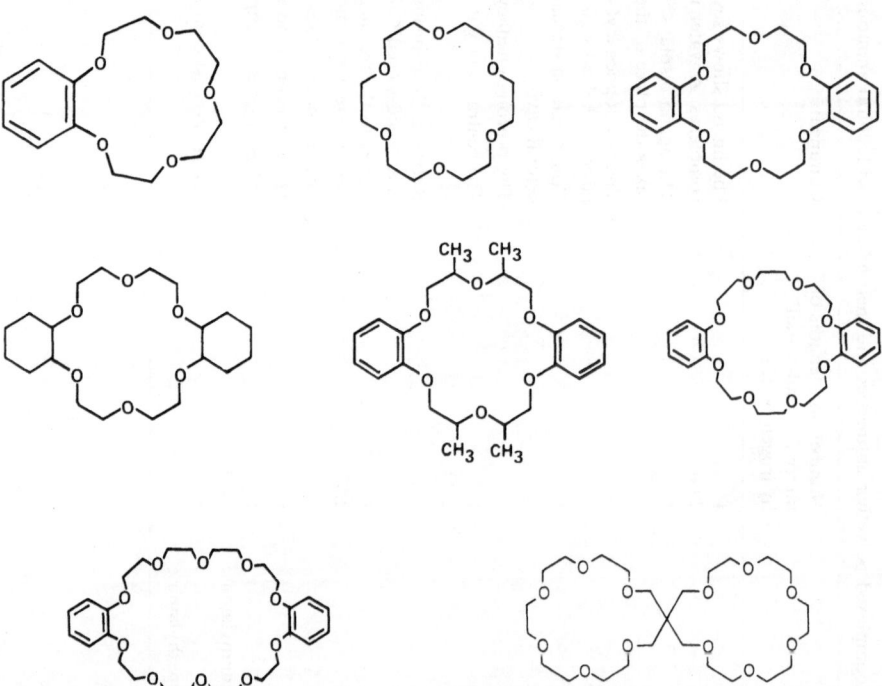

Fig. 22. Structural formulae of crown ethers. From left to right and top to bottom: benzo[15]crown-5, [18]crown-6; dibenzo[18]crown-6, dicyclohexano[18]crown-6; tetramethyldibenzo[18]crown-6, dibenzo[24]crown-8; dibenzo[30]crown-10, spiro-bis[19]crown-6

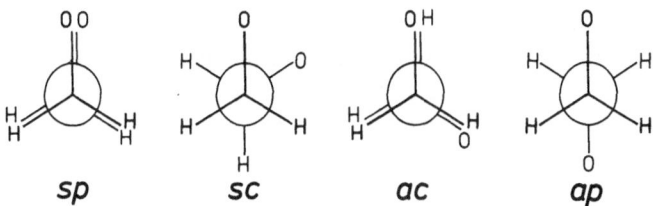

Fig. 23. Definition of torsion angles shown for the O—CH$_2$—CH$_2$—O unit. From left to right: *synperiplanar (sp), synclinal (sc), anticlinal (ac), antiperiplanar (ap)*

The latter process may cause major distortions of the torsion angles about C-O bonds whereas in the complexes, C-C bonds are essentially invariant in being synclinal ("all-gauche conformation") in order to place the ether oxygens close to the cation. It is noteworthy that if C-O bonds are forced into the synclinal conformation they usually adopt torsion angles larger than 70° to avoid short 1,4 CH ... HC contacts [257].

Interestingly, aliphatic C-C bonds are found to be systematically short in all the crystal structures of cyclic and open-chain oligoethers [240, 241, 330] and in their complexes with cations [244] as well as with neutral guests [340, 341]. Thus, the mean for these bond lengths is 1.507 Å in the uncomplexed hexaether 18-crown-6 [240] and even shorter by 0.03 Å in its Cs$^+$ complex [198] whereas the usually quoted value for a C(sp^3)-C(sp^3) bond is 1.537 Å [256]. Formerly, this effect was believed to be an artifact caused by inadequate treatment of thermal vibrations during the crystallographic refinement procedure [240]. Recently, however, Dunitz and coworkers repeated the structural analysis of 18-crown-6 at a temperature of 100 K so as to diminish thermal motions, and still, they found the mean C-C bond length (1.512 Å) to be only slightly increased with respect to the room temperature structure [241]. From this it may be concluded that the effect is at least partially real. A possible reason for the short C-C bonds might be the slightly polarized character of the adjacent C-O bonds which causes partially positive charges on the carbon atoms. By application of non-empiric SCF calculations to the hydrogen molecule it could be shown that the H-H bond length would slightly decrease if the protons were fractionally more positive than unity [342].

The KNCS complex of the hexaether [18]crown-6 [192] is highly ordered inasmuch as all the C—C—O—C and C—O—C—C torsion angles are close to 180° and all the O—C—C—O torsions are close to 60°. As a result, the average O ... O distance is as short as 2.82 Å which actually represents a Van der Waals contact. This unfavorable interaction is more than compensated for by the ion-dipole attraction induced by the complexed cation. The potassium ion is located at the center of the macrocycle and coordinated to the six ether oxygens which are all in a planar arrangement or, to be more exact, lie alternately above and below their mean plane by 0.2 Å (Fig. 24). Additionally, there is weak interaction between the metal ion and two disordered SCN$^-$ ions in the crystal lattice.

The conformation found in the K$^+$ complex of [18]crown-6 differs from that adopted by the free ligand, because in the uncomplexed molecule unfavorable alignment of dipoles is not compensated by a cation. Rather, the free coronand actually attains a somewhat elliptical conformation which allows for 1,5- and 1,8-CH ... O interactions [240, 241] (see Fig. 25). This is achieved by adjustment of several torsion angles including two C-O and two C-C bonds forced into synclinal (80°) and antiperiplanar (174°) conformation, respectively. As a result, not all the oxygen atoms point to the interior of the molecule as they do in the potassium complex. This, together with the elliptical shape of the molecule, is a general feature common to most of the uncomplexed polyethers.

The conformation of uncomplexed [18]crown-6 found in the crystalline state has been shown by IR spectroscopy to be predominant also in solution [258], and force-field calculations have essentially confirmed the same structure [241, 259]. Interestingly, it was found that the "ordered" D_{3d} [18]crown-6 conformation observed in the potassium complex would be lower in energy by $7 \text{ kJ} \cdot \text{mol}^{-1}$ were it not for the aforementioned unfavorable electrostatic interactions [259]. The tendency to remove the latter by complexation of a cation and thus enable the ligand to adopt a low-energy conformation might contribute considerably to the high affinity of [18]crown-6 to potassium ions.

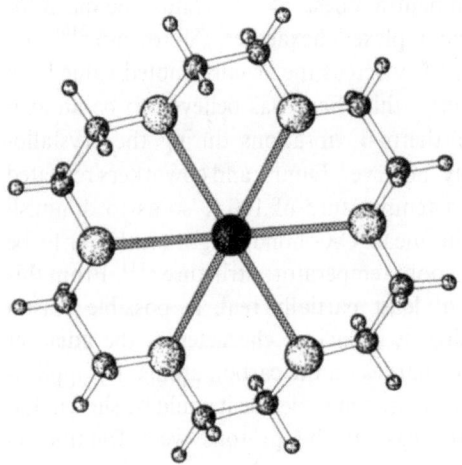

Fig. 24. Structure of the [18]crown-6 complex with potassium isothiocyanate. The anion is not shown

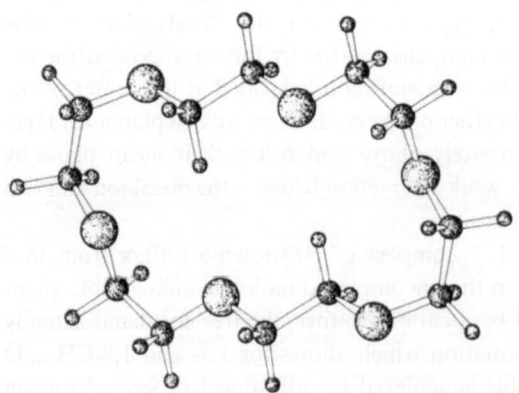

Fig. 25. Conformation of the free hexa-ether [18]crown-6. Note the elliptical shape of the molecule

5.2 The Ion-Cavity Concept

Examination of the diameters of ligand cavities and of the diameters of alkali and alkaline earth ions as given in Table 9 clearly shows that either the metal ion is too small to fill the cavity, or too large to fit in it, or it just meets the cavity size. The

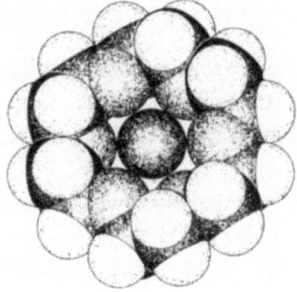

Fig. 26. Space-filling model of [18]crown-6-K$^+$

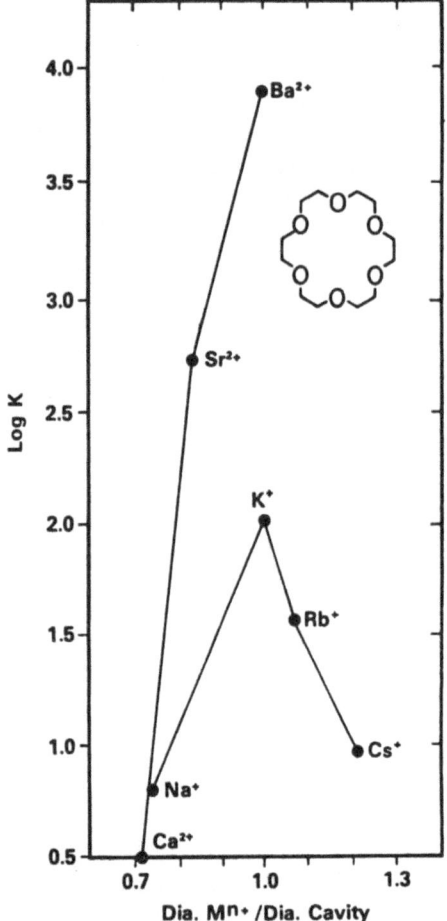

Fig. 27. Variation of the stability constant in water for [18]crown-6 complexes with alkali and alkaline ether cation depends on the degree of fit between host and guest. The ratio of cation size to crown cavity size is depicted on the abscissa (from Ref. [344])

Table 9. Diameters of crown cavities calculated from crystallographic data (after ref. [244]). Diameters of some alkali cations are given for comparison (all values in Å units)

[12]crown-4	1.2	Li^+	1.20
[15]crown-5	1.72–1.84	Na^+	1.90
[18]crown-6	2.67–2.86	K^+	2.66

latter is the case for [18]crown-6 and K^+, as can be seen from the space-filling model in Fig. 26, and in fact, potassium is the one most strongly complexed among the alkali metal ions [260]. The relationship between complex stability and the degree of fit between ligand and cation is depicted in Fig. 27.

Both Rb^+ and Cs^+ are too large to be accommodated into the 18-crown-6 cavity. Therefore, they occupy a site somewhat distant from the plane of the ether oxygens. Rb^+ is situated by 1.19 Å and Cs^+ even by 1.44 Å above this plane, which leads to a less favorable interaction with the ligand donor atoms, thus explaining the diminished stabilities of the [18]crown-6 complexes with these ions. The crown ether conformation does not alter in comparison to the K^+ complex. Coordination of the metal ion is completed by contacts with the bridging SCN^- ions, thus resulting in the formation of a 2:2 dimeric structure (Fig. 28) [197, 198].

On the other hand, Na^+ is too small to completely fill the ligand cavity. To render a sufficient interaction with all the possible ether oxygens, the ligand wraps around the sodium ion, one of the donor atoms thus occupying an apical position of the coordination sphere while the remaining five oxygens lie approximately in a plane. The metal ion is also coordinated to a water molecule. The SCN^- anion

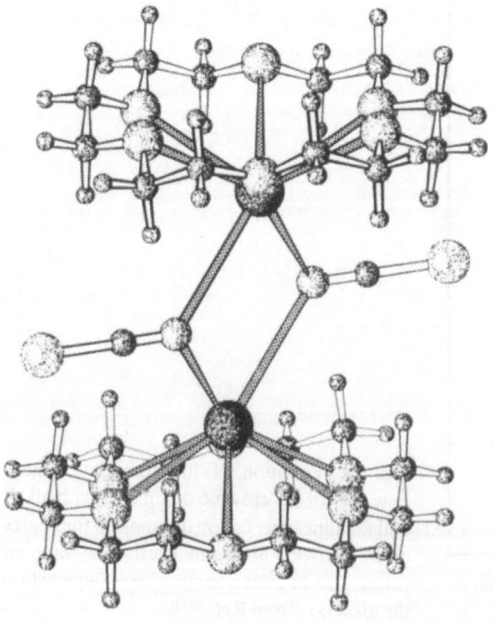

Fig. 28. Spatial structure of the 2:2 complex between [18]crown-6 and CsNCS

does not interact directly with the Na$^+$ but is hydrogen-bonded to the H$_2$O molecule (see Fig. 29). The perturbation of the crown-ether conformation is indicated by four torsion angles about C-O bonds severely deviating from the expected range [190].

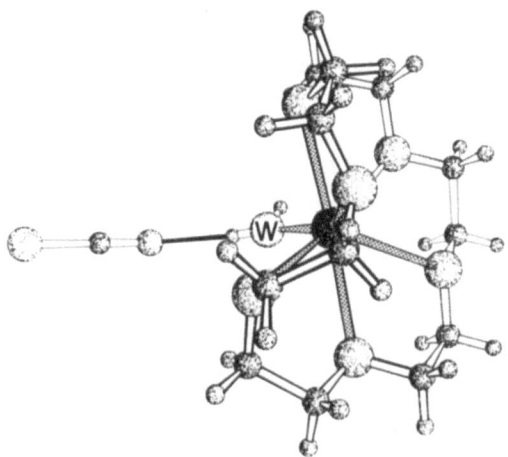

Fig. 29. The [18]crown-6 complex with NaNCS. "Double action" water molecule indicated by "W"

5.3 Effects of Anion and Cation Type on Complex Structure

Though the cavity-metal ion size relationship is very plausible, it should however be emphasized that this concept is oversimplified. Actually, there are other important factors ruling the complex structure such as ligand substituents and the anion involved. Thus, the ligand dibenzo[18]crown-6, which is less flexible than the unsubstituted hexaether, does not show any distortion in its complex with NaBr [208]. The sodium occupies a site slightly distant from the center of the macrocycle with contacts to the ether oxygens considerably longer than in the [18]crown-6-NaSCN complex (mean 2.71 Å, as compared to 2.55 Å). This is also true for a 1:1:1 complex between 18-crown-6, NaP(CN)$_2$, and tetrahydrofuran (THF). Its X-ray analysis [191] revealed two distinct complex units one of which consists of the macrocycle with Na$^+$ inside the cavity and the oxygens of two THF molecules occupying the apical coordination sites. The other also contains the sodium ion complexed inside the macrocyclic cavity, two dicyanophosphide ions coordinating to the metal ion *via* one cyano nitrogen. The latter unit is actually the first known example of a crown ether complex with negative overall charge. In both complex ions, [18-crown-6 · Na · (THF)$_2$]$^+$ and [18-crown-6 · Na · (P(CN)$_2$)$_2$]$^-$, the macrocycle attains the completely regular conformation also found in the [18]crown-6 complexes with K$^+$, Rb$^+$, and Cs$^+$. This can be explained by the availability of suitable solvent molecules or anions to occupy the apical coordination sites and thus avoiding unfavorable distortions of the ligand conformation. Evidence is added to this view by the fact that in the [18]crown-6 complex with Ca(NCS)$_2$ [199], the Ca^{2+} ion is located in the center of the (undistorted) crown though the ionic radius of Ca^{2+} ion is very similar to that of Na$^+$ (see

Table 1). This is due to the presence of *two* coordinating anions which lead to a sufficiently occupied coordination sphere of the metal ion.

If the anion is strongly coordinated to the metal ion to form an ion pair, it tends to pull the latter out of the crown ether ring. Thus, whereas K$^+$ is located exactly in the ether oxygen plane in the [18]crown-6 complex with KNCS, it is 0.9 Å above this plane, if the anion employed is the chelating ethyl acetoacetate [193]. As a result, the K ... O (crown) distances are considerably increased (2.83–3.02 Å) as compared to the KNCS complex (2.77–2.83 Å).

This effect is even more pronounced in the complex between benzo-15-crown-5 and calcium picrate [189]. In this structure, the metal ion though smoothly fitting into the cavity of the 15-membered ring, does not have any direct contact to the latter at all. Rather, it prefers chelation by the two picrate ions each of which provides its phenolate oxygen and an orthonitro oxygen. The remaining coordination sites are occupied by three water molecules. Two of these are hydrogen-bonded to adjacent crown ether molecules and thus stabilize the structure (see Fig. 30). Interestingly, though being uncomplexed, the polyether ligand adopts a conformation distinct from that of the free benzo[15]crown-5 [239], but similar to that in other benzo[15]crown-5 complexes with metal ions. From this it may be concluded that hydrogen bonding from water molecules is able to induce the same conformational

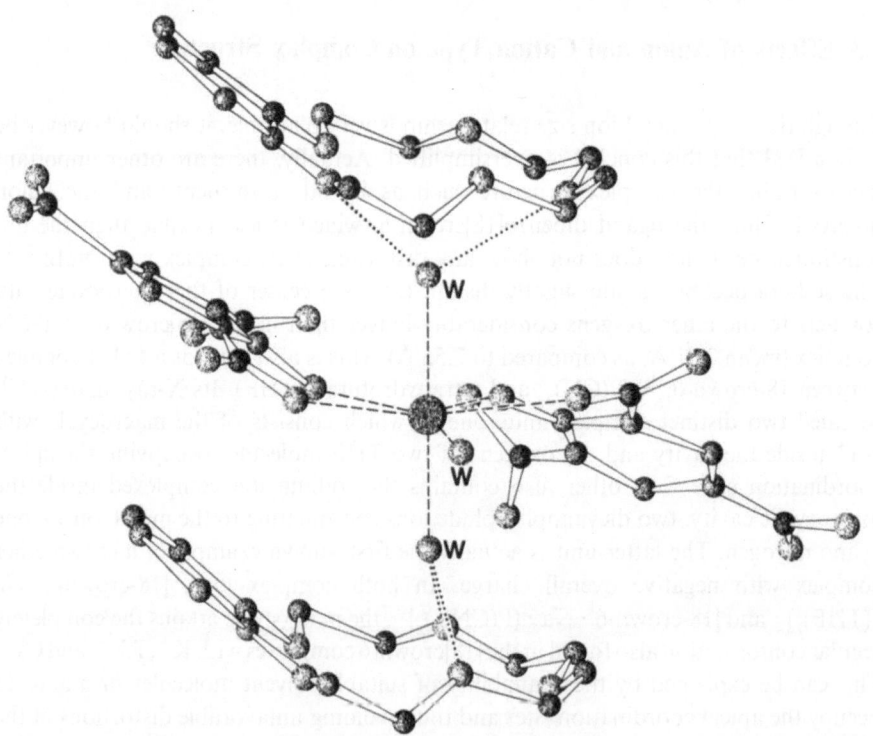

Fig. 30. Structure of the benzo[15]crown-5 complex with calcium picrate and three water molecules "W". Hydrogen atoms have been omitted for clarity

changes of the ligand as do metal ions, suggesting rather shallow energy barriers between different conformational states.

Clearlx, the above complex structure is dictated by the bulky picrate anions the o-nitro groups of which preclude further approach of the crown molecule toward the Ca^{2+} ion. When picrate is replaced by 3,5-dinitrobenzoate which lacks an o-nitro group, the Ca^{2+} is ninefold coordinated to the carboxylate oxygens and to all of the five crown ether oxygens but displaced by 1.38 Å from their mean plane [188]. Additionally, the crystal structure contains an uncomplexed molecule of benzo[15]-crown-5 which is hydrogen-bonded to water.

The anion involved not only shows an effect on the distance between the metal ion and the ligand but also on the ligand conformation itself. Two crystal structures of a benzo[15]crown-5 complex with $Ca(NCS)_2$ have been published [186, 187], one being a water and the other the corresponding methanol solvate. The conformations of the cyclic ether in both these structures are very similar to each other, the Ca^{2+} being situated by 1.22 Å above the plane of the donor atoms. However, the ligand conformation differs considerably from that found in the complex with calcium 3,5-dinitrobenzoate [188]. This may be explained by the different coordination number (8 in the SCN^- complexes, and 9 in the 3,5-dinitrobenzoate complex) which influences the effective charge density of the cation, and in turn, together with the altered distance between the crown and the Ca^{2+}, results in a different polarization of the ligand by the cation. Furthermore, in the 3,5-dinitrobenzoate complex one of the bulky aromatic anions is stacked above the benzene ring of the ether.

The dependence of the complex structure on the anion involved is particularly pronounced in the case alkaline earth cations. Due to their relatively high charge densities (see Table 1), these ions are *always ion-paired* in their crown ether complexes. Only two exceptions to this rule are known: in the 12-crown-4 complexes with $MgCl_2$ [178] and $CaCl_2$ [179] the metal ions are heavily hydrated. These water molecules expel the chloride ions and, in the case of the Mg complex, also the crown molecule. The H_2O molecules display a "double action" [261] because they stabilize (1) the cation by coordination and (2) the anion by hydrogen bonding, thus diminishing the nucleophilic power of the latter.

Actually, highly nucleophilic anions such as Cl^- are deleterious for the synthesis of crown ether complexes, if they are not effectively stabilized by hydrogen bonding from protic solvents or other proton donors such as picric acid [262]. Anions that have been found useful for crown ether complex preparation are of soft HSAB character, e.g. SCN^-, ClO_4^-, Br^-, I^-, and picrate.

The difference in charge density between alkali and alkaline earth cations of similar radius, e.g. Na^+ and Ca^{2+}, causes some structural differences of their crown ether complexes which again demonstrate that the ion-cavity relation is not the only factor crucial for the complex structure. In the benzo-15-crown-5 complex with NaI [184], cation and anion are charge separated, only being connected by a H_2O molecule which again displays a "double action", as described above. The Na^+ ion lies only 0.75 Å above the crown, in comparison to 1.22 Å in the case of the corresponding complex with $Ca(NCS)_2$ [186, 187]. Moreover, there are severe differences in ligand conformation: One C—C—O—C torsion angle differs by as much as 90°.

5.4 Simultaneous Complexation of Metal Ion and Water or of two Metal Ions

Hitherto, we have seen two different structural possibilities for a crown ether complex with a cation the radius of which is too small to fill the ligand cavity: Either the ligand wraps around the ion as in [18]crown-6-NaSCN, or the crown is unchanged with the metal ion at its center assuming longer than optimum distances to the donor atoms as in [18]crown-6-NaP(CN)$_2$-THF. Yet another possibility is filling the host cavity by simultaneous complexation of a metal ion and a H$_2$O molecule. An example for this was recently found by Czugler and Kálmán in the structure of a lithium iodide complex of a multiloop crown ether in which two [19]crown-6 units are attached to each other by a spiro junction [232]. Each loop contains one Li$^+$ and one water molecule. The latter donates two hydrogen bonds to two ether oxygen atoms and interacts also with the lithium ion (Fig. 31). Besides to this water molecule, the metal ion is coordinated to three ether oxygens and to a second water molecule outside the ligand cavity. Thus, one of the ligand donor atoms interacts neither with H$_2$O nor Li$^+$, and it is just in this part of the coronand that large deviations from the normal torsion angles occur.

A somewhat similar situation has been found in the dibenzo-24-crown-8 complex with barium picrate [227]. The Ba^{2+} does only fill one compartment of the ring whilst the other is occupied by a water molecule which is hydrogen-bonded to the receptor and simultaneously coordinated to the Ba^{2+} ion. Interestingly, the situation is completely changed when ClO$_4^-$ is employed as anion instead of picrate. In this complex, the polyether wraps about the Ba^{2+} ion which is coordinated to all the ether oxygens and to the two perchlorate anions [226]. This is another example for the structurally decisive influence of the anion.

The ionic radius of K$^+$ is very close to that of Ba^{2+} but its charge density is much lower (see Table 1). As a result, dibenzo[24]crown-8 is able to form a complex with 1:2 stoichiometry (crown:metal ion) with KNCS in which *two* potassium ions

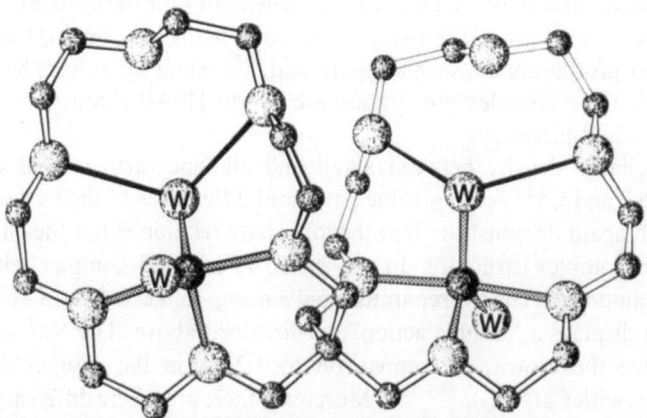

Fig. 31. 1:2 LiI complex of the bicyclic ligand spiro-bis[19]crown-6. Water molecules are indicated by "W". Neither iodide anions nor hydrogen atoms are included in the figure

are simultaneously complexed in the macrocyclic cavity and connected by bridging SCN⁻ ions [224, 225]. A somewhat similar arrangement was detected by X-ray crystallography in the 1:2 complex between the same ligand and sodium *o*-nitrophenolate [223] (Fig. 32).

Similar to dibenzo[24]crown-8, the much larger dibenzo[30]crown-10 complexes two Na⁺ ions simultaneously [228]. With potassium and rubidium, however, only 1:1 complexes are formed [229−231]. This is because the ligand completely wraps around

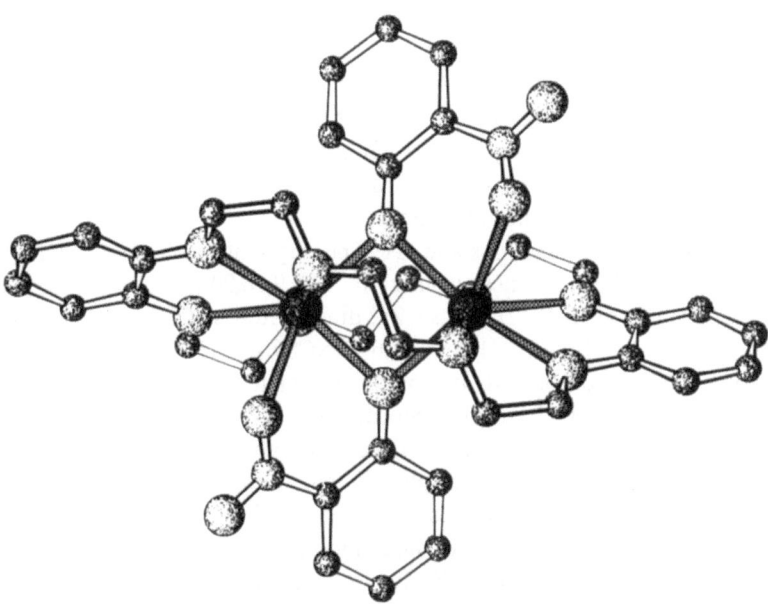

Fig. 32. Structure of the 1:2 dibenzo[24]crown-8 complex with sodium *o*-nitrophenolate. Hydrogens are not shown

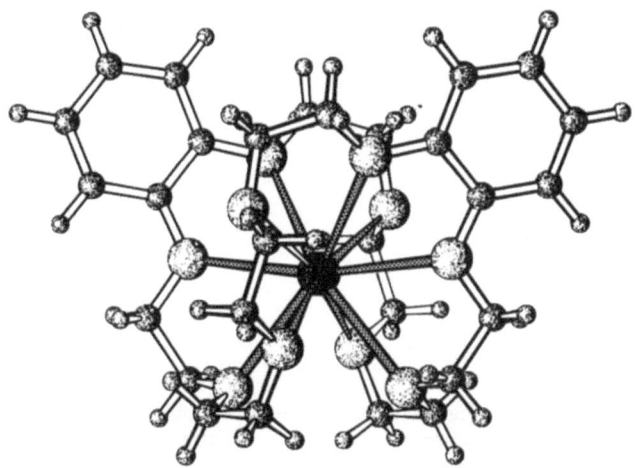

Fig. 33. Three-dimensional structure of the dibenzo[30]-crown-10 complex with K⁺

the complexed ion leading to a very effective three-dimensional encapsulation (Fig. 33) similar to that achieved by valinomycin.

5.5 Sandwich Formation

The cations Rb^+ and Cs^+ are too large in size for the [18]crown-6 cavity and lie above it. However, ^1H-, ^{13}C- and ^{133}Cs-NMR studies of the [18]crown-6- and dibenzo[18]crown-6-Cs^+ systems [263-266] indicated that at low temperatures formation of *2:1 sandwich complexes* occurs, but it should be emphasized that the anions involved in these investigations were BPh_4^- and BF_4^- which do not coordinate to cations and thus favor sandwich formation. Nonetheless, sandwich complexes are a common structural feature in many crown complexes with cations that are too large to fit into the macrocyclic cavity. The first crown sandwich whose structure was determined by X-ray crystallography was the KI complex of benzo[15]crown-5 [185]. In this complex, the potassium ion is simply coordinated by all of the ten ether oxygens, thus being completely shielded from the anion (Fig. 34).

The complex stoichiometry may, in some cases, be very sensitive to subtle changes in ligand configuration. Thus, the meso isomer "F" of 7,9,18,20-tetramethyldibenzo-[18]crown-6 (for formula see Fig. 22) forms a 2:2 dimer with CsNCS by bridging iso-thiocyanate ions [214], as was found in [18]crown-6-CsNCS itself. However, if the optically active isomer "G" is the ligand, use of the racemate leads to a 2:1 charge-separated sandwich in which Cs^+ is 12fold coordinated [214, 215].

Finally, it should be mentioned that sandwich formation is not a privilege of crown ether complexes with metal ions displaying ionic radii which are too large to fit into the ligand cavity. Infrared spectroscopic studies revealed that benzo[15]crown-5 forms a 2:1 sandwich with Na^+ (which could smoothly fit into the cavity), if the anion is tetraphenyl borate [267]. This is due to the inability of the anion to provide donor atoms for the Na^+ ion which thus requires a second crown for a sufficient coordination.

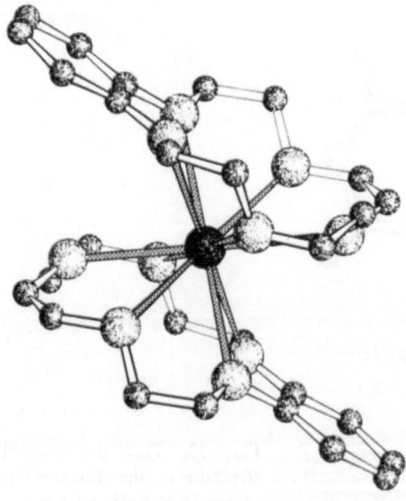

Fig. 34. 2:1 Sandwich-type complex between benzo-[15]crown-5 and K^+. For clarity, hydrogen atoms have been omitted

5.6 Ternary Crown Ether Complexes

In 1970, Pedersen reported on the formation of ternary adducts between dibenzo-[18]crown-6, potassium thiocyanate or iodide, *and* thiourea [268]. The stoichiometries varied between 1:1:1 and 1:1:6. One of these "supercomplexes", a 1:1:1 adduct between the crown ether, KI, and thiourea, has recently been the subject of an X-ray structural analysis [212]. It revealed that the potassium ion located in the center of the cyclic hexaether is coordinated to all the ether oxygens and to the iodide anion as was also found in the crystal structure of the simple 1:1 complex between the same ligand and KI [211]. Thiourea is not involved in the complexation of the metal ion, nor does it have any contact to the polyether ligand. Instead, it forms polymeric, hydrogen-bonded sheets. One hydrogen atom of each amide group is in contact with the sulfur of an adjacent thiourea molecule whereas the other one is involved in an N—H ... I hydrogen bond to the iodide anion (see Fig. 35). As a result of this involvement of the anion in hydrogen bonding, the K^+ ... I^- distance is slightly larger than the one found in the simple KI complex. Apparently, thiourea stabilizes the anion by hydrogen bonding as found for H_2O in many other crown ether complexes, thus facilitating complexation of the metal ion by the coronand. The resulting adduct might be called a *doubly wrapped salt* [268].

6 Macropolycyclic Host Molecules: Cryptands and their Cation Complexes

Macropolycyclic ligands, commonly referred to as *cryptands*, contain intramolecular cavities of three-dimensional shape ("*crypts*"). In their complexes ("*cryptates*") with alkali and alkaline earth cations, they display considerably enhanced stabilities with respect to crown ethers ("*cryptate effect*") [13, 289]. Thus, the K^+ complex of [2.2.2] cryptand (for nomenclature see Fig. 36) is by a factor of 10^5 more stable than the corresponding diaza[18]crown-6 complex and even by four orders of magnitude compared with the valinomycin potassium complex [293]. Furthermore, the smaller cryptands exhibit pronounced peak selectivity for alkali or alkaline earth cations (see Table 12) which agrees well with the ion-cavity size concept.

In bicyclic oligoethers which are usually designated as [2]-cryptates, two nitrogen atoms serve as bridgeheads. Each of these may be oriented either inward or outward with respect to the central cavity, leading to three possible stereoisomeric forms: *in-in*, *in-out*, and *out-out* (Fig. 37) [13]. Due to the electronic lone pairs on the nitrogen atoms pointing toward the metal ion, the most favorable isomer for complex formation is the *in-in* form which was actually found by X-ray crystallographic methods in all the cation complexes of cryptands and also for the uncomplexed [2.2.2] cryptand [291].

The *in-out* and *out-out* forms have been detected in the crystal structures of N-borane-[1.1.1] [292] and N,N-diborane-[2.2.2] [291] where BH_3 groups are attached to the amine nitrogens.

As in the case of monocyclic polyether ligands, the preferred conformations about C-C and C-O bonds are *synclinal* and *antiperiplanar*, respectively. The C—N—C—C, C—C—N—C torsion angles may lie in either of these low-energy ranges.

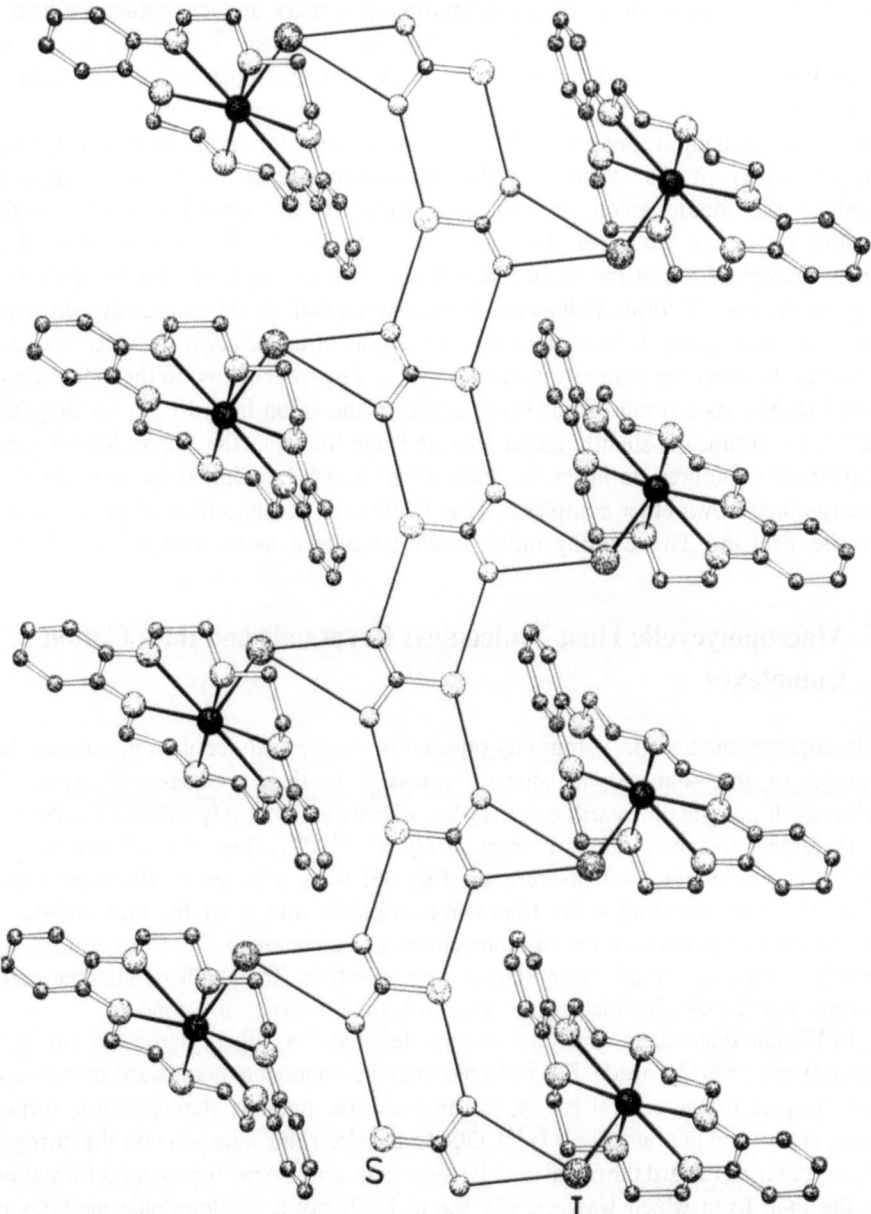

Fig. 35. Polymeric structure of the ternary "super" complex between dibenzo[18]crown-6, potassium iodide, and thiourea. Note the endless sheet of thiourea molecules. Iodide and sulfur indicated by I and S, resp.

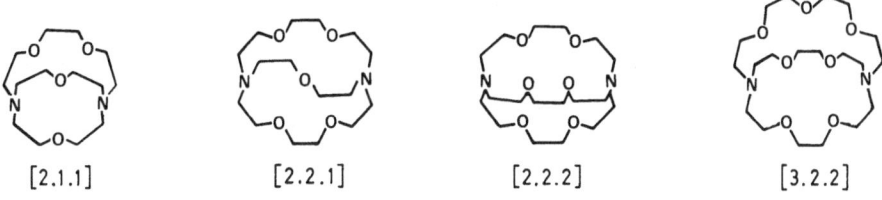

Fig. 36. Structural formulae of some cryptands

[2.1.1]   [2.2.1]   [2.2.2]   [3.2.2]

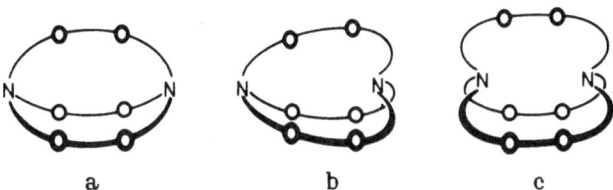

a   b   c

Fig. 37a—c. Cryptands: a *in-in*, b *in-out*, and c *out-out* stereoisomers (from Ref. [15])

A large number of [2.2.2] complexes have been subjected to X-ray analyses (see Table 13). The cation-ligand size relation for this cryptand has its optimum for potassium as the complexed species (cavity diameter $\approx$ 2.80 Å, ionic diameter 2.66 Å). However, the cryptand is flexible enough to accommodate the undersized Na$^+$ ion. This is accompanied by significant twisting of the ligand as expressed by some less favorable torsion angles [272]. Moreover, whereas Na$^+$ is fairly regularly eightfold coordinated in the [2.2.2]-NaI cryptate [272], somewhat different coordination modes have been detected in the crystal structure of ([2.2.2]Na)$_2$ Fe$_2$(CO)$_6$(μ_2-PPh$_2$)$_2$ [274] and ([2.2.2]Na)$_4$Sn$_9$ [277]. In these cases, the coordination number of sodium is decreased to seven and six, respectively. On the other hand, the twisting of the cryptand in both these structures is less pronounced. Apparently, an interplay of less favorable coordination of the guest ion and conformational strain is operative in [2.2.2] complexes with Na$^+$.

The flexibility of the [2.2.2] cryptand is also evident from its complexes with larger ions such as Rb$^+$ and Cs$^+$. If going from K$^+$ to Rb$^+$ to Cs$^+$, the distance between the two bridgehead nitrogen atoms which is significant for the cavity size increases from 5.75 to 6.00 to 6.07 Å, accompanied by a change of the mean torsion angle about C-C bonds from 54 to 67 to 70° [279]. Moreover, a N ... N separation as short as 4.92 Å has been found in [2.2.2]-Ag$^+$ [293], whereas it reaches its maximum of 6.87 Å in the free host molecule [291]. The latter is of an elongated shape as described in Chapter 5 for uncomplexed crown ethers.

In the [2.2.2] complex with cesium, the Cs$^+$ ... O distances are somewhat smaller than the corresponding sum of the Van-der-Waals radius of oxygen and the ionic radius of Cs$^+$, as was similarly observed in the Cs$^+$ complex of nonactin (see Chap. 3). This indicates that the cation presses heavily on the ligand. In fact, it was concluded from ^{133}Cs-NMR data that in solution, [2.2.2] forms an *exclusive complex* with Cs$^+$, i.e. the metal ion does not occupy the central ligand cavity but is only partly

Table 10. Complexes of cyclic crown ethers

Compound	Coord. no.	Bond distances			Stoichiometry and type of structure	Ref.
		M—O	M-anion	M—O (solv.)		
[12]crown-4 complexes						
[Na · L₂]⁺ [Cl · (H₂O)₅]⁻	8	2.47–2.52	—	—	2:1 sandwich	176)
[Na · L₂]⁺ [OH · (H₂O)₈]⁻	8	2.44–2.51	—	—	2:1 sandwich	177)
L · [Mg(H₂O)₆]²⁺ (Br)₂⁻	6	—	—	2.05–2.08	Mg²⁺ not encapsulated in the crown	178)
[Ca · L · (H₂O)₄]²⁺ (Cl)₂	8	2.51–2.54	—	2.38–2.40	1:1; Ca²⁺ 1.63 Å above the crown plane	179)
[Cu · L · Cl₂]	6	2.11–2.40	2.21, 2.23	—	1:1	180)
[(UO₂) · L · (H₂O)₂]²⁺ (NO₃⁻)₂	6	1.77–2.05	—	2.37, 2.37	1:1; UO₂²⁺ encapsulated in the crown	181)
[15]crown-5 complexes						
[Ba · L₂]²⁺ [Br₂(H₂O)₂]²⁻	10	2.75–2.88	—	—	2:1 sandwich; crowns are apparently disordered	182)
L · [Cu(H₂O)₂Br₂]	4	—	2.35, 2.37	1.95, 2.00	Cu²⁺ not encapsulated in the crown	183)
benzo[15]crown-5 complexes						
[Na · L · H₂O]⁺ I⁻	6	2.35–2.43	—	2.29	1:1; I⁻ hydrogen-bonded to water	184)
[K · L₂]⁺ I⁻	10	2.77–2.96	—	—	2:1 sandwich	185)
[Mg · L · (NCS)₂]	7	2.17–2.20	2.06	—	1:1	186)
[Ca · L · (NCS)₂ · H₂O]	8	2.46–2.61	2.42, 2.43	2.40	1:1; Ca²⁺ 1.23 Å above the crown	186,187)
[Ca · L · (NCS)₂ · MeOH]	8	2.51–2.55	2.40, 2.49	2.38	1:1; very similar to the former complex	186,187)
[Ca · L · (dinitrobenzoate)₂] · [L · (H₂O)₃]	9	2.52–2.78	2.46–2.47	—	one of the two crowns does not interact with Ca²⁺	188)
L · [Ca(pic)₂(H₂O)₃]	7	—	2.28–2.59	2.24–2.38	Ca²⁺ not encapsulated in the crown	189)

[18]crown-6 complexes

Compound	CN				Comments	Ref
[Na · L · H₂O]⁺ NCS⁻	7	2.45–2.62	—	2.32	1:1; water displays a double action	190)
[Na · L · (THF)₂]⁺	8	2.71–2.79	—	2.36, 2.36	first example of a negatively charged crown unit	191)
[Na · L · (P(CN)₂)₂]⁻	8	2.73–2.78	2.44, 2.48	—		191)
[K · L]⁺ NCS⁻	6(8)	2.77–2.83	3.19, 3.19	—	1:1; very weak interaction with the disordered anion	192)
[K · L · (ethyl acetoacetatoenolate)]	8	2.83–3.02	2.65, 2.73	—	1:1; K⁺ 0.9 Å above the oxygen plane	193)
[K · L · (tosylate)]	8	2.78–2.94	2.69, 2.93	—	1:1; crown disordered	194)
[K · L · (H₂O)₂]⁺	8	2.78–3.05	—	2.78, 2.82	1:1; both the K⁺ ions displaced	195)
[K · L · (H₂O) · MoO₄]⁻	8	2.76–2.99	2.79	2.81	from the mean crown planes by 0.92 and 0.78 Å, resp.	195)
[K₂ · L₂ · (H₂O)₂ · Mo₆O₁₉]	8	2.72–2.88	2.70, 2.72	2.89, 2.93	2:2; two crown-K⁺ units linked by the anion	196)
[Rb · L · (NCS)]₂	8	2.93–3.15	3.23, 3.31	—	2:2; formation of dimers by bridging SCN⁻ anions	197)
[Cs · L · (NCS)]₂	8	3.03–3.27	3.30, 3.32	—	2:2; bridging SCN⁻ ions	198)
[Ca · L · (NCS)₂]	8	2.56–2.74	2.35, 2.35	—	1:1; Ca²⁺ in the center of the ether oxygen plane	199)
(L)₂ [Mn(H₂O)₆]²⁺ (ClO₄⁻)₂	6	—	—	2.14–2.22	Mn²⁺ not encapsulated in the crown ether	200)
(L)₂ [Co(H₂O)₆]²⁺ [CoCl₄]²⁻	6/4	—	—	2.04–2.12	both the Co²⁺ ions not complexed by the crowns	201)
([UCl₃ · L]⁺)₂ [UO₂Cl₃(OH) (H₂O)]²⁻	9	2.48–2.61	—	—	1:1; the published coordinates are suspicious	202)
L · (H₂O)₃ · [UO₂(H₂O)₂(NO₃)₂]	8	—	2.48–2.49	2.43, 2.43	UO₂²⁺ not encapsulated by the crown ether	

benzo(L')- and 4-nitrobenzo(L'')/[18]crown-6 complexes

Compound	CN				Comments	Ref
[Rb · L' · (NCS)]₂	8	2.91–3.13	3.04, 3.05	—	2:2; formation of 5imers by bridging SCN⁻ ions	204)
[Sr · L' · (H₂O)₃]²⁺ (ClO₄⁻)₂	9	2.66–2.72	—	2.55–2.58	1:1	205)
[Ba · L' · (H₂O)₂ (ClO₄)₂]	10	2.80–2.85	2.79, 2.94	2.78, 2.84	1:1	205)
[Rb · L'' · (NCS)]	8	2.95–3.08	2.90	—	1:1; Rb⁺ interacts with the nitro group of an adjacent ligand molecule	206)

Table 10. (continued)

Compound	Coord. no.	Bond distances			Stoichiometry and type of structure	Ref.
		M—O	M—anion	M—O (solv.)		
[Cs · L'' · (NCS)]	8	3.04–3.25	Cs-N : 3.44 Cs-S : 3.68	—	1:1; Cs$^+$ interacts with nitro group as in the former complex	[207]
dibenzo[18]crown-6 complexes NaBr · L · 2 H$_2$O:						
Molecule A: [Na · L · H$_2$O · Br]	8	2.54–2.89	2.82	2.35	1:1; two distinct complexes with different coordination. A is a complexed ion pair and B a complexed cation	[208]
Molecule B: [Na · L · (H$_2$O)$_2$]$^+$ Br$^-$	8	2.63–2.82	—	2.27, 2.31		
[Na · L]$^+$ NCS$^-$	6	2.74–2.89	—	—	1:1; occupancy of cation site is 45% for Na$^+$ and 55% for Rb$^+$ (see below)	[209, 210]
KI · L · 1/2 H$_2$O: molecule A: [K · L · I]	7	2.73–2.79	3.52	—	1:1; two distinct complex molecules with different modes of coordination	[211]
molecule B: [K · L · H$_2$O]$^+$ I$^-$	7	2.73–2.79	—	2.72		
[K · L · I · (thiourea)]	7	2.71–2.80	3.57	—	1:1:1 complex between KI, crown, and thiourea; polymeric	[212]
[Rb · L · (NCS)]	7	2.86–2.94	2.94	—	1:1; occupancy of cation site is 55% for Rb$^+$ and 45% for Na$^+$ (see above)	[209, 210]
[Sm · L · (ClO$_4$)$_3$]	10	2.41–2.59	2.36–2.64	—	1:1	[213]
tetramethyldibenzo[18]crown-6 complexes (isomer "F" = L': methyl groups cis, anti, cis cis; isomer "G" = L'': methyl groups trans, anti, trans, trans)						
[Cs · L' · (NCS)]$_2$	8	3.07–3.34	3.30, 3.32	—	2:2; formation of dimer by bridging SCN$^-$ ions	[214]
[Cs · (L'')$_2$]$^+$ NCS$^-$	12	3.12–3.36	—	—	2:1 charge-separated sandwich	[214, 215]

dicyclohexano[18]crown-6 complexes						
[Na · L · (H₂O)₂]⁺ Br⁻	8	2.68–2.97	—	2.35, 2.35	1:1; ligand is the cis-anti-cis isomer	216, 217)
[Ba · L · (NCS)₂ · (H₂O)]	9	2.80–2.91	2.88, 2.88	2.80	1:1; ligand is the cis-syn-cis isomer	218)
[La · L · (NO₃)₃]	12	a)	a)	—	1:1; ligand is the cis-syn-cis isomer	219)
([UCl₃ · L]⁺)₂ UCl₆²⁻	9	2.47–2.65	—	—	1:1; cis-syn-cis isomer	220)
2,3-naphtho[20]crown-6 complex						
[K · L · (NCS)]	7	2.73–2.88	3.26	—	1:1; K⁺ interacts only weakly with disordered SCN⁻	221)
4,18-dioxo bezo[21]crown-7 complex						
[K · L · (NCS)]	8	2.77–3.07	N: 2.80 S : 3.49	—	1:1; K⁺ interacts also with S of adjacent SCN⁻	222)
dibenzo[24]crown-8 complexes						
[Na₂ · L · (o-dinitrophenolate)₂]	6	2.47–2.62	2.30–2.40	—	1:2; each Na⁺ coordinated to 3 ether oxygens	223)
[K₂ · L · (NCS)₂]	7	2.73–2.98	2.87, 2.88	—	1:2; both the SCN⁻ bridge metal ions	224, 225)
[Ba · L · (ClO₄)₂]	10	2.76–3.04	2.72, 2.79	—	1:1	226)
[Ba · L · (H₂O)₂ · (pic)₂]	10	2.86–3.00	2.67–3.09	2.73, 2.77	1:1; ligand cavity occupied by Ba²⁺ *and* a water molecule	227)
dibenzo[30]crown-10 complexes						
[Na₂ · L · (NCS)₂]	7	2.40–2.59	2.36	—	1:2	228)
[K · L]⁺ I⁻	10	2.85–2.93	—	—	1:1	229)
[K · L]⁺ NCS⁻	10	2.84–2.96	—	—	1:1	230)
[Rb · L]⁺ [(NCS) · (H₂O)]⁻	10	2.96–3.19	—	—	1:1	231)
spiro-bis[18, 18'-19-crown-6] complex						
[Li₂ · L · (H₂O)₄]²⁺ (I⁻)₂	5	1.93–2.21	—	1.93, 1.96	1:2; each loop contains a Li⁺ and a water molecule	232)

Table 10. (continued)

Compound	Coord. no.	Bond distances			Stoichiometry and type of structure	Ref.
		$M-O^b$	M-anion	M-O (solv.)		
1,10-diaza[18]crown-6 complexes^b						
[K · L · (NCS)]	7	O: 2.82–2.84 N: 2.86, 2.86	3.33	—	1:1; SCN⁻ ions link complex molecules in the crystal	233)
[Pb · L · (SCN)₂]	8	O: 2.79–2.88 N: 2.75, 2.75	2.89, 2.89	—	1:1; Pb²⁺ coordinated to sulfur atoms of SCN⁻ ions	234)
[Cd · L · I₂]	8	O: 2.81–2.84 N: 2.48, 2.48	2.83, 2.83	—	1:1	235)
[Hg · L · L₂]	8	O: 2.91–2.94 N: 2.72, 2.72	2.68, 2.68	—	1:1	235)
[Cu · L · Cl₂]	6	O: 2.71, 2.75 N: 2.03, 2.04	2.28, 2.34	—	1:1; Cu²⁺ coordinated to two O and two N of ligand	236)
[Cu · L · Br₂]	6	O: 2.77, 2.77 N: 2.01, 2.01	2.44, 2.51	—	1:1; similar to the former complex	237)

a coordinates not available
b distances between metal ion and ligand nitrogen are given separately ("N")
L = ligand; refers to the crown ether quoted in the corresponding headline
pic = picrate

Table 11. Published X-ray structures of uncomplexed crown molecules

Compound	Comments, if any	Ref.
[12]crown-4		238)
benzo[15]crown-5		239)
[18]crown-6		240, 241)
dibenzo[18]crown-6		209, 210, 242)
dicyclohexano[18]crown-6		243, 244)
tetramethyldibenzo[18]crown-6		245)
2,6-dioxo[18]crown-6	ordered structure, does not show the usual elliptical shape	246)
[18]crown-6 with two attached sugar moieties		247)
[20]crown-6 derivative	partial conformational disorder	248)
2,6-dimethylbenzoic acid-[18]crown-5		249)
dimethyldibenzo[21]crown-7		250)
2,4-dioxo[16]crown-5	all the ether oxygens are directed to the interior of the cavity	251)
bis-2,4-pyrimidino[20]crown-6		252)
dibenzo[24]crown-8		253)
dibenzo[30]crown-10		229)
1,10-diaza[18]crown-6	cavity nearly circular	254)
1,10-dithia[18]crown-6	sulfur atoms point out of cavity	255)
1,4-dithia[18]crown-6	sulfur atoms point out of cavity	255)

Table 12. Stability constants (log K) for cryptates (data taken from Ref. [296])

Ligand (see Fig. 36)	Stability constants in water: log K for								
	Li^+	Na^+	K^+	Rb^+	Cs^+	Mg^{2+}	Ca^{2+}	Sr^{2+}	Ba^{2+}
[2.1.1]	4.3	2.8	<2.0	<2.0	<2.0		2.8	<2.0	<2.0
[2.2.1]	2.5	5.4	4.0	2.6	<2.0	<2.0	7.0	7.4	6.3
[2.2.2]	<2.0	3.9	5.4	4.4	<2.0	<2.0	4.4	8.0	9.5
[3.2.2]	<2.0	<2.0	2.2	2.1	2.2	<2.0	≈ 2.0	3.4	6.0

encapsulated by the cryptand, thus still being able to interact with the anion or the solvent [265, 294, 295].

An exclusive complex of this type was actually found also in the crystalline state for the KNCS complex of the smaller [2.2.1] cryptand [270]. From molecular models, the cavity radius of this ligand was estimated to be 1.1 Å which is too small for potassium to enter. As a result, K^+ occupies a site in the 18-membered ring rather than in the central cavity (see Fig. 38), thus resembling the coordination of K^+ by [18]crown-6. Additionally, the potassium ion is bonded to the isocyanate anion whereas anions are generally not coordinating to alkali metal ions in *inclusive* cryptates.

In contrast to K^+, Na^+ is of optimum size for the [2.2.1] ligand, and consequently, it is completely entrapped in the three-dimensional cage [270] (Fig. 39). From this it might be considered plausible that [2.2.1] exhibits selectivity for Na^+ and against K^+, but

Table 13. Coordination distances found in cryptates by X-ray crystallographic studies. The N ⋯ N separation is also given to demonstrate the relative flexibility of the cryptand cages

Compound	Coord. no.	Bond distances (Å)				N ⋯ N separation (Å)	Comments, if any	Ref.
		M–O	M–N	M-anion	M–OH₂			
[2.1.1]-complexes								
[Li · L]⁺ I⁻	6	2.08–2.17	2.29, 2.29	—	—	4.21		269)
[2.2.1]-complexes								
[Na · L]⁺ SCN⁻	7	2.45–2.52	2.59, 2.70	—	—	4.94	Na⁺ occupies a central position	270)
[K · L · (NCS)]	8	2.70–2.87	2.90, 2.92	2.78	—	5.14	K⁺ lies in 18-membered ring	270)
[Co · L]²⁺ [Co(SCN)₄]²⁻	7	2.10–2.22	2.20, 2.24	—	—	4.20	Co²⁺ occupies a central position	271)
[2.2.2]-complexes								
[Na · L]⁺ I⁻	8	2.57–2.58	2.72, 2.78	—	—	5.50		272)
[(Na · L)⁺]₂ [Fe(CO)₄]²⁻	8	av. 2.53	av. 2.80	—	—	av. 5.60		273)
[(Na · L)⁺]₂ [Fe₂(CO)₆ (μ₂ – PPh₂)₂)]²⁻	7	2.50–2.75	2.69, (3.14)ᵃ	—	—	5.84		274)
[Na · L]⁺ Na⁻	8	av. 2.57	av. 2.72	—	—	5.43		275)
[Na · L]⁺]₃ Sb₇³⁻	8	2.40–2.71	2.83, 2.94	—	—	av. 5.84		276)
[Na · L]⁺]₄ Sm₄⁴⁻: Molecule A:	6	2.47–2.57	2.90, 2.93	—	—	5.92		277)
Molecules B, C, D:	8	2.41–2.76		—	—	6.07		
[K · L]⁺ I⁻	8	2.78–2.79	2.87, 2.87	—	—	5.75		278)
[Rb · L]⁺ [(SCN) (H₂O)]⁻	8	2.88–2.93	2.99, 3.01	—	—	6.00	water hydrogen-bonded to SCN⁻	279)
[Cs · L]⁺ [(SCN) (H₂O)]⁻	8	2.96–2.97	3.02, 3.05	—	—	6.07	water hydrogen-bonded to SCN⁻	279)
[Ca · L · (H₂O)]²⁺ (Br⁻)₂	9	2.49–2.55	2.72, 2.72	—	2.42	5.44	water displays a double action	280)

Compound							Comments	Ref.
[Ba · L · (NCS) (H₂O)]⁺ NCS⁻								
Molecule A:	10	2.75–2.82	2.94, 3.00	2.91	2.88	5.94	In both molecules, H₂O displays a double action	281)
Molecule B:	10	2.74–2.89	2.99, 3.00	2.88	2.84	5.99		
[Tl · L]⁺ [(HCOO) (H₂O)]⁻	8	2.90–2.91	2.95, 2.95	—	—	5.89	water hydrogen-bonded to HCOO⁻	282)
[Pb · L · (NCS) (SCN)]	10	2.73–2.98	2.86, 2.91	N: 2.64 S: 3.12	—	5.76		283)
([La · L · (NO₃)₂]⁺)₃ [La(NO₃)₆]³⁻	12	2.64–2.74	2.81–2.85	2.63–2.69	—	av. 5.62		284)
[Eu · L · (ClO₄)]²⁺ (ClO₄⁻)₂	10	2.44–2.52	2.64, 2.70	2.67, 2.71	—	5.34	one ClO₄⁻ acts as a bidentate anion	285, 286)
[Sm · L · (NO₃)]²⁺								
[Sm(NO₃)₅(H₂O)]²⁻	10	2.44–2.57	2.75, 2.78	2.48, 2.50	—	5.53	NO₃⁻ acts as a bidentate	287)
[Ag · L]⁺ [Ag₃(SCN)₄]⁻	8	2.66–2.85	2.40–2.50	—	—	4.92	Ag—N bonds are partially covalent	293)
[3.2.2]-complexes								
[Ba · L · (H₂O)₂]²⁺ (NCS⁻)₂	11	2.80–3.09	3.08, 3.18	—	2.81, 2.87	6.10	water molecules display a double action	288)

L = ligand; refers to the cryptand quoted in the corresponding headline
a) not considered a binding distance

101

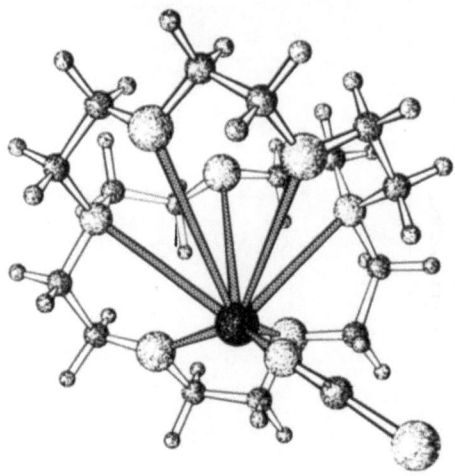

Fig. 38. Structure of the "exclusive" [2.2.1] complex with potassium isothiocyanate

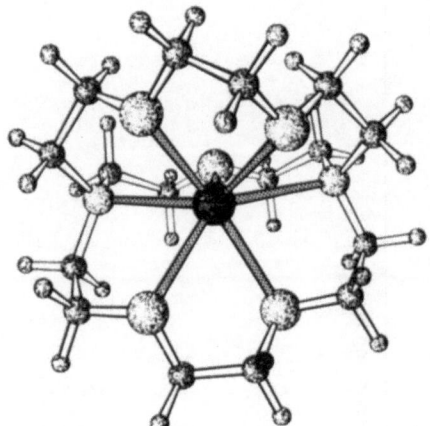

Fig. 39. The "inclusive" [2.2.1] —Na⁺ complex

thermodynamic studies revealed that this selectivity is of entropic rather than enthalpic origin [296]. This suggests that the ion-cavity radius concept though being more generally valid for cryptands than for crown ethers, again is only one criterion among others that determine the complex stability.

7 Open-Chain Polyethers: Wrapping of Metal Ions

7.1 Podands with Aromatic Donor End Groups

When going from the cyclic [18]crown-6 to its open-chain analog *pentaglyme* (Fig. 40) as a ligand for K⁺, the complex stability decreases by factor as high as 10^4 [324] although both these ligands offer the same number of donor atoms. The enhanced stability of cyclic crown ether complexes with respect to those of corresponding linear polyethers (*podands*) is attributed to a *macrocyclic effect* [325–328] which is most likely of entropic origin [326, 327].

Because of this, open-chain polyethers were believed until recently not to be capable of forming crystalline complexes with alkali and alkaline earth metal ions. However, in 1977 Vögtle and coworkers reported on considerable enhancement of complex stabilities by attaching rigid aromatic donor end groups to the oligo-(ethylene glycol) backbone as in podands *1–10* (Fig. 41) [14,329]. A series of rubidium iodide complexes with ligands *1–4* of increasing length containing 8-quinolinole moieties as rigid end groups has been subjected to X-ray analyses [330]. As a rule, it was found that *all* the heteroatoms of the ligands are coordinated to the Rb$^+$ ion. Furthermore, the steric preferences described in Chapters 5 and 6 for coronand and cryptand complexes are valid also for podands, i.e. torsion angles about C—O bonds are generally *antiperiplanar* (ap), those about C—C bonds *synclinal* (sc). Again, deviations from this rule reflect special steric requirements.

The short ligand *1* (five heteroatoms) wraps about the Rb$^+$ ion in a circular arrangement with all the torsion angles in the usual range [297]. While the cation is shielded

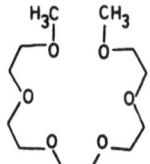

Fig. 40. Pentaglyme

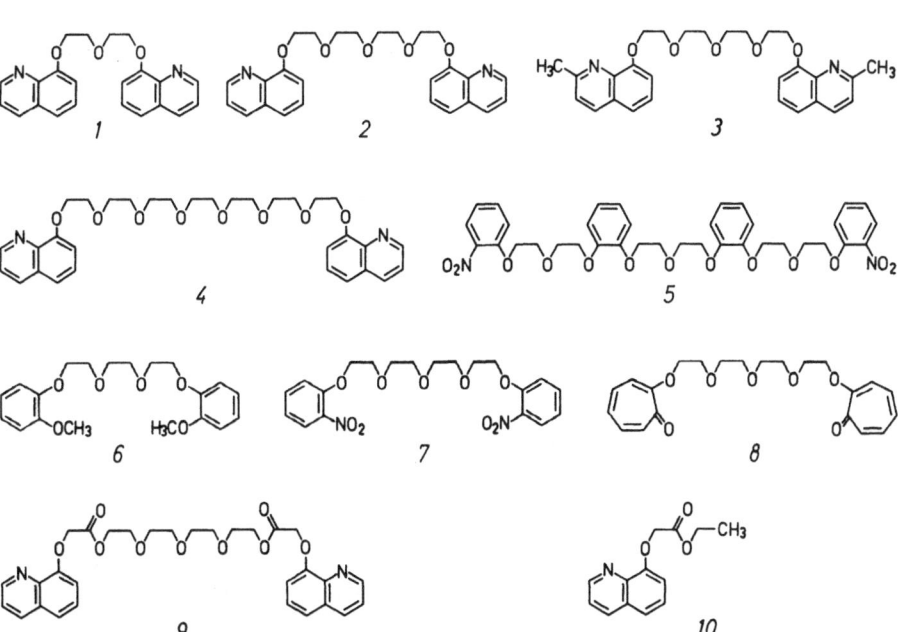

Fig. 41. Structural formulae of synthetic open-chain polyether ligands containing aromatic donor end groups

by the oligoether from one side, it is free to coordinate to two (crystallographically equivalent) iodide anions on the opposite side (Fig. 42). If podand *1* is extended by two ethylene glycol units as in *2*, it does not fit circularly around the rubidium ion but has to addopt a helical structure [298, 299]. Thus, this *achiral* ligand forms a *chiral* complex. Starting from one heteroaromatic moiety, the polyether wraps around the Rb$^+$ ion in a planar arrangement, the metal ion being positioned 0.75 Å above this plane. However, in order to avoid intramolecular collision between the two quinoline systems, a sharp turn (indicated by an arrow in Fig. 43) is introduced in the smooth wrapping of the ligand around the equatorial coordination sphere of the cation. This "kink" is achieved by a rotation of one C—O torsion angle from

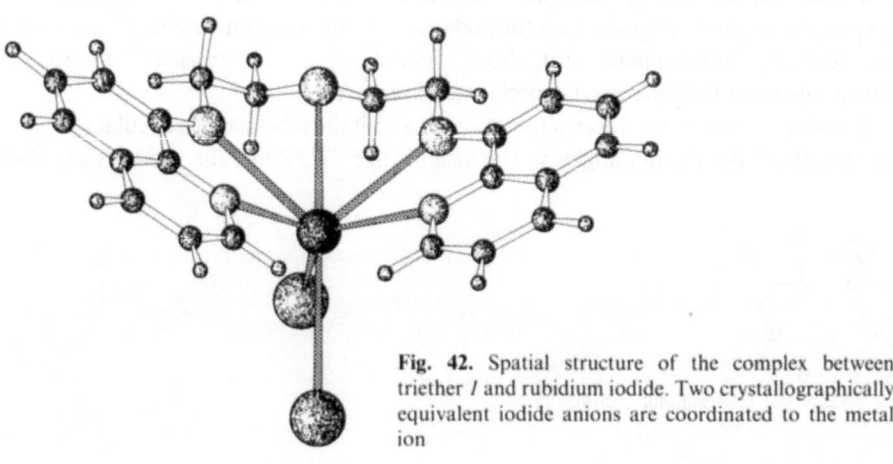

Fig. 42. Spatial structure of the complex between triether *1* and rubidium iodide. Two crystallographically equivalent iodide anions are coordinated to the metal ion

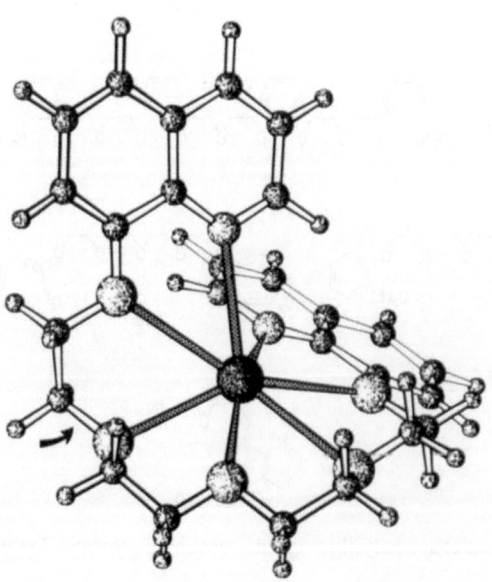

Fig. 43. Helical structure of the Rb$^+$ complex of podand *2*. The ligand "kink" with *sc* instead of *ap* orientation of torsion angle C—O—C—C is indicated by an arrow

antiperiplanar to synclinal. As a result, the nitrogen atom of the second hetero-aromatic moiety is placed by 3 Å above the equatorial ligand plane.

Interestingly, the Rb^+ ... O distances to the *aromatic* oxygen atoms are significantly longer (3.07 and 3.09 Å) than those to sp^3 oxygens. This is due to the diminished basicity of the former [298], probably associated with an unfavorable orientation of the electron lone pairs with respect to the metal ion.

If quinaldine rather than quinoline moieties are introduced as ligand end groups, an entirely different complex structure results. Podand *3* displays a helical configuration with both heterocycles stacked parallel to each other at 3.4 Å [301] (Fig. 44). The helix is somewhat more continuous than in the quinoline complex and does not display an abrupt kink. Apparently, this is accomplished by slight deviations of *all* the torsion angles from their ideal values, which is particularly true for the aliphatic C—C bonds, the mean torsion angle about which is 71° (range: 65–75°) as compared to 62° (range: 60–67°) in the quinoline complex. In order to facilitate π—π interaction between the heterocycles, the torsion angle about one of the two "aromatic" C—O bonds is in the anticlinal (128°) rather than in the more common antiperiplanar range.

The interaction between Rb^+ and the aromatic nitrogens (distances 3.04 and 3.11 Å) is weaker compared to the quinoline complex (2.93 and 2.96 Å). This is compensated for by additional coordination to the iodide anion and by the energy gained from stacking interactions.

A plausible explanation for the structural differences between the RbI complexes with ligands as similar to each other as the quinoline pentaether *2* and its methylated analog *3* is evident if one assumes the rigid donor end groups to be potential centers of nucleation of the binding process [331]. In the case of 8-quinolinole this is particularly striking as this compound itself is a strong chelating ligand for alkali and alkaline earth cations [332]. This view is supported by the short distances displayed

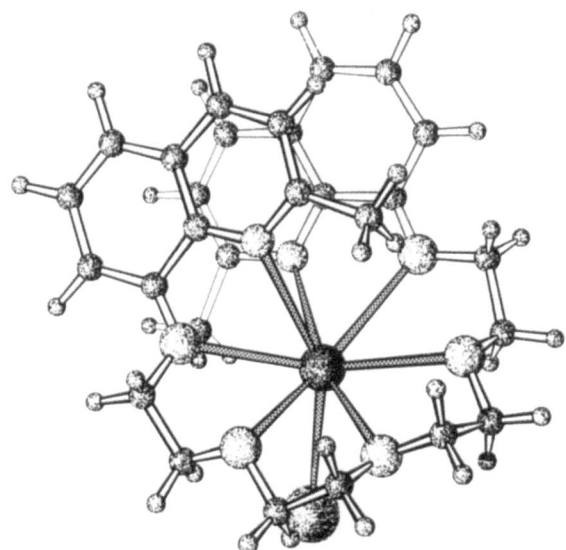

Fig. 44. Spatial structure of the complex between quinaldine ligand *3* and RbI

between Rb$^+$ and the quinoline nitrogen atoms which are actually even shorter than the sum of the corresponding atomic and ionic radii.

In contrast, 8-hydroxyquinaldine cannot function as "recognizable" site for the metal ion. The latter is not able to approach the donor atom sufficiently because of the presence of the 2-methyl group as reflected by the significantly elongated Rb$^+$ ... N distances.

range [302, 303]. All the heteroatoms may be considered coordinating if Rb$^+$ ... N

Ligand *4* with ten heteroatoms wraps smoothly around the Rb$^+$ ion with only one C—O torsion angle (127°) deviating largely from the expected antiperiplanar and Rb$^+$ ... O distances of 3.37 and 3.15 Å, respectively, are tolerated. The cation is completely shielded by the ligand (see Fig. 45) and therefore is unable to contact the anion.

Employing spectrophotometric titrations, Tümmler et al. [331] found noncyclic polyether ligands containing aromatic donor end groups to be strong but relatively nonselective host molecules for alkali cations; thus, in the case of ligand *2* log K was found to be 3.22, 3.51, and 3.06 for complexation of Na$^+$, K$^+$ and Rb$^+$, respectively. The authors explained this behavior by the high flexibility of the polyether chain which is able to easily adjust to the size of the metal ion. However, a recent X-ray study of the *2* · KSCN complex suggests another explanation [300]. Since K$^+$ is significantly smaller in size compared to Rb$^+$, one might expect the kink in ligand conformation, i.e. the deviation from planar surrounding of the cation, to occur one or two bonds earlier than in the Rb$^+$ complex. This, however, is not the

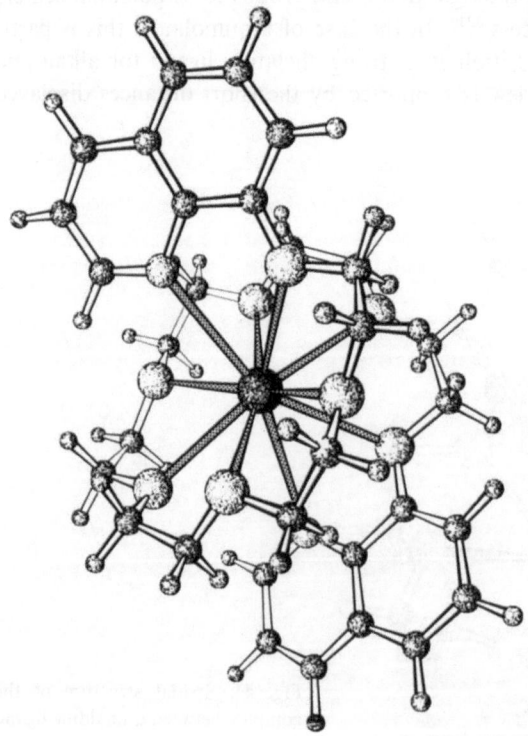

Fig. 45. Spherical wrapping of a metal ion: Podand *4*-Rb$^+$

case. Rather, the abnormal torsion angle is actually connected with the *same* C—O bond as in the Rb⁺ complex, and its value is almost identical with that found in the latter (75° as compared to 69°). Furthermore, though smotthly fitting into the cavity made up by the planar part of the ligand, the potassium ion is still placed 0.66 Å above it, i.e. it occupies a site nearly identical to that of the Rb⁺ ion. As a result, the K⁺ ... O distances are in the range 2.80–2.93 Å (average 2.86 Å), which is significantly *longer* than those observed in dibenzo[18]crown-6-K⁺ (2.71–2.80 Å, mean 2.76 Å) where the potassium ion lies within the plane formed by the ether oxygens [211,212]. On the other hand, the K⁺ ... N distances (mean 2.81 Å) are slightly *shorter* than the theoretical value (2.83 Å). This is accounted for by the strong interaction with both the terminal groups, and, in fact, these donor sites mainly determine the complex stabilities rather than the overall conformation of the polyether chain. Thus, the low selectivity of ligands of this type is obviously due to the predominant donor ability of the heteroaromatic end groups whereas the discriminating function of the ligand pseudo-cavity as complexing site is lost.

7.2 Complexes of Podands Containing End Groups Capable of Hydrogen Bonding

The ligands discussed in Section 7.1. are sometimes quoted as model compounds for polyether antibiotics (see Chap. 4). However, this does not exactly meet the situation because they lack the intramolecular hydrogen bond of the head-to-tail type which is characteristic of the polyether bioionophores. Furthermore, the latter

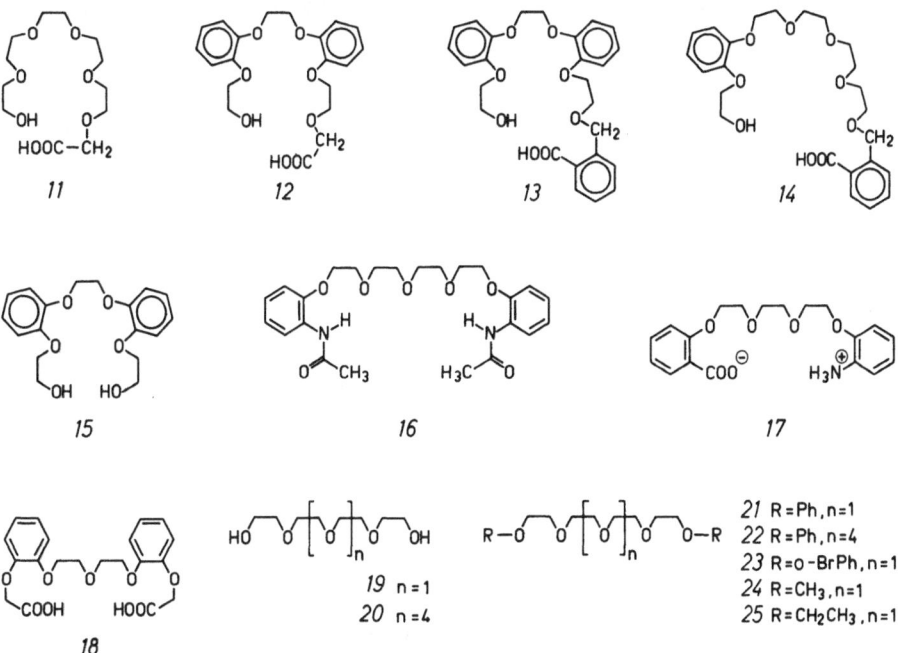

Fig. 46. Structural formulae of linear polyethers lacking rigid donor end groups

do not display helical structures in their complexes, as is observed with many synthetic podands.

Yamazaki et al. [333] were able to show that the hydrogen bonding is obviously a prerequisite to the ability of these ligands to transport metal ions across (artificial) membranes. These authors investigated the carrier properties of the open-chain polyethers depicted in Fig. 46 and found that *12* selectively transported K^+, as did *13* and *14* for Rb^+, whereas *11* and *15* were unable to transport alkali metal ions. This is in agreement with X-ray structural studies of both the NaNCS and KNCS complexes of *15* [314] which revealed that they do not contain the internal hydrogen bond which is essential for membrane transport. Rather, *two* ligand molecules are involved in the coordination of one metal ion in both complexes. As a result, the NaNCS complex is polymeric but of 1:1 stoichiometry whereas the KNCS complex consists of distinct 2:1 (ligand: metal ion) units (Fig. 47).

Although not containing rigid donor end groups as the ligands discussed in Section 7.1., these podands are still capable of forming alkali metal ion complexes due to their hydroxy end groups. In all the structures highlighted here, the latter display a "*double action*" similar to that described for water molecules in Section 5.3., i.e. they are coordinated to the cation and simultaneously stabilize the anion by hydrogen bonding, for example of the $O—H ... N—C—S^-$ type. This interaction prevents the anion from interacting too strongly with the metal ion and therefore, these polyethers containing hydroxy or carboxy groups are called "*double action ligands*" [261].

Recently, the first X-ray study on a metal ion complex of a synthetic oligoether ligand containing an internal head-to-tail hydrogen bond was completed [311]. The α-carboxy-ω-amino tetraether *17* (see Fig. 46 for formula) surrounds the Na^+ ion in a more or less planar arrangement. The terminal groups, COO^- and NH_3^+,

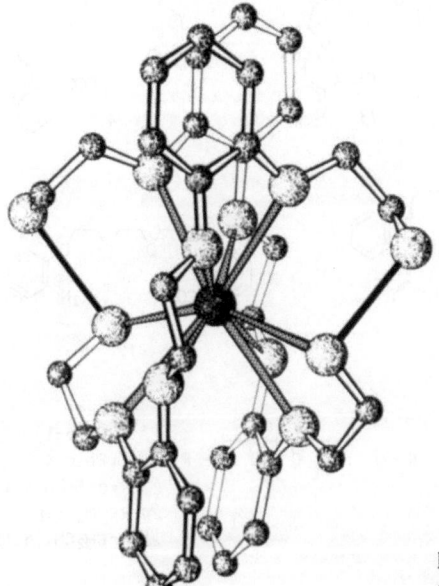

Fig. 47. Structure of the 2:1 complex between dihydroxypodand *15* and K^+. Hydrogen atoms are not shown

are linked to each other by a strong intramolecular N—H ... O hydrogen bond (Fig. 48). Presumably, the latter is to a much greater extent responsible for the pseudocyclic structure of the ligand than are the ion-dipole forces between Na$^+$ and the ether oxygens. This is also reflected by the fact that only three of the four ether oxygens contribute to the coordination of the sodium ion. A similar situation has been found, for example, in the Tl$^+$ complex of the polyether antibiotic Ionomycin [153, 154]. ^{1}H-NMR studies suggest that the N—H ... O hydrogen bond is also present in the free ligand leading to a pseudocyclic configuration as well [334]. This is exactly what is found in naturally occurring polyether antibiotics so that it seems justified to consider *17* as a suitable analog of the latter, especially for the pyrrole ethers A23187 and X-14547 A both of which contain an intramolecular N ... H—O hydrogen bond. But still, there are important structural differences.

In the crystalline state, the synthetic α-carboxy-ω-amino ligand forms a 2:2 dimer by formation of hydrogen bonds with the ClO$_4^-$ anions (see Fig. 48). This arrangement is somewhat similar to that found in the 2:2 lasalocid complex with Na$^+$ [134] if one neglects the presence of the perchlorate anions.

Unfortunately, most of the uncomplexed podands are oils so that crystallographic studies are not possible. The only free linear polyether whose structure was elucidated

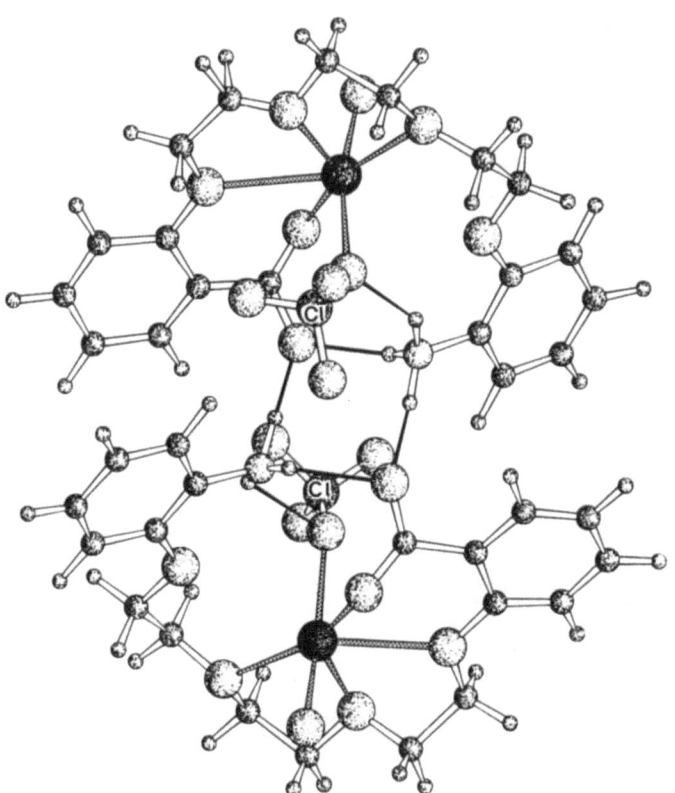

Fig. 48. 2:2 Dimer of the NaClO$_4$ complex of the α-carboxy-ω-amino ligand *17*

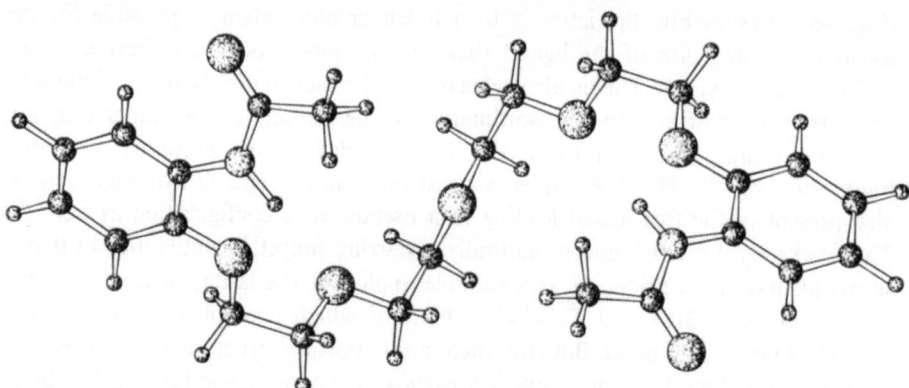

Fig. 49. S-like configuration of the uncomplexed amide ligand *16*

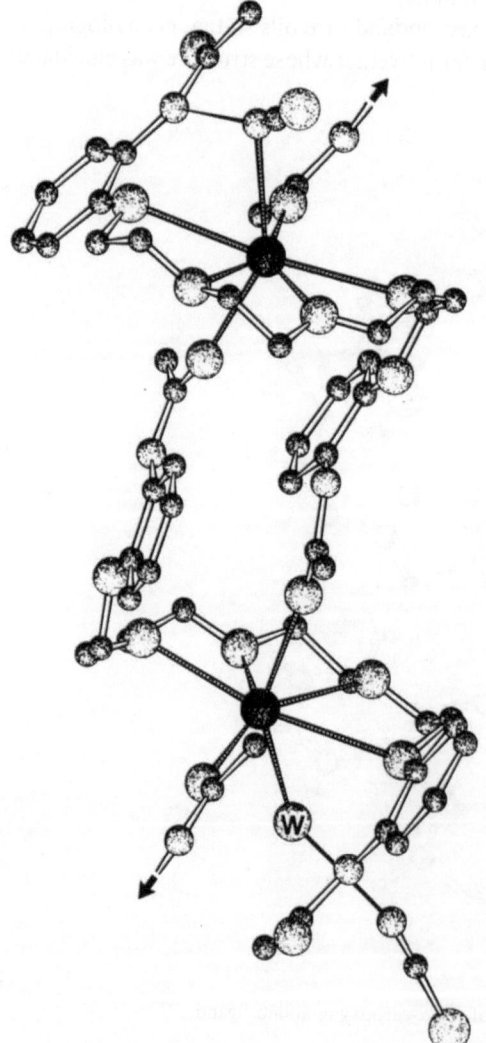

Fig. 50. Polymeric structure of the complex between polyether *16* and potassium isothiocyanate. Hydrogen atoms have been omitted for clarity. "W" indicates a water molecule. Continuation of adjacent molecules indicated by arrows

by X-ray methods is the amide ligand 16 [310] (for formula see Fig. 46). The molecule adopts an S-like configuration as is shown in Fig. 49. The torsion angles follow the general scheme being ($\pm$) synclinal for O—C—C—O but antiperiplanar for C—O—C—C and C—C—O—C.

Assuming that in solution ligand 16 adopts, on the time average, a similar configuration as found in the crystal structure, we can infer from the structure of the complex between 16 and K$^+$ that binding of the cation is initiated by recognition of one of the loops of this "empty" ligand by the cation, followed by coordination. This might be associated with a major conformational adjustment to allow for wrapping of the remaining donor atoms around the K$^+$ in a helical manner. These structural changes of the ligand are obvious from an inspection of torsion angles along the oligoether chain: Whereas C—O torsions remain antiperiplanar and those about C—C bonds synclinal, some of the latter change sign [310]. From this it can be concluded that rotation about C—C bonds contributes to the free activation energy ΔG^+ of complex formation whereas unfavorable conformations around C—O bonds induced by sterical strain, if present, affect the free reaction enthalpy $\Delta G°$.

The 16 · KNCS complex forms a complicated polymeric structure as depicted in Fig. 50 [310]. An important feature of this complex is the classical "double action" displayed by a water molecule which is hydrogen-bonded to an SCN$^-$ anion, thus shielding it from the cation. The other isothiocyanate anion accepts a hydrogen bond from the amide N—H grouping.

The formation of this polymer is in good agreement with the high entropy of complexation ($\Delta S° \approx -200 \text{ J} \cdot \text{K}^{-1} \cdot \text{mol}^{-1}$) found for the corresponding NaClO$_4$ complex by ^{23}Na-NMR techniques [335]. In fact, polymerization is a process accompanied by large negative entropy changes. This agreement between the results provided by crystallographic and spectroscopic studies suggests that the polymeric structure revealed in the crystalline state is probably also present in (concentrated) solutions.

7.3 Linear Polyethers without Donor End Groups

The previous two sections suggest that either strong aromatic donors or protic groups capable of displaying double action have to be employed as end groups for a potent noncyclic oligoether ligand. Recently, however, Vögtle and his coworkers succeeded in obtaining crystalline *alkaline earth* complexes of ligands without any terminal donor groups, including the very simple *glymes* (glycol dimethyl ethers) [336].

Ligands of this type are of great interest, as they, though forming much less stable complexes than podands containing donor end groups, exhibit considerably enhanced *selectivities* with respect to the latter [331]. Thus, the ethylene glycol diphenyl ether 21 shows the highest peak selectivity for K$^+$ of all the podands hitherto tested. Substitution of one or both phenoxy groups by the 8-quinolinol moiety results in highly increased complex stabilities associated with dramatically diminished selectivity, as can be seen from Fig. 51.

The low stabilities of the oligo (ethylene glycol) diphenyl ether complexes are accounted for by unfavorable entropy changes upon complexation which are no longer important after attachment of aromatic donor end groups to the ligand

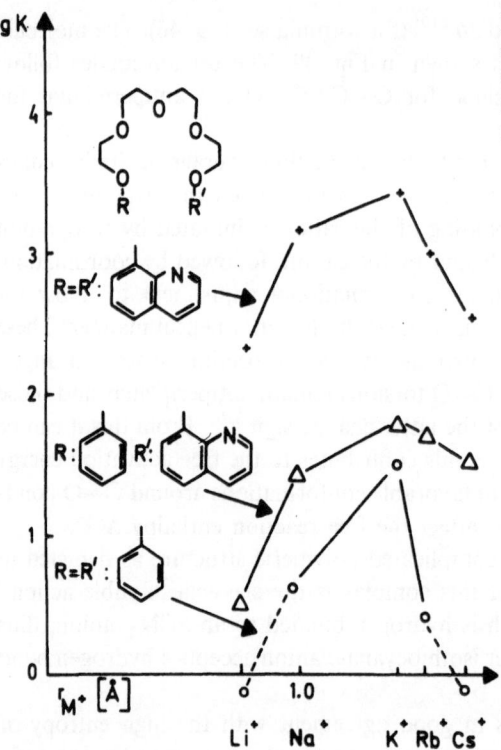

Fig. 51. Stabilities of complexes between open-chain pentaethers and alkali metal ions: Enhancement of stabilities by attaching quinoline moieties as donor end groups (from Ref. [331])

chain [331]. Until recently, this behavior together with the pronounced selectivity was not explainable. However, the first X-ray crystallographic study of a complex between a polyether lacking any donor end groups and a metal ion [317] provided much insight in the structural features that govern the selectivity pattern. Interestingly, the *octa*-ether *22* behaves only as a *hepta*dentate in its $Ba(NCS)_2$ complex (Fig. 52).

As found for the polyethers containing aromatic donor groups, the ligand wraps around the ion in a helical manner including a region of a more or less planar arrangement of the donor atoms in the equatorial zone of the coordination sphere. However, whereas the former ligands display *one* "kink" introduced by *one* abnormal torsion angle if necessary to avoid collision of the terminal chain segments, the structure described here contains *two* of these turns involving *four* C—O torsion angles being shifted to the *anticlinal* range. These conformational abnormalities, which are indicated by arrows in Fig. 52, locate the two phenyl groups below and above the mean ligand plane, respectively and suggest the following mechanism of cation inclusion [337].

Substitution of the Ba^{2+} solvation shell starts with the central ligand oxygens, and the steric hindrance of both the terminal segments of the ligand chain is overcome by a *cooperative mechanism*. Both the chain ends escape each other by undergoing unfavourable changes in conformation. This may be described by the term "binding of central portion followed by binding of floppy ends". In contrast, the mechanism postulated for podands containing donor end groups [298, 331] is different: Starting from one end group, the ligand wraps around the metal ion

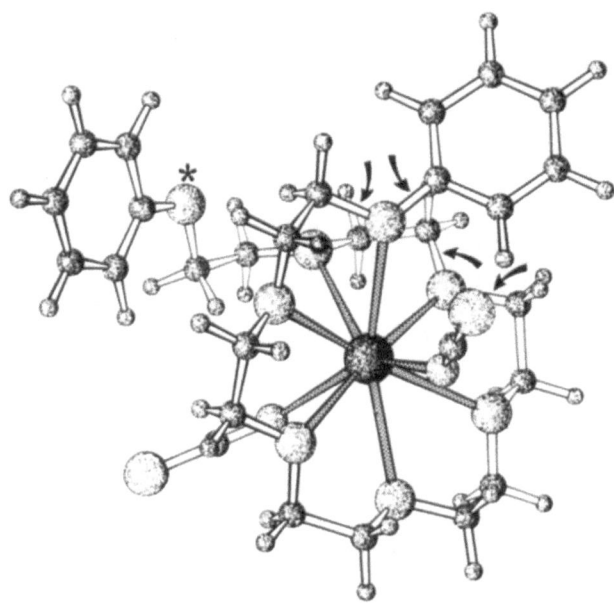

Fig. 52. X-ray structure of a Ba(NCS)$_2$ complex of the octaether *22*. Arrows indicate conformational changes from normally preferred *ap* into *ac* in the polyether chain. One of the oxygen atoms (marked by an asterisk) is not coordinated to the Ba^{2+} ion

in a regular arrangement until continuation of this process in hampered by intramolecular collision which forces the ligand into an unusual conformation. This is best designated as "binding of one end followed by binding of the other end".

Clearly, the fixation of the very flexible ligand *22* in a sterically disadvantageous conformation is entropically unfavorable. Furthermore, since the metal ion is located at a position coplanar with the central donor atoms, this ligand conformation is very sensitive to a change of the cation radius, thus providing an explanation for the pronounced peak selectivities exhibited by ligands without donor end groups.

7.4 Tripod Ligands

In 1977, it was found that the attachment of additional ligand atoms to linear polyethers resulted in a large enhancement of complex stabilities [338]. These *polypodands* exhibit a cryptand-like behavior also in their phase-transfer properties.

The first tripodand structure elucidated by X-ray crystallography was the KNCS complex of ligand *26* (see Fig. 53 for formula) [322]. The cation was found to be located in the center of a three-dimensional cavity and coordinated to all the available donor atoms (Fig. 54), reminiscent of prey cought by a three-legged octopus. From this structure, the designation of this tripodand as a "noncyclic cryptand" [338] appears to be justified.

This does not hold strictly, however, for tripodand *27* complexed with RbI. The metal ion is too big to be located in the central cavity. Therefore, it occupies a site in a two-dimensional cavity built up by two of the three octopus arms and distances to the donor atoms of the third arm are significantly longer [323]. Additionally, there is an interaction with the anion. As a whole, the binding situation is reminis-

Table 14. Geometrical features of podand complexes in the solid state

Compound	Coord. no.	M–O	M–N	M-anion	M–OH$_2$	Comments, if any	Ref.
Complexes of podands containing rigid donor end groups							
$[1 \cdot RbI]$	7	3.06–3.17	2.97, 2.97	3.69, 3.90	—	Rb$^+$ coordinated to two crystallographically equivalent I$^-$ ions	297)
$[2 \cdot Rb]^+ I^-$	7	2.89–3.08	2.93, 2.96	—	—	helical structure	298,299)
$[2 \cdot K]^+ SCN^-$	7	2.80–2.93	2.81, 2.82	—	—	helical structure	300)
$[3 \cdot RbI]$	8	2.84–3.12	3.05, 3.11	3.63	—	helical structure	301)
$[4 \cdot Rb]^+ I^-$	10	2.95–3.12	3.16, 3.36	—	—	helical structure	302,303)
$[5 \cdot K_2 \cdot (SCN)_2]$	8	2.69–3.18	a	2.33	—	K$^+$ interacts weakly with SCN$^-$	304,305)
$[6 \cdot Na(NCS)]$	7	2.35–2.54	a	2.72	—	helical structure	306)
$[7 \cdot K(NCS)]$	8	2.79–3.22	a	3.67	—	helical structure	307)
$[8 \cdot RbI]$	8	2.94–3.18	a	—	—	planar structure	308)
$[9 \cdot Ba(NCS)_2]$	11	2.79–3.17	2.92, 2.92	2.87, 2.87	—	helical; carbonyl oxygens do not coordinate	309)
$[10 \cdot Ba(NCS)_2]$	9	2.76–3.01	2.89, 2.92	2.87, 2.91	2.73	2:1 complex; carbonyl oxygens are coordinated to the cation	309)
Complexes of podands containing hydroxy, carboxy, or amide end groups							
$[15 \cdot Na(NCS)]$	7	2.42–2.57	a	2.41	—	polymeric structure	314)
$[(15)_2 \cdot K]^+ SCN^-$	10	2.86–3.05	a	—	—	2:1 complex	314)
16 · KSCN:							
Molecule A:							
$[16 \cdot K \cdot (H_2O)]^+ NCS^-$	7	2.71–3.09	—	—	2.79	polymeric structure	310)
Molecule B:							
$[16 \cdot K \cdot (NCS)]$	7	2.61–2.97	—	2.84	—		
$[17 \cdot Na \cdot ClO_4]_2$	6	2.25–2.74	—	2.38, 2.41	—	polymeric structure	311)
$([18 \cdot K]^+)_2 \cdot (pic^-)_2$	8	2.73–2.90	a	2.33, 2.62	—	dimeric structure	312,313)
$[19 \cdot Ca \cdot (pic)(H_2O)]^+ pic^-$	8	2.39–2.50	a	—	2.38	water and ligand itself display a double action	315)
$[20 \cdot Sr(NCS)]^+ SCN^-$	9	2.56–2.73	a	2.57	—		316)

Complexes with podands containing no donor end groups							
[22 · Ba(NCS)$_2$(H$_2$O)]	10	2.79–3.11	a	2.77, 2.81	3.08	only 7 of the 8 ligand oxygens coordinated to Ba^{2+}	317)
[23 · (HgBr$_2$)]	7	2.72–3.06	a	(2.39, 2.41)	—		318)
[24 · (HgCl$_2$)]	7	2.78–2.96	a	(2.29, 2.31)	—		319)
[24 · (CdCl$_2$)$_2$]	6	2.41–2.74	a	2.42–2.68	—	two Cd^{2+} ions connected with each other by bridging Cl anions	321)
[25 · (HgCl$_2$)$_2$]	6	2.66–2.91	a	2.38	—		320)
Complexes of tripod ligands							
[26 · K]$^+$ SCN$^-$	10	2.77–3.00	2.94	—	—		322)
[27 · RbI]	8	3.03–3.06	3.00–3.20	3.76	—		323)

a does not apply; pic = picrate; The ligand numbering refers to the formulae depicted in Figs. 41 and 46

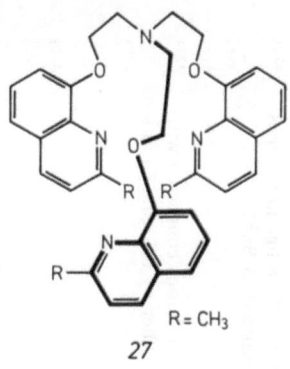

Fig. 53. Structural formulae of tripod ligands

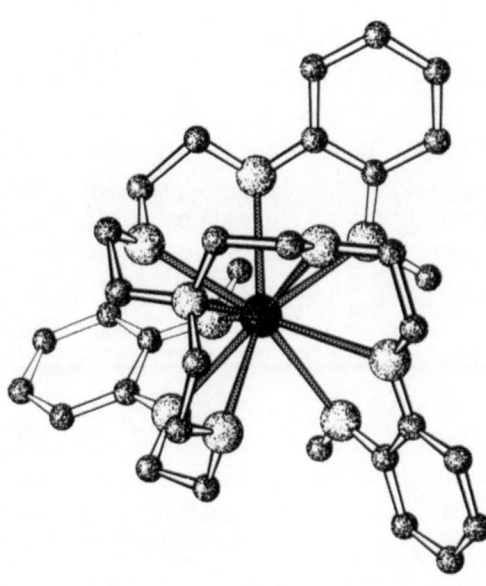

Fig. 54. Spatial structure of the 26-K$^+$ complex

cent of the "exclusive" cryptates, e.g. [2.2.1] · KNCS, which we have discussed in Chap. 6.

For the sake of completeness, mention should be made of a water complex of the same tripod ligand 27 studied by X-ray diffraction recently [339]. The water molecule is complexed by only one of the three ligand loops employing two N ... H—O hydrogen bonds.

8 Concluding Remarks

Looking backwards, one has to agree that it is indeed mainly the results obtained from X-ray crystallographic studies that the present knowledge on the structural foundations of selective complexation is based on. What is needed now is refinement of this knowledge toward the goal of being able to explain the mechanisms of host-guest complex formation with metal ions in greater detail than is possible today.

A promising approach to this is the application of X-ray crystallography in a highly systematic way. This is to say, emphasis should not be put on the investigation of as many new ligands as possible but rather on complexes of a given host molecule with various guest entities and, possibly, crystallized from solvents of different polarity. From a precise knowledge of the conformational changes of the ionophore molecule it is possible to propose detailed mechanisms for both ion capture and release, as we have seen in Chapter 4 for monensin and lasalocid. These studies should, of course, be complemented by spectroscopic, thermodynamic and kinetic investigations.

9 Acknowledgement

Most of the figures were drawn by the UNIVAC 1100/82 computer of the Gesellschaft für wissenschaftliche Datenverarbeitung (GWD), Göttingen, employing the plot program SCHAKAL which was written by Dr. Egbert Keller, Freiburg, and generously placed at our disposal. We are grateful to the GWD operators for their efforts in obtaining high-quality plots.

We would like to thank Drs. Galena Tishchenko, Moscow, and Matyas Czugler, Budapest, for communicating their results prior to publication. We are also indebted to Mr. Manfred Steifa for carrying out calculations using the Cambridge Crystallographic Data File [343], to Mrs. Anna-Liisa Peltola-Hilgenfeld for help in the literature work, to Mr. Ludwig Kolb for drawing several figures, and to Miss Petra Große for typing the manuscript.

10 References

1. Berger, J. et al.: J. Am. Chem. Soc. 73, 5295 (1951)
2. Mueller, P., Rudin, D. O.: Biochem. Biophys. Res. Commun. 26, 398 (1967)
3. Moore, C., Pressman, B. C.: Biochem. Biophys. Res. Commun. 15, 562 (1964)
4. Wipf, H. K., Simon, W.: Helv. Chim. Acta 53, 1732 (1970)
5. Wipf, H. K., Olivier, A., Simon, W.: Helv. Chim. Acta 53, 1605 (1970)
6. Ivanov, V. T.: Ann. N.Y. Acad. Sci. 264, 221 (1975)
7. Gramicidin A has been subject of an (uncompleted) X-ray study: Koeppe, R. E., Hodgson, K. O., Stryer, L.: J. Mol. Biol. 121, 41 (1978); Koeppe, R. E. et al.: Nature 279, 723 (1979) A crystallographic study of alamethicin is in progress: Fox, R., private communication
8. Lardy, H.: Fed. Proc. 27, 1278 (1968)
9. Tosteson, D. C.: Fed. Proc. 27, 1269 (1968)
10. Eisenman, G., Ciani, S. M., Szaba, G.: Fed. Proc. 27, 1289 (1968)
11. Harris, E. J. et al.: Arch. Biochem. Biophys. 182, 311 (1977)
12. Pedersen, C. J.: J. Am. Chem. Soc. 89, 7017 (1967)
13. Lehn, J. M.: Acc. Chem. Res. 11, 49 (1978)
14. Vögtle, F., Weber, E.: Angew. Chem. 91, 813 (1979); Angew. Chem. Int. Ed. Engl. 19, 753 (1979)
15. Ovchinnikov, Yu. A., Ivanov, V. T., Shkrob, A. M.: Membrane-active complexones. Amsterdam: Elsevier 1974
16. Burgermeister, W., Winkler-Oswatitsch, R.: Top. Curr. Chem. 69, 91 (1977)

17. Ovchinnikov, Yu. A.: FEBS Lett. *44*, 1 (1974)
18. Pinkerton, M., Steinrauf, L. K., Dawkins, P.: Biochem. Biophys. Res. Commun. *35*, 512 (1969)
19. Neupert-Laves, K., Dobler, M.: Helv. Chim. Acta *58*, 432 (1975)
20. Venkatachalam, C. M.: Biopolymers *6*, 1425 (1968)
21. Steinrauf, L. K., Sabesan, M. N.: Computer simulation studies on the ion-transporting antibiotics. In: Metal-ligand interactions in organic chemistry and biochemistry. Pullman, B., Goldblum, N. (eds.), part 2, pp. 43–57. Dordrecht: D. Reidel 1977
22. Ivanov, V. T. et al.: Biochem. Biophys. Res. Commun. *34*, 803 (1969)
23. Bystrov, V. F. et al.: Eur. J. Biochem. *78*, 63 (1977)
24. Grell, E., Funck, T., Sauter, H.: Eur. J. Biochem. *34*, 415 (1973)
25. Krishna, N. R. et al.: Biophys. J. *24*, 791 (1978)
26. Urry, D. W., Kumar, N. G.: Biochemistry *13*, 1829 (1974)
27. Ohnishi, M. et al.: Biochem. Biophys. Res. Commun. *46*, 312 (1972)
28. Fisher, I. M., Rothschild, K. J., Stanley, H. E.: J. Mol. Biol. *89*, 205 (1974)
29. Mayers, D. F., Urry, D. W.: J. Am. Chem. Soc. *94*, 77 (1972)
30. Duax, W. L. et al.: Science *176*, 911 (1972)
31. Duax, W. L., Hauptman, H.: Acta Cryst. *B28*, 2912 (1972)
32. Karle, I. L.: J. Am. Chem. Soc. *97*, 4379 (1975)
33. Smith, G. D. et al.: J. Am. Chem. Soc. *97*, 7242 (1975)
34. Smith, G. D., Duax, W. L.: Crystallographic studies of valinomycin and A23187. In: Metal-ligand interactions in organic chemistry and biochemistry. Pullmann, B., Goldblum, N. (eds.), part 1, pp. 291–315. Dordrecht: D. Reidel 1977
35. Davies, D. B., Abu Khaled, M.: J. Chem. Soc. Perkin Trans. II, *1976*, 1327
36. Ovchinnikov, Yu. A., Ivanov, V. T.: Tetrahedron *30*, 1871 (1974)
37. Urry, D. W., Kumar, N. G.: Biochemistry *13*, 1829 (1974)
38. Patel, D. J., Tonelli, A. E.: Biochemistry *12*, 486 (1973)
39. Patel, D. J.: Biochemistry *12*, 496 (1973)
40. Glickson, J. D. et al.: Biochemistry *15*, 5721 (1976)
41. Servis, K. L., Patel, D. J.: Tetrahedron *31*, 1359 (1975)
42. Rothschild, K. J. et al.: J. Am. Chem. Soc. *99*, 2032 (1977)
43. Feigenson, G. W., Meers, P. R.: Nature *283*, 313 (1980)
44. Grell, E., Funck, T.: J. Supramol. Struct. *1*, 307 (1973)
45. Ivanov, V. T. et al.: Biochem. Biophys. Res. Commun. *34*, 803 (1969)
46. Maigret, B., Pullman, B.: Theoret. Chim. Acta *37*, 17 (1975)
47. Pletnev, V. Z. et al.: Biopolymers *18*, 2145 (1979)
48. Pletnev, V. Z. et al.: Biopolymers *19*, 1517 (1980)
49. Rothschild, K. J. et al.: Science *182*, 384 (1973)
50. Asher, I. M. et al.: J. Am. Chem. Soc. *99*, 2024 (1977)
51. Shemyakin, M. M. et al.: Biochem. Biophys. Res. Commun. *29*, 834 (1967)
52. Hamill, R. L. et al.: Tetrahedron Lett. *1969*, 4255
53. Plattner, P., Nager, U.: Experientia *3*, 325 (1947)
54. Ovchinnikov, Yu. A. et al.: Int. J. Peptide Protein Res. *6*, 465 (1974)
55. Ovchinnikov, Yu. A et al.: Biochem. Biophys. Res. Commun. *37*, 668 (1969)
56. Dobler, M., Dunitz, J. D., Krajewski, J.: J. Mol. Biol. *42*, 603 (1969)
57. Vainshtein, B. K. et al.: 4th Eur. Cryst. Meet., Oxford 1977
58. Estrada-O., S., Gomez-Louero, C., Montal, M.: Bioenergetics *3*, 417 (1972)
59. Roeske, R. W. et al.: Biochem. Biophys. Res. Commun. *57*, 554 (1974)
60. Yafuso, M. et al.: Fed. Proc. Abstracts *33*, 1258 (1974)
61. Braden, B. et al.: J. Am. Chem. Soc. *102*, 2704 (1980)
62. Prince, R. C., Crofts, A. R., Steinrauf, L. K.: Biochem. Biophys. Res. Commun. *59*, 697 (1974)
63. Hamilton, J. A., Steinrauf, L. K., Braden, B.: Biochem. Biophys. Res. Commun. *64*, 151 (1975)
64. Geddes, A. J., Akriegg, D.: 4th Eur. Cryst. Meet., Oxford 1977
65. Geddes, A. J., Akrigg, D.: Acta Cryst. *B32*, 3164 (1976)
66. Popov, E. M. et al.: Khim. Priv. Soedin. *5*, 616 (1970)

67. Hamilton, J. A. et al.: Biochem. Biophys. Res. Commun. *80*, 949 (1978)
68. Zhukhlistova, N. E., Tishchenko, G. N.: Kristallografiya, submitted (1981)
69. Karle, I. L. et al.: Proc. Natl. Acad. Sci. USA *70*, 1836 (1973)
70. Karle, I. L.: J. Am. Chem. Soc. *96*, 4000 (1974)
71. Smirnova, V. I., Tishchenko, G. N., Vainshtein, B. K.: Dokl. Akad. Nauk. SSSR, submitted (1981)
72. Karaulov, A. I., Tishchenko, G. N.: Cryst. Struct. Commun., submitted (1981)
73. Tishchenko, G. N. et al.: 6th Eur. Cryst. Meet., Barcelona 1980
74. Tishchenko, G. N., Karaulov, A. I., Karimov, Z.: Cryst. Struct. Commun., submitted (1981)
75. Dunitz, J. D., Dobler, M.: Structural studies of ionophores and their ion-complexes. In: Biological aspects of inorganic chemistry. Addison, A. W. et al. (eds.), pp. 113–140. New York: J. Wiley 1977
76. Shishova, T. G., Simonov, V. I.: Kristallografiya *22*, 515 (1977)
77. Karle, I. L. et al.: Proc. Natl. Acad. Sci. USA *76*, 1532 (1979)
78. Karle, I. L. et al.: Proc. Natl. Acad. Sci. USA *73*, 1782 (1976)
79. Karle, I. L.: J. Am. Chem. Soc. *99*, 5152 (1977)
80. Karle, I. L., Duesler, E.: Proc. Natl. Acad. Sci. USA *74*, 2602 (1977)
81. Dobler, M.: Helv. Chim. Acta *55*, 1371 (1972)
82. Nawata, Y., Sakamaki, T., Iitaka, Y.: Acta Cryst. *B30*, 1047 (1974)
83. Nawata, Y. et al.: Chem. Lett. *1980*, 315
84. Kilbourn, B. T. et al.: J. Mol. Biol. *30*, 559 (1967)
85. Dobler, M., Dunitz, J. D., Kilbourn, B. T.: Helv. Chim. Acta *52*, 2573 (1969)
86. Dobler, M., Phizackerley, R. P.: Helv. Chim. Acta *57*, 664 (1974)
87. Sakamaki, T., Iitaka, Y., Nawata, Y.: Acta Cryst. *B32*, 768 (1976)
88. Iitaka, Y., Sakamaki, T., Nawata, Y.: Chem. Lett. *1972*, 1225
89. Sakamaki, T., Iitaka, Y., Nawata, Y.: Acta Cryst. *B33*, 52 (1977)
90. Neupert-Laves, K., Dobler, M.: Helv. Chim. Acta *59*, 614 (1976)
91. Nawata, Y., Sakamaki, T., Iitaka, Y.: Chem. Lett. *1975*, 151
92. Nawata, Y., Sakamaki, T., Iitaka, Y.: Acta Cryst. *B33*, 1201 (1977)
93. Pioda, L. A. R. et al.: Helv. Chim. Acta *50*, 1373 (1967)
94. Morf, W. E., Simon, W.: Helv. Chim. Acta *54*, 2683 (1971)
95. Szabo, G., Eisenman, G., Ciani, S.: J. Membr. Biol. *1*, 346 (1969)
96. Eisenman, G., Krasne, S., Ciani, S.: Ann. N.Y. Acad. Sci. *264*, 34 (1975)
97. Kyogoku, Y. et al.: Biopolymers *14*, 1049 (1975)
98. Anteunis, M. J. O., De Bruyn, A.: Bull. Soc. Chim. Belg. *86*, 445 (1977)
99. Phillies, G. D. J., Asher, I. M., Stanley, H. E.: Biopolymers *14*, 2311 (1975)
100. Prestegard, J. H., Chan, S. I.: Biochemistry *8*, 3921 (1969)
101. Prestegard, J. H., Chan, S. I.: J. Am. Chem. Soc. *92*, 4440 (1970)
102. Ivanov, V. T. et al.: FEBS Lett. *30*, 199 (1973)
103. Asher, I. M., Phillies, G. D. J., Stanley, H. E.: Biochem. Biophys. Res. Commun. *61*, 1356 (1974)
104. Asher, I. M. et al.: Biopolymers *16*, 157 (1977)
105. Dunitz, J. D. et al.: Helv. Chim. Acta *54*, 1709 (1971)
106. Marsh, W., Dunitz, J. D., White, D. N. J.: Helv. Chim. Acta *57*, 10 (1974)
107. Gertenbach, P. G., Popov, A. I.: J. Am. Chem. Soc. *97*, 4738 (1975)
108. Ward, D. L. et al.: Acta Cryst. *B34*, 110 (1978)
109. Westley, J. W.: Ann. Rep. Med. Chem. *10*, 246 (1975)
110. Westley, J. W.: Adv. Appl. Microbiol. *22*, 177 (1977)
111. Ebata, E. et al.: J. Antibiot. *28*, 118 (1975)
112. Liu, W.-C. et al.: J. Antibiot. *31*, 815 (1978)
113. Liu, C., Hermann, T. E.: J. Biol. Chem. *253*, 5892 (1978)
114. Liu, C. et al.: J. Antibiot. *32*, 95 (1979)
115. Lutz, W. K., Wipf, H.-K., Simon, W.: Helv. Chim. Acta *53*, 1741 (1970)
116. Pressman, B. C., Haynes, D. H.: Ionophore agents as mobile ion carriers. In: The molecular basis of membrane function. Tosteson, D. C. (ed.), pp. 221–246. Englewood Cliffs, N. Y.: Prentice Hall 1969
117. Pressman, B. C.: Ann. Rev. Biochem. *45*, 4925 (1976)

118. Agtarap, A. et al.: J. Am. Chem. Soc. *89*, 5737 (1967)
119. Pinkerton, M., Steinrauf, L. K.: J. Mol. Biol. *49*, 533 (1970)
120. Duax, W. L., Smith, G. D., Strong, P. D.: J. Am. Chem. Soc. *102*, 6725 (1980)
121. Alléaume, M., Hickel, D.: J.C.S. Chem. Commun. *1970*, 1422
122. Alléaume, M., Hickel, D.: J.C.S. Chem. Commun. *1972*, 175
123. Lutz, W. K., Winkler, F. K., Dunitz, J. D.: Helv. Chim. Acta *54*, 1103 (1971)
124. Haynes, D. H., Pressman, B. C., Kowalsky, A.: Biochemistry *10*, 852 (1971)
125. Anteunis, M. J. O.: Bull. Soc. Chim. Belg. *86*, 367 (1977)
126. Anteunis, M. J. O., Rodios, N. A.: Biorg. Chem. *7*, 47 (1978)
127. Briggs, R. W., Hinton, J. F.: Biochemistry *17*, 5576 (1978)
128. Painter, G., Pressman, B. C.: Biochem. Biophys. Res. Commun. *91*, 1117 (1979)
129. Young, S. P., Comperts, B. D.: Biochim. Biophys. Acta *469*, 281 (1977)
130. Westley, J. W., Evans, R. H., Blount, J. F.: J. Am. Chem. Soc. *99*, 6057 (1977)
131. Johnson, S. M. et al.: J.C.S. Chem. Commun. *1970*, 72
132. Johnson, S. M. et al.: J. Am. Chem. Soc. *92*, 4428 (1970)
133. Maier, C. A., Paul, I. C.: J.C.S. Chem. Commun. *1971*, 181
134. Schmidt, P. G., Wang, A. H. J., Paul, I. C.: J. Am. Chem. Soc. *96*, 6189 (1974)
135. Bissell, E. C., Paul, I. C.: J.C.S. Chem. Commun. *1972*, 967
136. Patel, D. J., Shen, C.: Proc. Natl. Acad. Sci. USA *73*, 1786 (1976)
137. Shen, C., Patel, D. J.: Proc. Natl. Acad. Sci. USA *73*, 4277 (1976)
138. Chiang, C. C., Paul, I. C.: Science *196*, 1441 (1977)
139. Friedman, J. M. et al.: J. Chem. Soc. Perkin Trans. II *1979*, 835
140. Smith, G. D., Duax, W. L., Fortier, S.: J. Am. Chem. Soc. *100*, 6725 (1978)
141. Chaney, M. O. et al.: J. Am. Chem. Soc. *96*, 1932 (1974)
142. Chaney, M. O., Jones, N. D., Debono, M.: J. Antibiot. *29*, 424 (1976)
143. Smith, G. D., Duax, W. L.: J. Am. Chem. Soc. *98*, 1578 (1976)
144. Perlman, R. L., Cossi, A. F., Role, L. W.: J. Pharmacol. Exp. Ther. *213*, 241 (1980)
145. Toeplitz, B. K. et al.: J. Am. Chem. Soc. *101*, 3344 (1979)
146. Barrans, Y., Alléaume, M.: Acta Cryst. *B36*, 936 (1980)
147. Steinrauf, L. K., Pinkerton, M., Chamberlin, J. W.: Biochem. Biophys. Res. Commun. *33*, 29 (1968)
148. Shiro, M., Koyama, H.: J. Chem. Soc. (B) *1970*, 243
149. Kubota, T. et al.: J.C.S. Chem. Commun. *1968*, 1541
150. Geddes, A. J.: Biochem. Biophys. Res. Commun. *60*, 1245 (1974)
151. Riche, C., Pascard-Billy, C.: J.C.S. Chem. Commun. *1975*, 951
152. Yamazaki, K., Abe, K., Sano, M.: J. Antibiot. *29*, 91 (1976)
153. Otake, N., Koenuma, M.: Tetrahedron. Lett. *1975*, 4147
154. Otake, N. et al.: J. Chem. Soc. Perkin Trans. II *1977*, 494
155. Blount, J. F., Westley, J. W.: J.C.S. Chem. Commun. *1971*, 927
156. Blount, J. F., Westley, J. W.: J.C.S. Chem. Commun. *1975*, 533
157. Alléaume, M. et al.: J.C.S. Chem. Commun. *1975*, 411
158. Czerwinski, E. W., Steinrauf, L. K.: Biochem. Biophys. Res. Commun. *45*, 1284 (1971)
159. Blount, J. F. et al.: J.C.S. Chem. Commun. *1975*, 853
160. Koyama, H., Utsumi-Oda, K.: J. Chem. Soc. Perkin Trans. II *1977*, 1531
161. Jones, N. D. et al.: J. Am. Chem. Soc. *95*, 3399 (1973)
162. Otake, N. et al.: J.C.S. Chem. Commun. *1977*, 590
163. Nakayama, H. et al.: J. Chem. Soc. Perkin Trans. II *1979*, 293
164. Shiro, M. et al.: J.C.S. Chem. Commun. *1978*, 682
165. Otake, N. et al.: J.C.S. Chem. Commun. *1978*, 875
166. Otake, N. et al.: J.C.S. Chem. Commun. *1975*, 92
167. Koenuma, M., Kinashi, H., Otake, N.: Acta Cryst. *B32*, 1267 (1976)
168. Alléaume, M.: 2nd. Eur. Cryst. Meet., Keszthely, Hungary, 1974
169. Kinashi, H. et al.: Tetrahedron. Lett. *1973*, 4955
170. Kinashi, H. et al.: Acta Cryst. *B31*, 2411 (1975)
171. Petcher, T. J., Weber, H. P.: J.C.S. Chem. Commun. *1974*, 697
172. Smith, G. D., Strong, P. D., Duax, W. L.: Acta Cryst. *B34*, 3436 (1978)
173. Westley, J. W. et al.: J. Antibiot. *27*, 597 (1974)

174. Westley, J. W. et al.: J. Am. Chem. Soc. *100*, 6784 (1978)
175. Westley, J. W. et al.: J. Antibiot. *32*, 100 (1979)
176. Van Remoortere, F. P., Boer, F. P.: Inorg. Chem. *13*, 2071 (1974)
177. Boer, F. P. et al.: Inorg. Chem. *13*, 2826 (1974)
178. Neuman, M. A. et al.: Inorg. Chem. *14*, 734 (1975)
179. North, P. P. et al.: Acta Cryst. *B32*, 370 (1976)
180. Van Remoortere, F. P., Boer, F. P., Steiner, E. C.: Acta Cryst. *B31*, 1420 (1975)
181. Armağan, N.: Acta Cryst. *B33*, 2281 (1977)
182. Feneau-Dupont, J. et al.: Acta Cryst. *B35*, 1217 (1979)
183. Arte, E. et al.: Acta Cryst. *B35*, 1215 (1979)
184. Bush, M. A., Truter, M. R.: J. Chem. Soc. Perkin Trans. II, *1972*, 341
185. Mallinson, P. R., Truter, M. R.: J. Chem. Soc. Perkin Trans. II, *1972*, 1818
186. Owen, J. D.: J. Chem. Soc. Dalton Trans. *1978*, 1418
187. Owen, J. D., Wingfield, J. N.: J.C.S. Chem. Commun. *1976*, 318
188. Cradwick, P. D., Poonia, N. S.: Acta Cryst. *B33*, 197 (1977)
189. Bhagwat, V. W., Manohar, H., Poonia, N. S.: 11th Internat. Congr. Cryst., Warsaw, 1978; J. Inorg. Nucl. Chem. Lett. *16*, 373 (1980)
190. Dobler, M., Dunitz, J. D., Seiler, P.: Acta Cryst. *B30*, 2741 (1974)
191. Sheldrick, W. S. et al.: Angew. Chem. *91*, 998 (1979); Angew. Chem. Int. Ed. Engl. *18*, 934 (1979)
192. Seiler, P., Dobler, M., Dunitz, J. D.: Acta Cryst. *B30*, 2744 (1974)
193. Riche, C. et al.: J.C.S. Chem. Commun. *1977*, 183
194. Groth, P.: Acta Chem. Scand. *25*, 3189 (1971)
195. Nagano, O.: Acta Cryst. *B35*, 465 (1979)
196. Nagano, O., Sasaki, Y.: Acta Cryst. *B35*, 2387 (1979)
197. Dobler, M., Phizackerley, R. P.: Acta Cryst. *B30*, 2746 (1974)
198. Dobler, M., Phizackerley, R. P.: Acta Cryst. *B30*, 2748 (1974)
199. Dunitz, J. D., Seiler, P.: Acta Cryst. *B30*, 2750 (1974)
200. Vance, T. B. et al.: Acta Cryst. *B36*, 153 (1980)
201. Vance, T. B. et al.: Acta Cryst. *B36*, 150 (1980)
202. Bombieri, G., De Paoli, G., Immirzi, A.: J. Inorg. Nucl. Chem. *40*, 1889 (1978)
203. Bombieri, G. et al.: Inorg. Chim. Acta *18*, L23 (1976)
204. Hašek, J., Huml, K.: Acta Cryst. *B34*, 1812 (1978)
205. Hughes, D. L., Mortimer, C. L., Truter, M. R.: Inorg. Chim. Acta *29*, 43 (1978)
206. Hlavatá, D., Hašek, J., Huml, K.: Acta Cryst. *B34*, 416 (1978)
207. Hašek, J., Hlavatá, D., Huml, K.: Acta Cryst. *B33*, 3372 (1977)
208. Bush, M. A., Truter, M. R.: J. Chem. Soc. (B) *1971*, 1440
209. Bright, D., Truter, M. R.: Nature *225*, 176 (1970)
210. Bright, D., Truter, M. R.: J. Chem. Soc. (B) *1970*, 1544
211. Myskiv, M. G. et al.: 11th Internat. Congr. Cryst., Warsaw 1978
212. Hilgenfeld, R., Saenger, W.: 6th Eur. Cryst. Meet., Barcelona 1980; Angew. Chem., in the press
213. Ciampolini, M. et al.: J. Chem. Soc. Dalton Trans. *1979*, 1983
214. Mallinson, P. R.: J. Chem. Soc. Perkin Trans. II, *1975*, 261
215. Layton, A. J. et al.: J.C.S. Chem. Commun. *1973*, 694
216. Fenton, D. E., Mercer, M., Truter, M. R.: Biochem. Biophys. Res. Commun. *48*, 10 (1972)
217. Mercer, M., Truter, M. R.: J. Chem. Soc. Dalton Trans. *1973*, 2215
218. Dalley, N. K. et al.: J.C.S. Chem. Commun. *1972*, 90
219. Harman, M. E. et al.: J.C.S. Chem. Commun. *1976*, 396
220. De Villardi, G. C. et al.: J.C.S. Chem. Commun. *1978*, 90
221. Ward, D. L., Brown, H. S., Sousa, L. R.: Acta Cryst. *B33*, 3537 (1977)
222. Czugler, M.: private communication
223. Hughes, D. L.: J. Chem. Soc. Dalton Trans. *1975*, 2374
224. Fenton, D. E. et al.: J.C.S. Chem. Commun. *1972*, 66
225. Mercer, M., Truter, M. R.: J. Chem. Soc. Dalton Trans. *1973*, 2469
226. Hughes, D. L., Mortimer, C. L., Truter, M. R.: Acta Cryst. *B34*, 800 (1978)
227. Hughes, D. L., Wingfield, J. N.: J.C.S. Chem. Commun. *1977*, 804
228. Owen, J. D., Truter, M. R.: J. Chem. Soc. Dalton Trans. *1979*, 1831

229. Bush, M. A., Truter, M. R.: J. Chem. Soc. Perkin Trans. II *1972*, 345
230. Hašek, J., Hlavatá, D., Huml, K.: Acta Cryst. *B36*, 1782 (1980)
231. Hašek, J., Huml, K., Hlavatá, D.: Acta Cryst. *B35*, 330 (1979)
232. M. Czugler and E. Weber: J.C.S. Chem. Commun. *1981*, 472
233. Moras, D. et al.: Bull. Soc. Chim. France *1972*, 551
234. Metz, B., Weiss, R.: Acta Cryst. *B29*, 1088 (1973)
235. Malmsten, L.-Å.: Acta Cryst. *B35*, 1702 (1979)
236. Herceg, M., Weiss, R.: Acta Cryst. *B29*, 542 (1973)
237. Herceg, M., Weiss, R.: Rev. Chim. Min. *10*, 509 (1973)
238. Groth, P.: Acta Chem. Scand. *A32*, 279 (1978)
239. Hanson, I. R.: Acta Cryst. *B34*, 1026 (1978)
240. Dunitz, J. D., Seiler, P.: Acta Cryst. *B30*, 2739 (1974)
241. Maverick, E. et al.: Acta Cryst. *B36*, 615 (1980)
242. Truter, M. R.: Effects of cations of groups IA and IIA on crown ethers. In: Metal-ligand interactions in organic chemistry and biochemistry. Pullman, B., Goldblum, N. (eds.), part. 1, pp. 317–335. Dordrecht: D. Reidel 1977
243. Dalley, N. K. et al.: J.C.S. Chem. Commun. *1975*, 43
244. Dalley, N. K.: Structural studies of synthetic macrocyclic molecules and their cation complexes. In: Synthetic multidentate macrocyclic compounds. Izatt, R. M., Christensen, J. J. (eds.), pp. 207–243. New York: Academic Press 1978
245. Mallinson, P. R.: J. Chem. Soc. Perkin Trans. II *1975*, 266
246. Dalley, N. K., Larson, S. B.: Acta Cryst. *B35*, 1901 (1979)
247. Czugler, M.: private communication
248. Goldberg, I.: Acta Cryst. *B34*, 2224 (1978)
249. Goldberg, I.: Acta Cryst. *B32*, 41 (1976)
250. Owen, J. D., Nowell, I. W.: Acta Cryst. *B34*, 2354 (1978)
251. Dalley, N. K., Larson, S. B.: Acta Cryst. *B35*, 2428 (1979)
252. Fronczek, F., Nayak, A., Newkome, G. R.: Acta Cryst. *B35*, 775 (1979)
253. Hanson, I. R., Hughes, D. L., Truter, M. R.: J. Chem. Soc. Perkin Trans. II *1976*, 972
254. Herceg, M., Weiss, R.: Bull. Soc. Chim. Fr. *1972*, 549
255. Dalley, N. K. et al.: J.C.S. Chem. Commun. *1975*, 84
256. Tables of interatomic distances and configurations in molecules and ions, London: The Chemical Society 1960
257. Dale, J.: Tetrahedron Lett. *30*, 1683 (1974)
258. Dale, J., Kristiansen, P. O.: Acta Chem. Scand. *26*, 1471 (1972)
259. Bovill, M. J. et al.: J. Chem. Soc. Perkin Trans. II *1980*, 1529
260. Frensdorff, H. K.: J. Am. Chem. Soc. *93*, 600 (1971)
261. Poonia, N. S., Bajaj, A. V.: Chem. Rev. *79*, 389 (1979)
262. Poonia, N. S. et al.: Ind. J. Chem. *19A*, 37 (1980)
263. Live, D., Chan, S. I.: J. Am. Chem. Soc. *98*, 3769 (1976)
264. Mei, E., Dye, J. L., Popov, A. I.: J. Am. Chem. Soc. *98*, 1619 (1976)
265. Popov, A. I.: Multinuclear NMR studies of crown and cryptand complexes. In: Stereodynamics of molecular systems. Sarma, R. H. (ed.), pp. 197–207. New York: Pergamon Press 1979
266. Krane, J., Dale, J., Daasvatn, K.: Acta Chem. Scand. *B34*, 59 (1980)
267. Parsons, D. G., Truter, M. R., Wingfield, J. N.: Inorg. Chim. Acta *14*, 45 (1975)
268. Pedersen, C. J.: J. Org. Chem. *36*, 1690 (1971)
269. Moras, D., Weiss, R.: Acta Cryst. *B29*, 400 (1973)
270. Mathieu, F. et al.: J. Am. Chem. Soc. *100*, 4412 (1978)
271. Mathieu, F., Weiss, R.: J.C.S. Chem. Commun. *1973*, 816
272. Moras, D., Weiss, R.: Acta Cryst. *B29*, 396 (1973)
273. Teller, R. G. et al.: J. Am. Chem. Soc. *99*, 1104 (1977)
274. Ginsburg, R. E. et al.: J. Am. Chem. Soc. *101*, 6550 (1979)
275. Tehan, F. J., Barnett, B. L., Dye, J. L.: J. Am. Chem. Soc. *96*, 7203 (1974)
276. Adolphson, D. G., Corbett, J. D., Merryman, D. J.: J. Am. Chem. Soc. *98*, 7234 (1976)
277. Corbett, J. D., Edwards, P. A.: J. Am. Chem. Soc. *99*, 3313 (1977)
278. Moras, D., Metz, B., Weiss, R.: Acta Cryst. *B29*, 383 (1973)
279. Moras, D., Metz, B., Weiss, R.: Acta Cryst. *B29*, 388 (1973)

280. Metz, B., Moras, D., Weiss, R.: Acta Cryst. *B29*, 1377 (1973)
281. Metz, B., Moras, D., Weiss, R.: Acta Cryst. *B29*, 1382 (1973)
282. Moras, D., Weiss, R.: Acta Cryst. *B29*, 1059 (1973)
283. Metz, B., Weiss, R.: Inorg. Chem. *13*, 2094 (1974)
284. Hart, F. A. et al.: J.C.S. Chem. Commun. *1978*, 549
285. Ciampolini, M., Dapporto, P., Nardi, N.: J.C.S. Chem. Commun. *1978*, 788
286. Ciampolini, M., Dapporto, P., Nardi, N.: J. Chem. Soc. Dalton Trans. *1979*, 974
287. Burns, J. H.: Inorg. Chem. *18*, 3044 (1979)
288. Metz, B., Moras, D., Weiss, R.: Acta Cryst. *B29*, 1388 (1973)
289. Lehn, J. M.: Molecular receptors, carriers, and catalysts: Design, scope, and prosepcts. In: Bioenergetics and thermodynamics: Model Systems. Braibanti, A. (ed.), pp. 455–461. Dordrecht: D. Reidel 1980
290. Lehn, J. M., Sauvage, J. P.: J. Am. Chem. Soc. *97*, 6700 (1975)
291. Metz, B., Moras, D., Weiss, R.: J. Chem. Soc. Perkin Trans. II *1976*, 423
292. Metz, B., Weiss, R.: Nouv. J. Chim. *2*, 615 (1978)
293. Metz, B., Moras, D., Weiss, R.: 2nd Eur. Cryst. Meet., Keszthely, Hungary, 1974
294. Mei, E., Popov, A. I., Dye, J. L.: J. Am. Chem. Soc. *99*, 6532 (1977)
295. Mei, E. et al.: J. Solution Chem. *6*, 771 (1977)
296. Kauffmann, E., Lehn, J. M., Sauvage, J. P.: Helv. Chim. Acta *59*, 1099 (1976)
297. Saenger, W., Reddy, B. S.: Acta Cryst. *B35*, 56 (1979)
298. Saenger, W. et al.: X-ray structure of a synthetic, non-cyclic, chiral polyether complex as analog of nigericin antibiotics. In: Metal-ligand interactions in organic chemistry and biochemistry. Pullman, B., Goldblum, N. (eds.), part 1, pp. 363–374. Dordrecht: D. Reidel 1977
299. Saenger, W., Brand, H.: Acta Cryst. *B35*, 838 (1979)
300. Hilgenfeld, R., Saenger, W.: unpublished results
301. Weber, G., Saenger, W.: Acta Cryst. *B35*, 1346 (1979)
302. Weber, G. et al.: Angew. Chem. *91*, 234 (1979); Angew. Chem. Int. Ed. Engl. *18*, 226 (1979)
303. Weber, G., Saenger, W.: Acta Cryst. *B35*, 3093 (1979)
304. Weber, G., Saenger, W.: Angew. Chem. *91*, 237 (1979); Angew. Chem. Int. Ed. Engl. *18*, 227 (1979)
305. Weber, G., Saenger, W.: Acta Cryst. *B36*, 61 (1980)
306. Suh, I.-H., Weber, G., Saenger, W.: Acta Cryst. *B34*, 2752 (1978)
307. Suh, I.-H., Weber, G., Saenger, W.: Acta Cryst. *B36*, 946 (1980)
308. Chacko, K. K., Saenger, W.: Z. Naturforsch. *35b*, 1533 (1980)
309. Czugler, M.: private communication
310. Suh, I.-H. et al.: Z. Naturforsch. *35b*, 352 (1980)
311. Chacko, K. K., Saenger, W.: Z. Naturforsch. *36b*, 102 (1981)
312. Hughes, D. L. et al.: Inorg. Chim. Acta *21*, L23 (1977)
313. Hughes, D. L., Mortimer, C. L., Truter, M. R.: Inorg. Chim. Acta *28*, 83 (1978)
314. Hughes, D. L., Wingfield, J. N.: J.C.S. Chem. Commun. *1978*, 1001
315. Singh, T. P., Reinhardt, R., Poonia, N. S.: Inorg. Nucl. Chem. Lett. *16*, 293 (1980)
316. Ohmoto, H. et al.: Bull. Chem. Soc. Jpn. *52*, 1209 (1979)
317. Hilgenfeld, R., Saenger, W.: Z. Anal. Chem. *304*, 277 (1980)
318. Weber, G.: Acta Cryst. *B36*, 2779 (1980)
319. Iwamoto, R.: Bull. Chem. Soc. Jpn. *46*, 1114 (1973)
320. Iwamoto, R.: Bull. Chem. Soc. Jpn. *46*, 1123 (1973)
321. Iwamoto, R., Wakano, H.: J. Am. Chem. Soc. *98*, 3764 (1976)
322. Saenger, W., Suh, I.-H.: unpublished results
323. Weber, G., Sheldrick, G. M.: Inorg. Chim. Acta *45*, L35 (1980)
324. Izatt, R. M., Eatough, D. J., Christensen, J. J.: Struc. Bonding *16*, 161 (1973)
325. Cabbiness, D. K., Margerum, D. W.: J. Am. Chem. Soc. *91*, 6540 (1969)
326. Kodama, M., Kimura, E.: Bull. Chem. Soc. Jpn. *49*, 2465 (1976)
327. Fabbrizzi, L., Paoletti, P., Lever, A. B. P.: Inorg. Chem. *15*, 1502 (1976)
328. Hancock, R. D., McDougall, G. J.: J. Am. Chem. Soc. *102*, 6551 (1980)
329. Vögtle, F., Sieger, H.: Angew. Chem. *89*, 410 (1977); Angew. Chem. Int. Ed. Engl. *16*, 396 (1977)
330. Saenger, W., Suh, I.-H., Weber, G.: Isr. J. Chem. *18*, 253 (1979)
331. Tümmler, B. et al.: J. Am. Chem. Soc. *101*, 2588 (1979)

332. Hughes, D. L., Truter, M. R.: J. Chem. Soc. Dalton Trans. *1979*, 520
333. Yamazaki, N. et al.: Tetrahedron Lett. *1978*, 2429
334. Sieger, H., Vögtle, F.: Liebigs Ann. Chem. *1980*, 425
335. Grandjean, J. et al.: Angew. Chem. *90*, 902 (1978); Angew. Chem. Int. Ed. Engl. *17*, 856 (1978)
336. Sieger, H., Vögtle, F.: Angew. Chem. *90*, 212 (1978); Angew. Chem. Int. Ed. Engl. *17*, 198 (1978)
337. Hilgenfeld, R., Saenger, W.: unpublished results
338. Vögtle, F. et al.: Angew. Chem. *89*, 564 (1977); Angew. Chem. Int. Ed. Engl. *16*, 548 (1977)
339. Weber, G., Sheldrick, G. M.: Acta Cryst. *B36*, 1978 (1980)
340. Goldberg, I.: Acta Cryst. *B31*, 754 (1975)
341. Hilgenfeld, R., Saenger, W.: Z. Naturforsch. *36 b*, 242 (1981)
342. Dunitz, J. D., Ha, T. K.: J.C.S. Chem. Commun. *1972*, 568
343. Allen, F. H. et al.: Acta Cryst. *B35*, 2331 (1979)
344. Christensen, J. J.: Transport of metal ions by liquid membranes containing macrocyclic carriers. In: Bioenergetics and thermodynamics: Model systems. Braibanti, A. (ed.), pp. 111 to 126. Dordrecht: D. Reidel 1980
345. Vögtle, F., Sieger, H., Müller, W. M.: Top. Curr. Chem. *98*, 107 (1981)
346. Cram, D. J., Trueblood, K. N.: Top. Curr. Chem. *98*, in the press

Concept, Structure, and Binding in Complexation

Donald J. Cram, Kenneth N. Trueblood

Department of Chemistry, University of California at Los Angeles, Los Angeles, California 90024, USA

Table of Contents

1 Introduction

Structural molecular complexation is central to biological phenomena. Enzymic catalysis and inhibition, immunological response, storage and retrieval of genetic information, replication, biological regulatory function, drug action, and ion transfer all involve structural recognition in complexation. At least one of the partners in most of the complexes of evolutionary chemistry is large and complicated enough to inhibit studies of its detailed structure and an analysis of the forces that control its shape.

The structures of the simple compounds supplied by nature have long inspired and challenged organic chemists to develop laboratory syntheses of the same compounds. A newer challenge provided by the biotic world is the design and synthesis of abiotic systems that mimic some of the properties of biotic systems. Any response to this challenge depends directly on an understanding of structural recognition in complexation. The potential fruits of the development of a field of synthetic organic complexation are obvious. Analytical and separation science have already benefited. The design and synthesis of nonpeptide organic catalysts with the specificity and turnover of enzyme systems are particularly exciting possibilities.

This paper is concerned with applications of the techniques of organic chemistry and crystallography to the study of structural recognition in complexation between partners, at least one of which is a designed synthetic-organic compound.

1.1 Definitions

A *complex* is composed of two or more distinct molecules held together in a definable structural relationship. The binding forces are of a pole-pole, pole-dipole, or dipole-dipole nature. More specifically, complexes are held together by hydrogen bonding, ion pairing, metal ion to ligand attractions, π-acid to π-base attractions, van der Waals attractive forces, solvent-liberation driving forces, or partially made and broken covalent bonds (transition states).

Multiple binding sites usually are required to provide high structural organization to ground state complexes involving organic compounds since binding energies at any one site are small compared to those provided by covalent bonds. For complexation to occur, the binding sites and steric barriers in potential partners must be complementary to one another in electronic character and geometric arrangement. Thus *structural recognition* in complexation depends on the degree of unique complementary relationships between potential complexing partners.

In biotic systems. complexing partners are named according to their functions. Example are: *enzymes*, *substrates* or *inhibitors*; *antibodies*, *antigens*; *ionophores*, *metal ions*. In medicinal chemistry, the terms *receptor* and *drug* have evolved. In inorganic chemistry, the terms *ligand* and *metal ion* differentiate the parts of complexes. We have adopted the terms *host* and *guest* to name the complexing partners in synthetic organic chemical studies.

A highly structured molecular complex is composed of at least one host and one guest that possess complementary stereoelectronic binding sites. The host is defined as an organic molecule or ion whose *binding sites converge in the complex*. The guest is defined as a molecule or ion whose *binding sites diverge in the complex*. Guests may be organic compounds or ions, metals or metal ions, or metal-ligand assemblies. Bonds or binding sites of single nuclei naturally diverge. Hosts are usually larger than guests since positioning of convergent binding sites involves support structures not required for guests. However, large nonbinding parts can be attached to either the hosts or the guests to manipulate their size and, thus, their properties. Simple guests are abundant, but hosts usually must be designed and synthesized [1].

Hosts are open-chain, cyclic, bicyclic, or polycyclic compounds that frequently contain repeating units. The classic examples are polyethylene glycols [H(OCH$_2$CHa)$_n$OH] and polyethylenediamines [H(NHCH$_2$CH$_2$)$_n$NH$_2$], which when cyclized provide crowns [(CH$_2$CH$_2$O)$_n$] and azacrowns [(CH$_2$CH$_2$NH)$_n$], respectively [2]. Cryptands were introduced when polyethyleneoxy units were used to make bridges between nitrogen bridgeheads in bicyclic and polycyclic hosts (e.g., N[CH$_2$(CH$_2$OCH$_2$)$_n$-CH$_2$]$_3$N) [3]. Many hosts have been designed and studied in which a variety of other units have been substituted for the CH$_2$CH$_2$, CH$_2$OCH$_2$, N(CH$_2$)$_3$, or CH$_2$OCH$_2$CH$_2$OCH$_2$ units of crowns or cryptands. When the crown or cryptand basic structures dominate, it is convenient to refer to such hosts as modified crowns or cryptands. When other units dominate structures and dictate the properties of hosts, other names are appropriate. Examples are the spherands [4] and the hemispherands [5]. Spherands are defined as hosts conformationally organized prior to complexation. Thus, the spherands and the host parts of their complexes are conformationally the same. Spherands must be composed of units much more rigid than those of crowns or cryptands. Hemispherands contain contiguous rigid parts sufficient in number to dominate the general shapes of hosts. Other parts are conformationally mobile.

We have found it convenient to classify complexes according to the degree to which a host envelops a guest in a complex. Terms borrowed from crystallography are inappropriate since many crystalline complexes depend for their existence on the mutual packing interactions of many structural units. In crystals, both host and guest compounds are completely enveloped. We suggest the term *perching complex* to characterize structures in which well under half of the guest surface contacts the host. The term *nesting complex* is applied to structures in which over half of the guest surface contacts the host. The term *capsular complex* is reserved for those structures in which the guest surface is covered well enough by the host to prevent external ligand or solvent from contacting the guest.

1.2 Scope, Themes, and Organization

This paper is not a general review. General [6-9] and specialized reviews [10,11] already exist. Rather, we develop themes that have been particularly useful to us in the development of host-guest complexation chemistry. The most important of these illustrates how the techniques of X-ray crystallography and space-filling molecular model building have been combined with those of organic chemistry to provide guidance in the design of complementary host-guest relationships. A second theme deals with the question, "To what degree does a guest reorganize its host upon complexation?" A third topic involves the relationships between structure and free energy of binding of complexing partners.

2 Types of Host-Guest Complexes

2.1 Complexes with Crown Hosts

Pedersen's macrocyclic polyether "crown" compounds have provided hosts for the most studied complexes [2]. Formula *1* points to a prototype complex in which a potassium ion as guest is complexed in a nesting arrangement by the six oxygens of the host, [18]crown-6. The drawing of the Corey-Pauling-Koltun (CPK) molecular model of the complex (*2*) illustrates the matching diameters of the potassium ion and the cavity it neatly occupies. The complex resembles a disk with two equivalent faces. The drawing disregards the more subtle facts that the ethylene glycol units are gauche and two sets of alternate oxygens occupy two close parallel planes. A second type of complex is illustrated with *3*, in which three acidic hydrogens of methylammonium ion guest are hydrogen bonded to three alternate oxygens of [18]crown-6 in a tripod arrangement. The nitrogen and its attached methyl group protrude above one face of the host to provide a perching type of host-guest relationship. The Newman projection formula (*4*) of the complex suggests that the six hydrogens possess a staggered arrangement and that the C—N bond is normal to the best plane of the crown. One-to-one perching complexes lack equivalent faces. Drawing *5* of a CPK model of *4* is a view of the complex from the methyl side, and *6* is a view from the opposite side. In *6* the three nonhydrogen bonded oxygens contact the N^+ and probably provide approximately a third of the total binding free energy [12].

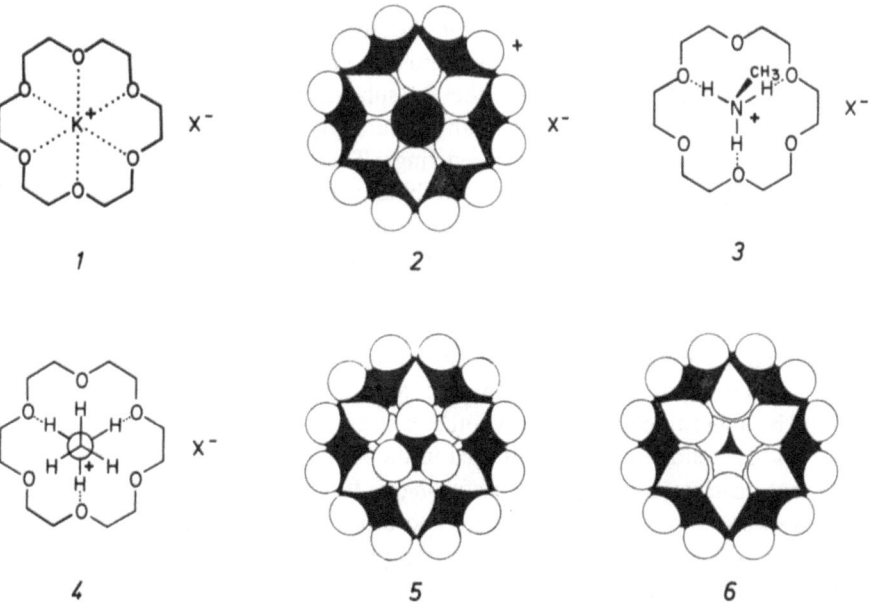

<table>
<tr><td align="center">*1*</td><td align="center">*2*</td><td align="center">*3*</td></tr>
<tr><td align="center">*4*</td><td align="center">*5*</td><td align="center">*6*</td></tr>
</table>

Four other types of guests complex [18]crown-6 and its analogues. Hydronium salt complexes of dicyclohexano[18]crown-6 have been characterized and appear to resemble the ammonium salt complexes as envisioned in structure 7 [13]. Aryldiazonium salts complex[18]crown-6 and other macrocyclic polyethers with large enough potential cavities [1,14]. In these complexes the N_2^+ group projects into the hole and the complex is considerably more stable and soluble in organic media than is the uncomplexed salt in water. Substituents in the *ortho* position of the aryldiazonium salts complex [18]crown-6 and other macrocyclic polyethers with large

of the $\overset{+}{N} \equiv N$ group into the potential cavity of hosts [1,14]. Although crystal structures of 8 and the dicyclohexano derivative of 7 have been cited [15], no structures had been published at the time of this writing. Hydrazinium salts complex [18]crown-6 (see sect. 3.3.) by full insertion into the hole of the host. Guanidinium salts are complexed and rendered soluble in organic media by benzo[27]crown-9 and other hosts [1,16]. The wreath-shaped structure of 9 is envisioned for the crystalline complex isolated [1].

7   8   9

2.2 Units in Hosts that Structure Complexes

Various units have been substituted for the ethylene glycol, 1,2-cyclohexanediol, or catechol units of the simpler crown hosts [2]. For example, the hosts of complexes *10–13* contain, respectively, the 2,6-substituted pyridyl [17], 2,5-substituted furanyl [12,18], the 2,5-substituted tetrahydrofuranyl [18], and 1,3-substituted benzo units [19] incorporated into 18-membered rings. These units place heteroatoms with unshared electron pairs in positions to bind metal or alkylammonium anions. Except for those containing the furanyl unit, hosts containing these units arraned in suitable fashion are good complexing agents. The unshared electron pairs on the furanyl oxygen are too delocalized to provide good binding [12]. In molecular models of complex *13* with X = CO_2CH_3, the carbonyl oxygen of the ester group possesses an enforced conformation and contacts the N^+ through the hole. The RNH_3^+ group binds the host from the side opposite the CO_2CH_3 group [19]. It would be interesting to test this hypothesis by crystal structure determination.

The 1,1'-dinaphthyl group substituted in the 2,2'-positions with oxygens provides a unit that possesses a particularly useful shape. It is chiral, configurationally stable, and possesses a C_2 axis that can render two faces of an attached macroring stereochemically equivalent to each other. Appropriate substituents in the 3- and 3'-positions of the unit provide "arms", one that extends across one face and one that

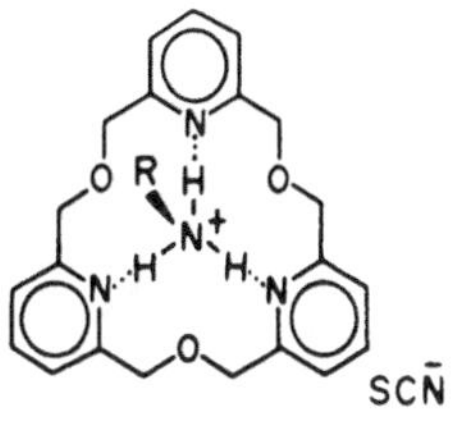

10 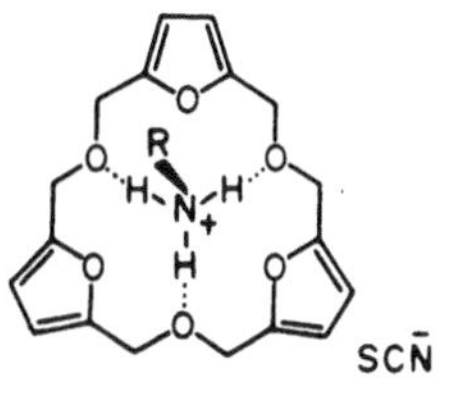   11 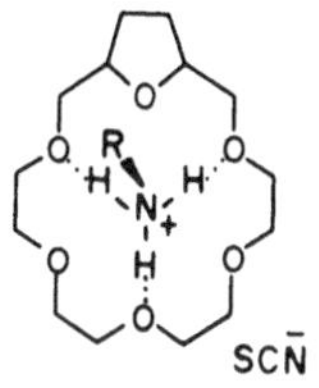   12 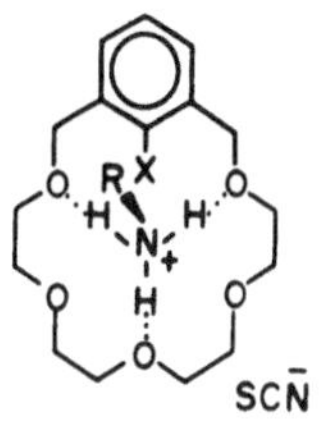   13

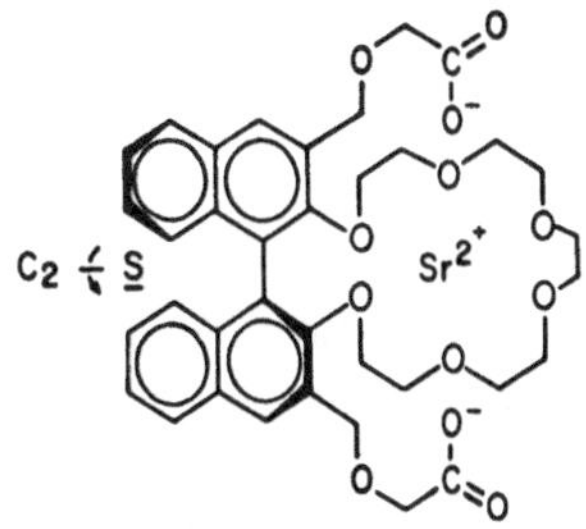

14 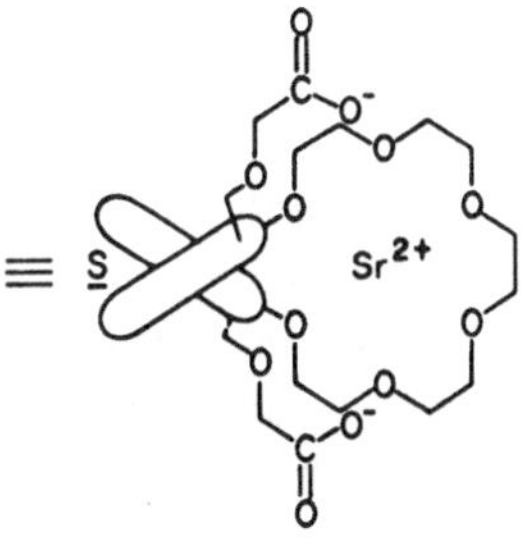   15 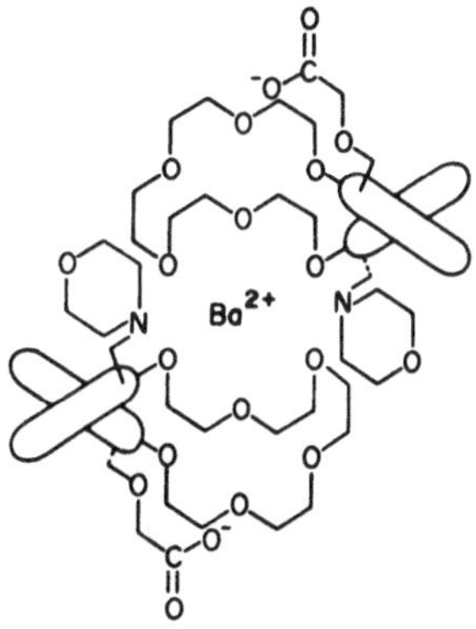   16

extends across the other face of the macroring. In structures such as *15*, these naphthyl units are tangent to the macroring. Formula *15* provides a cross-sectional view of *14* in which the dinaphthyl group, observed from the aryl-aryl bond axis, resembles an open pair of scissors. The 3,3′-substituents extend the chiral barrier of the dinaphthyl group and can contain additional binding sites. In molecular models of *14*, the anions that terminate these side chains can center on two faces of the macroring and contact the bound Sr^{2+}. These anions counter the two positive charges of the guest [20]. In molecular models of *16*, two ligand systems, each carrying one counterion, cluster around Ba^{2+} which is too large to nest in either macroring but can perch on each at the same time to give a capsular complex. In molecular models, the carboxylate oxygens appear best able to contact the Ba^{2+} through the holes of the macrorings [20]. Hosts containing the similarly shaped 1,1′-ditetralyl unit behave like those containing the 1,1′-dinaphthyl unit [21]. The chiral 9,9′-bifluorene unit has been used by Prelog for many of the same purposes for which we have employed the 1,1′-dinaphthyl unit [22].

2.3 Complexes of Cryptands, Hemispherands, and Spherands

The chemistry of Lehn's cryptands and their complexes has been reviewed [6]. They are exemplified by the potassium ion complex of [2.2.2]cryptand (*17*) in which the electron pairs of eight heteroatoms contact the potassium ion. The third polyethyleneoxy bridge attached to the two nitrogens of *17* turns what would otherwise be a perching or nesting complex into a capsular complex.

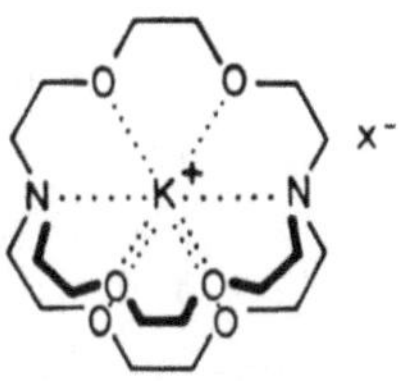

17 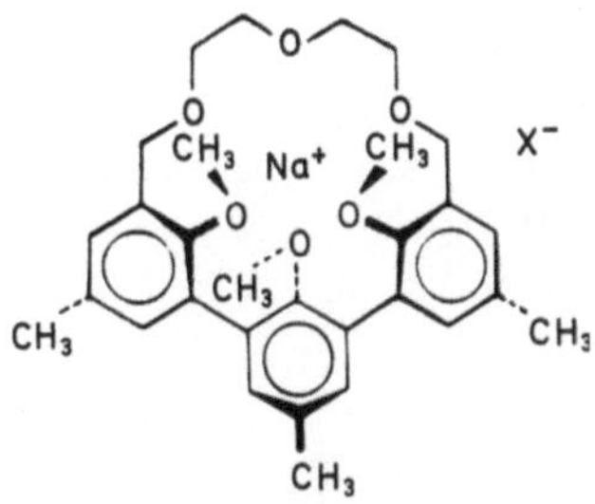   18 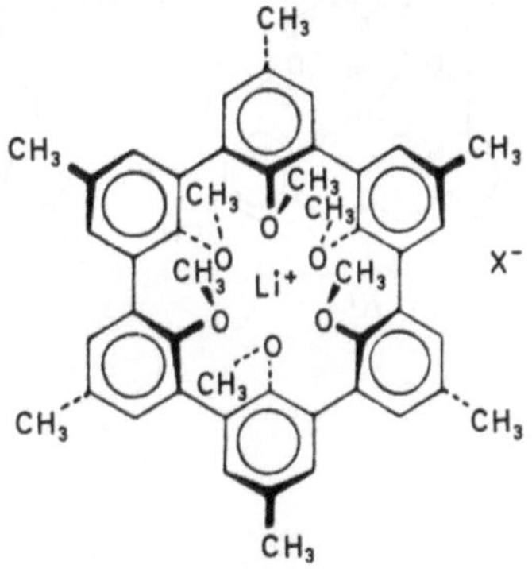   19

Complexes of the hemispherands [5] and the spherands [4] are illustrated with structures *18* and *19*, respectively. In CPK molecular models of *18*, the *o,o'*-terphenyl unit possesses a conformation enforced by its incorporation into a ring system. The two peripheral methoxyl groups protrude from the near face, and the central methoxyl group protrudes from the far face of the macroring in structure *18*. The three aryl oxygens form a partial nest for the Na$^+$, and the diethyleneoxy bridge completes it with three additional binding sites. The bridge also draws the twelve unshared electrons of the three aryl oxygens into proximity with one another [5]. In the lithiospherium complex *19*, the six oxygens possess an approximately octahedral arrangement with their unshared electron pairs focused on the guest [4]. Structure *19* is a good example of a capsular complex.

3 Structures of Complexes

3.1 The Molecular-Crystal Structure Connection

The general structures of the first host-guest complexes were inferred from their elemental compositions, the structures of their components, their solubilities, and by examination of CPK molecular models of possible structures [6-9]. The X-ray studies of the crown-alkali metal salt complexes by Truter [7] and Dunitz [23] and of the cryptand-alkali metal complexes by Weiss [24] provided a firm basis for the structures of these complexes in the crystalline state. The spectral results of Dale [25] and of Chan [26] on both the hosts and their metal salt complexes allowed structures in solution to be compared with those in the solid state.

In this section we compare the structures of complexes determined by X-ray diffraction with those predicted by CPK molecular model examination which led to the design and synthesis of the complexes. Figure 1 outlines the steps in the evolution of the role played by space-filling models in the derivation and prediction of chemical structures of complexing partners, first in biotic and then in abiotic chemistry.

| Crystal structures of simple biotic compounds (Corey, Pauling, etc.) | → | Corey-Pauling-Koltun (CPK) molecular models | → | Molecular models of biotic systems such as proteins, DNA, and their complexes |

↓ ↓

| Crystal structures of abiotic systems and their complexes | ← | Molecular models of abiotic systems and their complexes | | Crystal structures of biotic systems such as proteins, DNA, and their complexes |

Fig. 1. The molecular model-crystal structure connection

3.2 Crown-Alkylammonium Complexes

In correlations between structures and complexing affinities of hosts, 2,3-naphtho-[18]crown-6 has been used as a standard host and $(CH_3)_3CNH_3X$ has been used as a standard guest. Formula *20* indicates the perching structure of the complex expected on the basis of CPK molecular model examination, and *21* is that observed by crystal structure determination [27]. The gross features that provide structure to this complex are the tripod arrangement of the three $^+NH \cdots O$ hydrogen bonds, the all-gauche arrangements of the OCH_2CH_2O units, the staggered conformations for the $H—N—C—CH_3$ groups, and the near 0° angle (actually 1°) between the normal to the best plane of the six oxygens and the axis of the C—N bond. Interestingly, the best plane of the six oxygens and that of the naphthalene ring are nearly parallel (dihedral angle of 6°), which indicates little repulsion between the aryl group and the methyl that projects toward that group. The ^1H NMR spectrum of the complex in $CDCl_3$ indicates that some of the CH_3 hydrogens are in the shielding region of the aryl group and, accordingly, are moved upfield. This property of the complex in solution is compatible with the structure in the crystal.

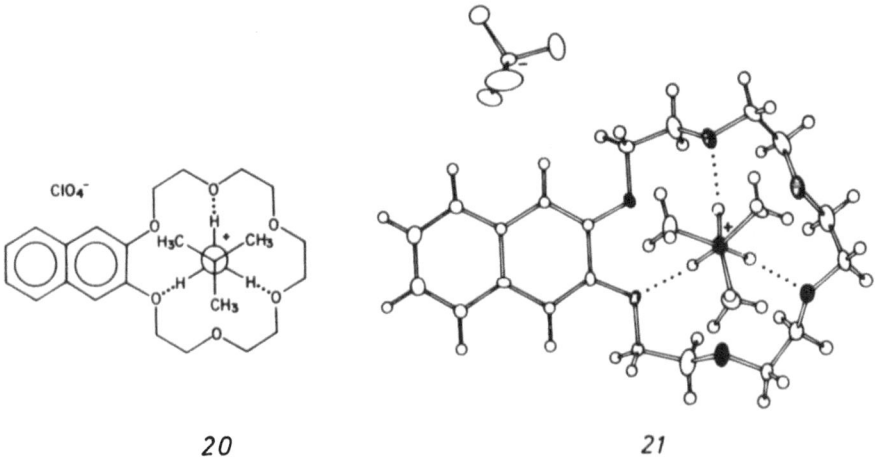

20 *21*

Host *22* was designed to complex guests containing two or more functional groups attached to one another by $(CH_2)_4$, $(CH_2)_5$, m-C_6H_4, or p-C_6H_4 units [28]. For example, models indicated that a complex between *22* and tetramethylene diammonium ion should have structure *23*. Accordingly, the bis-hexafluorophosphate salt of the complex was prepared and found by X-ray crystal analysis [29] to have a perching structure of which *24* and *25* are two different views.

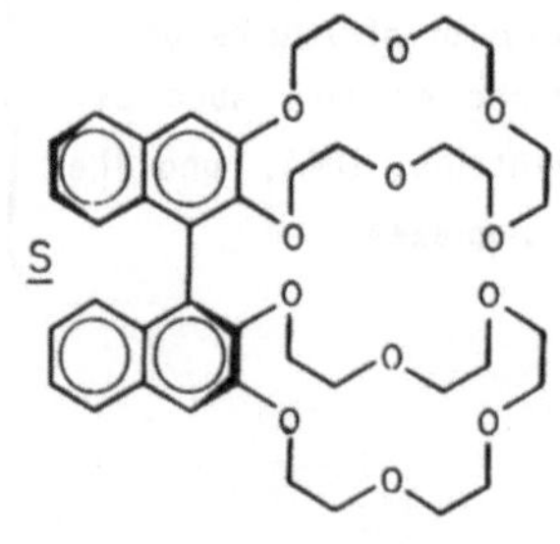

22

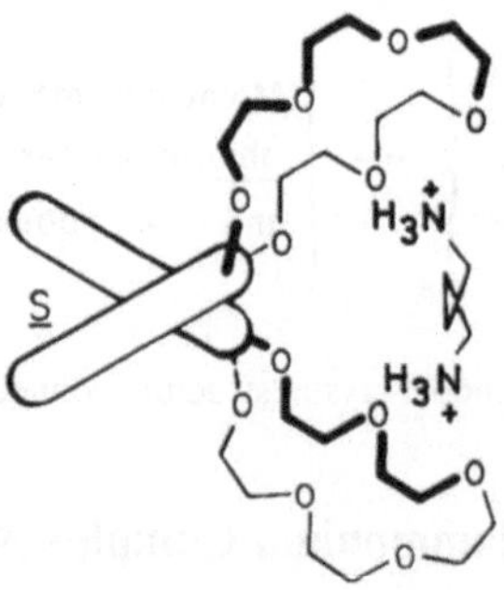

23

In *24*, the upper "jaw" of the host complexes one NH_3^+ group and the lower "jaw" complexes the other to give a complex with a C_2 axis. One of the NH^+ hydrogen bonds in each jaw appears bifurcated, being directed about midway between two ArO oxygens, as indicated in *25* [29]. As expected from molecular model examination, *26* in $CDCl_3$ allowed each of the four salt guests (formulated collectively as *27*) to be extracted on a roughly equal basis from D_2O into the organic phase, presumably by engagement of all of the binding sites in both host and guest [28].

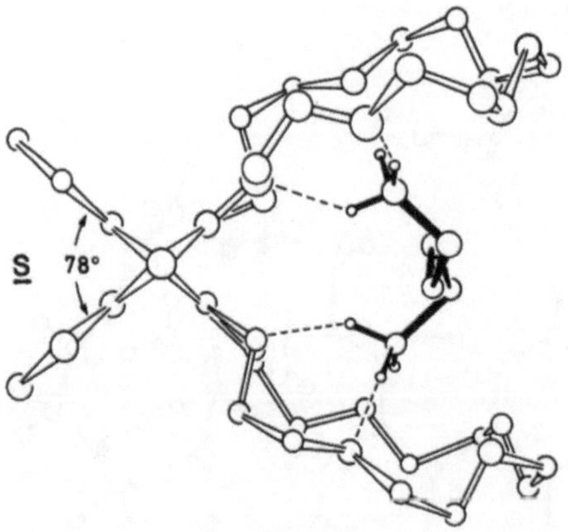

24

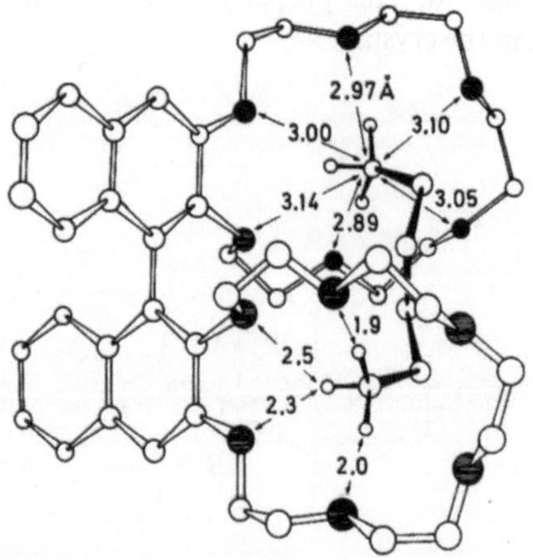

25

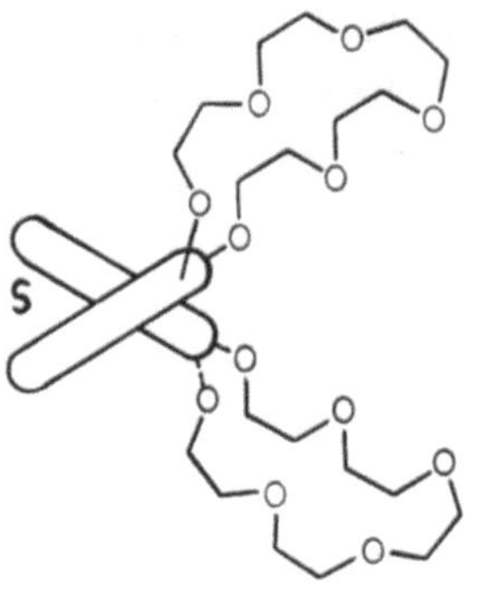

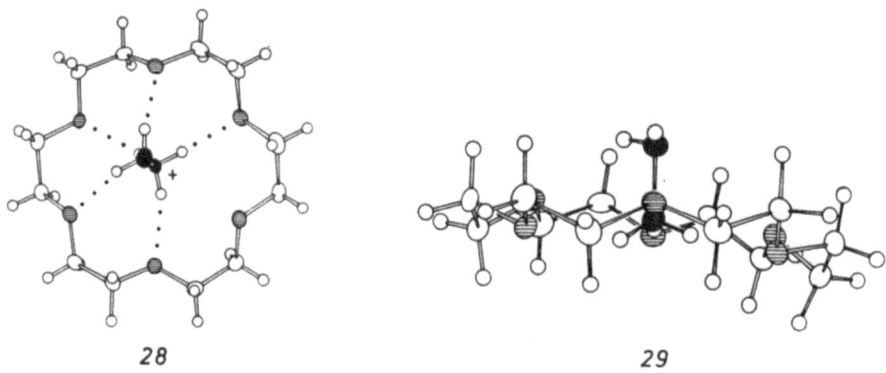

26 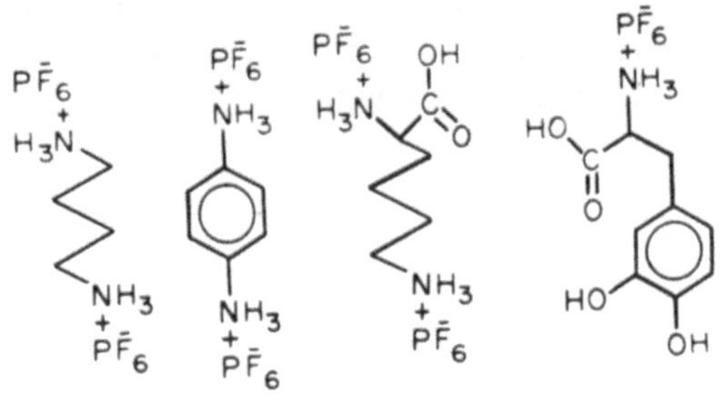   27

3.3 Crown-Hydrazinium Complex

The crystal structure of the complex [30] between [18]crown-6 and H_2N—$NH_3^+ClO_4^-$ has been determined. Structures 28 and 29 provide two different views of the complex. The host in 28 is beautifully regular in shape. Its complex possesses a filled-hole diameter of 2.90 Å if the van der Waals radius of oxygen is assumed to be 1.40 Å. The H—N—N—H dihedral angles are $60 \pm 4°$. View 29 shows that the guest is fully inserted into the hole so that all five hydrogens of the guest hydrogen bond five different oxygens of the host. The sixth oxygen is somewhat more distant from the NH_2 group (3.33 Å) than are the two oxygens hydrogen bonded to that group (3.05 Å). The N—N axis is tipped 6° relative to the normal to the best oxygen plane in the direction of the oxygen atoms bonded to the NH_2 group. One of these hydrogen bonds is bifurcated, a perchlorate oxygen being involved as well (at 3,21 Å). The three $^+N \cdots O$ hydrogen bonds average 2.84 Å, whereas the two from the NH_2 group are longer and less linear. This structure is a particularly good example of nesting complexation [30].

28   29

3.4 Heterocycles as Parts of Macrocyclic Hosts and Their Complexes

Substitution of appropriate heterocycles for CH_2OCH_2 units in [18]crown-6 leads to hosts with powerful binding abilities. For example, hosts such as 30–32 have

135

been prepared and found to be about equal in their abilities to complex K^+ or RNH_3^+ as [18]crown-6 [12,17]. Host *31* is distinguished by its synthesis from cellulose as the sole organic starting material [12].

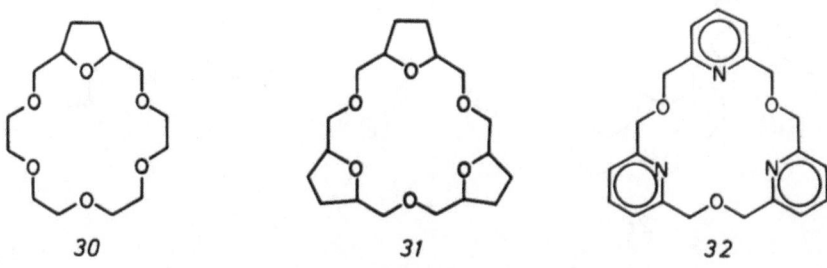

30   31   32

The interesting question of whether the pyrido nitrogens or the ether oxygens are the better hydrogen bonding sites in hosts such as *32* was answered provisionally by comparisons of the free energies of association of a series of hosts with $(CH_3)_3CNH_3^+SCN^-$ in $CHCl_3$ [12]. Conclusive evidence regarding the answer to this question was gained from the crystal structure of the *tert*-butylammonium perchlorate complex of pyrido[18]crown-6 [31]. Molecular model examination suggested *33* as the probable structure (see *34* for another view). Two similar views of the structure as found in crystals are shown in *35* and *36*.

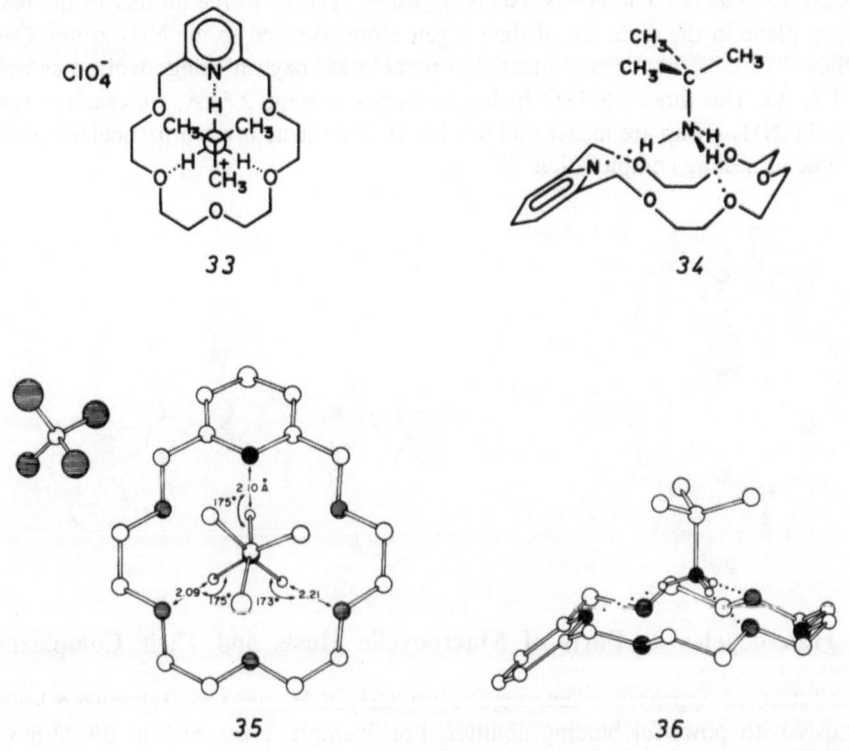

33   34

35   36

3.5 The Complex Designed Not to Form

Chiral hosts containing 1,1'-dinaphthyl units were designed and prepared to separate by extraction the enantiomers of racemic amino ester and amine salts through chiral recognition in complexation [32]. One diastereomeric complex was designed to be significantly more stable in solution than the other. Although differences in free energy between diastereomeric complexes of as much as about 2 kcal mol^{-1} were realized, this difference represented only about one-third of the total binding (free energy) for the more stable diastereomer [32]. Fortuitously, the diastereomeric complex between (SS)-host 37 (38 in cross section) and (D)-C$_6$H$_5$CH(CO$_2$CH$_3$)NH$_3$PF$_6$ crystallized, although in solution the complex between (SS)-37 and the (L) enantiomer was the more stable. The crystal structure of this less stable complex (39) is informative because it was designed not to form [32]. Formula 40 is a drawing of 39 abstracted from a CPK molecular model of the crystal structure [33].

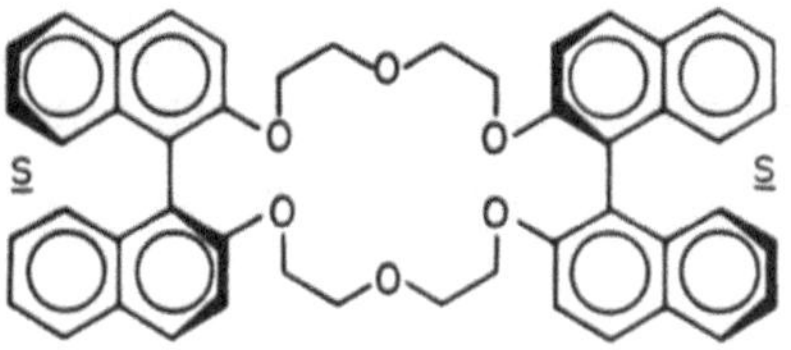

37

38

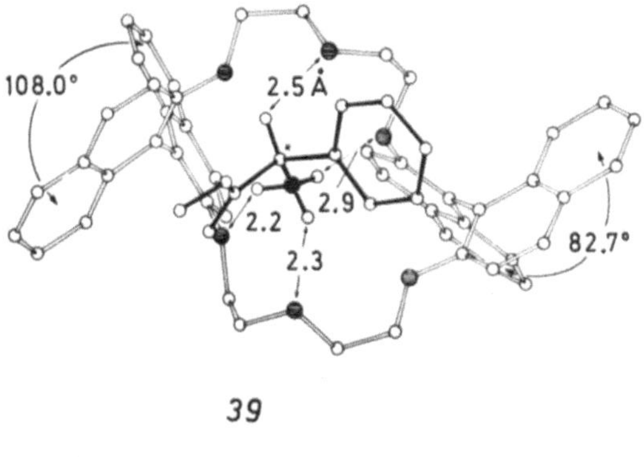

39

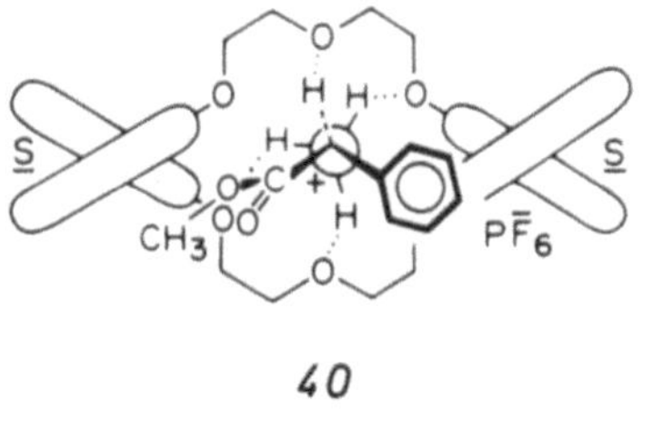

40

To reduce the effects of steric compression designed into the complex, the host and guest in 38 have adapted to each other in several ways. 1) The dihedral angle of one dinaphthyl group has opened to 108°, which is about 30° greater than that observed in most hosts or complexes containing the dinaphthyl group. 2) The hydrogen bonds are far from linear. 3) A weak nonlinear C—H ··· O interaction appears to be present, involving the C̊—H group of the guest and one oxygen of the host. This interaction tilts the guest with respect to the host so that the C$_6$H$_5$ group can escape collision with a naphthyl steric barrier. The C̊—H hydrogen atom is greatly acidified by the attached NH$_3^+$, CO$_2$CH$_3$, and C$_6$H$_5$ groups. 4) The planes of one naphthyl and the ester group are roughly parallel and only 3.45 Å apart. These groups contact one another in what appears to be a *pi*-acid to *pi*-base attractive interaction. 5) Although electron pairs of all six oxygens of the host are oriented generally toward the N^+ of the guest, only five of them appear close enough to provide much binding. Thus the complex reconciles binding with steric repulsion. 6) In the crystalline complex, a molecule of chloroform appears to be hydrogen bonded weakly to a PF_6^- ion, and helps to fill voids in the structure, thus providing additional stabilization [33].

3.6 A Crown-Water Complex

When host *41* was prepared, it was found to cling to a mole of water even at elevated temperatures [28]. Molecular model examination of potential complexes indicated that *41* and H_2O posses a complementary relationship. Thus three well-aligned hydrogen bonds might hold the complex together, as in *42*. One hydrogen might be donated by the ArOH group, and the other two by the water. The crystal structure of the complex (*43*) shows that the water molecule nests in the host, the three hydrogen bonds holding the complex together. The other OH group of the host is involved in an intermolecular hydrogen bond [34].

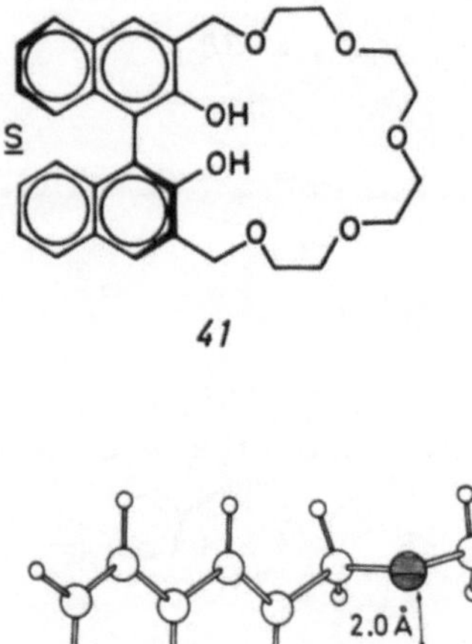

41

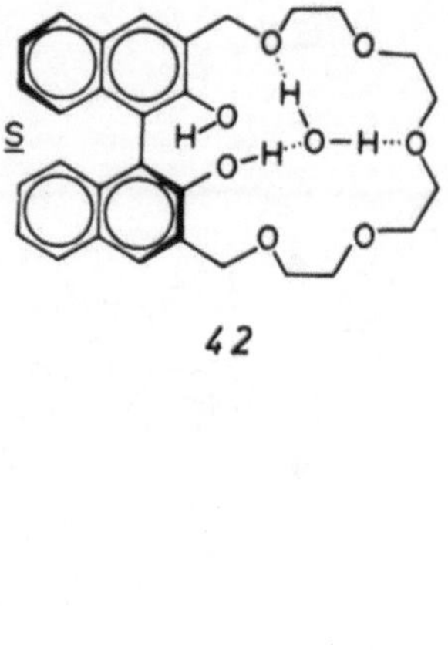

42

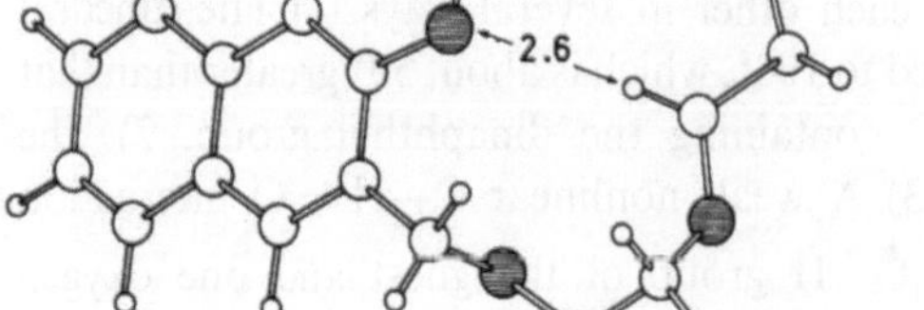

43

3.7 A Crown-Methyl Ester Complex

In contrast to the host-guest complexes thus far discussed, this inclusion complex of an ester and a crown appears to be held together mainly by van der Waals forces. When CH_3—O_2C—$C\equiv C$—CO_2—CH_3 and [18]crown-6 were mixed in equal proportions in a number of solvents, a one-to-one complex crystallized [12]. The same was true when monofurano[18]crown-6 was substituted for [18]crown-6 [12]. In the crystal structure of the former complex, portrayed as *44*, each CH_3 group of the guest perches on a crown whose oxygens are all turned inward [35]. Two of the hydrogens of each methyl group appear to be weakly hydrogen bonded to two of the oxygens of the guest, as shown in part structure *45*.

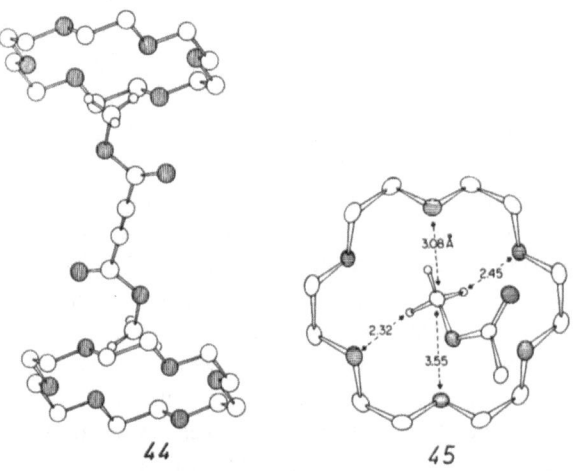

44   *45*

3.8 Intramolecular Complexes

The primary and secondary structures of proteins are determined by the patterns of covalent bonds and of intramolecular hydrogen bonds between neighboring amino acid residues. Two intramolecular complexes, designed with the help of CPK molecular models, show similar features. Electronically complementary units are held by support structures in positions relative to one another to maximize their attractive interactions, which involve hydrogen bonds and other dipole-dipole forces.

46   *47*

In *46*, substituted benzoic acid and pyridine units are transannularly incorporated in a ring system so that the CO_2H group converges on the :N. The compound was synthesized [36], and its structure is depicted in drawing *47* [37]. As expected from its design, the nonbonded distances between the C=O carbon and the flanking ether oxygens (2.84 to 2.87 Å) are considerably shorter than the normal van der Waals distance (≈ 3.1 Å). This fact suggests the presence of O $\cdots$ C=O attractive forces in *46*. Such interactions have been much studied in other systems [38]. The anticipated $CO_2H \cdots N$ hydrogen bond is short and linear (N to O is 2.66 Å). The hydrogen is on the carboxyl group and the structure is not zwitterionic. This compound is an analogue of potential proton-relay systems in which carboxyl and imidazole units are similarly situated, as in some of the proteases.

In molecular models of the intramolecular complex *48*, two carboxyl groups converge on each other to provide an ideal hydrogen bonding arrangement. The compound was prepared [39], and its crystal structure (*49*) [40] was found to be as anticipated. The two carboxyl groups are coplanar and form an eight-membered heterocyclic ring, including two nearly linear O—H $\cdots$ O units. The macrocycle contains four equivalent O=C $\cdots$ O interactions at a distance of 2.97 Å. As in the pyrido complex *47*, at least one electron pair of each ether oxygen of *49* faces inward toward the cavity filled by the complexing groups. The O—CH_2—CH_2—O groups, which provide some of the support structure, are gauche and the CH_2—O—CH_2—CH_2 groups are trans. The planes of the two aryl groups in *47* are nearly parallel, but are mutually displaced by about 1.2 Å to provide more effective hydrogen bonding. In *49* the aryl planes are twisted in opposite directions relative to the plane of the carboxylic acid dimer. The dihedral angle between the aryl planes is 103° [40]. These results demonstrate the general compatibility between expectation and reality concerning the structures of host-guest complexes. Physical organic concepts applied to molecular models derived from crystal structures can be used to design complexes which, when prepared in the laboratory, are frequently found to have the expected structures. Thus a cycle of operations is completed in which one starts and ends with crystal structures.

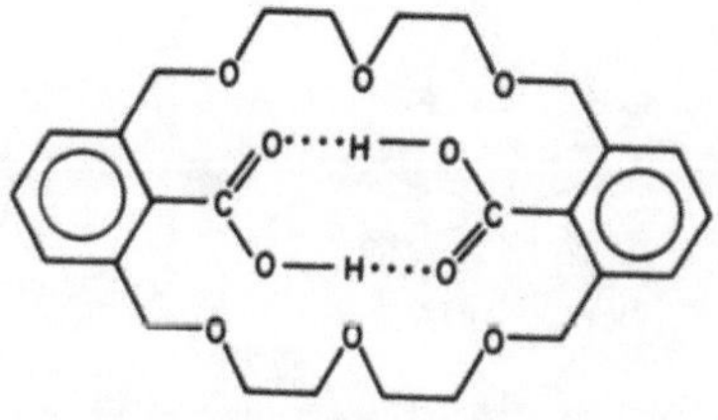

48

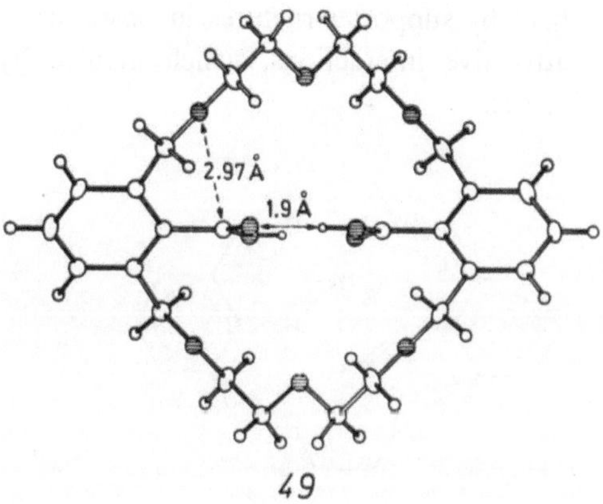

49

4 Comparisons of Structures of Noncomplexed and Complexed Hosts

4.1 The Organizational Question

A central theme in host-guest complexation involves the varieties of answers that can be given to the question, "To what extent does a guest organize its host upon complexation?" A lithium ion in water gathers the oxygens of up to six water molecules. The CH_2CH_2 groups of polyethylene glycol ethers connect and provide spacers for oxygen atoms capable of complexing cations. Although the oxygens of such hosts are collected, they are disorganized in a conformational sense until they complex a suitable guest. When the ends of a polyethylene glycol chain are connected to form a ring, as in the crown ethers, the number of possible conformations of the chains is greatly diminished. Accordingly, the crown ethers form complexes that are more stable than those of their open-chain counterparts by several kcal mol^{-1}. The cryptands' bicyclic structures and the hemispherands' accumulation of rigid units further constrain the number of conformations available to uncomplexed hosts. They do not, however, eliminate conformations having little or no complexing ability. Only in the spherands are the conformations of the free hosts and their complexes essentially the same. The next sections provide structural comparisons between hosts and host-guest complexes, as found in crystals, that illustrate these points.

4.2 Crown and Cryptand Reorganization

Dunitz's crystal structures of [18]crown-6 (*50*) and those of its K^+SCN^- complex (*51*) [23] indicate that the host and its complex have different conformational organizations. The potential crown cavity of the pure host is filled with two inward-turned CH_2 groups, and the electron pairs of two oxygens face outward and away from the best center of the roughly rectangular structure. Thus the free host does not have a "crown shape" nor a cavity. Only when the oxygens become engaged with a guest such as K^+ or the CH_3 groups of $CH_3O_2CC\equiv CCO_2CH_3$, as in *44*, does a *filled cavity* develop. Only the presence of a guest in the complex induces the electron pairs to converge on the center of a crown-shaped object. In other words, the *guest conformationally reorganizes the host upon complexation*. Many solvents are probably able to play the same role as a guest.

A similar reorganization occurs when Lehn's [2.2.2]cryptand complexes KI. Weiss's crystal structure of the free host *52* [24] shows that two methylene groups occupy the middle of the bicyclic structure and that the lone pairs of two oxygens face outward. Upon formation of the capsular complex *53* (K^+ as guest), the host reorganizes so that a potassium ion-occupied cavity develops, lined with 28 electrons. Thus with both the crown and the cryptand, the guest, upon complexation, substitutes for the two methylene groups in the middle of the free host and focuses unshared electron pairs on the middle of the occupied cavity.

Other types of reorganizations accompany complexation of cyclic polyethers containing dinaphthyl units. Thus in the crystal structure 55 [41] of host 54 [20] the center of the macroring system is occupied by two hydrogens of an inward-turned CH_2 group. When 54 was treated with $(CH_3)_3CNH_3^+ClO_4^-$, 56 was envisioned as the complex formed [42]. Its structure is shown in 57 [41]. In 57 all oxygens are turned inward toward the NH_3^+ group, and the C—N bond is nearly perpendicular to the best plane of the macroring. As expected, the H—C—N—H torsion angles are about 60°, and a methyl of the host contacts and fits between two methyls of the guest in a "gear" type of arrangement. Three linear ^+N—H $\cdots$ O hydrogen bonds in a tripod arrangement hold this perching complex together, along with three $N^+ \cdots O$ interactions [41].

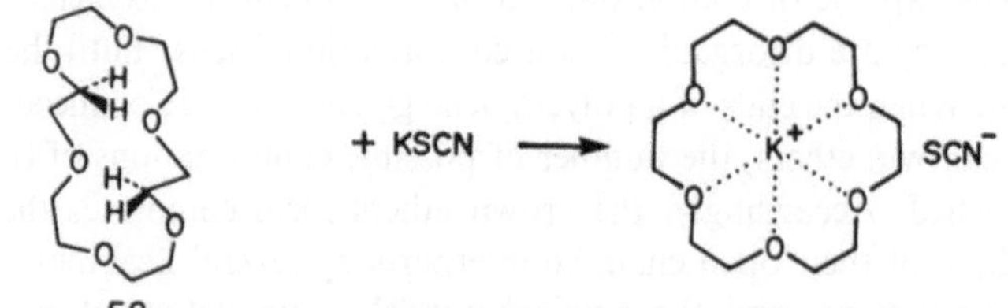

50   + KSCN   51

Crystal Structures

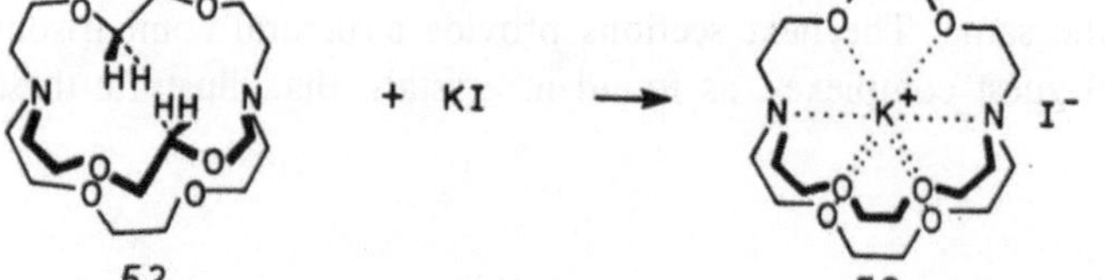

52   + KI   53

Crystal Structures

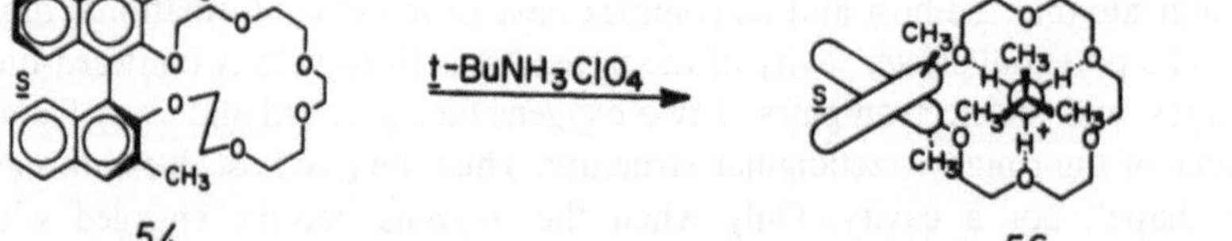

54   t-BuNH₃ClO₄   56

Molecular model of host   Molecular model of complex

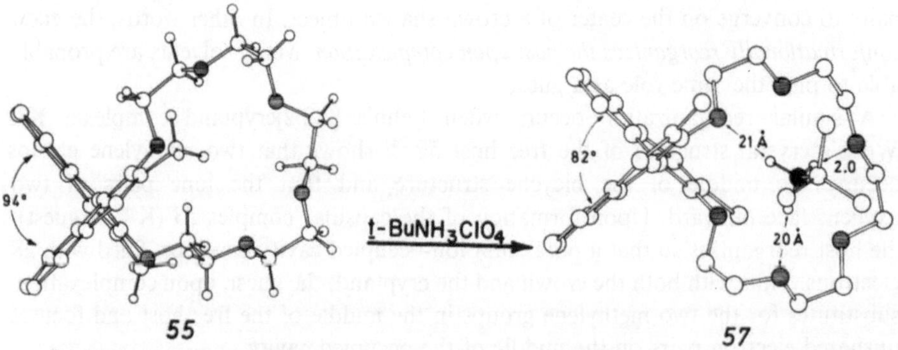

55   t-BuNH₃ClO₄   57

Crystal structure of host   Crystal structure of complex

142

4.3 Hinged Host Reorganization

In molecular models of host *58*, two macrocyclic polyether systems are hinged by an aryl-aryl bond to provide a "jaws" type of arrangement [28]. The crystal structure of host *58* is shown in *59* [43]. In *59* the jaws are wide open and the oxygens' electrons face in a variety of directions. In molecular models of the complex of *58* with KSCN, all ten oxygens gather around the K^+ ion and focus their electrons on the K^+ ion, as in *60* [28]. In the crystal structure of the complex shown in *61*, the jaws have closed on the K^+ ion to provide a capsular structure [43]. All ten oxygens contact the guest and focus their electrons on the ion. No SCN^- ions are liganded to the K^+. The two closest anions, shown in stereoview *62*, are on the outside of the complex more than 6 Å distant from any cation. The "skin" of the cavity in *61* and *62* is completely lipophilic. Again, the guest organizes the cavity of the host during complexation.

In crystals of *58*, which contain both enantiomers of the structure shown in *59*, the [16]crown-5 ring on each jaw of each molecule is juxtaposed close to a similar ring on the other jaw of a neighboring enantiomeric molecule. Each molecule "shakes hands" with two neighbors, producing endless chains ascending through the crystal, as shown in stereoview *63* [43].

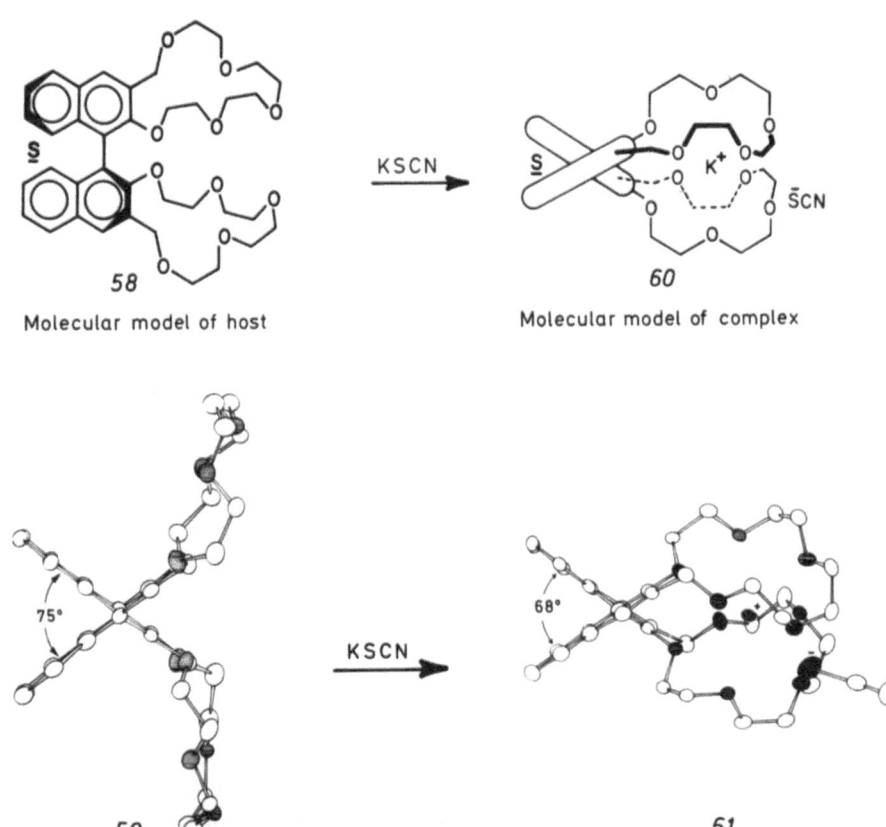

58	*60*
Molecular model of host	Molecular model of complex
59	*61*
Crystal structure of host	Crystal structure of complex

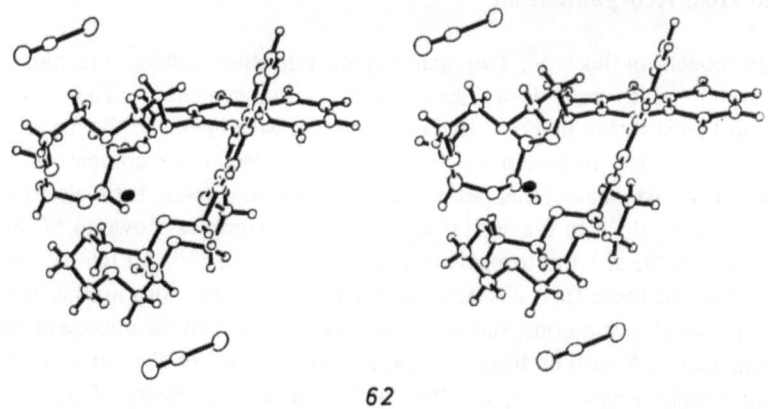

62

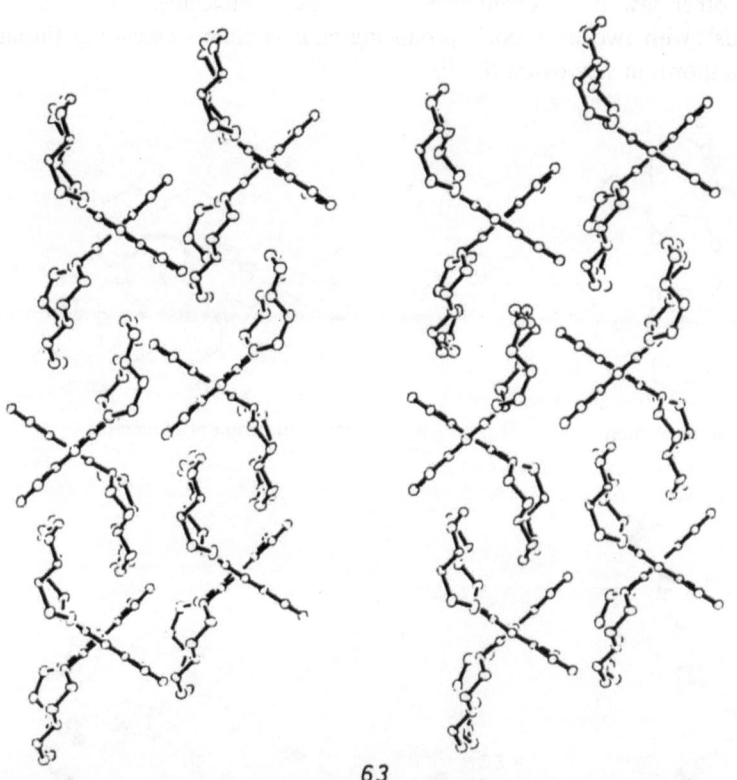

63

4.4 Hinged and Belted Host Reorganization

Host *64* resembles *54* except that two methyl groups attached to the 1,1'-dinaphthyl unit in the 3,3'-positions of *54* have been replaced by a bridging $(CH_2OCH_2)_5$ group in *64*[28]. In molecular models of *64*, as well as in the crystal structure *65*[43], the electron pairs of the oxygens are unfocused and the center of the potential complexing ring is filled with the hydrogens of an inward-turned CH$_2$ group. In the KSCN complex of the host, as represented by molecular model *66* [28)] and its crystal structure *67* [43)], the electrons of seven oxygens are focused on the K$^+$ guest. Six of these oxygens are part of the bridge attached to the 2,2'-positions of the 1,1'-dinaphthyl system and the seventh provides a floor to the potassium ion-filled cavity in the nesting complex. This seventh oxygen is in the "belt" that spans the 3,3'-positions of the dinaphthyl group. The remainder of that bridge only holds this one oxygen in place. The SCN$^-$ ligates the K$^+$ almost end-on through the N atom from the same side of the macroring as the seventh oxygen. The K$^+$ ⋯ $^-$NCS distance is 2.76 Å, shorter than all but three of the K$^+$ ⋯ O distances [43)].

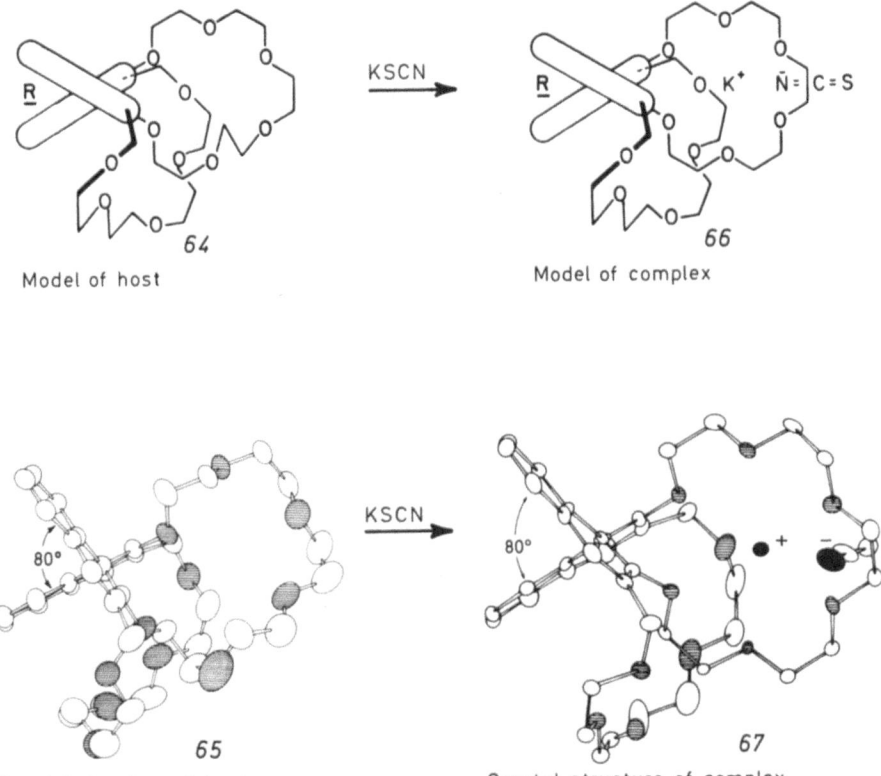

64
Model of host

KSCN →

66
Model of complex

65
Crystal structure of host

KSCN →

67
Crystal structure of complex

4.5 Intramolecular to Intermolecular Complex Reorganization

Compound *68* is an intramolecular complex whose carboxyl group occupies the cavity of the macrocycle [44)]. The OH ⋯ O hydrogen bond and short O ⋯ C=O distances in its crystal structure (*69*) [45)] confirm the features anticipated from molecular model examination. When *68* was treated with (CH$_3$)$_3$CNH$_2$, an intermolecular complex formed which molecular models suggested might be *70* [44)]. The structure of the substance is shown in *71* [46)]. Complex *70* is a salt. Structure *71* shows the usual tripod arrangement of three nearly linear N$^+$—H ⋯ O hydrogen bonds. One of the three hydrogen-bonded oxygens is provided by the carboxylate anion, which is able to engage the $^+$N—H part of the guest by assuming a position perpendicular to the aryl group, which itself turns more than 90° about its two Ar—CH$_2$ bonds (compare *69* and *71*). Thus when treated with an external guest, *68* conformationally reorganizes to a structure complementary to that of the guest.

In *71* the RNH$_3^+$ guest is in a perching position relative to the host. As observed in the other complexes of (CH$_3$)$_3$CNH$_3^+$, the C—N bond is nearly perpendicular to the best plane of the oxygens. The C—C—N—H torsion angles are close to 60°, as are all of the C—C—C—H torsion angles of the (CH$_3$)$_3$C group. In other words, the conformations about the C—N and C—C bonds in (CH$_3$)$_3$CNH$_3^+$ are all staggered. If the ion is viewed down the C—N bond, three of the nine methyl hydrogens are pseudo-axial (directed away from the nitrogen), and the other six are pseudo-equatorial (see *71*) [46]. The same conformation for the (CH$_3$)$_3$CNH$_3^+$ group is observed in structures *21*, *36*, *57*, and *75*. It is also found in the crystal structure of (CH$_3$)$_3$CNH$_3$Cl itself [47]. Therefore this conformation is inherent in this guest. Interestingly, the (CH$_3$)$_3$CNH$_3^+$ ion is isosteric with (CH$_3$)$_4$C, whose conformations are probably similarly staggered.

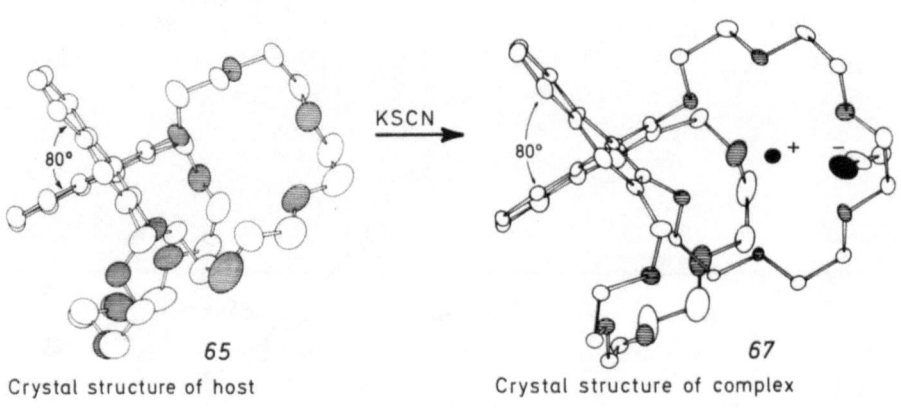

65
Crystal structure of host

67
Crystal structure of complex

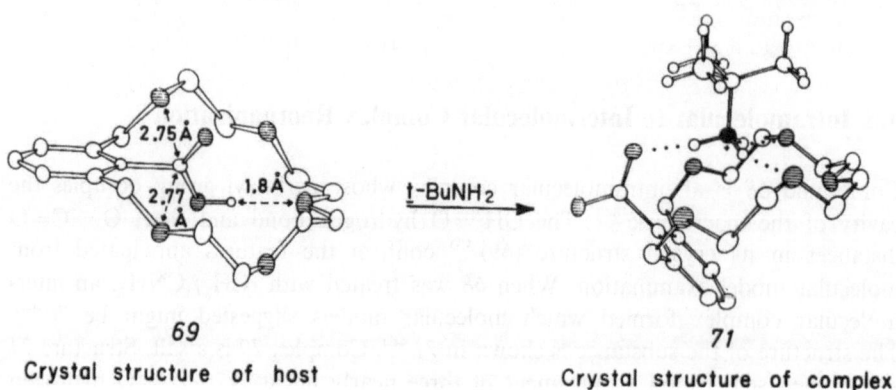

69
Crystal structure of host

71
Crystal structure of complex

4.6 Hemispherand Partially Organized Prior to Complexation

In molecular models of hemispherand 72[5], three methoxyl groups are conformationally constrained as in the structure drawn. However, the methylene groups of the diethylene glycol bridge are conformationally mobile so that the oxygen atoms can either face inward or outward. For example, in structure 72 the hydrogens of two methylene groups occupy the center of the system and the unshared electron pairs of the bridge are oriented outward. In the structure of 73 in crystals[48], the unshared electrons on the three aryl oxygens face one another, the two O—CH₂— —CH₂—O groups possess an *anti* arrangement, methylene hydrogens fill portions of the cavity, and, at the same time, the oxygen atoms' unshared electron pairs turn outward.

When treated with $(CH_3)_3CNH_3ClO_4$, 72 forms a complex that, in molecular models, could have structure 74 or, alternatively, one whose guest was hydrogen bonded to the opposite face of the macroring[5]. The crystal structure (75)[43] demonstrates that the $(CH_3)_3CNH_3^+$ group is bonded to the face from which the two methoxyl groups protrude in a perching arrangement. Two $^+$N—H $\cdots$:O hydrogen bonds are essentially linear and involve the oxygens of the diethylene glycol bridge. The third $^+$N—H hydrogen approaches the three ArOCH₃ oxygens to provide what at first N appears to be a novel *trifurcated hydrogen bond*, but which is better described as a bifurcated hydrogen bond. Structure 76 provides a stereoview of the complex[43].

A comparison of some of the molecular dimensions of host and complex is informative (see Fig. 2). The pseudo-ortho CH₃O oxygen distances change very little upon complexation, but the pseudo-meta oxygen distance increases from 3.56 to 3.96 Å. The O—CH₂—CH₂—O oxygen distances change from 3.61 Å in the free host to 2.89 Å in the complex. The two aryl groups that flank the central aryl are more planar in the free host than in the complex. The angle of fold of these outer aryl rings about their CH₃C₆H₂O axes increases from 3° to 7° upon complexation. Thus the aryls are significantly more deformed in the complex than in the free host.

The binding free energy in CHCl₃ solution of complex 75 is about 1 kcal mol⁻¹ greater than that of 21 (complex of M₃CNH₃ClO₄ and 2,3-naphtho[18]crown-6[5]). greater than that of 21 (between $(CH_3)_3CNH_3ClO_4$ and 2,3-naphtho[18]crown-6[5]). In spite of the aryl folding that accompanies the formation of 74, this complex is

72

Molecular model of host

74

Molecular model of complex

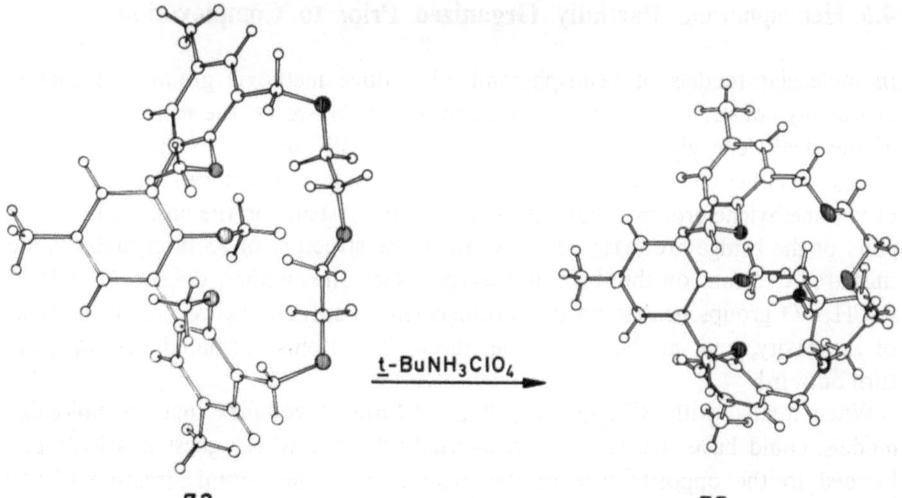

73

Crystal structure of host

$\xrightarrow{t\text{-BuNH}_3\text{ClO}_4}$

75

Crystal structure of complex

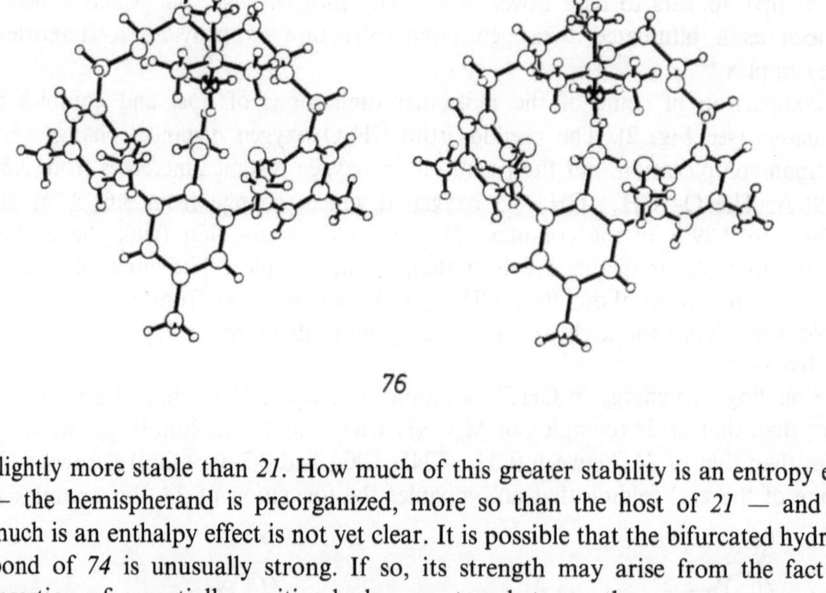

76

slightly more stable than *21*. How much of this greater stability is an entropy effect — the hemispherand is preorganized, more so than the host of *21* — and how much is an enthalpy effect is not yet clear. It is possible that the bifurcated hydrogen bond of *74* is unusually strong. If so, its strength may arise from the fact that insertion of a partially positive hydrogen atom between three oxygen atoms compensates somewhat for their lone-pair repulsions. It may also be relevant that the pseudo-meta oxygens in the hemispherand are significantly (about 20%) closer to one another than are the next nearest oxygen atoms in *21* and other crowns.

4.7 Spherand Organized Prior to Complexation

At one extreme, hosts are rigidly organized prior to complexation. If such hosts are designed to complex metal cations, they should contain enforced, spherical

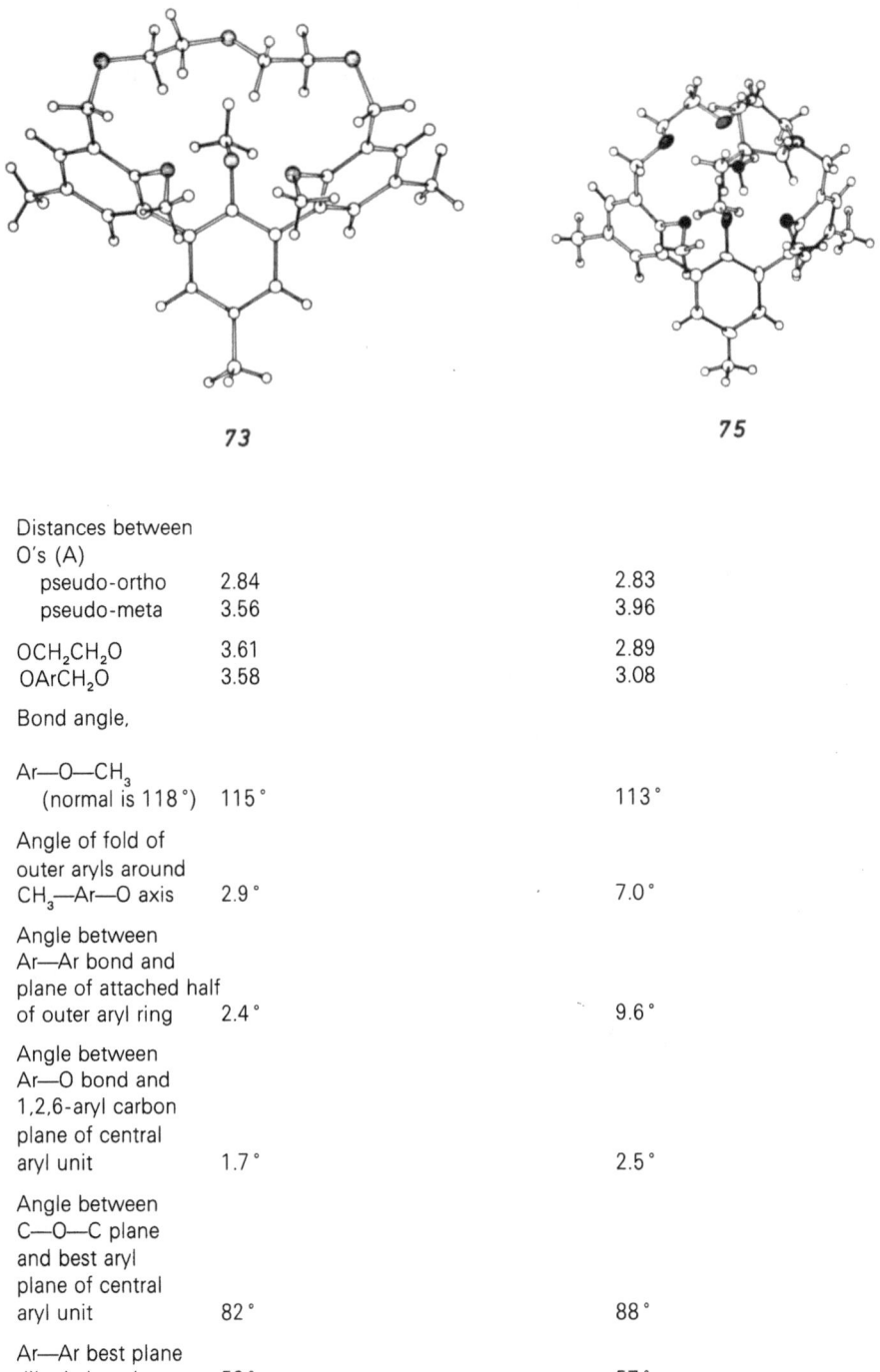

73

75

Distances between O's (A)		
pseudo-ortho	2.84	2.83
pseudo-meta	3.56	3.96
OCH₂CH₂O	3.61	2.89
OArCH₂O	3.58	3.08

Distances between
O's (A)
    pseudo-ortho   2.84   2.83
    pseudo-meta   3.56   3.96

OCH_2CH_2O 3.61   2.89
$OArCH_2O$ 3.58   3.08

Bond angle,

$Ar—O—CH_3$
  (normal is 118°)   115°   113°

Angle of fold of
outer aryls around
$CH_3—Ar—O$ axis   2.9°   7.0°

Angle between
Ar—Ar bond and
plane of attached half
of outer aryl ring   2.4°   9.6°

Angle between
Ar—O bond and
1,2,6-aryl carbon
plane of central
aryl unit   1.7°   2.5°

Angle between
C—O—C plane
and best aryl
plane of central
aryl unit   82°   88°

Ar—Ar best plane
dihedral angles   58°   57°

Fig. 2. Average Molecular Dimensions of Hemispherand and Complex

cavities lined with electron pairs of heteroatoms so that upon complexation no conformational reorganization occurs.

Host *77* is a prototype spherand composed of six methoxyl groups mounted on six benzene rings, which in turn are attached to one another at their 2,6-positions. In *77* the cyclohexaarylene system provides a support structure for the six heteroatoms, which converge on one another. The only CPK molecular model of *77* that can be assembled is one in which the oxygens are octahedrally arranged around a cavity lined with 24 electrons. The size of the cavity is limited on the lower side by collision of the oxygens with one another, and on the higher side by the tendency of the benzene rings and the atoms bonded directly to them to be coplanar. The cavity is too small to accommodate either solvent molecules or inward-turned methyl groups attached to the oxygens. Consequently, the methoxy methyl groups are directed away from the center of the cycles, three in one direction and the other three in the opposite direction. The model of *76* possesses D_{3d} symmetry, similar to that of chair cyclohexane. In a sense, the six methyls attached to the oxygens are axially arranged, and the six methyls attached to the para positions of the benzene rings are equatorially arranged. In *77* the oxygens are buried in a skin of C—H bonds. Their unshared electrons are unavailable for interacting with solvent or with potential guest cations unless such guests enter the cavity. Thus *only capsular complexes are possible with this particular spherand*[4].

The cavity diameters in molecular models of *76* appear to range from approximately 1.3 to 2.0 Å, depending on the aryl-aryl dihedral angles and the degree of deformation of normal bond angles and planarities. With diameters in excess of 2.0 Å, the electron pairs of the oxygens start to focus not on the interior of the cavity, but along its surface or above the cavity. The lower range includes diameters of Li^+ and Na^+, so it was anticipated that complexes such as lithiospherium chloride (*79*) would be preparable. Compounds *77* and *79* were synthesized, and their ^1H NMR spectra were found to be consistent with their anticipated D_{3d} structures[4]. Their respective structures, as found in crystals, are shown in *78* and *80*[43]. The corresponding sodiospherium salt was also prepared[4]. Its crystal structure was determined, and found to generally resemble that of the lithiospherium salt[43]. Notice the resemblance of these structures to snowflakes.

The molecular dimensions taken from the crystal structures of prototype spherand (*78*) and of lithiospherium chloride (*80*) are given in Fig. 3. If one assumes the oxygen-atom radius in these compounds is 1.40 Å, the hole diameter is 1.62 Å in the free ligand system and 1.48 Å in the Li^+ complex. The diameter of the hole occupied by the Li^+ ion in the [2.1.1]cryptand complex of LiI is also 1.48 Å if the same oxygen-atom radius is assumed[24]. Each of the six aryl groups is folded about the O—Ar—CH_3 axis by 6.3° in *78* and by only 2.6° in *80*. The six oxygens are displaced 0.20 Å from their attached aryl planes in *78* but only 0.07 Å in complex *80*. Thus the six aryl groups of the noncomplexed ligand system are distinctly more strained than those in the complexed ligand. Aryl deformation strain is released upon complexation, and this release provides some of the driving force for complexation. The strain is undoubtedly associated with oxygen-oxygen electron repulsion in the free spherand. To increase their distance from one another, the oxygens induce in the free system deformations of the aryl rings and of the exocyclic bond angles. When the hole is filled with a cation, the electron-electron repulsion

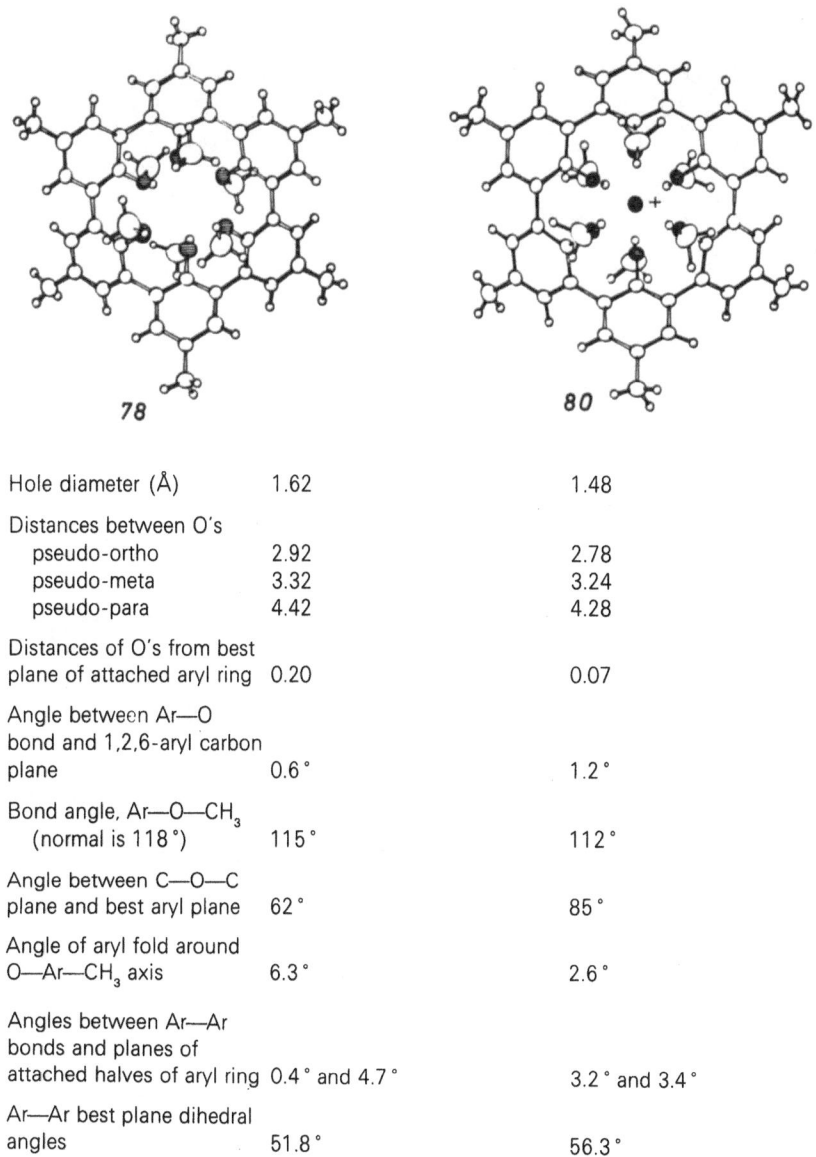

Hole diameter (Å)	1.62	1.48
Distances between O's		
pseudo-ortho	2.92	2.78
pseudo-meta	3.32	3.24
pseudo-para	4.42	4.28
Distances of O's from best plane of attached aryl ring	0.20	0.07
Angle between Ar—O bond and 1,2,6-aryl carbon plane	0.6°	1.2°
Bond angle, Ar—O—CH_3 (normal is 118°)	115°	112°
Angle between C—O—C plane and best aryl plane	62°	85°
Angle of aryl fold around O—Ar—CH_3 axis	6.3°	2.6°
Angles between Ar—Ar bonds and planes of attached halves of aryl ring	0.4° and 4.7°	3.2° and 3.4°
Ar—Ar best plane dihedral angles	51.8°	56.3°

Fig. 3. Molecular Dimensions of Prototype Spherand and Its Lithiospherium Chloride

is not removed, but is somewhat compensated for by attractive forces, the oxygens come closer together, and the aryls and their attached oxygen and aryl carbons approach their lower energy, coplanar geometry.

The crystal packing arrangement of the prototype lithiospherium chloride *80* also is interesting [43]. Each Li^+ is surrounded by eight Cl^- and each Cl^- is surrounded by eight Li^+, with spherand shells packed between the ions. The chloride ions are centered between two sets of three methyl groups attached to

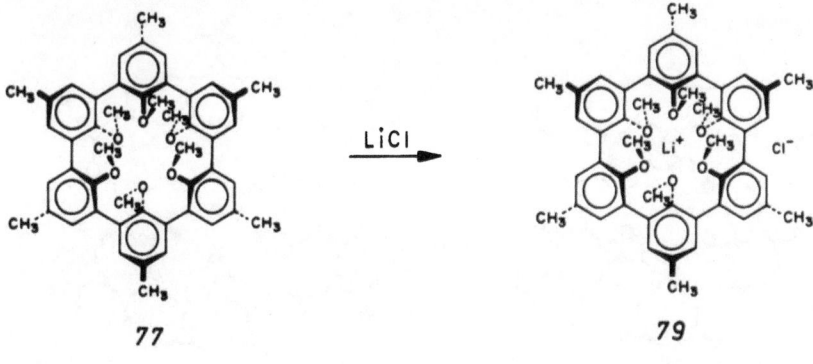

77   79

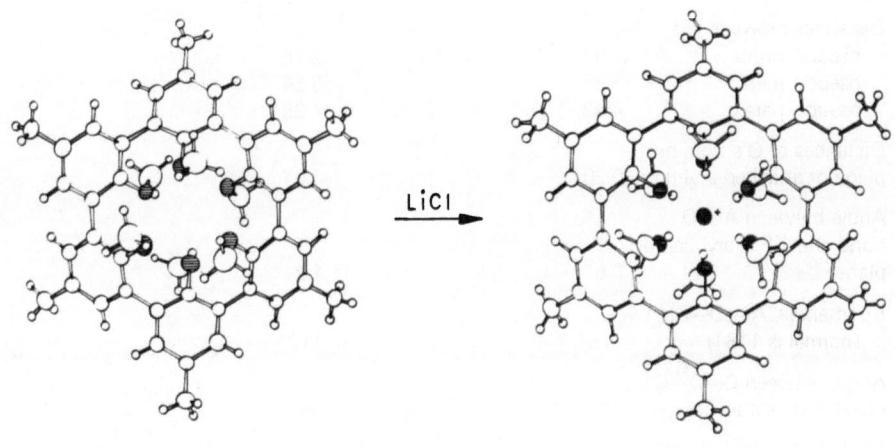

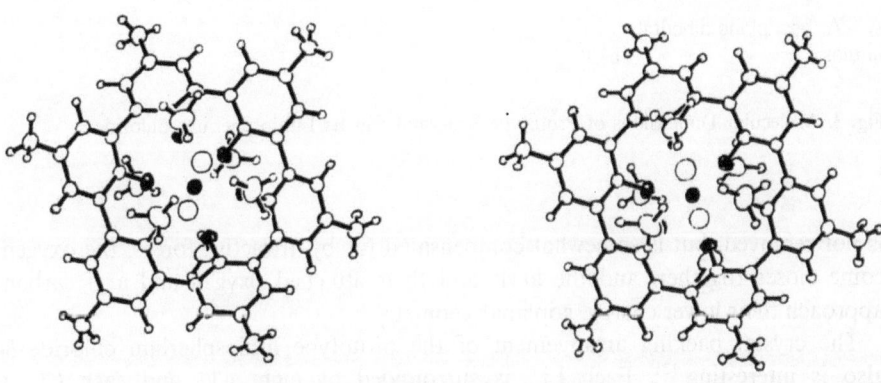

81

oxygens to provide parallel stacks of molecules. Within each stack, the molecular axes are colinear. These relationships are visible in stereodrawing *81* of complex *80*, viewed 10° from the $\bar{3}$ axis of the molecule. When *80* is viewed from the side, each column resembles a stack of "hourglasses" piled on top of one another, with the $(CH_3)_3Cl^-(CH_3)_3$ parts as the "waists" and the $(ArO)_3Li^+(OAr)_3$ parts as the "shoulders". In the packing of parallel stacks, each waist is surrounded by three shoulders and each shoulder by three waists. The distance between the closest Li^+ and Cl^- ions along the $[(CH_3)_3Cl^-(CH_3)_3(ArO)_3Li^+(OAr)_3]_n$ axes is 5.26 Å, and that between the interaxial "near" Li^+ and Cl^- ions is 10.73 Å. The packing of the Li^+ and Cl^- ions reflects the $\bar{3}$ symmetry. The spatial relationships in the crystal (and of each molecule in it) of one Li^+ ion surrounded by eight Cl^- ions and one Cl^- ion surrounded by eight Li^+ ions is sketched roughly in *82*. Drawing *83* represents cross sections of seven contiguous stacks of molecules in the crystal viewed along their $[(CH_3)_3Cl^-(CH_3)_3(ArO)_3Li^+(OAr)_3]$ axes. The three heavy circles represent cross sections above the plane; the ordinary circle is in the plane, and the three dashed circles are below the plane of the page. In the center of each circle is a Li^+ [43].

The diameter of Na^+ in the crystal structure of the sodiospherium cation of spherand *78* is 1.76 Å. To accommodate the Na^+ ion, the spherand cavity must expand from 1.62 Å in the free host to 1.76 Å in the complex, which introduces additional strain into an already highly strained system. The Ar—Ar best plane dihedral angles expand from about 52° in the free host to about 60° in the Na^+ complex to accommodate this larger diameter [43].

82   83

4.8 Host that Resists Reorganization

Host *84* was prepared, since in molecular models it appeared to have a potential cavity large enough to accommodate two Na^+ or two K^+ ions at the same time [28]. Structure *85* represents the compound drawn in cross section with all oxygens turned inward so their electron pairs line an egg-shaped cavity. The structure of the compound in crystals is shown in *86* [48], and demonstrates again the tendency of hosts to fill their own holes when possible. In *86* the electrons of five of the ten oxygens face away from the center of the molecule, and the potential cavity is filled with inward-turned hydrogens of CH_2 groups.

The compound is a poor complexing agent for alkali metal ions [28], and no complex has been crystallized. Apparently the energy costs are too high for guests to reorganize this host so that its oxygens could turn their electron pairs inward to line a cavity filled with one or two metal cations. Complexes formed by *84* are probably perching or, at best, nesting, but not capsular.

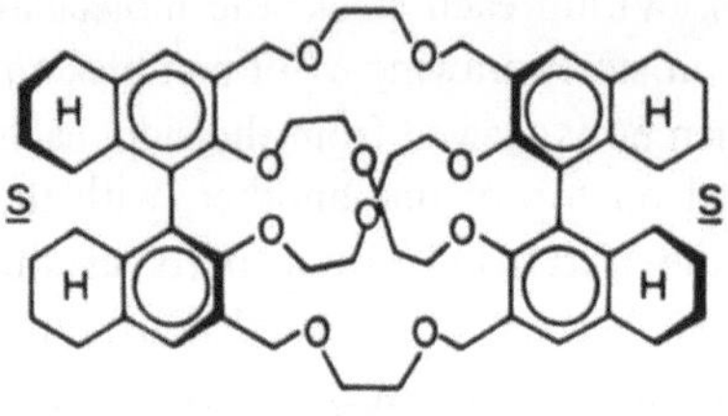

84

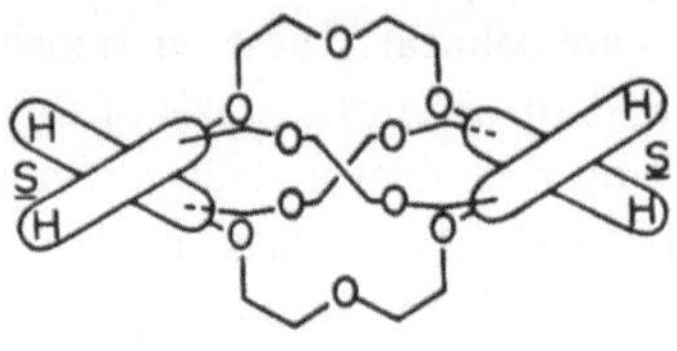

85

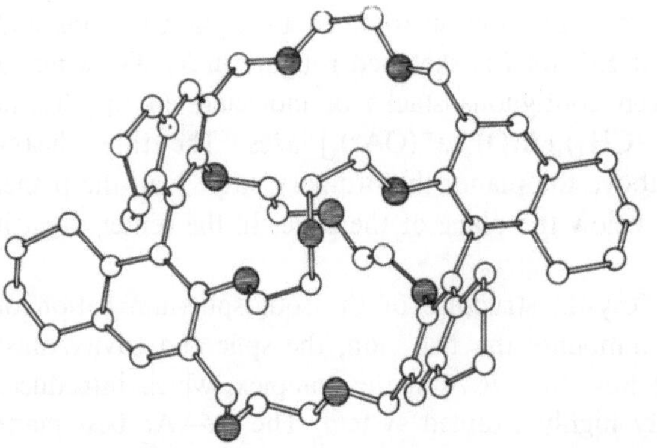

86

5 Detailed Structural Comparisons

5.1 Geometry of 1,1′-Diaryl Units

Many of the hosts and complexes whose X-ray structures have been determined possess 1,1′-diaryl units with oxygens attached at the 2,2′-positions, and frequently with carbons attached at the 3,3′-positions. This section deals with the similarities and differences in the geometries of these units. For comparing hosts, guests, and complexes, it is convenient to write shorthand formulas. In these, M stands for methyl; D for 1,1′-dinaphthyl substituted in the 2,2′-positions by oxygen (O), and in the 3,3′-positions by H, CH_3, CH_2, or O; E for CH_2CH_2; Nap for 2,3-disubstituted naphthalene; Pyr for 2,6-disubstituted pyridine; BCO_2H for 2,6-disubstituted benzoic acid; and MBOM for 2,6-disubstituted 4-methylanisole.

Table 1 records values for some features of the molecular geometries of hosts and their complexes. The Ar—Ar dihedral angles of the dinaphthyl unit range from a low of 68° to a high of 108°. The low value of 68° is associated with the "closed jaws" complex, $D[CH_2(OEOE)_2O] \cdot KNCS$ (60), and appears due to K^+ pulling the jaws together. The free host 59 has a value of 75°, close to that of several others. The high value of 108° involves one of the two dinaphthyl units of the complex $D(OEOEO)_2D \cdot C_6H_5CH(CO_2M)NH_3PF_6$ (39), into which were designed noncomplementary configurational relationships between host and guest. The large dihedral angle appears associated with the adaptation of the host to its poor fit to the complexing guest. The values for the other hosts and complexes range from 75° to 94°. Of the three cases for which crystal structures of both host and complex are available, in two the dihedral angles are lower in the complex and in one the dihedral angles are the same.

The Ar—Ar dihedral angles of the diphenyl units in Table 1 vary from a low of 52° to a high of 58° in the hemispherand and spherand and their complexes. Complexation of $(MBOM)_3(CH_2OE)_2O$ with $M_3CNH_3ClO_4$ produces no overall change in dihedral angles which average about 58° in both host (73) and complex (75). Complexation of $(MBOM)_6$ with LiCl increases the six equivalent dihedral angles from 52° to 56°.

The aryl-aryl bonds are all somewhat deformed. The deformation is expressed by the values for "bend" in Table 1. Bend is the sum of the absolute deviations from 90° of the angles between the aryl-aryl bond and the normal to the best plane of each aryl of the bonding pair. The bend values vary only from 0.2 to 4.3° in hosts and complexes such as 55, 57, 24, and 59, whose macroring systems in CPK molecular models appear strain-free. In the "strapped" host $O(EOEOCH_2)_2$-$D(OEOEO)_2E$ (65) and its KNCS complex (67), these bends are 12.1 and 14.1, respectively. The bridge spanning the 3,3'-positions of the dinaphthyl unit is the smallest that could be synthesized readily [28], and in models, this "strap" strains the system. Hydrate complex 43 also contains the strap, but not the central macroring of 65 which must be circumvented by its strap. The bond in 43 is only 7.3°.

In models, $(MBOM)_6$ (78) and its LiCl complex (80), both appear strained, and the bend values are 10.4° and 8.6°, respectively. Thus a slight relaxation of this type of strain accompanies complexation. In contrast, upon complexation with $M_3CNH_3 \cdot ClO_4$, the bend in the hemispherand $(MBOM)_3(CH_2OE)_2O$ (73) increases from an average of 6.8° to an average of 17.4°. Thus an increase in strain appears to accompany complexation in this system, although some of this apparent increase only represents a redistribution of strain. In the complex, the $(CH_2OCH_2)_3$ units possess gauche conformations.

All of the hosts of Table 1 contain O—Ar—Ar—O units in which the oxygens are substituted in the 2,2'-positions. All but the spherands contain $O—CH_2—CH_2—O$ units as well. Comparisons of distances between the nearest oxygens in hosts and complexes are interesting because they provide one measure of how much the host is organized *for* complexation prior *to* complexation. Three values of nonbonded O ⋯ O distances provide useful calibration points. The normal van der Waals O ⋯ O distance is usually assumed to be 2.80 Å, although it is clear that oxygen atoms are compressible and can approach more closely when forced together sufficiently. The average O ⋯ O distance in the gauche $O—CH_2—CH_2—O$ units of twelve of the crystal structures discussed here is 2.85 Å, while the average O ⋯ O distance in the few anti $O—CH_2—CH_2—O$ units is 3.60 Å.

In the three examples in which Table 1 lists both host and complex containing 1,1'-dinaphthyl units, the distances between the 2,2'-oxygens decrease from an average of 3.62 Å in the host to be an average of 3.13 Å in the complex. This average decrease of about 0.5 Å compares with an average decrease of about 0.75 Å in oxygen-to-oxygen distance when an anti $O—CH_2—CH_2—O$ changes

Donald J. Cram and Kenneth N. Trueblood

Table 1. Geometry of 1,1'-Dinaphthyl and 1,1'-Diphenyl Units, and their 2,2'- and 3,3'-Atoms in

Complex[a] between				Ar-Ar angle (°)	
Host	Guest	Counterion	No.	Dihedral[b]	Bend[c]
$M_2D(OEOEO)_2E$	none	none	55	94	3.3
$M_2D(OEOEO)_2E$	$M_3CNH_3^+$	ClO_4^-	57	82	2.0
$O(EOEOCH_2)_2D(OEOEO)_2E$	none	none	65	80	12.1
$O(EOEOCH_2)_2D(OEOEO)_2E$	K^+	NCS^-	67	80	14.1
$D[CH_2(OEOE)_2O]_2$	none	none	59	75	1.7
$D[CH_2(OEOE)_2O]_2$	K^+	NCS^-	61	68	5.1
$D[(OEOEO)_2E]_2$	$^+H_3N(CH_2)_4NH_3^+$	PF_6^-	24	78	0.2
$O(EOEOCH_2)_2D(OH)_2$	H_2O	none	43	79	7.3
$D(OEOEO)_2D$	$C_6H_5CH(CO_2M)NH_3^+$	PF_6^-	39	108	8.2
				83	4.3
$(MBOM)_6$	none	none	78	52	10.4
$(MBOM)_6$	Li^+	Cl^-	80	56	8.6
$(MBOM)_3(CH_2OE)_2O$	none	none	73	58	6.8
$(MBOM)_3(CH_2OE)_2O$	$M_3CNH_3^+$	ClO_4^-	75	57[j]	17.4[j]

[a] D is 1,1'-dinaphthyl substituted in its 2,2'-positions by O's and 3,3'-positions by H, CH_3, CH_2, or O; E is CH_2CH_2; M is CH_3; B is benzene substituted, respectively, in its 1-, 2-, 3-, and 5-positions by B, by OM, by B or CH_2, and by M. [b] Angle between the normals to the least squares planes of the aryl rings. [c] Sum of the absolute deviations from 90° of the angles between the Ar-Ar bond and the normals to the least squares planes. [d] Between O's attached to 2,2'-C's of 1,1'-aryls. [e] + means bent toward, and − means away from best center of host cavity. [f] Hydro-

to a gauche, a change that occurs in some units upon complexation of most crown hosts.

Complexation of the hemispherand and the spherand involves contrasting movements of near oxygens. The distance between nearest oxygens decreases from 2.92 Å for the spherand to 2.78 Å in the LiCl complex, a difference of only 0.14 Å. In the hemispherand, the near aryl oxygen distances decrease from 2.84 Å to 2.83 Å in the $M_3CNH_3ClO_4$ complex, an insignificant difference of only 0.01 Å. The near oxygens of the O—CH_2—CH_2—O units move from a separation of 3.61 Å in the host to 2.89 Å in the complex, a difference of 0.72 Å. Thus it appears that when at least three aryls are bound to one another their near oxygens become highly organized prior to complexation. This important conclusion becomes apparent upon examination of CPK molecular models of the hemispherands.

The aryl atoms and their attached atoms in hosts and complexes do not lie in a single plane. Tabel 1 records the average and the maximum deviations of the aryl carbons from their least squares planes. For the four sets of host and complex pairs other than the spherands that contain di- or triaryl groups (*55* and *57*; *65* and *67*; *59* and *61*; *73* and *75*) the average deviations increase by 0.01 or 0.02 Å in passing from guest to complex, and the maximum deviations by 0.03 or 0.05 Å. This increased aryl deformation presumably reflects the effects of the guest's reorganization of the host on complexation. The maximum deviation in any of the above eight crystal

Hosts and Complexes

O ... O distance (Å) of O-Ar-Ar-O[d]	Deviations from least squares aryl planes						Ref.
	Aryl C's (Å)		Distance (Å) attached atom[e]				
	Average Δ	Maximum Δ	O at 2	O at 2'	C at 3	C at 3'	
3.76	0.02	0.04	+0.01	+0.02	−0,07	+0.20	41)
3.26	0.03	0.09	+0.31	+0.07	+0.13	+0.17	41)
3.72	0.02	0.06	+0.07	−0.21	+0.18	−0.23	43)
3.15	0.04	0.11	+0.42	+0.05	−0.20	+0.30	43)
3.38	0.01	0.03	+0.09	+0.08	+0.10	0.00	43)
2.98	0.03	0.06	+0.09	−0.01	+0.16	+0.28	43)
3.11	—	—	—	—	—	—	29)
3.48	0.02	0.06	+0.13[f]	−0.07[g]	+0.02	+0.11	34)
3.87	—	—	—	—	—	—	?
3.29	—	—	—	—	—	—	33)
2.92	0.03	0.05	+0.20	—	−0.16[h]	+0.14[i]	43)
2.78	0.01	0.02	+0.07	—	−0.13[h]	−0.01[i]	43)
2.84	0.01	0.02	—	+0.04[j]	−0.11[k]	0.04[i]	48)
2.83[j]	0.03[m]	0.06	—	+0.15[j,n]	−0.36[j,p]	+0.10[i]	43)

gen bonded in cavity. [g] Hydrogen bonded intermolecularly. [h] Carbons of attached benzene rings. [i] CH_3's para to OCH_3 groups, averaged in hemispherands. [j] Two kinds of Ar's. Values for each are close and averaged. [k] Attached CH_2 and Ar's C's. [m] Flanking Ar's only. Central aryl maximum deviation is 0.00 Å. [n] Flanking Ar's only. Central Ar, +0.07 Å. [p] Flanking Ar's, ArC's only. Attached CH_2's, −0.15 Å. Central Ar attached Ar'sC's, +0.12 Å.

structures is observed to be 0.11 Å in complex $O(EOEOCH_2)_2D(OEOEO)_2E \cdot KNCS$ (67). In both this complex and its host (65), the $O(EOEOCH_2)_2$ strap spanning the 3,3'-positions appears tight enough to deform the aryls more than usual. The value of 0.09 Å maximum deviation observed in complex $M_2D(OEOEO)_2E \cdot M_3NH_3ClO_4$ (57) is the next highest, and may reflect accommodation to the bulky $(CH_3)_3C$ group of the guest by the Ar—CH_3 groups of the host. The averages of the deviations of the four atoms attached to the two aryls in the same eight crystal structures also increase in passing from host to complex. When $M_2D(OEOEO)_2E$ (55) complexes $M_3CNH_3ClO_4$, the average increases from 0.08 to 0.17 Å. When $O(EOEOCH_2)_2D(OEOEO)_2E$ (65) complexes KNCS, the average increases from 0.17 to 0.24 Å. When $D[CH_2(OEOE)_2O]_2$ (59) complexes KNCS, the average increases from 0.07 to 0.14 Å. When $(MBOM)_3(CH_2OE)_2O$ (73) complexes $M_3CNH_3ClO_4$, the average increases from 0.07 to 0.18 Å. Thus deformations of aryls and their attached atoms all *increase* upon complexation for the four hosts 55, 65, 59, and 73.

When spherand 78 complexes LiCl to form $(MBOM)_6 \cdot LiCl$ (80), these deformations of aryls and their attached atoms all *decrease*. In 78 the average deviation of the aryl carbons from their least squares plane (all aryls are equivalent) is 0.03 Å. In complex 80, this average decreases to 0.01 Å. In host 78, the four atoms attached to each aryl ring are out of the aryl plane by an average of

0.17 Å. In complex *80*, this average decreases to 0.08 Å. Thus deformation of aryls and their attached atoms *decreases* upon complexation. These deformations *favor* complexation of the spherand, but *disfavor* complexation in the other systems.

5.2 Substituted Ammonium and Oxygen Binding Sites

The crystal structures of eight complexes between RNH_3^+ and hosts containing six binding heteroatoms are available for comparisons. Table 2 provides the distances and angles that answer the most interesting structural questions about binding

Table 2. Geometric Description of Interactions Between RNH_3^+ and Oxygens in Complexes

Complex[a] between			Com-plex No.	Ave. devn. O's from their best plane (Å)	Dist. N$^+$ to best O plane (Å)	Angle C—N bond to normal of plane (°)	Ave. devn. C—C—N—H torsion angle from 60° (°)
Host	Guest	Coun-terion					
Nap(OEOEO)$_2$E	M$_3$CNH$_3^+$	ClO$_4^-$ *21*	0.18	0.99	1	7
Pyr(CH$_2$OEOE)$_2$O	M$_3$CNH$_3^+$	ClO$_4^-$ *35*	0.19	1.14	1	8
BCO$_2^-$(CH$_2$OEOE)$_2$O M$_3$CNH$_3^+$		CO$_2^-$ *71*	0.46	0.98	3	0
M$_2$D(OEOEO)$_2$E	M$_3$CNH$_3^+$	ClO$_4^-$ *57*	0.45	1.22	2	7
(MBOM)$_3$(CH$_2$OE)$_2$O M$_3$CNH$_3^+$		ClO$_4^-$ *75*	0.65(0.27^f)	1.37(1.05^f)	12(3^f)	4
D(OEOEO)$_2$D	C$_6$H$_5$CH(CO$_2$M)NH$_3^+$ PF$_6^-$	*39*	—^{h}	1.11	26^h	12
D[(OEOEO)$_2$E]$_2$	$^+$H$_3$N(CH$_2$)$_4$NH$_3^+$	PF$_6^-$ *24*	0.36	1.26	28	10
E(OEOEO)$_2$E	H$_2$NNH$_3^+$	ClO$_4^-$ *28*	0.24	0.11	6^i	4^j

[a] Nap is 2,3-disubstituted naphthalene. M is methyl. E is CH$_2$CH$_2$. Pyr is 2,6-disubstituted pyridine. B is benzene, substituted respectively in its 1-, 2-, 3-, and 5-positions by B, OM, B or CH$_2$, and M. BCO$_2$H is 2,6-disubstituted benzoic acid. D is 1,1′-dinaphthyl substituted in the 2,2′-positions with O's, and in the 3,3′-positions with O's or C's. [b] Difference between observed N ⋯ O vector

between complexing partners of this type. In the comparisons, $E(OEOEO)_2E$ (18-crown-6) and $Nap(OEOEO)_2E$ (2,3-naphtho[18]crown-6) serve as prototype hosts, and $M_3CNH_3ClO_4$ serves as a prototype guest.

The average deviations of the six binding heteroatoms from their least squares planes are listed in Table 2. The average deviations for complexes $Nap(OEOEO)_2E \cdot$ $\cdot M_3CNH_3ClO_4$ (21), $Pyr(CH_2OEOE)_2O \cdot M_3CNH_3ClO_4$ (35) and $E(OEOEO)_2E \cdot$ $\cdot H_2NNH_3ClO_4$ (28) are 0.18, 0.19, and 0.24 Å, respectively. These complexes are probably typical when the host is [18]crown-6 or one of its analogues. The BCO_2^- and O—Ar—Ar—O units in $BCO_2^-(CH_2OEOE)_2O \cdot M_3CNH_3^+$ (71), M_2D-$(OEOEO)_2E \cdot M_3CNH_3ClO_4$ (57), and $(MBOM)_3(CH_2OE)_2O \cdot M_3CNH_3ClO_4$ (75)

H_3^+ interactions with C—O—C

					Nonhydrogen bonded			
Hydrogen bonded								
H ··· O		Distance N ··· O	Angle[b] from N ··· O to		Distance N ··· O	Angle[b] from N ··· O to		
angle	Distance H ··· O (Å)	(Å)	tetrahedral at O (°)	trigonal at O (°)	(Å)	tetrahedral at O (°)	trigonal at O (°)	Ref.
	2.0	2.96	12	43	3.00	28[c]	27[c]	27)
	2.0	2.96	10	48	3.02	10	45	
	2.0	2.97	37[c]	18[c]	3.02	5	50	
	2.1[d]	3.00[d]	38[d]	17[d]	3.05	0	55	
	2.1	2.94	32	23	3.06	3	55	31)
	2.2	3.03	26	28	3.07	4	51	
	1.7[e]	2.68[e]	—[e]	—[e]	3.48	3	57	
	2.2	3.11	20	35	3.11	8	47	46)
	2.2	3.11	20	35	3.48	3	57	
	2.0	3.00	34[c]	21[c]	3.01	4	54	
	2.1	2.91	26	28	3.12	4	56	41)
	2.1	2.95	18	39	2.92	52[c]	3[c]	
	2.0	2.96	18	37	—	—	—	
	2.1	2.89	22	33	2.87	6	49	43)
104[g]	2.2, 2.5[g]	2.84, 2.87[g]	50[c], 51[c]	5[c], 4[c]	3.10	54[c]	2[c]	
	2.1	2.90	42[c]	13[c]	3.27	20[c]	67[c]	
	2.3	2.94	30	32	3.15	14	44	33)
	2.2	2.90	26[c]	29[c]	3.66	6[c]	55[c]	
	2.0	2.97	18	37	3.10	12	42	
	1.9	2.89	30	25	3.05	8	47	29)
127[g]	2.3, 2.4[g]	3.00, 3.14[g]	21[c], 6[c]	34[c], 50[c]	—	—	—	
	2.1	2.87	6	50	2.86	8	46	
	2.0	2.84	6	49	2.88	6	49	30)
	2.1	2.82	12	42	2.86	5	50	

and direction of unshared electron pairs on oxygen had they been tetrahedrally or trigonally oriented. [c] ArO's are involved. [d] $N^+ \cdots N$. [e] N—H $\cdots O_2C$. [f] Best five of six O's used to calculate best plane and its normal. [g] Bifurcated hydrogen bond. [h] Complex designed not to form turned out badly deformed. [i] N—N bond. [j] H—N—N—H torsion angle.

159

perturb the crown structure sufficiently to increase the average deviations to 0.46, 0.45, and 0.65 Å, respectively. The binding oxygens in these complexes are more three-dimensionally disposed than those in the complexes of the normal crowns. If the five most bound oxygens in *75* are used in calculating the plane, the average deviation drops to 0.27 Å. The complex in which the guest fills the open jaws of the host, $D[(OEOEO)_2E]_2 \cdot PF_6H_3N(CH_2)_4NH_3PF_6$ (*24*), gives a value of 0.36 Å. The detailed structure of this bidentate complex (*24*) appears constrained by an imperfectly complementary relationship of host and guest. The tetramethylene bridge does not appear long enough to permit as effective a centering of the NH_3^+ groups in the faces of the crown rings as is observed in the monodentate systems. Therefore, the NH_3^+ groups cannot organize the oxygens as completely as in simpler complexes, such as $Nap(OEOEO)_2E \cdot M_3CNH_3ClO_4$ (*21*).

A second organizational question deals with the angle which the C—N bond makes with the normal to the best plane of the heteroatoms. Table 2 lists these values. In complexes $Nap(OEOEO)_2 \cdot M_3CNH_3ClO_4$, $Pyr(CH_2OEOE)_2O \cdot M_3CNH_3 \cdot ClO_4$, $BCO_2^-(CH_2OEOE)_2O \cdot M_3CNH_3^+$, and $M_2D(OEOEO)_2E \cdot M_3CNH_3ClO_4$ angles are 1°, 1°, 3°, and 2°, respectively. The large value of 26° that is observed in $D(OEOEO)_2D \cdot C_6H_5CH(CO_2CH_3)NH_3PF_6$ reflects the noncomplementary character of host and guest in this complex [33]. In $(MBOM)_3(CH_2OE)_2O \cdot M_3CNH_3ClO_4$, 12° is observed for the best six-oxygen plane and 3° for the best five-oxygen plane. The effect of the less than perfect length of the tetramethylene bridge in the guest of bidentate complex $D[OEOEO)_2E]_2 \cdot PF_6H_3N(CH_2)_4NH_3PF_6$ (*24*) is visible in the value of 28° for this angle. A value of 6° is observed for the angle between the N—N bond and the normal to the best oxygen plane in $E(OEOEO)_2E \cdot H_2NNH_3ClO_4$ (*28*). As mentioned earlier, this slight tilt seems associated with the hydrogen bonds from the NH_2 group to the ring oxygens.

A third organizational feature of the complexes is the tendency for the guests to assume a staggered conformation about their C—N bonds (N—N bond in *28*). Table 2 lists values for the average absolute deviation of the C—C—N—H torsion angles from the ideal of 60°. For the eight complexes, these deviations varied between 0° and 12°, the average being $\approx 7°$. The complex "designed not to form", $D(OEOEO)_2D \cdot C_6H_5CH(CO_2M)NH_3PF_6$ (*39*), exhibited the highest deviation, 12°.

A fourth structural feature is the binding of RNH_3^+ to the six oxygens of the host. Table 2 provides the N—H $\cdots$ O bond angles, the H $\cdots$ O distances, and the N $\cdots$ O distances, both directly and through the hydrogen bond.

The N—H $\cdots$ O bond angles range from a high of 175° to a low of 104° for the 26 listed, the average being $157 \pm 20°$. If the bifurcated hydrogen bonds are omitted, the average of the remaining 22 angles is $163 \pm 13°$. The average of the H $\cdots$ O distances, not including the bifurcated bonds, is 2.1 ± 0.1 Å. The shortest N $\cdots$ O and H $\cdots$ O distances are in $^+N—H \cdots O^-$ of complex $BCO_2(CH_2OEOE)_2O \cdot M_3CNH_3^+$ (*71*), the shortness being a consequence of the charge on the acceptor oxygen atom. In general, the longer the H $\cdots$ O distance, the greater the deviation of the N—H $\cdots$ O bond angle from 180°.

The most linear hydrogen bonds are those of complexes $Pyr(CH_2OEOE)_2O \cdot M_3CNH_3ClO_4$ (*35*) and $Nap(OEOEO)_2E \cdot M_3CNH_3ClO_4$, averaging 174° and 172°, respectively. As molecular models of complex *35* suggest, substitution of a 2,6-disubstituted pyrido group for a CH_2OCH_2 group in a crown does little to disturb the structure of the complex.

The two kinds of N ⋯ O interactions, those that are along or near the hydrogen bond directions and those that are between the hydrogen bonds (omitting the bifurcated bonds) provide another interesting indicator of the binding interactions of host and guest. The 26 N ⋯ O distances of Table 2 that involve the hydrogen bond average 2.94 ± 0.10 Å. The 22 direct interactions average 3.10 ± 0,21 Å in length. Clearly those attractions acting through the hydrogen bonds produce substantially shorter distances on the average than those that act directly. In most individual comparisons within a given complex, the same thing is true. The hydrogen bonded distances are shorter because all but *28* are perching complexes.

Table 2 lists the distances between the N^+ of the guest and the best plane of the oxygens (or heteroatoms) of the host for eight complexes. This distance provides a rough measure of the perching vs. nesting character of the NH_3^+ groups. These distances in Å increase in the following order: $E(OEOEO)_2E \cdot H_2NNH_3ClO_4$, 0,11; $BCO_2^-(CH_2OEOE)_2O \cdot M_3CNH_3^+$, 0,98; $Nap(OEOEO)_2E \cdot M_3CNH_3ClO_4$, 0.99; $D(OEOEO)_2D \cdot C_6H_5CH(CO_2M)NH_3PF_6$, 1.11; $Pyr(CH_2OEOE)_2O$ $\cdot M_3CNH_3ClO_4$, 1.14; $M_2D(OEOEO)_2E \cdot M_3CNH_3ClO_4$, 1.22; $D[(OEOEO)_2E]_2$ $\cdot PF_6NH_3(CH_2)_4NH_3PF_6$, 1.26; $(MBOM)_3(CH_2OE)_2O \cdot M_3CNH_3ClO_4$, 1.37. In complex $E(OEOEO)_2E \cdot H_2NNH_3ClO_4$, the guest nests in the cavity to provide a N^+ to plane distance about 10% as long as those for the other seven complexes, which are all perching. The largest values are observed when the complexes involve bifurcated hydrogen bonds. The N^+ to plane distances do not appear to correlate either with the sums of the N ⋯ O distances, or with complex stabilities (see sect. 6).

Another interesting question is, "What is the geometry of the interactions at the oxygen binding sites?" We have analyzed data in terms of the tetrahedral and trigonal directions implicit in each observed C—O—C group, as we and others have done earlier [31,41,49,50]. The tetrahedral directions are defined as those that make an angle of 109.5° with each other and equal angles with the two O—C bonds. The trigonal direction lies in the plane of the C—O—C group, directed away from the two O—C bonds and making equal angles with them. Table 2 lists the values for eight complexes of the angles between the O ⋯ N vector and the closet tetrahedral direction, and the angles between the O ⋯ N vector and the trigonal direction. The average difference in angle from the tetrahedral direction for the sites that involve hydrogen bonds is 24 ± 13°. The average difference in angle from the tetrahedral for the sites not involving hydrogen bonds is 12 ± 15°.

The bias toward a tetrahedral direction when the oxygens are attached to only aliphatic carbons is greater than when the oxygens are attached to one aryl and one aliphatic carbon. The differences in direction from tetrahedral in the eight complexes are as follows: for $(CH_2)_2O \cdots HN^+$, 19 ± 7°; for $(CH_2)_2O \cdots N^+$, 6 ± 3°; for $CH_2(Ar)O \cdots HN^+$, 33 ± 12°; for $CH_2(Ar)O \cdots N^+$, 32 ± 17°. The differences in direction from trigonal are: for $(CH_2)_2O \cdots HN^+$, 36 ± 7°; for $(CH_2)_2O \cdots N^+$, 50 ± 4°; for $CH_2(Ar)O \cdots HN^+$, 22 ± 12°; for $CH_2(Ar)O \cdots N^+$, 31 ± 24°. Thus when the oxygens are attached to one aryl, the directions show essentially no average bias toward either tetrahedral or trigonal. This result suggests that some delocalization of the unshared electron pairs into attached aryl groups occurs in these complexes, accompanied by rehybridization at oxygen toward sp^2.

5.3 Metal Cations Bound to Oxygen

Two of the complexes discussed here contain KNCS as guest. In the "closed jaws" complex $D[CH_2(OEOE)_2O]_2 \cdot KNCS$ (60), the K^+ ion is surrounded by ten oxygens. Although the K^+ perches on each macroring, the complex overall is capsular. The ten oxygens all turn their electron pairs toward the K^+ ion and the $K^+ \cdots O$ lines lie, on the average, closer to a tetrahedral direction for an electron pair on each oxygen ($20°$) than to a trigonal direction ($36°$). The $K^+ \cdots O$ distances are 2.90, 2.78, 2.81, 2.92, 3.40, 2.83, 2.76, 2.95, 2.87, and 3.30 Å to provide an average distance of 2.95 ± 0.22 Å. This average for K^+ bound to ten oxygens is larger than the average of 2.80 ± 0.03 Å for the $K^+ \cdots O$ distance in $E(OEOEO)_2E \cdot KNCS$. This difference is not surprising, since enveloping K^+ with ten contacting oxygens and their support structures is sterically much more demanding than forming a nest for K^+ with six oxygens and their support structure. The average of the eight shortest $K^+ \ldots O$ distances in (60) is 2.85 ± 0.07 Å. In 60, the closest approach of any NCS^- to K^+ is 6.18 Å. Thus the anion is not liganded to the cation [43].

In the "strapped complex" $O(EOEOCH_2)_2D(OEOEO)_2E \cdot KNCS$ (67), the K^+ is surrounded on all but one side by oxygens (seven in all). These generally turn their electron pairs toward the K^+ to provide a deeply nesting complex. The $K^+ \cdots O$ lines again lie somewhat closer to a tetrahedral direction for an electron pair on each oxygen ($23°$) than to a trigonal direction ($37°$). The $K^+ \cdots O$ distances are 3.04, 2.70, 2.87, 2.76, 2.91, 2.74, and 3.02 Å, to give an average of 2.86 ± 0.14 Å. This average is closer to that for $E(OEOEO)_2E \cdot KNCS$ with six oxygens (2.80 ± 0.03 Å) than that for $D[CH_2(OEOE)_2O]_2 \cdot KNCS$ with ten oxygens (2.95 ± 0.22 Å), as might be expected on steric grounds. In 67 the N of the NCS^- anion is liganded to the K^+ at a distance of 2.76 Å. It is on the same side of the K^+ as the binding oxygen of the belt, opposite to the six-oxygen ring. The axis of the NCS^- anion of 67 is $12°$ from the normal to the best oxygen plane of the binding macroring [43]. This orientation is significant because it is the only model available for carbanion orientation in reactions of $K^+CR_3^-$ (carbonide salts) that are catalyzed by chiral hosts.

The six $Li^+ \cdots O$ distances in lithiospherium chloride $(MBOM)_6 \cdot LiCl$ (80) are all 2.14 Å. Interestingly, the $Li^+ \cdots O$ lines lie much nearer to a trigonal direction for an electron pair on each oxygen ($9°$) than is observed for either the $N^+ \cdots O$ or $K^+ \cdots O$ lines [43]. This orientation may, however, be merely an artifact of the many forces that compose this capsular complex, and not reflect an effect intrinsic to the $Li^+ \cdots O$ interaction.

5.4 Heavy-Atom Distances to Binding Sites

The complexes $D[CH_2(OEOE)_2O]_2 \cdot KNCS$ (60), $O(EOEOCH_2)_2D(OEOEO)_2E \cdot KNCS$ (67) and $E(OEOEO)_2E \cdot KNCS$ taken together provide a total of 23 oxygen binding sites for K^+ in hosts with quite different support structures, steric requirements, and counterion placements. The average $K^+ \cdots O$ distance is 2.81 ± 0.11 Å, and the extremes are 2.70 and 3.40 Å [43]. Complexes $Nap(OEOEO)_2E \cdot M_3CNH_3ClO_4$ (21), $Pyr(CH_2OEOE)_2O \cdot M_3CNH_3ClO_4$ (35), $M_2D(OEOEO)_2E \cdot M_3CNH_3ClO_4$

(57), (MBOM)$_3$(CH$_2$OE)$_2$O · M$_3$CNH$_3$ClO$_4$ (75), D(OEOEO)$_2$D · C$_6$H$_5$CH(CO$_2$M) · NH$_3$PF$_6$ (39), and D[(OEOEO)$_2$E]$_2$ · PF$_6$H$_3$N(CH$_2$)$_4$NH$_3$PF$_6$ (24) taken together have 41 N$^+$ ··· O sites involved in binding which provide an average N$^+$ ··· O distance of 3.02 ± 0.09 Å, with extremes of 2.84 and 3.66 Å (Table 2). The average of the eight shortest host-guest C ··· O interactions in the nonsalt complex E(OEOEO)$_2$E · CH$_3$O$_2$CC=CCO$_2$CH$_3$ · E(OEOEO)$_2$E (45) is 3.37 ± 0.13 Å, with extremes of 3.08 and 3.55 Å [35]. The six Li$^+$ ··· O distances in (MBOM)$_6$ · LiCl (80) are all the same, 2.14 Å [43].

5.5 Bond and Torsion Angles to Oxygens in Complexes

Aryloxy units have been extensively used to bind and shape complexes, and one structural parameter that has shown some systematic variation is the Ar—O—C bond angle. The free hosts and their complexes provide 18 values for the Nap—O—CH$_2$ bond angle, with an average of 117.4 ± 3.8° and a range of 111 to 128°. This average compares well with the usual Ar—O—CH$_3$ angle of 117–118°, and is appreciably greater than the average C—O—C angle for the CH$_2$—O—CH$_2$ parts of ten different related crown compounds of 112.6° [51], similar to the average CH$_2$—O—CH$_2$ angle in the present compounds.

The 18 B—O—CH$_3$ bond angles in the spherand, hemispherand, and their complexes provide an average of 113.4 ± 1.8° with extremes of 111.6° and 117.4° [43]. This average is significantly lower than that for the Nap—O—CH$_2$ bond angles. In molecular models of spherand (MBOM)$_6$ and its complex (MBOM)$_6$ · LiCl, as well as in their crystal structures, the methyls of the OCH$_3$ groups in an apparent geared arrangement barely contact one another, the shortest H ... H distance being 2.4 Å. Thus intramolecular steric interactions probably do not affect the B—O—CH$_3$ bond angles. Either packing interactions or rehybridization effects at oxygen to undo interoxygen electron-electron repulsion might be responsible.

The tendency of complexed crowns to assume gauche (synclinal) O—C—C—O and anti (antiperiplanar) C—O—C—C conformations is well-documented [7,23,24,51]. The same tendencies are exhibited in the complexed "part crown" hosts of this paper. Only a few of the more interesting examples of the O—C—C—O torsion angles are discussed here. The averages quoted are those of absolute values. The standard complex, Nap(OEOEO)$_2$E · M$_3$CNH$_3$ClO$_4$ (21) has five gauche torsion angles that average 66° [27]. The pyridine-containing complex Pyr(CH$_2$OEOE)$_2$O · M$_3$CNH$_3$ClO$_4$ (35) has four gauche torsion angles averaging 62° [31]. The zwitterionic complex BCO$_2^-$(CH$_2$OEOE)$_2$O · M$_3$CNH$_3^+$ 71 has four gauche torsion angles averaging 67° [46]. The dinaphthyl-containing complex M$_2$D(OEOEO)$_2$E · M$_3$CNH$_3$ClO$_4$ (57) has five gauche torsion angles that average 62° [41]. The hemispherand complex (MBOM)$_3$(CH$_2$OE)$_2$O · M$_3$CNH$_3$ClO$_4$ (75) has two gauche torsion angles that average 70° [43]. The bidentate complex D[(OEOEO)$_2$E]$_2$ · PF$_6$H$_3$N(CH$_2$)$_4$NH$_3$PF$_6$ (24) has ten gauche torsion angles averaging 68° [29]. The "closed jaws" complex D[CH$_2$(OEOE)$_2$O]$_2$ · KNCS (61) has eight gauche torsion angles that average 63° [43]. The "strapped" complex O(EOEOCH$_2$)$_2$D(OEOEO)$_2$E · KCNS (67) has five gauche torsion angles in the binding macroring that average 64° [43].

These complexes involve a wide variety of rigid or semirigid units designed not to disturb greatly the general dimensions or geometries of the crown parts of complexes. The overall average of these gauche torsion angles is $\approx 65°$. This compares with an overall average for the twelve gauche O—C—C—O torsion angles of $72°$ for the two complexes that involve [18]crown-6 as host, anemly, E(OEOEO)$_2$E · H$_2$NNH$_3$ClO$_4$ (28)[30] and E(OEOEO)$_2$E · CH$_3$O$_2$CC≡CCO$_2$CH$_3$ (45)[35]. The similarity between the overall averages of the torsion angles for complexes of [18]crown-6 and those of part crowns containing a variety of other units attests to the utility of the CPK molecular models that were used to select those other units that would least disturb the crown structure.

6 Correlation of Structure with Free Energies of Complexation

6.1 Extraction Methods of Determining Free Energies of Complexation

Scales provide a basis for correlation of structure with free energies of complexation of salt guests with hosts of widely varying structure. Ideal scales should have various characteristics to take into account different properties of hosts, guests, and complexes. 1) Complexes differ in their stabilities by more than 16 kcal mol^{-1}. Scales should cover as wide a range as possible. 2) Salt guests are hydrophilic and hosts range from being highly lipophilic to somewhat hydrophilic. Scales should be applicable to a wide variety of solubility properties of hosts, guests, and complexes. 3) Many of the uses to which hosts are put involve guest transfer from water to organic media. Scales developed in organic media saturated with water more closely approximate laboratory applications than scales involving either water or methanol as the main medium or dry organic solvents. The naturally occurring ionophores transport guests through lipophilic media from aqueous phase to aqueous phase. The presence of water has been referred to as a disadvantage[52]. We consider it an advantage. 4) Frequently only small amounts of hosts are available. Scales should depend on analytical techniques that are as sensitive and simple to apply as possible.

Semiquantitative scales that largely satisfy the above criteria have been developed[12,19,42]. They involve distributing salt guests between D$_2$O and CDCl$_3$ in the absence and presence of hosts. The amounts of guest drawn into the organic layer by the lipophilic host are determined by UV or ^{1}H NMR spectroscopy. The results provide association constants (Ka) defined in equation 1. The free energies of complexation ($-\Delta G°$) are defined in equation 2, and apply to CDCl$_3$ saturated with D$_2$O at about 25 °C. The picrate salts of Li$^+$, Na$^+$, K$^+$, Rb$^+$, Cs$^+$, NH$_4^+$, CH$_3$NH$_3^+$, and (CH$_3$)$_3$CNH$_3^+$ have served as standard guests for development of *picrate extraction scales*[42]. Scales were developed initially that depended on extractions of (CH$_3$)$_3$CNH$_3^+$X$^-$ where X$^-$ = SCN$^-$, ClO$_4^-$, Cl$^-$, or Pic$^-$. The $-\Delta G°$ values of these scales are linearly related[19], and those for the SCN$^-$ and Pic$^-$ are the same. The $-\Delta G°$ values (kcal mol^{-1}) quoted in the next sections are accurate enough only for the "coarse-grained" comparisons made there.

$$\text{host} + \text{guest} \cdot (\text{H}_2\text{O})_m^+ \cdot \text{X}(\text{H}_2\text{O})_n^- \overset{Ka}{\rightleftarrows} (\text{host} \cdot \text{guest})^+ \cdot \text{X}(\text{H}_2\text{O})_n^- + m\text{H}_2\text{O} \qquad (1)$$

$$-\Delta G° = RT \ln Ka . \qquad (2)$$

6.2 Ring Organizational Effects in Hosts

Below each formula in this section is listed the compound number, followed by the $-\Delta G°$ value in kcal mol^{-1} at ~25 °C in CDCl$_3$ saturated with D$_2$O for the host binding $(CH_3)_3CNH_3^+SCN^-$. Unlike the metal salts, this guest interacts even with the less lipophilic hosts such as [18]crown-6 to form complexes distributed mainly in the CDCl$_3$ layer, making determinations possible [12].

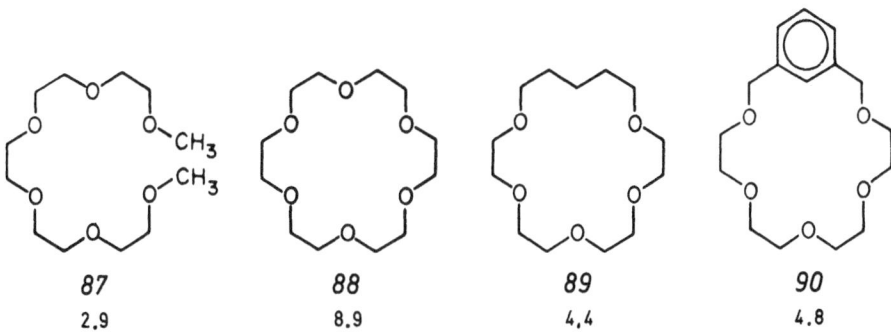

87	88	89	90
2.9	8.9	4.4	4.8

Open-chain polyether *87* and cycle *88* differ in molecular composition only by two hydrogen atoms and yet they differ in binding free energies by 6 kcal mol^{-1}. Although the binding sites are collected in *87*, they are both collected and partially organized for a tripod type of binding in *88*. Molecular models (CPK) of hosts *89* and *90* can be organized similarly to *88* but they lack one oxygen. The structural changes reduce the binding power from that of *88* by 4.5 kcal mol^{-1} for *89*, and by 4.1 kcal mol^{-1} for *90* [12].

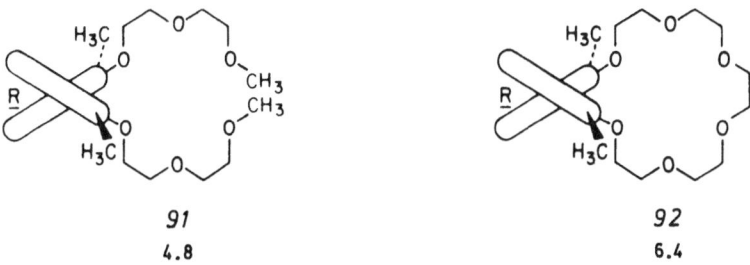

91	92
4.8	6.4

Similarly, *91* and *92* differ in molecular composition by two hydrogen atoms, but differ in binding free energies by 1.6 kcal mol^{-1}. Lack of a larger difference between cycle *92* and open-chain analogue *91* is attributed to the conformational constraints that the rigid 3,3′-dimethyl-1,1′-dinaphthyl unit imposes on its attached polyethyleneoxy "arms". Host *91* has many less possible conformations than host *87*. Although many of the naturally occurring ionophores are open-chain, they are largely composed of rigid units that limit the numbers of conformations of the whole structure [42].

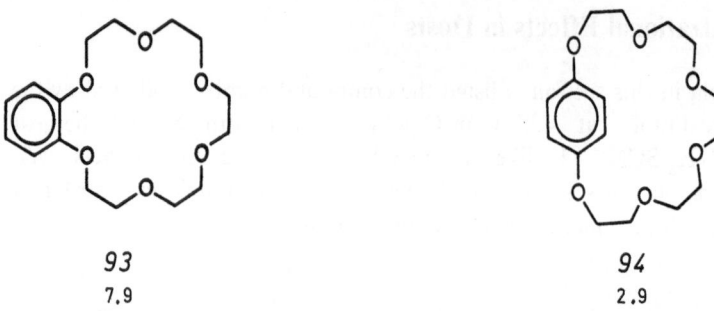

93
7.9

94
2.9

Comparisons of the isomeric hosts *93* and *94* illustrate the importance of placement of binding sites in cycles so they can act cooperatively. In *93* all six oxygens can bind simultaneously $M_3CNH_3^+$, and $-\Delta G°$ is 7.9 kcal mol^{-1}. In *94* only three or four oxygens can bind at the same time and the binding free energy drops to 2.9 kcal mol^{-1}, a difference of 5.0 kcal mol^{-1} [12].

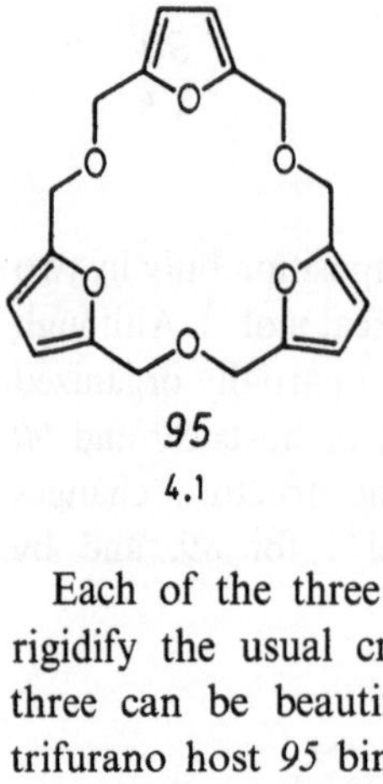

95
4.1

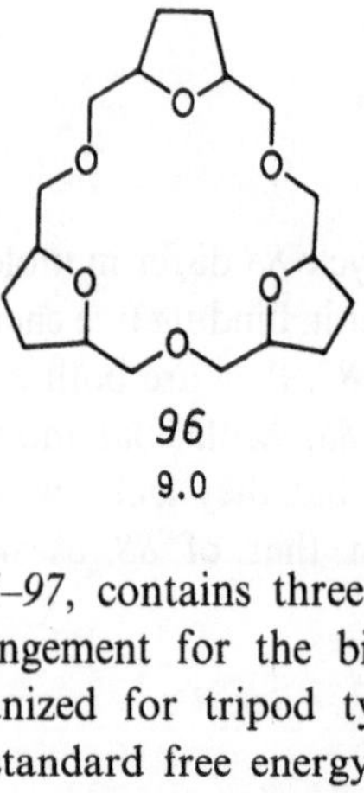

96
9.0

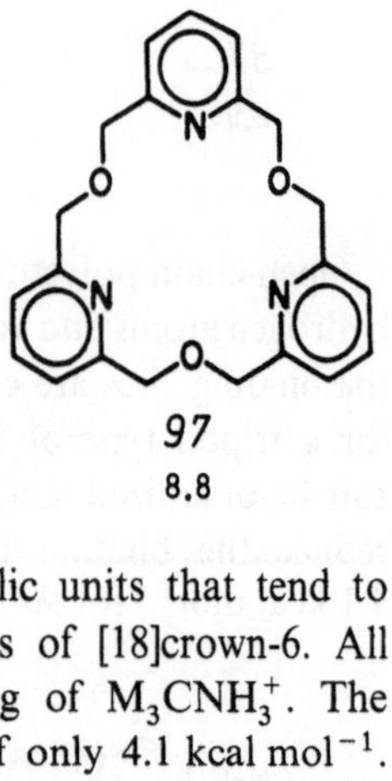

97
8.8

Each of the three hosts, *95–97*, contains three heterocyclic units that tend to rigidify the usual crown arrangement for the binding sites of [18]crown-6. All three can be beautifully organized for tripod type binding of $M_3CNH_3^+$. The trifurano host *95* binds by a standard free energy change of only 4.1 kcal mol^{-1}. The partial positive change on each of the three furano oxygens undoubtedly depresses its binding potential. When the three furano units are reduced to tetra-hydrofuran units as in *96*, the binding energy increases to 9.0 kcal mol^{-1}, essentially the same as that of [18]crown-6 (8.9 kcal mol^{-1}) [12]. Trispyrido host *97* provides the similar value of 8.8 kcal mol^{-1} [17]. The crystal structure of Pyr(CH$_2$OEOE)$_2$O · M$_3$CNH$_3$ClO$_4$ (*35*) [31] suggests that the complex of *97* contains three $^+$NH $\cdots$ N hydrogen bonds, rather than the alternative with three $^+$NH $\cdots$ O hydrogen bonds.

6.3 Dinaphthyl and Ditetralyl Units in Hosts

The binding free energies ($-\Delta G°$ in kcal mol^{-1}) for KPic and NH$_4$Pic complexed to hosts containing the 1,1′-dinaphthyl or 1,1′-ditetralyl units in CDCl$_3$ saturated with D$_2$O at 25 °C are listed below the formulas of the hosts. Included for comparison are standard [18]crown-6 hosts (*98* and *99*) made lipophilic with hydrocarbon substituents. The complexes of [18]crown-6 itself are too water-soluble for direct measurement.

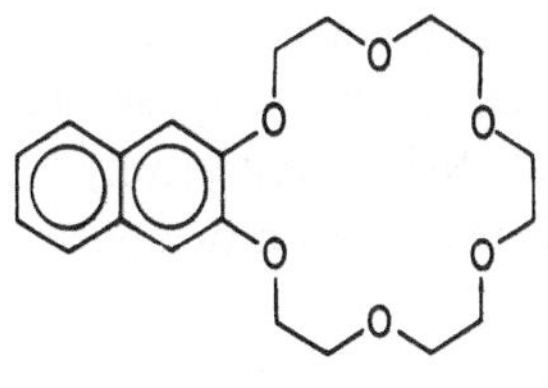

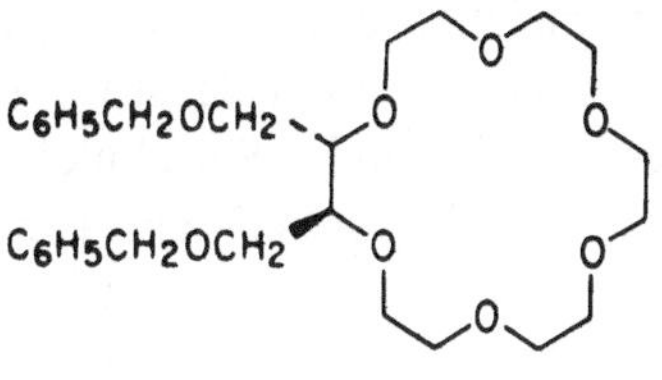

98

K$^+$, 10.8

NH$_4^+$, 9.5

99

11.4

10.1

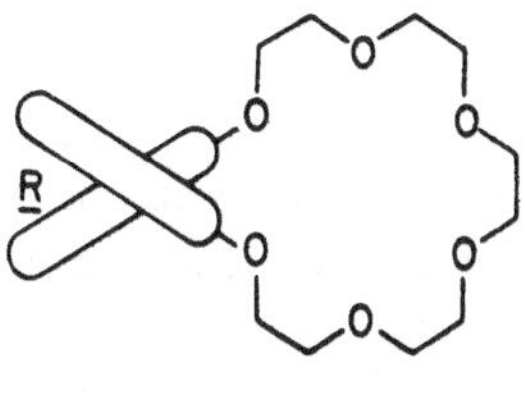

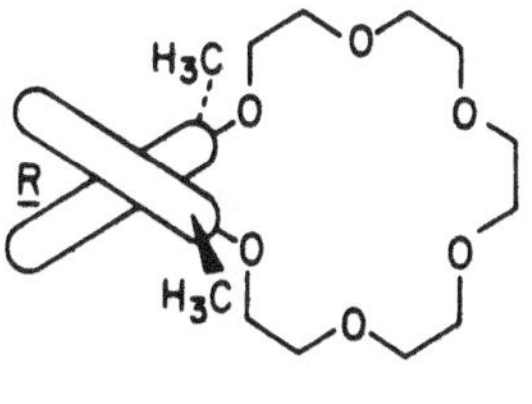

100

K$^+$, 8.8

NH$_4^+$, 7.8

92

10.4

8.9

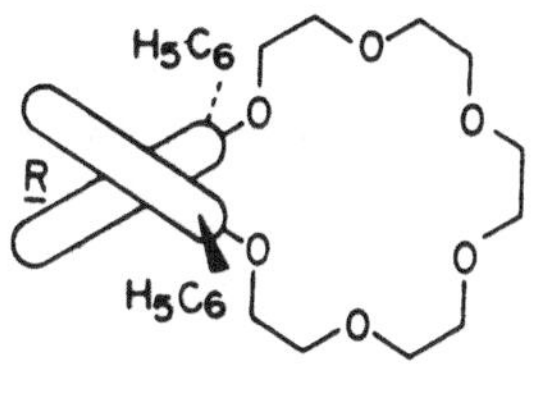

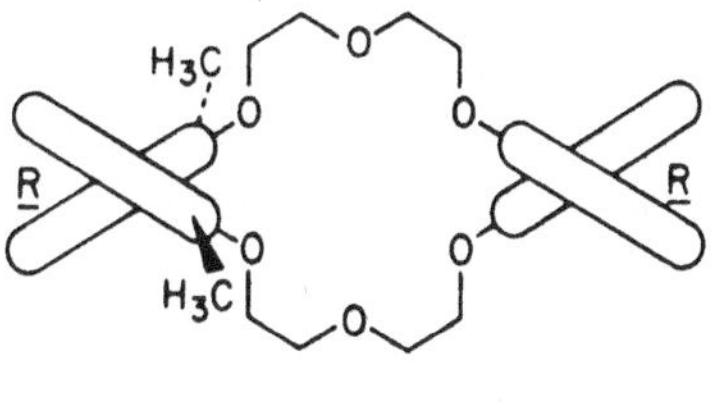

101

K$^+$, 9.1

NH$_4^+$, 8.0

102

8.1

7.4

The standard [18]crown-6 hosts *98* and *99* show the usual strong binding of the K$^+$ ion with $-\Delta G°$ values of 10.8 and 11.4 kcal mol^{-1}, respectively. The diameter of K$^+$ matches the diameters of the host cavities. The values for NH$_4^+$ are each about 1.3 kcal mol^{-1} below those for K$^+$ [42]. These [18]crown-6 cavities are well adapted to the tripod arrangement of hydrogen bonds (e.g., see crystal structure *21* [27]).

In hosts *92* and *100* to *102* containing dinaphthyl units, a similar pattern of

167

$-\Delta G^{\circ}$ values is maintained as in the [18]crown-6 hosts, except that the K^+ and NH_4^+ values are somewhat lower. Host *92*, with methyl groups in the 3,3′-positions, provides values of 10.4 and 8.9 kcal mol^{-1} for K^+ and NH_4^+, respectively [42], close to those of *98*, which also contains two aryloxy binding sites. With two phenyl groups in the 3,3′-positions, as in *101*, these values drop somewhat to 9.1 and 8.0, respectively [53]. With H's in the 3,3′-positions, as in *100*, the values drop further to 8.8 and 7.8 kcal mol^{-1}, respectively [42]. They drop further to 8.1 and 7.4 kcal mol^{-1}, respectively, when two dinaphthyl units are introduced [54] as in *102*.

The crystal structure of *92* and of its $M_3CNH_3^+$ complex [41] show that the CH_3 groups in the 3,3′-positions of the 1,1′-dinaphthyl unit limit the conformations available for the two ArOCH$_2$ groups in particular and force the two oxygens of the free host into conformations similar to those found in the complex. This enforced preorganization of two binding sites (which is due to the methyl groups) appears to be worth 1.1 to 1.6 kcal mol^{-1} in binding free energy. The two phenyl groups of *101* exhibit the same effect at a lower level. Molecular model (CPK) examination of complexes of *101* indicate that the phenyl groups are large enough to partially inhibit contact ion pairing between bound K^+ and picrate ion [53]. The same may be true of the methyl groups in *102*. Hosts *101* and *102* provide the highest chiral recognition yet observed in complexation of amino acids and esters [32, 53]. Hosts *92* and *102* provide the highest asymmetric induction observed for crowns (at the time of this writing) in the catalysis of organic reactions [55].

The binding powers of hosts *64*, *103*, *58*, *26*, *85*, and *104* illustrate the effects of introducing more binding sites into potential complexes [28]. In host *64*, one potentially binding oxygen is located in the $CH_2O(EO)_4CH_2$ bridge that extends between the 3,3′-positions of the dinaphthyl group. The crystal structure of the KNCS complex of *64* (see *67*) shows that this oxygen ligates the K^+, as does the NCS^- counterion [43]. This extra oxygen brings the binding power of *64* for K^+ and NH_4^+ to 11.1 and 9.4 kcal mol^{-1}, respectively. These values are slightly higher than those of either *98* or *92*, which also contain two ArO binding sites, and are comparable to those of *99*, which contains only aliphatic oxygen binding sites. Host *64* should have interesting properties as a chiral catalyst.

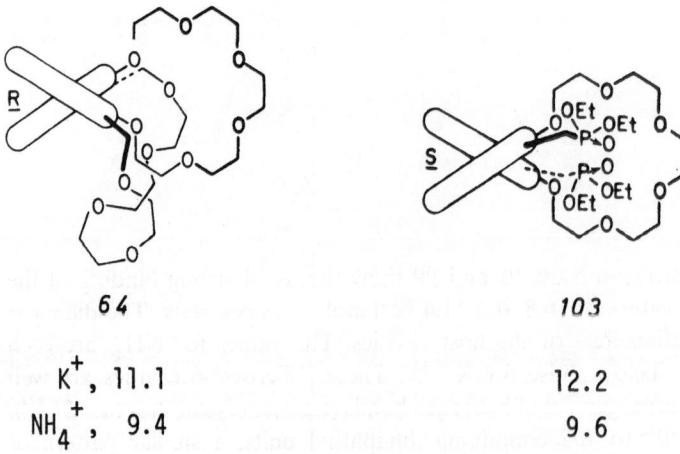

64	*103*
K^+, 11.1	12.2
NH_4^+, 9.4	9.6

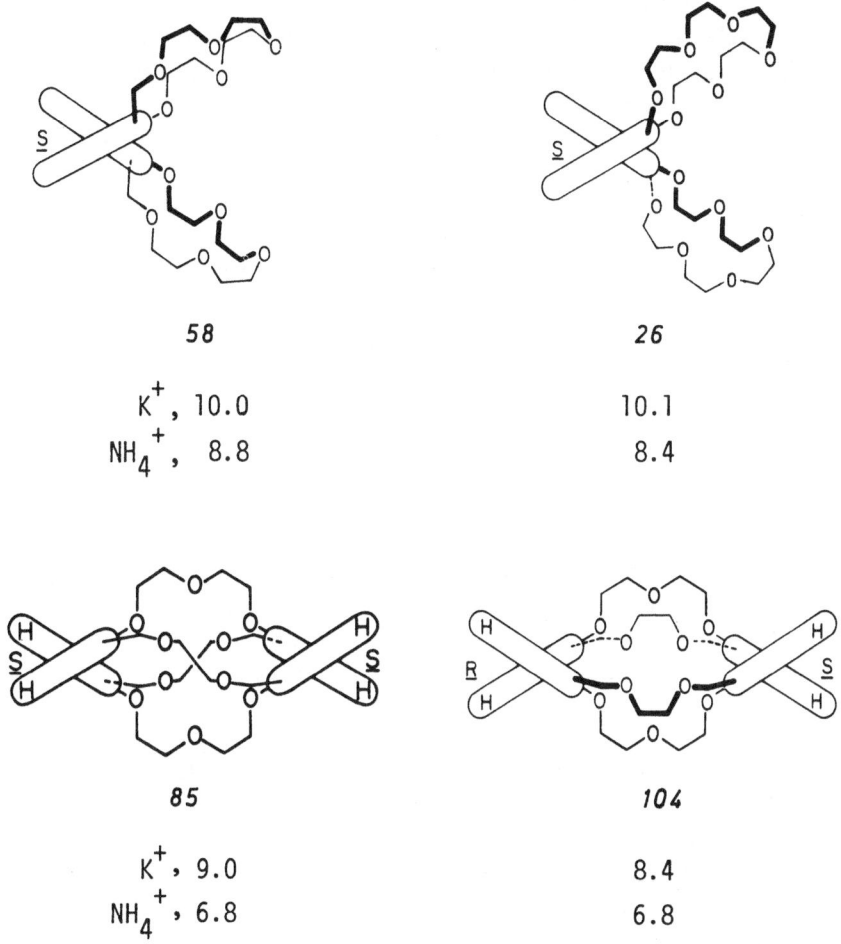

58

K$^+$, 10.0

NH$_4^+$, 8.8

26

10.1

8.4

85

K$^+$, 9.0

NH$_4^+$, 6.8

104

8.4

6.8

Host *103* contains a CH$_2$PO(OEt)$_2$ group in each of its 3,3'-positions. Molecular models (CPK) show that in reasonable conformations of a K$^+$ complex, the P → O oxygens can ligate the guest to give a hexagonal-bipyramidal arrangement involving eight oxygens. The crystal structure of a KNCS complex of *103* was recently solved [43]. Host' *103* is the most powerful binder of K$^+$ of the hosts yet examined that contain the 1,1'-dinaphthyl unit. The values are 12.2 and 9.6 kcal mol^{-1} for K$^+$ and NH$_4^+$, respectively. As expected the P → O oxygens ligate K$^+$, which can nest in the macroring. One oxygen might be hydrogen bonded to the NH$^+$ in the ammonium complex, and the other might have a direct N$^+$ ⋯ O interaction with no intervening proton.

The jaws type hosts, *58* and *26*, potentially provide ten or twelve oxygen ligands to a guest. The crystal structure of the complex of *58* with KNCS (see *61*) shows all ten oxygens to be ligated and the NCS$^-$ anions to be excluded [43]. In *61* the K$^+$ ion perches on each 16-membered macroring but is fully capsulated in a cavity defined by the ten oxygens. The free energies of complexation by *58* of K$^+$ and NH$_4^+$ are 10.0 and 8.8 kcal mol^{-1}, respectively, just below those of model crown *98*. The closing of the jaws on the guest requires considerable conformational organi-

169

zation. Host *26* gave values of only 10.1 and 8.4 kcal mol^{-1} for binding K^+ and NH_4^+, respectively. These values are slightly lower than those for model crown host *98*, which suggests that if the two macrorings are cooperating in binding these guests, the costs of conformational organization are high.

Diastereomeric, polycyclic hosts *85* and *104* involve two 1,1'-ditetralyl units between which are strung two OEOEO bridges in their 2,2'-positions, and two CH_2OEOCH_2 bridges in their 3,3'-positions. The letter H in the formulas denotes the tetralyl units. The crystal structure of *85* (see *86*) shows the cavity is filled with inward-turned CH_2 groups [48]. The binding power of the two complexes toward both ions is well below that of all the other hosts except *100*, whose value for K^+ is comparable. These facts suggest that *85* and *104* only form perching complexes, and the potential guests never reach the interior of the host, probably because the free energy costs of conformational reorganization are too high.

6.4 Hemispherands as Hosts

The cumulative effects of the partially enforced conformational organization of binding sites in the hemispherands is shown in the free energies of complexation of the series of hosts shown in Fig. 4 [5, 56]. The Li^+, Na^+, K^+, Rb^+, Cs^+, and NH_4^+ picrates served as guests. The standard crown host, *99*, is included for comparisons.

In passing from crown *99* to hemispherand *72*, 2,6-disubstituted 4-methylanisole units are introduced successively into the macrorings in positions adjacent to one another in place of the usual crown units. These four hosts all contain six oxygens in their macrorings, which are 18-membered. Introduction of the first anisyl unit to give *105* depresses the binding power as compared to the parent crown *99* for all six ions. The first anisyl unit appears to partially disrupt the organization of binding sites characteristic of the crown host. Introduction of the second anisyl unit as in *106* restores much of the binding power of the system, particularly toward the smaller ions. This increase is attributed to the ligand organization that the two attached anisyl units force on each other, which compensates for the loss in crown organization. Introduction of the third anisyl unit provides hemispherand *72*, whose binding power toward all of the guests but NH_4^+ substantially exceeds that of crown *99*. The crystal structure of *72* (see *73*) shows that the three methoxyl oxygens form a well-organized partial nest lined with unshared electron pairs, whereas the oxygens of the crown portion of the host turn their electron pairs away from this nest [48]. Thus the hemispherands are distinguished from the crowns by the presence in the former of at least three adjacent binding sites that are intrinsically organized for binding during synthesis and need not be organized by the guest. In hemispherand *107*, the organization of the free ligand system is further enforced by an extra EOE bridge whose oxygen, models suggest, has no choice but to turn its unshared electrons toward the nest. This extra bridge further increases the binding power of the system for the larger ions, in particular, K^+ and Rb^+ [56].

The patterns of binding free energies for different hosts with a given ion may be seen by scanning the columns of Fig. 4. Hemispherand *72* exceeds standard crown host *99* in binding power by the following $-\Delta(\Delta G°)$ values in kcal mol^{-1}: for Li^+, 0.9; for Na^+, 3.8; for K^+, 0.3; for Rb^+, 0.6; for Cs^+, 0.5. The two hosts are comparable binders of NH_4^+. Hemispherand *107* exceeds *99* by greater values: for Li^+, 1.0; for Na^+, 4.3; for K^+, 2.6; for Rb^+, 1.9; for Cs^+, 1.2; for NH_4^+, 0.9 kcal mol^{-1}. Thus, the hemispherands are substantially better general binders than the ordinary crowns of which *99* is among the best, particularly for the alkali metal cations. They also show a selectivity pattern which is different

	Li^+	Na^+	K^+	Rb^+	Cs^+	NH_4^+
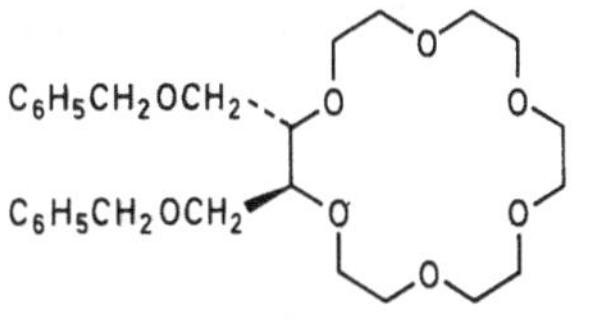 *99*	6.3	8.4	11.4	9.9	8.5	10.1
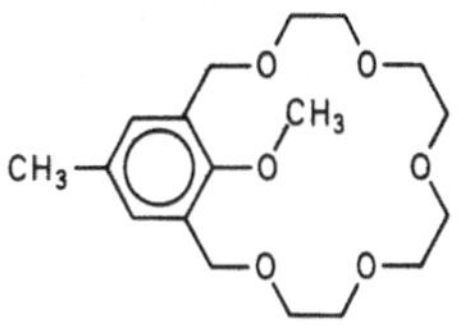 *105*	5.5	6.4	8.5	7.5	6.9	7.6
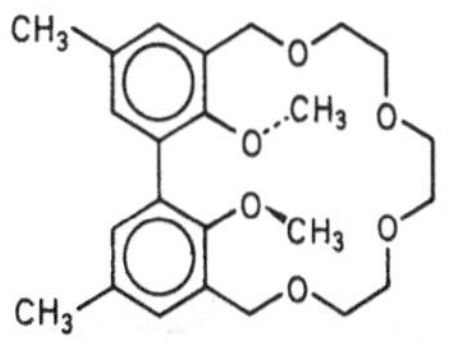 *106*	6.5	8.7	9.8	8.6	7.8	7.9
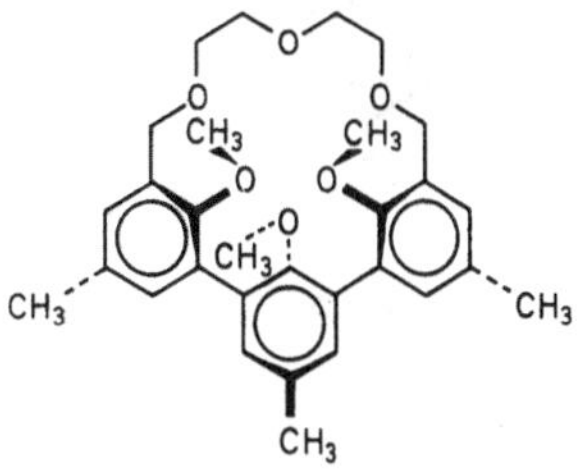 *72*	7.2	12.2	11.7	10.5	9.0	9.8
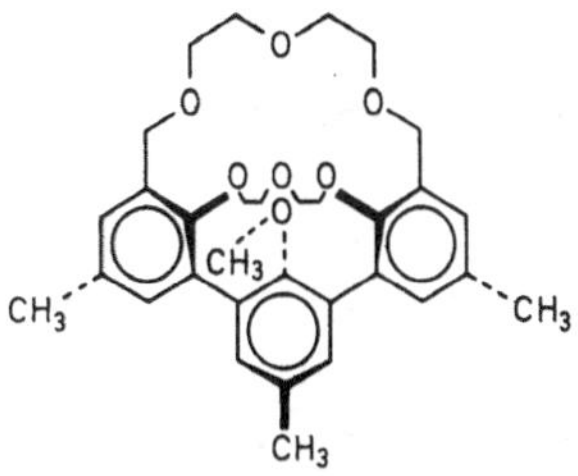 *107*	7.3	12.7	14.0	11.8	9.7	11.0

Fig. 4. Hemispherand and Model Complexation Free Energies ($-\Delta G°$, kcal mol^{-1} at 25 °C in CDCl$_3$ Saturated with D$_2$O) of Picrate Salts

171

from that of the [18]crown-6 crowns. The latter tend to bind K^+ considerably better than Na^+, whereas the hemispherands tend to bind Na^+ and K^+ almost equally well [5,56].

The hemispherands appear to owe their superior binding power to an enforced arrangement of at least three adjacent binding sites such that no conformational reorganization of them occurs during complexation. Molecular model examination suggests that many rigid binding units that resemble the anisyl in shape might substitute for one or more anisyl units in *72* or *107*. Such units include appropriately substituted methoxycyclohexane, pyridine oxide, cyclic urea, benzoquinone, cyclohexanone, cyclic sulfoxide, cyclic phosphine oxide, or cyclic amide. Similarly, appropriately substituted pyridine or tetrahydrofuran units might replace the central CH_2OCH_2 units in *72* or *107* without destroying the ability of the bridges to adapt to a spherical cavity.

Hemispherands *108* and *109* illustrate such substitutions. In *108* a cyclic urea unit has been substituted for the central anisyl unit of *72*. In *109* a pyridyl unit has been substituted for the central CH_2OCH_2 unit of *72*. Compounds *108* and *109* have been prepared and been found to exhibit the same patterns of superior binding properties characteristic of *72* [56,57].

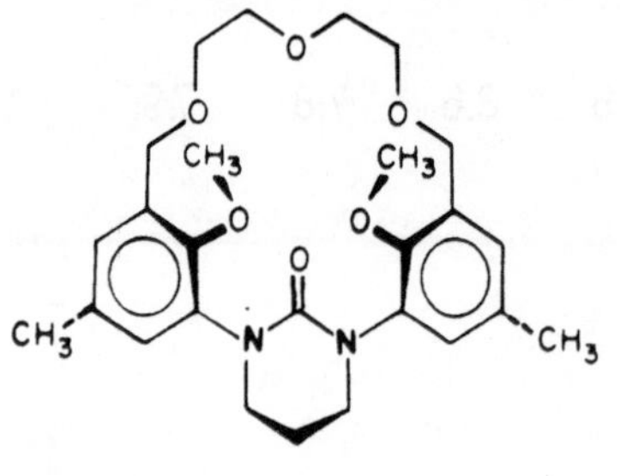

108

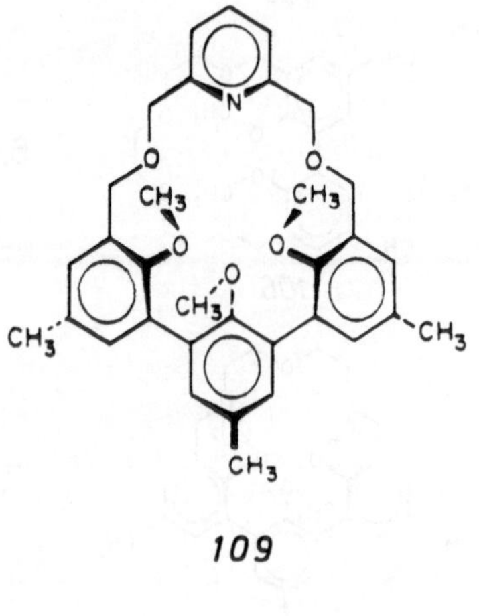

109

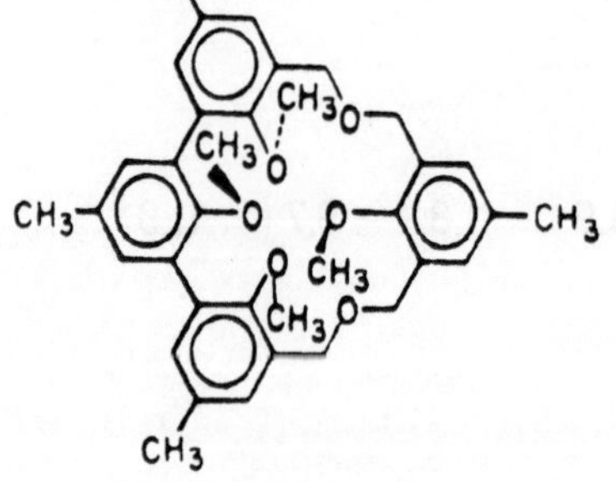

110

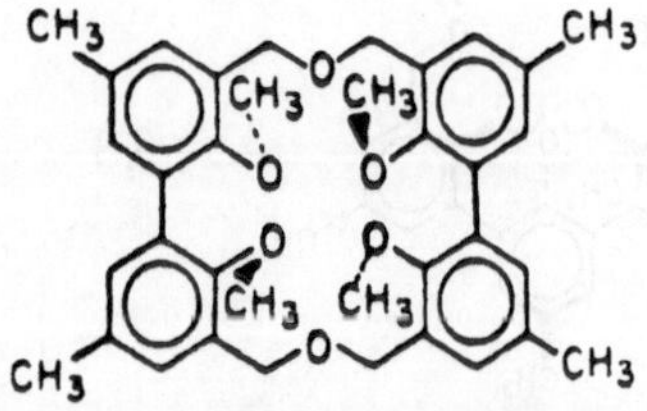

111

The relative positions of rigid and nonrigid units are critical in the design of hemispherands. For example, macrocycles *110* and *111* both contain four anisyl and two CH_2OCH_2 units incorporated into 18-membered rings. Both compounds are exceptionally poor hosts for the six ions of Fig. 4 [5]. Molecular models of *110* indicate that it possesses a conformation in which the CH_3 group of the bridging anisyl unit occupies the nest formed by the three contiguous anisyl groups. Conformational reorganization of *110* by a potential guest introduces more strain than is evident in models of hemispherand *72*. Molecular models of *111* indicate that although the four oxygens of the methoxyl groups form a nest, the two CH_2OCH_2 oxygens are too distant from the center of the cavity to cooperate in the binding of a potential guest. Thus the hemispherand concept appears to be general, but it must be applied with care to the design of hosts.

6.5 Spherands as Hosts

Because spherands are a relatively new class of compound and their syntheses are unconventional, the critical ring-closing steps in the syntheses of *79* [4] and *113* [58] are illustrated. This reaction is new and involves oxidation of aryllithiums with $Fe(AcAc)_3$ to give what are probably aryl radicals that couple. In the cyclization reaction, yields increase with high dilution and decrease in the presence of good H-donors, either in reactants, products, or solvents. The exothermicity of the aryl-aryl radical coupling explains why highly strained systems can be assembled with this reaction.

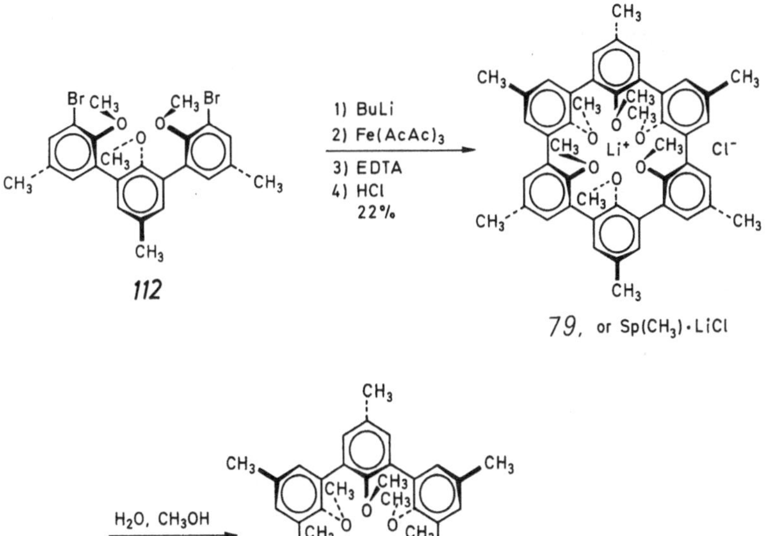

112

1) BuLi
2) Fe(AcAc)₃
3) EDTA
4) HCl
22%

79, or Sp(CH₃)·LiCl

H₂O, CH₃OH
150°   80%

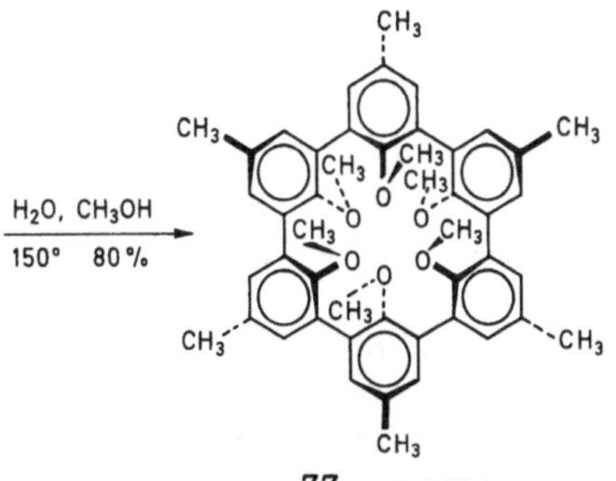

77, or Sp(CH₃)

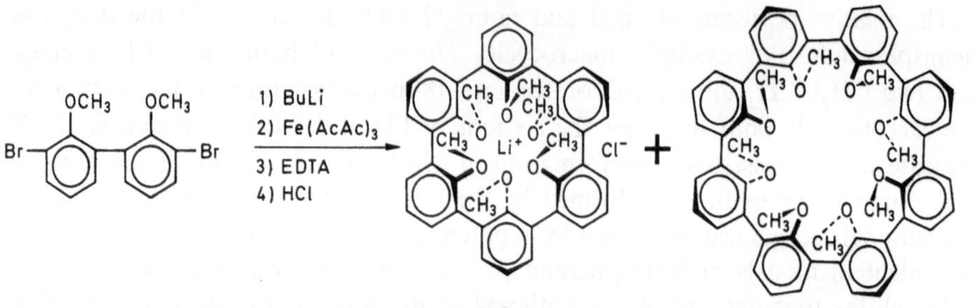

113, ~8 %   +   *114,* ~4 %

The smaller spherands always contain complexed Li$^+$, which can be removed by heating the complexed salt in water-methanol. The decomplexation reaction of *79* takes several days at 150 °C and appears to be driven by the crystallization of free spherand *77*. The overall yield of spherand *77* from p-cresol is 9.2 %[4].

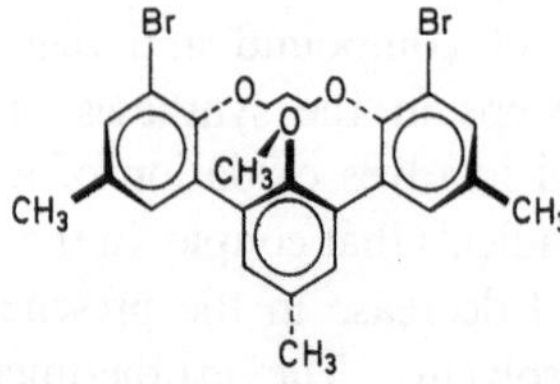

115

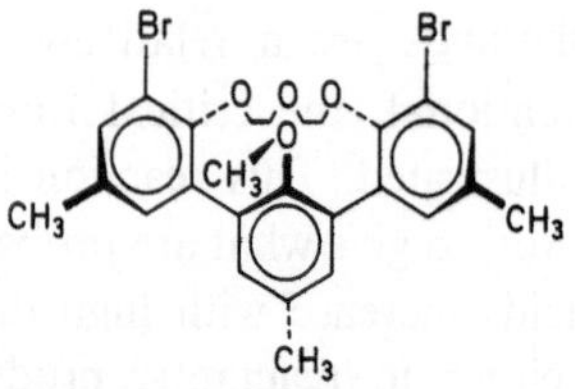

116

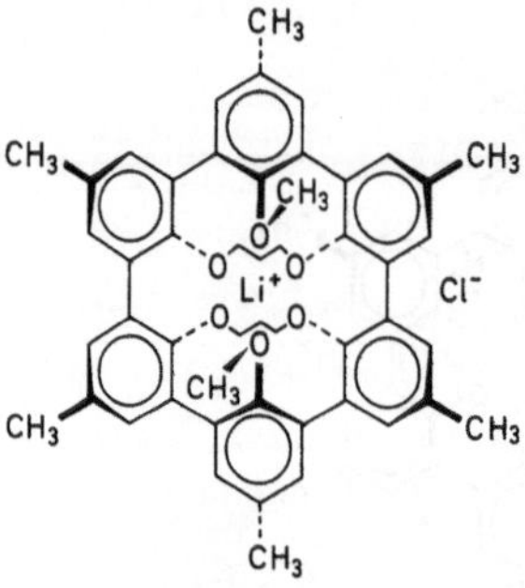

117

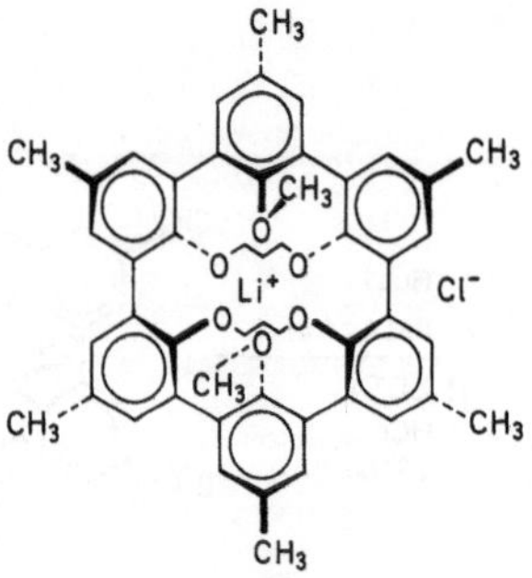

118

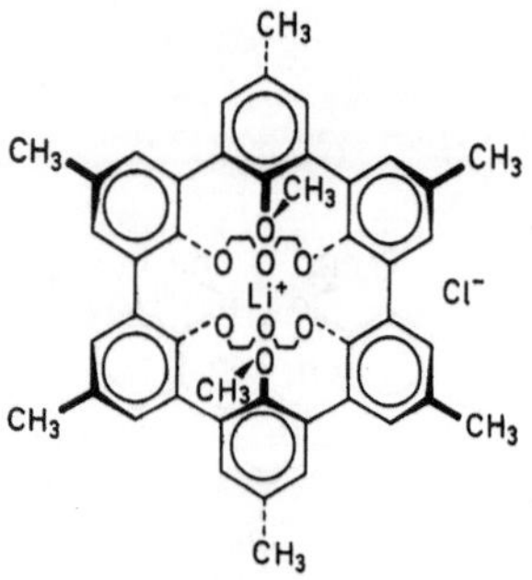

119

Dibromides *115* and *116* were ring closed by treatment with first butyllithium and then Fe(AcAc)$_3$ to give complexes *117* and *119*, respectively[4,59]. The structures for these complexes were established by crystal structure determination[43]. Complex *117* was erroneously assigned isomeric structure *118* when first isolated[4]. This error illustrates an important point. Structure *118* is similar to *79* in the sense that the alkyl groups attached to pseudo-ortho oxygens are oriented anti to one another. Although CPK molecular models of *118* can be assembled, models of *117* *cannot* because of interference between the two sets of pseudo-ortho oxygens attached to the two (CH$_2$)$_3$ bridges. For similar reasons, CPK models of *119* *cannot* be made, although its isomer with the two bridges on the opposite side is easy to assemble. In the preliminary crystal structures of *117* and *119*, these pseudo-ortho oxygen-to-oxygen distances are ≈2.51 Å and ≈2.56 Å, respectively, about 10 % less than the normal 2.80 Å. The complexes are severely strained in other ways as well. Thus although CPK molecular models provide a reasonable guide for making complexes from hosts and guests, they fail badly when complexes are assembled by making covalent bonds from high-energy starting materials. The reactions leading to *117* and *119* probably involve aryl to aryl radical coupling in the ring-closing step, and thus plenty of energy is available to make highly strained complexes. Probably the diradicals leading to isomers *117* and *119* are already complexed to Li$^+$, and the reactions are templated. Possibly the diradicals that would lead to the isomers of *117* and *119* bind Li$^+$ less well, and hence these isomers are not produced.

When heated in water-methanol at 100°, complexes *117* and *119* decomplexed to give the free hosts *120* and *121*, respectively. The driving forces for the decomplexations are the precipitations of the free spherands from the medium, in which they are insoluble.

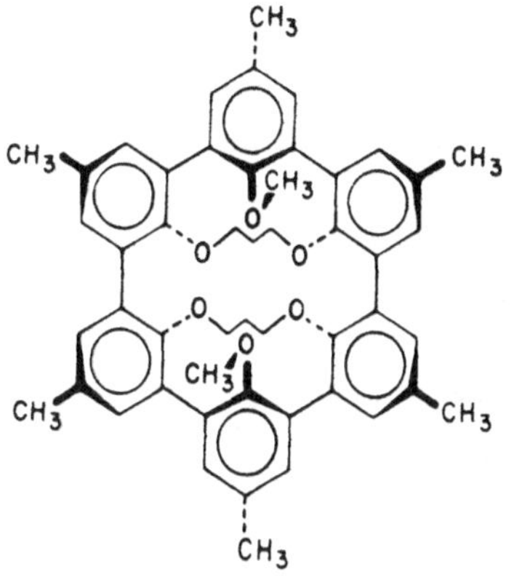

120, or Sp(CH₂)

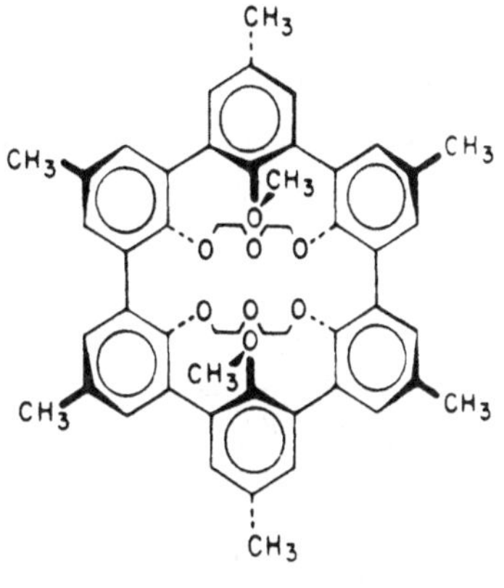

121, or Sp(EOE)

When heated to 200° in pyridine, complex *113* lost one methyl group to give the lithium phenoxide salt complex, which when heated in acid gave the free spherand *122*. This phenol when oxidized with $Tl(NO_3)_3$ gave spherand *123* which contains a benzoquinone unit in place of one anisole unit of the other spherands. Both phenolic spherand *122* and quinone spherand *123* strongly complex Li$^+$, but *122* much less so than 77 [58].

The ^{1}H NMR spectrum of each spherand and each metallospherium salt gives characteristic chemical shifts for the CH_3O, ArH, and CH_3Ar protons. Hosts and their complexes at ambient temperature do not equilibrate rapidly enough to affect one another's spectra. These characteristics greatly facilitate their study. Spherands *77*, *120*, and *121* have been the most studied [56]. In the following discussion, *77* will be indicated by the notation, Sp(CH$_3$); *120* by Sp(CH$_2$); and *121* by Sp(EOE).

Phenolic spherand *124* was prepared from Sp(CH$_3$) · LiCl *79*), and was heated with $(CH_3)_2SO_4$ and either LiOH, NaOH, KOH, Ca(OH)$_2$, or MgH$_2$. The product in CH_2Cl_2 solution was washed with aqueous HCl. The metal cations of the first two bases were encapsulated to give Sp(CH$_3$) · LiCl (*79*) and Sp(CH$_3$) · NaCl (*125*), respectively, but the metal ions of the other bases gave only free Sp(CH₃) mixed with traces of Sp(CH₃) · NaCl, the Na⁺ being scavenged from the other bases [4].

Low concentrations of Sp(CH₃) in CDCl₃, and of HClO₄, LiClO₄, NaClO₄, KClO₄, RbClO₄, CsClO₄, Mg(ClO₄)₂, CaBr₂, or their hydrates in (CD₃)₂SO were mixed to give homogeneous solutions of potential complexing partners in host-to-guest molar ratios of about 2.2 to 1. By the time ¹H NMR spectra could be taken, all Li⁺ and Na⁺ guest present was complexed. Even after extended times, none of the other cations were complexed [4]. Solutions of bridged spherands Sp(CH₂) (*120*) and Sp(EOE) (*121*) behaved similarly [56]. These three hosts are highly ion-selective. Of those studied, monovalent cations of diameters that range from 1.4 to 1.80 Å when in the same solution with these spherands, are rapidly complexed.

Unlike all the other types of complexes studied, those of *77*, *120*, and *121* dissolved in CDCl₃ or CH₂Cl₂ release their cations when shaken with water only after many hours, if at all. The free spherands and their complexes are completely insoluble in water, and the complexes very rapidly exchange their anions when their CDCl₃ or CH₂Cl₂ solutions are shaken with aqueous solutions of a variety of salts. They resemble lipophilic quaternary ammonium salts in their abilities to act as anion exchange agents, and to lipophilize anions [4].

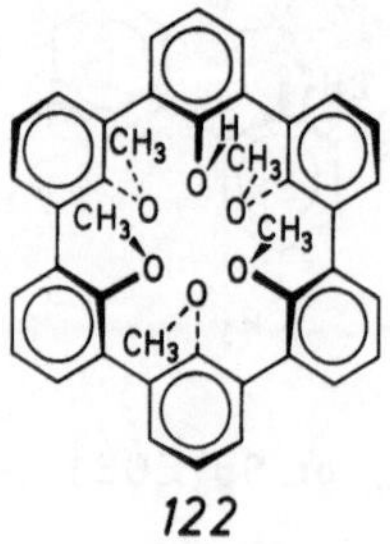

122

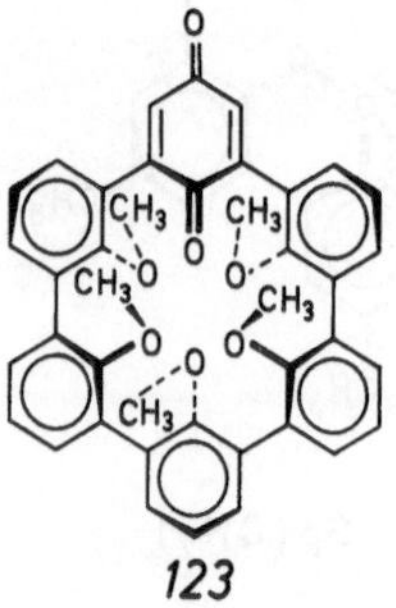

123

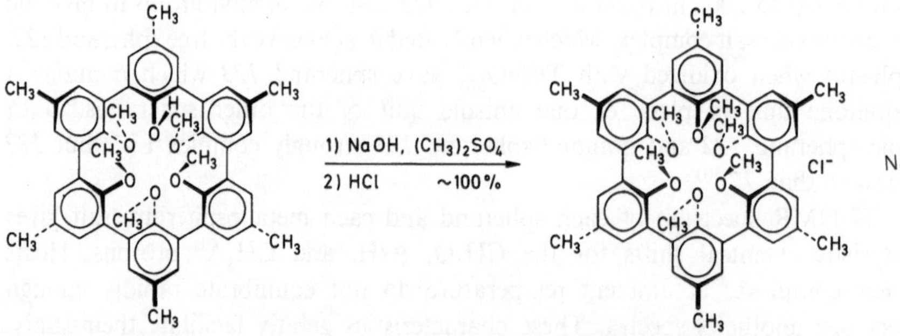

124

1) NaOH, (CH₃)₂SO₄
2) HCl   ~100%

125, or Sp(CH₃)·NaCl

When free spherands Sp(CH₃), Sp(CH₂), and Sp(EOE) in dilute CDCl₃ solution were shaken at 30 °C with 3–6 M solutions of HBr, LiBr, NaBr, KBr, RbBr, CsBr, MgBr₂, CaBr₂, or SrBr₂, the three hosts became completely complexed within hours only when the aqueous solutions contained LiBr or NaBr. With the other solutions, small amounts of Na⁺ present as an impurity were scavenged, but none of the other ions became complexed. When 0.0026 M solutions of the three spherands were shaken with equal volumes of 0.20 M solutions of LiCl or NaCl at 25 °C, equilibrium was reached with the LiCl solutions in about 700–800 hours, and with NaCl solutions in about 20–125 hours. When 0.0026 M solutions of the three LiCl and three NaCl complexes in CDCl$_3$ were shaken with equal volumes of D$_2$O at 25 °C, equilibrium was reached in 400–800 hours with the Li$^+$, and in 21–731 hours with the Na$^+$ complexes. From the concentrations of complex and free ligand present in each phase at equilibrium in these experiments, the differences in free energies have been estimated for each of the three spherands complexing LiCl and NaCl in D$_2$O-saturated CDCl$_3$ at 25 °C (Fig. 5) [58]. The host that showed the poorest complexing power toward each salt was in each case used as standard ($-\Delta(\Delta G^\circ) = 0$).

The picrate extraction scale for measuring the free energies of complexation has been applied to the widest variety of compounds. Values for $-\Delta G^\circ$ in kcal mol^{-1} for hosts complexing NaPic in CDCl$_3$ saturated with D$_2$O at 25 °C are listed in order of increasing binding power in Fig. 6. It proved feasible to determine the $-\Delta G^\circ$ values for spherand Sp(CH$_2$) (120) directly [56]. Values for Sp(CH$_3$) (77) and Sp(EOE) (121) proved to be too high for direct measurement, but were estimated as follows. The water solubility of [2.2.1]cryptand (126) (and its Na$^+$ complex) prevented its $-\Delta G^\circ$ value from being determined directly on the picrate extraction scale. Equilibration at 25 °C in CDCl$_3$ saturated with D$_2$O of NaPic between [2.2.1]cryptand and Sp(CH$_2$) favored the complexation constant of the cryptand over that of the spherand by a factor of $\sim$95. This value was used to estimate $-\Delta G^\circ$ for [2.2.1]cryptand on the picrate scale to be $\sim$16.3 kcal mol^{-1}. A similar equilibration was carried out with NaPic, [2.2.1]cryptand, and Sp(CH$_3$), and was found to favor the complexation constant of Sp(CH$_3$) over that of [2.2.1]cryptand by a factor of $\sim$100. Thus the $-\Delta G^\circ$ value for Sp(CH$_3$) is $\sim$19 kcal mol^{-1} [56]. Equilibration of NaPic, Sp(CH$_3$) (77), and Sp(EOE) (121) was too slow to observe. If the $-\Delta(\Delta G^\circ)$ values for these spherands are roughly anion-independent, the $-\Delta G^\circ$ value for Sp(EOE) complexing NaPic should be $\sim$3 kcal mol^{-1} higher than that for Sp(CH$_3$) (see Fig. 5), or $\sim$22 kcal mol^{-1}. Although the $-\Delta G^\circ$ values listed in Fig. 6 for the last three entries are only estimates, they differ by large enough amounts to provide informative correlations of structure with complexing power of hosts with Na$^+$ [56].

The striking generalization that correlates host structure with binding power is: *The larger the number of host ligating sites that are organized for maximum binding prior to complexation, the higher the free energy of complexation.* This generalization is intuitively obvious and is well-illustrated by the results listed in Fig. 6.

	$-\Delta(\Delta G^\circ)$ in kcal mol^{-1}		
	Sp(CH$_3$)	Sp(CH$_2$)	Sp(EOE)
LiCl	>10	1	0
NaCl	4	0	7

Fig. 5. Differences in complexation free energies in CDCl$_3$ saturated with D$_2$O at 25 °C

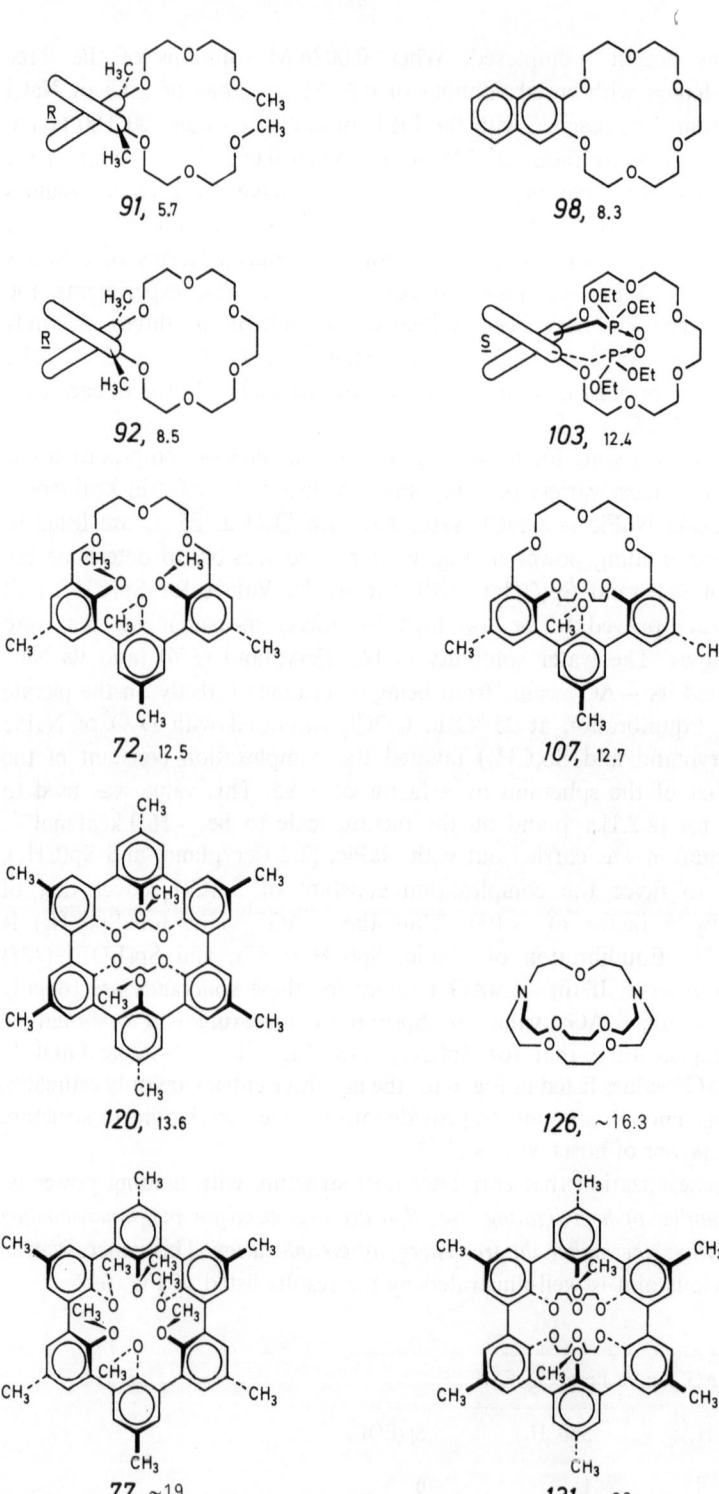

Fig. 6. Free energies of complexation ($-\Delta G°$, kcal mol^{-1}) of NaPic at 25 °C in CDCl$_3$ saturated with D$_2$O

Acyclic host *91* is at the bottom of the scale with a value of 5.7 kcal mol^{-1}. It is organized in the sense that six oxygens are collected by conformationally flexible spacer units *amenable to organization by a guest such as Na$^+$*. This host is disorganized in the sense that to bind Na$^+$, it must freeze out many conformational degrees of freedom and liberate the unshared electron pairs of the oxygens from attractive interactions with solvent and with the CH$_2$ groups of the spacer units.

In the three crown compounds *98*, *92*, and *103*, the $-\Delta G°$ values increase to 8.3, 8.5, and 12.4 kcal mol^{-1}, respectively. In these hosts, six oxygens are partially organized by their incorporation in a conformationally flexible macroring. They are disorganized for complexation in the sense that their unshared electron pairs are available for stabilizing interactions with solvent or CH$_2$ groups in some of the conformations possible for the macroring. The fact that the macroring is more organized than the open-chain system is worth about 3 kcal mol^{-1} in binding free energy. In crown *103*, two additional oxygen binding sites add almost 4 kcal mol^{-1} more in binding free energy. The bulk of the PO(OEt)$_2$ groups probably limits the number of conformations of the macroring that must be frozen out, and the number of solvent-electron pair interactions that must be destroyed during complexation.

In hemispherands *72* and *107*, the three CH$_3$O oxygens are rigidly organized for complexation with their unshared electron pairs well shielded from stabilizing interactions with solvents. However, the CH$_2$O(CH$_2$CH$_2$O)$_2$CH$_2$ bridge is still conformationally flexible. The three CH$_2$OCH$_2$ oxygens of this bridge can turn outward to interact with solvent, and the CH$_2$ groups can turn inward to lower the energy of the CH$_3$O oxygens through CH$_2 \cdots$ OCH$_3$ interactions. In *107* one additional, rigidly organized oxygen is present in the extra (shorter) bridge, and the conformational flexibility of the longer bridge remains. A study of molecular models of *107* and of the crystal structures of *72* (*73*) [48] and its M$_3$CNH$_3^+$ complex *75* [43] suggests that complexation is accompanied by introduction of considerable strain into host *107*, particularly with ions as small as Na$^+$. Should this strain be absent, *107* would have a higher $-\Delta G°$ value.

Spherand Sp(CH$_2$) (*120*) has a $-\Delta G°$ value of 13.6 kcal mol^{-1}, the lowest observed for the three spherands listed in Fig. 6. Molecular models (CPK) of it can be constructed only by shaving about 10% from the side of each of the four oxygens attached to the (CH$_2$)$_3$ bridges. This model is conformationally rigid in the sense that the unshared electron pairs of all six oxygens face inward toward the cavity. However, the oxygens are unevenly distributed around the cavity, with the four oxygens that terminate the bridges on one side and the two oxygens attached to the methyls on the other. Thus the *methyl* side is more open than the methylene side. Introduction of a sphere of 1.48 Å diameter into the modified model of Sp(CH$_2$) does not appear to further strain the system, but a sphere of 1.95 Å placed in the cavity spreads the oxygens and further deforms the support structure. This fact suggests Sp(CH$_2$) should bind Li$^+$ better than Na$^+$.

The structure of Sp(CH$_2$) · LiFeCl$_4$ (*127*) in its crystal almost contains a mirror plane passing through the Li$^+$ and the two CH$_3$O groups. Five of the Li$^+$ to O distances are between 2.0 and 2.1 Å, but one of the methoxy oxygens is 2.9 Å from the Li$^+$. The average Li$^+$ diameter based on the five oxygens is 1.26 Å; that based on all six is 1.54 Å. The best-plane dihedral angles about the Ar-Ar bonds are 28° and 34° for the two sets of aryls attached to the bridge, and are 46°, 51°,

51°, and 45° for the others (overall average of 42°). The aryls attached to the methoxy groups are folded around their O-Ar-CH$_3$ axes by values of 4° and 7°, and those attached to the bridges by 10°, 11°, 5°, and 10° (overall average of ~8°). The oxygens are bent out of the best planes of their attached aryls by 0.08 and 0.18 Å for the two also attached to CH$_3$, and by 0.17, 0.32, 0.21, and 0.26 Å for the four also attached to the bridges (overall average of 0.20 Å). The aryl carbons are bent out of the best planes of their attached aryls by an average of 0.29 Å, with extremes of 0.38 and 0.22 Å. The distances between the two sets of near oxygens that terminate the two trimethylene bridges are 2.50 and 2.52 Å, whereas the methoxy-to-methyleneoxy near distances are 2.68, 2.74, 2.72, and 2.72 Å. The oxygen-to-oxygen distances in the two OCH$_2$CH$_2$CH$_2$O bridges are 2.63 and 2.95 Å [43]. Thus seven of the eight near oxygen-to-oxygen distances in *127* are substantially below the normal van der Waals oxygen-to-oxygen contact distances of 2.80 Å, some of them very substantially so. The host in this lithiospherium complex is severely strained. The crystal structures of the free spherand Sp(CH$_2$) or of its sodium complex are not yet available. Structures *127* provide a stereoview of Sp(CH$_2$) · Li$^+$ (the FeCl$_4^-$ has been omitted) in which many of the deformations can be observed.

127

Host *126* is [2.2.1]cryptand, which is the strongest of the simple cryptands for binding Na$^+$ [6]. It has a $-\Delta G°$ value of ~16.3 kcal mol^{-1}, nearly halfway between the values for the two more weakly binding spherands [56]. Host *126* is composed of a 1,9-diaza[18]crown-6 containing a CH$_2$CH$_2$OCH$_2$CH$_2$ bridge that extends between the two nitrogens. This bridge adds a well-placed seventh binding site to the system, and eliminates many conformational degrees of freedom present in the crowns. Molecular models of *126* show that at any one time two of its methylene groups can turn inward to fill its own cavity, and interact with trans-annularly located oxygens, as in structure *52* of its [2.2.2]cryptand analogue [24]. Many of the electron pairs can also turn outward to interact with solvent.

Host Sp(CH$_3$) (*77*) provides a $-\Delta G°$ value for Na$^+$ of ~19 kcal mol^{-1} [56]. Molecular models suggest, and ^{1}H NMR spectra confirm that the six CH$_3$O groups are equivalent. The crystal structure *78* of the spherand shows the six oxygens to be octahedrally arranged around a near spherical hole lined with 24 electrons in an enforced conformation. The host appears to be free of any conformational degrees of freedom other than those of rotations of the methyl groups about their bonding axes to oxygen or carbon. The six methyl and six aryl groups shield the unshared electron pairs on oxygen from interactions with solvent, and the molecular organization eliminates all intramolecular van der Waals interactions between oxygens and any atoms other than the other oxygens. The hole diameter shrinks from 1.62 Å in the crystal structure of the free spherand to 1.48 Å in that of its LiCl complex (80) [43]. Complexation with LiCl is accompanied by relaxation of considerable bond angle and out-of-plane deformations induced in the free host by electron-electron repulsion between the neighboring oxygens.

The crystal structure of $Sp(CH_3) \cdot NaSO_4CH_3$ has been determined [43]. The molecule in the crystal has approximate $\bar{3}$ symmetry, and the oxygens possess a nearly octahedral arrangement around Na^+. The Na^+ diameter is 1.76 Å, considerably less than the 1.95 Å found in the crystal structure of benzo[15]crown-5 $\cdot H_2O \cdot NaI$ [60]. The average displacement of the six carbons of each benzene ring from their least squares plane is 0.02 Å in $Sp(CH_3) \cdot NaSO_4CH_3$, as compared to 0.03 Å in $Sp(CH_3)$. The average displacement of the attached aryl carbons from the best plane of each aryl ring is the same (0.16 Å) in the complex and free spherand. The distance of the oxygen from the best plane of its attached aryl group is 0.11 Å in the complex, as compared to 0.20 Å in the spherand. The angle of the aryl fold around the $O—Ar—CH_3$ axis is 4° in the complex and 6° in the spherand. The dihedral angle between adjacent aromatic rings is 60 ° in the complex and 52° in the free spherand. The distance of the pseudo-ortho oxygens from one another is 2.99 Å in the complex, as compared to 2.92 Å in the free spherand. Thus the strain inherent in spherand $Sp(CH_3)$ is redistributed somewhat upon complexation with Na^+ [43].

All of these effects are summarized by the statement that unlike any of the other hosts that have been designed and synthesized, the spherands are almost fully organized for complexation during their synthesis. No significant further organization need, or can, occur during their complexation. Aside from small bond angle and bond rotational adjustments induced in the host by complexation, the guest does not have to reorganize the host in order to complex. This preorganizational feature explains why spherand 77 with six binding sites is a more powerful binder than cryptand 126 or hemispherand 107 with seven binding sites, or augmented crown 103 with eight binding sites. Comparison of standard crowns 98 or 99 with six oxygen binding sites with spherand 77 with six oxygen binding sites indicates that the difference in organization of the two types of hosts is worth about 11 kcal mol^{-1} in binding Na^+. Much of this difference is attributed to the different states of pre-organization of the two types of hosts.

Spherand $Sp(EOE)$ (121) has a $-\Delta G°$ value of ~ 22 kcal mol^{-1} (Fig. 6), the highest that has been estimated for any host. Molecular models (CPK) of 121 can be constructed only by shaving about 10% from the sides of the four oxygens attached to the $CH_2CH_2OCH_2CH_2$ bridges. The model is conformationally rigid except for a very low mobility in the bridges. The six aryl oxygens are locked into a conformation in which the unshared electron pairs on oxygen are oriented toward the cavity, and the attached groups away from the cavity. The OCH_2CH_2 $\cdot OCH_2CH_2O$ bridges can contain only gauche conformations for their $O—C—C—O$

units. The resulting cavity in Sp(EOE) is lined with 32 electrons, but is ellipsoidal rather than spherical. A sphere of 1.95 Å diameter fits this cavity better than does one of 1.48 Å.

The structure of Sp(EOE) · LiCl (*119*) in the crystal also nearly contains a mirror plane. The seven oxygens ligated to Li^+ are at distances between 2.00 and 2.44 Å, to provide an average Li^+ diameter of 1.72 Å, considerably larger than the six-coordinate Li^+ of Sp(CH₃) · LiCl, which has an effective Li^+ diameter of 1.48 Å. The nonliganded methoxy oxygen is 3.47 Å from the Li^+. The best-plane dihedral angles about the Ar—Ar bonds are 40° and 44° for the two sets of aryls attached to the bridges, and are 50°, 46°, 59° and 60° for the others (overall average of 50°). The aryls attached to the methoxy groups are folded around their O—Ar—CH_3 axes by values of 2° and 4°, whereas those attached to the bridges are folded by 7°, 10°, 9° and 11° (overall average of ≈7°). The aryl oxygens are bent out of the best planes of their attached aryls by 0.05 Å and 0.14 Å for the two also attached to CH_3, and by 0.20, 0.34, 0.38, and 0.42 Å for the four also attached to the bridges (overall average of 0.27 Å). The aryl carbons are bent out of the best planes of their attached aryls by an average of 0.32 Å, with extremes of 0.06 and 0.53 Å, those attached to the rings joined to bridges being much more displaced. The distances between the two sets of near oxygens that terminate the two bridges are 2.57 and 2.58 Å, whereas the methoxy-to-methyleneoxy near distances are 2.80, 2.87, 2.90, 2.95, and 3.07 Å. The four sets of oxygen-to-oxygen distances in the two $OCH_2CH_2OCH_2CH_2O$ bridges are 2.70, 2.68, 2.64, and 2.75 Å [43]. Thus there are six oxygen-to-oxygen distances shorter than the normal van der Waals contact distances of 2.80 Å, some of them substantially so. Clearly the host in this lithiospherium complex is badly strained. The crystal structures of the free spherand Sp(EOE) and of its sodiospherium complex are not yet available. Structures *128* provide a stereoview of the crystal structure of Sp(EOE) · LiCl. The deformations of the aryl rings and of the bonds to their attached atoms are clearly visible in *128*.

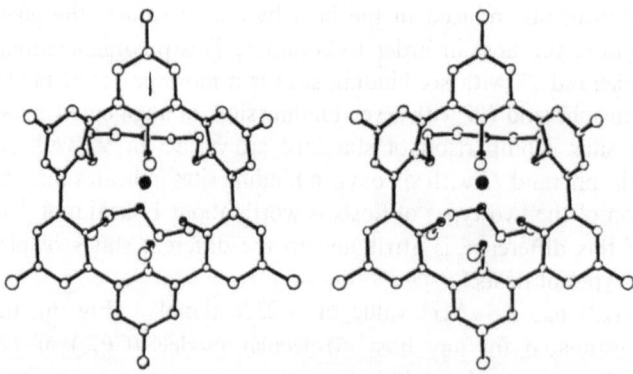

128

The effective diameters of five-, six-, and seven-coordinate Li^+ ion in Sp(CH₂)Li⁺, Sp(CH₃)Li⁺, and Sp(EOE)Li⁺ are 1.26 Å, 1.48 Å, and 1.72 Å, respectively. These values compare with the following averages calculated from distances in several crystal structures: four-coordinate, tetrahedral, 1.12 Å; five-coordinate, 1.36 Å; six-coordinate, octahedral, 1.50 Å [61].

The 3 kcal mol^{-1} increase in binding free energy substitute for Na$^+$ of Sp(EOE) over that of Sp(CH$_3$) is attributed to several factors. (1) Since Sp(EOE) · LiCl has seven oxygens that ligand Li$^+$, Na$^+$ in Sp(EOE) · NaCl must have at least seven oxygen ligands, at least one more than the Na$^+$ in Sp(CH$_3$) · NaCl. (2) There is probably a better match between the cavity diameter of Sp(EOE) and the Na$^+$ diameter than between the cavity diameter of Sp(CH$_3$) and the Na$^+$ diameter.

The difference in binding free energies of the host *most organized* for binding Na$^+$, Sp(EOE), and the host in which six oxygens are collected but are the *least* organized (open-chain host *91*) is about 16 kcal mol^{-1} (a factor of almost 10^{12} in Ka) [58]. Thus almost total organization of the host *during* its synthesis and prior to its complexation rather than during its complexation leads to the highest binding host on the NaPic binding scale.

The decomplexation rate constant for Sp(CH$_2$) · NaPic in CDCl$_3$ saturated with D$_2$O at 25 °C was determined by following with ^{1}H NMR the transfer of Na$^+$ from unlabeled Sp(CH$_2$) to Sp(CH$_2$) labeled with deuterium in its two methoxyl groups. A rate constant of about 2.2×10^{-4} sec^{-1} was obtained. The equilibrium constant for NaPic complexing Sp(CH$_2$) was determined by the picrate extraction method to be about 9.1×10^9 liter mol^{-1}. The complexation rate constant of NaPic by Sp(CH$_2$) must be about 2.0×10^6 liter mol^{-1} sec^{-1}, four powers of ten from a diffusion controlled rate constant [56].

Competition experiments for equimolar amounts of Sp(CH$_3$) and Sp(CH$_2$) complexing an equimolar amount of NaBi in CDCl$_3$ with 5% by volume of (CD$_3$)$_2$SO gave a ratio of rate constants of about unity. A similar experiment carried out with LiBr gave a 1.5 ratio of the rate constant for Sp(CH$_2$) over that for Sp(CH$_3$) complexing LiBr. In the same medium, when equimolar amounts of NaX and LiX were allowed to compete for an equimolar amount of any one of the three spherands, Sp(CH$_3$), Sp(CH$_2$), or Sp(EOE), the ratio of the complexation rate for Na$^+$ to that for Li$^+$ varied only from a minimum of two to a maximum of ten, depending on the spherand and the counterion (Pic$^-$, Br$^-$, or ClO$_4^-$). Thus the rate constants for any of the three spherands complexing either Na$^+$ or Li$^+$ are within a power of ten of each other. Consequently, changes in the equilibrium constants for complexation by the spherands for Na$^+$ and Li$^+$ appear to be associated more with changes in their decomplexation than with their complexation rate constants [56].

The question arises as to why the spherands take so long to reach equilibrium with their complexes when their CDCl$_3$ solutions are shaken with aqueous salt solutions. The crowns and hemispherands do not exhibit this property. The binding sites of the spherands, unlike those of other hosts, cannot sequentially displace solvating ligands such as H$_2$O or (CH$_3$)$_2$SO from Li$^+$ or Na$^+$. Molecular models indicate that to complex spherands, metal ions must pass through a lipophilic sleeve whose diameter is small enough to inhibit entry of ions containing more than about one ligand (see *81*). The principle of microscopic reversibility requires that in decomplexation the metal ions delivered to the medium must also initially have only one ligand. Probably the spherands cannot complex at the D$_2$O—CDCl$_3$ interface as is possible for hosts that can turn their electrons outward. The spherands' low solubility in D$_2$O prevents rapid complexation from occurring in the aqueous phase. Similarly, complexation in the $CDCl_3$ phase is very slow because concentrations of NaPic and LiPic are very low. Preliminary experiments indicate that rates of reaching equilibrium can be greatly increased by improved mixing of the phases. Phase transfer of components appears rate-limiting with the spherands presently at hand. Water-soluble spherands have been designed and are being prepared.

The little information available at the time of writing suggests that spherand $Sp(CH_2)$ complexes Li^+ better than Na^+. When $Sp(CH_2) \cdot NaPic$ was mixed with an equimolar amount of LiPic in $CDCl_3$ with 5% by volume of $(CD_3)_2SO$ at 25 °C, >95% of the Li^+ replaced Na^+ in the complex. Thus in this medium, $Sp(CH_2)$ binds LiPic >3.5 kcal mol^{-1} better than NaPic. The data of Fig. 5 indicate that in $CDCl_3$ saturated with D_2O, $Sp(CH_2)$ complexes LiCl 1 kcal mol^{-1} better than does Sp(EOE), and that $Sp(CH_3)$ complexes LiCl >9 kcal mol^{-1} better than does $Sp(CH_2)$. In another experiment, [2.1.1]cryptand $\cdot$ LiPic was equilibrated with an equimolar amount of spherand $Sp(CH_2)$ in $CDCl_3$ saturated with D_2O at 25 °C. The ratio of [2.1.1]cryptand $\cdot$ LiPic to $Sp(CH_2) \cdot$ LiPic at equilibrium was about unity. Of the cryptands, [2.1.1]cryptand is the best binder of Li^+ [6]. These data suggest that when the LiPic extraction scale is extended to the spherands, the strongest binder will be $Sp(CH_3)$, whose $-\Delta G°$ value in kcal mol^{-1} will be somewhere in the middle to high twenties, >9 kcal mol^{-1} higher than the best cryptand binder of Li^+, which also contains six binding sites [56].

The binding powers of the three spherands for LiCl fall in the order $Sp(CH_3)$ $\gg Sp(CH_2) > Sp(EOE)$ (see Fig. 5). This order correlates well with their structures. Comparisons of the crystal structures of $Sp(CH_3)$ (78) and of $Sp(CH_3) \cdot LiCl$ [43] indicate that much strain is released upon complexation (see Fig. 3). Comparison of the crystal structure of $Sp(CH_2) \cdot LiFeCl_4$ [43] with the modified CPK molecular models of $Sp(CH_2)$ suggest that there is little release of strain in the ligand support system upon complexation of Li^+. Comparison of the crystal structure of $Sp(EOE) \cdot LiCl$ [43] with the modified CPK molecular models of Sp(EOE) suggest that strain in the ligand support system might increase upon Li^+ complexation, since the cavity of Sp(EOE) appears too large for ordinary Li^+ binding [58].

6.6 Host-Metal Cation Selectivity

Many potential uses of hosts depend on their metal ion selectivity. The $-\Delta G°$ values of the picrate extraction scales provide convenient measures of ion selectivities of various hosts under conditions amenable for actual separations of guests. The scale as now applied has a lower limit of ~ 6.0 and an upper limit of about 16 kcal mol^{-1}. The scale can be extended by use of other counterions, or by equilibrations in the absence of an aqueous phase of guests between two hosts that differ in $-\Delta G°$ values by only a few kcal mol^{-1}. Hosts whose complexes are more soluble in water than in $CDCl_3$ saturated with water might also be brought onto this scale by similar equilibrations.

Scales might be extended to about 30 kcal mol^{-1}, but will be limited by rates of decomplexation. An example is $Sp(CH_3) \cdot LiCl$, which does not decomplex over periods of hundreds of hours when its solutions in $CDCl_3$ are shaken with water at 25 °C. Better methods of mixing and higher temperatures possibly will bring this decomplexation rate onto the human time scale. Ideally, the complexing abilities of all hosts should be graded on the same or related scales, as are the acidities of carbon acids.

Selectivities by the crown hosts for the alkali metal cations on the picrate extraction scale are typified by that of standard crown Nap(OEOEO)$_2$E (*98*). The $-\Delta G°$ values kcal mol^{-1} are as follows: Li$^+$, 5.9; Na$^+$, 8.3; K$^+$, 10.8; Rb$^+$, 9.6; Cs$^+$, 8.3. The maximum $-\Delta(\Delta G°)$ value in this series is ~ 5 kcal mol^{-1}, and involves Li$^+$ and K$^+$. This general pattern persists for a wide variety of modified crown compounds. These contain from six to twelve binding sites, two hinged macrorings in jaws-type arrangements of binding sites (e.g., *58*), belted macrorings such as *64*, and a wide variety of other hosts with other types of binding units. For most hosts dominated by OCH$_2$CH$_2$O binding units and possessing many conformational degrees of freedom, K$^+$ gives the highest $-\Delta G°$ values and Li$^+$ the lowest, with a maximum difference of 5 kcal mol^{-1} [28,42].

The hemispherands are better binders than the crowns for all alkali metal ions, the maximum $-\Delta(\Delta G°)$ value observed reaching about 7 kcal mol^{-1} for K$^+$ and Li$^+$ bound to *107* (Fig. 4). The pattern of $-\Delta G°$ values for the hemispherands differs from that of the crowns only in the sense that for the six-oxygen crowns (e.g., *98*) K$^+$ is bound better than Na$^+$ by ~ 2.5 kcal mol^{-1}, but in the six-oxygen hemispherand *72*, Na$^+$ binding exceeds that of K$^+$ by ~ 0.5 kcal mol^{-1} [56].

The cryptands are much more ion-selective than the crowns in methanol as solvent, but comparisons with the present compounds are difficult since few of the former have been put on the picrate extraction scales. In methanol at 25 °C [18]crown-6 and cyclohexano[18]crown-6 bind K$^+$ better than Na$^+$ by about 2.4 and 2.1 kcal mol^{-1}, respectively [6], close to the difference of 2.5 kcal mol^{-1} for Nap(OEOEO)$_2$ (*98*) in CDCl$_3$—D$_2$O. In methanol, [2.1.1]cryptand binds Li$^+$ better than Na$^+$ by >2.1 kcal mol^{-1}, and Na$^+$ better than K$^+$ by 5.1 kcal mol^{-1}. In the same solvent, [2.2.1]cryptand binds all three ions so well as to be off scale ($-\Delta G°$ is >8.3 kcal mol^{-1} for Li$^+$ and >10.8 kcal mol^{-1} for Na$^+$ and K$^+$), so comparisons cannot be made [6]. In water, [2.2.1]cryptand binds Na$^+$ and K$^+$ equally well (7.4 kcal mol^{-1}), and both ions are bound 3.9 kcal mol^{-1} better than Li$^+$. In 95% methanol — 5% water, [2.2.2]cryptand binds K$^+$ with $-\Delta G° = 12.9$ and Na$^+$ with 9.7 kcal mol^{-1}, so K$^+$ is better bound than Na$^+$ by 3.2 kcal mol^{-1} [6].

The crystal structures of complexes [2.1.1]cryptand · LiI, [2.2.2]cryptand · NaI, and [2.2.2]cryptand · KI show them all to be capsular, with enough adaptability in the guest and host cavity diameters to provide stability [24]. An example in which the guest is too large to fit into the host is provided by K$^+$ binding [2.1.1]cryptand ($-\Delta G° = 3.2$ kcal mol^{-1} in methanol). Molecular models (CPK) show the K$^+$ ion must perch on, rather than be encapsulated in this host. Such binding is dependent on the electron pairs being able to turn toward a noncentrally located guest.

The three spherands, Sp(CH$_3$), Sp(CH$_2$), and Sp(EOE) appear capable of binding *only* Na$^+$ and Li$^+$ among the alkali metal ions [4,56,59]. The crystal structure of Sp(CH$_3$) and molecular model of Sp(CH$_2$) show that perching type of binding is impossible since the electron pairs of the oxygens are completely shielded from guests by CH$_3$, C$_6$H$_2$, or CH$_2$ groups. Molecular models of Sp(EOE) show that only the two bridging oxygens can contact external guests, and then only one at a time. No binding of the three spherands to K^+ could be detected by the picrate extraction method, which can detect binding of ions only at a level of $-\Delta G° \geqq 6$ kcal mol^{-1}. Thus Sp(EOE), Sp(CH$_3$), and Sp(CH$_2$) bind Na$^+$ better than K^+ by >16, >13, and >8 kcal mol^{-1}, respectively in CDCl$_3$ saturated with D$_2$O at 25 °C. The fragmentary data available suggest that Sp(CH$_3$) binds Li$^+$ better than Na$^+$ by >7 kcal mol^{-1}, but that Sp(EOE) binds Na$^+$ better than Li$^+$ by ~ 8 kcal mol^{-1} [56].

Up to the time of writing a few crown hosts have been reported to bind Cs$^+$ better than K^+. Preliminary experiments show that the eight-oxygen host (C$_6$H$_3$OCH$_3$)$_8$ (*114*) binds CsPic several kcal mol^{-1} better on the picrate extraction scale than it binds either RbPic or KPic. Host *114* is shaped like a torus. When all eight oxygens are turned inward, they describe a square antiprism. Molecular models (CPK) indicate that, with all methyls turned outward, the diameter of the cavity changes with variation of the Ar—Ar dihedral angles over the range 3.4 Å to 5.2 Å. (A worn tennis ball just fits in the cavity of a CPK model of the host at the maximum allowable dihedral angle, which provides a maximum diameter for the cavity). Thus in models, all oxygens of *114* can touch at the same time a sphere with the effective diameter of Cs$^+$, which is 3.4 Å, whereas smaller ions would "rattle" in the hole. Thus with semirigid systems, it appears possible to design hosts that are ion-selective for guests larger than K^+ [58].

7 Acknowledgements

Support for parts of the X-ray crystal structure determinations by the National Science Foundation Grants 72-04385, GP-28248, and 77-18748 are gratefully acknowledged by K. N. Trueblood. The support for parts of the syntheses and determination of binding free energies by the Division of Basic Sciences of the Department of Energy, Contract (AT(04-3)34, P.A. 218) is gratefully acknowledged by D. J. Cram. Both authors warmly thank Drs. I. Goldberg, C. B. Knobler, and E. Maverick for much information in advance of publication concerning crystal structure determinations, and R. C. Helgeson, T. Kaneda, G. M. Lein, P. Cheng, G. D. Y. Sogah, I. Dicker, M. Newcomb, and T. L. Tarnowski for their many contributions in the organic chemical field.

8 References

1. Kyba, E. P., Helgeson, R. C., Madan, K., Gokel, G. W., Tarnowski, T. L., Moore, S. S., Cram, D. J.: J. Amer. Chem. Soc. *99*, 2564 (1977)
2. Pedersen, C. J.: J. Amer. Chem. Soc. *89*, 2495 (1967); Pedersen, C. J.: J. Amer. Chem. Soc. *89*, 7017 (1967); Richman, J. E., Atkins, T. J.: J. Amer. Chem. Soc. *96*, 2268 (1974)
3. Dietrich, B., Lehn, J.-M., Sauvage, J. P.: Tetrahedron Lett. 2885, 2889 (1969); Lehn, J.-M.: U.S. Pat. 3,888,877, June 10 (1975)
4. Cram, D. J., Kaneda, T., Helgeson, R. C., Lein, G. M.: J. Amer. Chem. Soc. *101*, 6752 (1979); Cram, D. J., Kaneda, T., Lein, G. M., Helgeson, R. C.: Chem. Commun. 948 (1979)
5. Koenig, K. E., Lein, G. M., Stückler, P., Kaneda, T., Cram, D. J.: J. Amer. Chem. Soc. *101*, 3553 (1979)

6. Lehn, J.-M.: Structure and Bonding, *16*, 1 (1973); Lehn, J.-M.: Pure and Applied Chemistry *50*, 871 (1978)
7. Truter, M. R.: Structure and Bonding *16*, 71 (1973)
8. Izatt, R. M., Christensen, J. J., (eds.): Synthetic Multidentate Macrocyclic Compounds. New York, Academic Press 1978
9. De Jong, F., Reinhoudt, D. N.: Adv. Physical Organic Chemistry (Gold, V. (ed.)), Vol. 17, pp. 279 London, Academic Press 1980
10. Cram, D. J., Cram, J. M.: Science, *183*, 803 (1974); Cram, D. J., Cram, J. M.: Acc. Chem. Res. *11*, 8 (1978); Cram, D. J., Synthetic Host-Guest Chemistry, Application of Biomedical Systems in Chemistry, Part II. (Bryan Jones, J. (ed.)), pp. 815–873. New York, Wiley 1976
11. Gokel, G. W., Dupont Durst, H.: Synthesis, 168 (1976)
12. Timko, J. M., Moore, S. S., Walba, D. M., Hiberty, P. C., Cram, D. J.: J. Amer. Chem. Soc. *99*, 4207 (1977)
13. Haymore, B. L., Huffman, J. C., presented at jount ACS-CIC meeting, Montreal, Canada, June 1977
14. Gokel, G. W., Cram, D. J.: Chem. Commun., 481 (1973)
15. Haymore, B. L., Huffman, J. C., presented at Fourth Symposium on Macrocyclic Compounds, Provo, Utah, August 1980
16. Madan, K., Cram, D. J.: Chem. Commun. 427 (1975)
17. Newcomb, M., Gokel, G. W., Cram, D. J.: J. Amer. Chem. Soc. *96*, 6810 (1974); Newcomb, M., Timko, J. M., Walba, D. M., Cram, D. J.: J. Amer. Chem. Soc., *99*, 6392 (1977)
18. Timko, J. M., Cram, D. J.: J. Amer. Chem. Soc. *96*, 7159 (1974)
19. Newcomb, M., Cram, D. J.: J. Amer. Chem. Soc. *97*, 1257 (1975); Moore, S. S., Tarnowski, T. L., Newcomb, M., Cram, D. J.: J. Amer. Chem. Soc. *99*, 6398 (1977)
20. Helgeson, R. C., Timko, J. M., Cram, D. J.: J. Amer. Chem. Soc. *95*, 3023 (1973); Cram, D. J., Helgeson, R. C., Koga, K., Kyba, E. P., Madan, K., Sousa, L. R., Siegel, M. G., Moreau, P., Gokel, G. W., Timko, J. M., Sogah, G. D. Y.: J. Org. Chem. *43*, 2758 (1978); Timko, J. M., Helgeson, R. C., Cram, D. J.: J. Amer. Chem. Soč. *100*, 2828 (1978)
21. Cram, D. J., Helgeson, R. C., Peacock, S. C., Kaplan, L. J., Domeier, L. A., Moreau, P., Koga, K., Mayer, J. M., Chao, Y., Siegel, M. G., Hoffmann, D. H., Sogah, G. D. Y.: J. Org. Chem. *43*, 1930 (1978)
22. Prelog, V.: Pure and Appl. Chem. *50*, 893 (1978)
23. Dunitz, J. D., Dobler, M., Seiler, P., Phizackerley, R. P.: Acta Cryst. *B30*, 2733 (1974); Dunitz, J. D., Seiler, P.: Acta Cryst. *B30*, 2739 (1974); Dobler, M., Dunitz, J. D., Seiler, P.: Acta Cryst. *B30*, 2741 (1974); Seiler, P., Dobler, M., Dunitz, J. D.: Acta Cryst. *B30*, 2744 (1974); Dobler, M., Phizackerley, R. P.: Acta Cryst. *B30*, 2746 (1974); Dobler, M., Phizackerley, R. P., Acta Cryst. *B30*, 2748 (1974)
24. Reviewed by Dalley, N. K.: Synthetic Multidentate Macrocyclic Compounds (Izatt, R. M., Christensen, J. J. (eds.)), pp. 209–216. New York, Academic Press 1978
25. Dale, J.: Tetrahedron *30*, 1683 (1974); Dale, J.: Israel J. Chem. *20* (*1–2*), 3, (1980)
26. Live, D., Chan, S. I.: J. Amer. Chem. Soc. *98*, 3769 (1976)
27. Knobler, C. B., Trueblood, K. N., Weiss, R. M.: unpublished
28. Tarnowski, T. L., Cram, D. J.: Chem. Commun. 661, (1976); Helgeson, R. C., Tarnowski, T. L., Cram, D. J.: J. Org. Chem. *44*, 2538 (1979)
29. Goldberg, I.: Acta Cryst. *B33*, 472 (1977)
30. Stevens, R. V., Lawrence, D., Knobler, C. B., Trueblood, K. N.: unpublished
31. Maverick, E., Grossenbacher, L., Trueblood, K. N.: Acta Cryst. *B35*, 2233 (1979)
32. Kyba, E. P., Koga, K., Sousa, L., Siegel, M. G., Cram, D. J.: J. Amer. Chem. Soc. *95*, 2692 (1973); Helgeson, R. C., Timko, J. M., Moreau, P., Peacock, S. C., Mayer, J. M., Cram, D. J.: J. Amer. Chem. Soc. *96*, 6762 (1974); Kyba, E. P., Gokel, G. W., De Jong, F., Koga, K., Sousa, L. R., Siegel, M. G., Kaplan, L., Sogah, G. D. Y., Cram, D. J.: J. Org. Chem. *42*, 4173 (1977); Kyba, E. P., Timko, J. M., Kaplan, L., De Jong, F., Gokel, G. W., Cram, D. J.: J. Amer. Chem. Soc. *100*, 4555 (1978); Peacock, S. C., Domeier, L. A., Gaeta, F. C. A., Helgeson, R. C., Timko, J. M., Cram, D. J.: J. Amer. Chem. Soc. *100*, 8190 (1978); Peacock, S. C., Walba, D. M., Gaeta, F. C. A., Helgeson, R. C., Cram, D. J.: J. Amer. Chem. Soc. *102*, 2043 (1980)
33. Goldberg, I.: J. Amer. Chem. Soc. *99*, 6049 (1977)

34. Goldberg, I.: Acta Cryst. *B34*, 3387 (1978)
35. Goldberg, I.: Acta Cryst. *B31*, 754 (1975)
36. Newcomb, M., Cram, D. J.: unpublished
37. Goldberg, I., Rezmovitz, H.: Acta Cryst. *B34*, 2894 (1978)
38. Bürgi, H. B., Dunitz, J. D., Shefter, E.: Acta Cryst. *B30*, 1517 (1974)
39. Cheng, P. G., Cram, D. J.: unpublished
40. Goldberg, I.: Acta Cryst. *B37*, 102 (1981)
41. Goldberg, I.: J. Amer. Chem. Soc. *102*, 4106 (1980)
42. Helgeson, R. C., Weisman, G. R., Toner, J. L., Tarnowski, T. L., Chao, Y., Mayer, J. M., Cram, D. J.: J. Amer. Chem. Soc. *101*, 4928 (1979)
43. Trueblood, K. N., Knobler, C. B., Maverick, E. F.: unpublished
44. Newcomb, M., Moore, S. S., Cram, D. J.: J. Amer. Chem. Soc. *99*, 6405 (1977)
45. Goldberg, I.: Acta Cryst. *B32*, 41 (1976)
46. Goldberg, I.: Acta Cryst. *B31*, 2592 (1975)
47. Trueblood, K. N.: unpublished
48. Goldberg, I.: Cryst. Struc. Comm. *9*, 1201 (1980); Goldberg, I.: Acta Cryst. *B36*, 2104 (1980)
49. Mercer, M., Truter, M. R.: J. Chem. Soc., Dalton Trans. 2215 (1973)
50. Trueblood, K. N., Maverick, E. F.: Paper E6, Meet. Amer. Crystallographic Assoc., Honolulu, March 26–30, 1979
51. Goldberg, I.: The Chemistry of Ethers, Crown Ethers, Hydroxyl Groups and their Sulfur Analogues. Supplement E1 (Ed. S. Patai), p. 175. London, J. Wiley and Sons (1980)
52. de Jong, F., Reinhoudt, D. N., Smit, C. J.: Tetrahedron Lett. 1371 (1976)
53. Lingenfelter, D. S., Helgeson, R. C., Cram, D. J.: J. Org. Chem. 46, 393 (1981)
54. Tarnowski, T. L., Cram, D. J.: unpublished
55. Sogah, G. D. Y., Cram, D. J.: unpublished
56. Lein, G. M., Cram, D. J.: unpublished
57. Dicker, I. B., Cram, D. J.: unpublished
58. Helgeson, R. C., Cram, D. J.: unpublished
59. Kaneda, T., Cram, D. J.: unpublished
60. Bush, M. A., Truter, M. R.: J. Chem. Soc., Perkin Transactions 2, 341 (1972)
61. Hermanson, H., Thomas, J. O., Olovsson, I.: Acta Cryst. *33*, 2859 (1977)

Analytical Applications of Crown Compounds and Cryptands

Ewald Blasius and Klaus-Peter Janzen

Fachrichtung „Anorganische Analytik und Radiochemie" der Universität,
D-6600 Saarbrücken, FRG

Table of Contents

1 Introduction

Crown compounds and cryptands have widely been used in analytical chemistry. Variable ring sizes, the type, the number, and the position of the donor atoms in the ether ring permit a selective adaptation to a certain cation. Because of the electroneutrality an anion is bound to a cation. The preference for a certain salt is expressed by the value of the selectivity constant, i.e. the ratio between two stability constants. When selectivity constants are high, separations and determinations of cations and anions are possible.

In solving such problems, both monomeric and polymeric cyclic polyethers are used. A survey is given in Table 1.

Table 1. Analytical Applications of Cyclic Polyethers

Monomeric and linear polymeric cyclic polyethers	Cross-linked polymeric cyclic polyethers
Separation methods:	Separation methods:
Masking	Column chromatography (low-pressure and high-pressure liquid chromatography, ion chromatography)
Extraction	
Extraction chromatography	Column electrophoresis
	Thin-layer chromatography
	Thin-layer electrophoresis
Determination methods:	Determination methods:
Photometry	Ion-sensitive Electrodes
Potentiometry	
Conductometry	
Polarography	
Voltammetry	
Ion-sensitive Electrodes	

2 Separation Methods

Monomeric cyclic polyethers are mainly used in the separation and determination of elements of the first and second main group of the periodic table. Linear polymeric cyclic polyethers are exclusively used for the extraction of salts of these elements. Cross-linked polymeric cyclic polyethers permit the separation and determination of many cations, anions and organic compounds.

2.1 Masking

1,4,7,10-Tetraazacyclododecane-N,N',N'',N'''-tetraacetic acid (1) forms the strongest complex ever recorded for Ca(II), with $pK = 15.85$ (compared to the EDTA-complex with $pK = 11.0$). The complexing agent is suitable for masking Ca(II) in the presence of Mg(II), Sr(II) and Ba(II). As for Sr(II), the complex with 1,4,7,10-tetra-azacyclotridecane-N,N',N'',N'''-tetraacetic acid (2) proves to be more stable than the complexes with the other alkaline earth ions, 1,4,8,11-Tetraazacyclotetradecane-

N,N',N'',N'''-tetraacetic acid (3) forms a complex with Ca(II) that is more stable than that with Mg(II) [1].

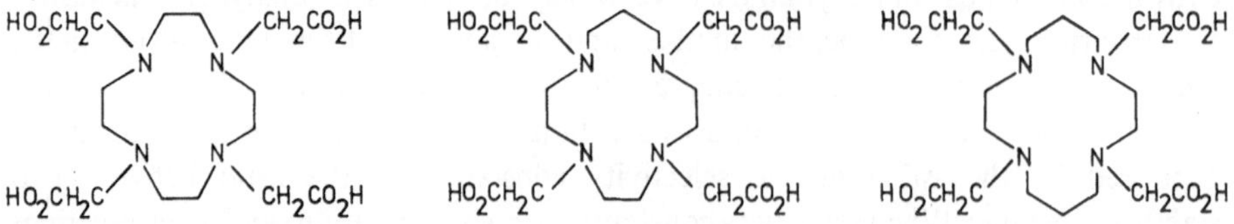

2.2 Extractions

For the selective extraction of inorganic salts monomeric and linear polymeric cyclic polyethers are used. With linear polymeric cyclic polyethers one can obtain higher extraction coefficients than with monomeric cyclic polyethers. Much research has been done on the extraction of alkali and alkaline earth salts. In contrast, only few extraction methods analytically suitable for heavy and transition metals or lanthanoids and actinoids have been described.

2.2.1 Extractions with Monomeric Cyclic Polyethers

The hardness of the anion of an extracted ion pair markedly affects the extraction coefficients. Therefore, many extractions are performed in the presence of the soft picrate anion. This method has the advantage that the extracted salts can be directly determined by photometry.

2.2.1.1 Organic Anions

Nonactin, monactin, dinactin and trinactin are characterized by a high selectivity to ammonium picrate and potassium picrate [2]. The highest selectivity to ammonium and potassium picrates is obtained with trinactin in dichloromethane (Table 2).

The selectivity to ammonium picrate is >10 with respect to potassium picrate and >1000 with respect to sodium picrate.

Alkali picrates are extracted by a solution of [18]crown-6 in benzene in the order K(I) > Rb(I) > Cs(I) > Na(I) [3]. The extraction coefficients for potassium picrate with dicyclohexano[18]crown-6 are different for the two isomers of the polyether [4]. The extraction coefficients for alkali picrates [5] as well as for thallium(I) and thallium(III) [6] are also known. If the crown compound is present in greater excess, Rb(I) and Cs(I) are also extracted. Mg(II) and Ca(II) do not interfere with up to a hundredfold excess [7].

Table 2. Extraction coefficients K_D for alkali and ammonium picrates with Trinactin in dichloromethane

Ion:	Li	Na	K	Rb	Cs	NH$_4$
K_D:	0.23	42	4000	1170	75	46000

The separation of Sr(II) from Mg(II) and Cs(II) in an acetate buffered solution (pH 4—7) is accomplished by using dicyclohexano[18]crown-6 in chloroform. Coextracted Ca(II) is removed by a picrate solution buffered with 0.015 mol/l NH_4Cl and Sr(II) is reextracted into an aqueous solution containing 2 mol/l HCl and 2 mol/l NH_4Cl [8].

In the system benzene/water the extraction coefficients of metal(I) picrates follow the order Ag(I) > Na(I) > Tl(I) ≫ K(I) > Rb(I) > Cs(I) > Li(I) for [15]crown-5; and Tl(I) > K(I) > Rb(I) > Ag(I) > Cs(I) > Na(I) ≫ Li(I) for [18]crown-6 [9]. For dibenzo[24]crown-8 are found the orders Tl(I) > Cs(I) > Ag(I) > Rb(I) > Na(I) ≫ Li(I) and Ba(II) ≫ Pb(II) ≫ Sr(II) > Ca(II) [10]. With metal(II) picrates the orders are Pb(II) > Sr(II) > Ba(II) > Ca(II) for [15]crown-5; Pb(II) ≫ Ba(II) = Sr(II) > Hg(II) > Ca(II) for [18]crown-6; and Pb(II) > Hg(II) > Sr(II) > Ca(II) for dibenzo[18]crown-6 [11].

One of the many applications is the sensitive determination of Na(I) after neutron activation [12]. For this purpose, Na(I) is extracted by dicyclohexano[18]-crown-6. The counter-ion is tetraphenyl borate.

For the photometrical determination of the K(I) content of serum [13] a 0.16% solution of [18]crown-6 in benzene and as reagent a 0.16% solution of bromocresol green in methanol/water (20:80 v/v), buffered with 0.2 mol/l acetic acid/lithium acetate are used. After combining aliquot parts of the bromocresol green solution and the serum, the mixture is centrifuged. The centrifugate is shaken with a little [18]crown-6 solution for 20 minutes and the indicator determined at 460 nm. Up to 500 ppm Na(I) does not interfere with the determination of 5 ppm K(I).

2.2.1.2 Inorganic Anions

Alkali salts with hard anions are only slightly soluble in most organic solvents. In order to improve the extraction of these salts by crown compounds, constituents which are able to form hydrogen bonds such as alcohols and phenols are added to the organic solvent [14, 15]. Dicyclohexano- and dibenzo[18]crown-6 are used as complexing agents. For potassium salts and dicyclohexano[18]crown-6, the extraction capability decreases in the order $F^- > CH_3COO^- > NO_3^- > I^- > Br^- > Cl^- > SO_4^{2-}$ [14].

An impressive example for the high Sr(II) selectivity is the diester of binaphthyl[18]-crown-6 hydrolyzed by a tenfold excess of Ba(OH)$_2$. Here, the selectivity of the crown compound is so high that only the 0.8% Sr(II) present in Ba(OH)$_2$ as impurity is extracted. The mass spectrum of the isolated crown compound contains only a molecular peak of the Sr(II)-complex [17].

The separation of Sr(II) and Ca(II) from milk is performed with dicyclohexano[18]crown-6 in chloroform. The method is used for the determination of ^{89,90}Sr(II) in milk [18].

3×10^{-4} mol/l Ba(II) can easily be separated from 3×10^{-4} up to 1.7×10^{-2} mol/l Sr(II) by extraction with a solution of dibenzo[24]crown-8 in nitrobenzene [19].

Synergistic effects cause a considerable increase in the percentage of extraction of $CsNO_3$ by nitrobenzene (Fig. 1).

The adduct dibenzo[24]crown-8/dodecatungstosilicic acid shows only the sum of the extraction capabilities of the two components. In contrast, the adduct dinonyl-dibenzo-[24]crown-8/dodecatungstosilicic acid in the ratio 4:1 displays a marked positive synergism.

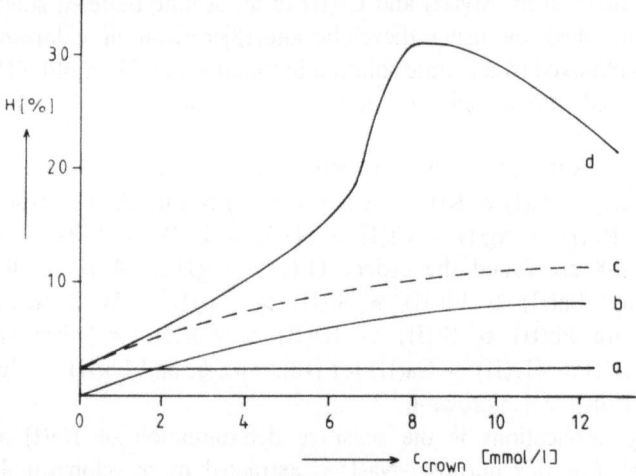

Fig. 1. Extraction of $CsNO_3$ from simulated middle-active waste solutions by mixtures of crown ethers and tungstosilicic acid in nitrobenzene.
a 2 mmol tungstosilicic acid, **b** bis(nonylbenzo)[24]crown-8 and dibenzo[24]crown-8, **c** dibenzo[24]-crown-8/tungstosilicic acid, **d** bis(nonylbenzo)[24]crown-8/tungstosilicic acid

This different behaviour is probably due to steric effects caused by the nonyl group. The adducts are suitable for the selective extraction of Cs(I) [20].

An example for the extraction of transition metals is the separation of Co(II) as $Co(SCN)_4^{2-}$ from ammonium thiocyanate solution by dicyclohexano[18]crown-6 or [18]crown-6 in dichloromethane [21].

Ni(II), Cu(II), Zn(II), Pb(II) and Sn(II) do not interfere. For the extraction of Ag(I) and Hg(I), thio-analogous crown compounds or O-, S-containing heterocycles are suitable [22]. $AgNO_3$ can also be extracted in the system water/chloroform [23].

2.2.2 Extractions with Linear Polymeric Cyclic Polyethers

Soluble polymeric crown compounds show extraction coefficients up to 250 times higher than those of the corresponding monomers [24]. Polyvinyl[15]crown-5 and polyvinyl[18]crown-6 are easily soluble in aliphatic hydrocarbons, toluene, chloroform, dichloromethane, tetrahydrofuran and nitrobenzene. In the system water/dichloromethane the extraction coefficients with polyvinyl[15]crown-5 are 5 times higher and those with polyvinyl[18]crown-6 250 times higher than those of the corresponding monomers. With polyvinyl[15]crown-5 one finds the following selectivity order for alkali and ammonium picrates:

$$K(I) > Rb(I) > Cs(I) > NH_4(I) > Na(I) > Li(I)$$

For polyvinyl[18]crown-6 the order is:

$$Cs(I) > K(I) > Rb(I) > NH_4(I) > Na(I) > Li(I).$$

2.3 Chromatography

In column chromatography, monomeric cyclic polyethers adsorbed on silica gel or dissolved in the eluent may be used in order to separate cations and optically active compounds. Polymeric cyclic polyethers are more widely applied. They allow the separation of cations, anions and organic compounds.

They are also used in thin-layer chromatography and thin-layer electrophoresis.

2.3.1 Column Chromatography

Monomeric cyclic polyethers can be employed in extraction and elution chromatography. Cross-linked polymeric cyclic polyethers are suitable for ion chromatography and other methods.

2.3.1.1 Extraction Chromatography

To separate radioactive alkali ions, silica gel is used as solid support for dibenzo[18]-crown-6. The elution is accomplished with 0.01 mol/l NH_4SCN in water at 40 °C and pH 7. The alkali ions appear in the eluate in the order Li(I), Na(I), Rb(I), K(I), Cs(I)[25]. Even the separation of $^{40}Ca(II)/^{44}Ca(II)$ is possible[26].

2.3.1.2 Elution Chromatography with Monomeric Cyclic Polyethers

The coefficients of the partition of alkali ions between Dowex 50 WX8 and an eluent consisting of crown compounds, HCl or HNO_3 and 80% v/v methanol/water are determined by radiometry[27].

Optically active crown compounds, e.g. di(binaphthyl)[18]crown-6 are of great interest[28].

Optically active primary amino ester salts are adsorbed on silicagel from an aqueous solution and eluted by a chloroform solution of different derivates of di(binaphthyl)[18]crown-6; they are completely separated in this process[29,30].

The chirospecific properties of some bicyclic and tricyclic cryptands in which the naphthyl group is bound via the bridging oxygen atoms can also be analytically exploited[31]. The ligands are deformed by the cations of the different alkali elements to different degrees so that the complex stabilities differ from each other. Binaphthyl derivatives of cyclic polyethers bearing carboxy groups at the crown ring are also of analytical interest. The stability constants of their alkali complexes are higher than those of the complexes with all the other polyethers[32].

[15]Crown-5, [18]crown-6 as well as the [2.1.2], [2.2.1] and [2.2.2] cryptands are suitable for the separation of cis- and trans-dimethylsemidion[33].

2.3.1.3 Elution Chromatography on Cross-Linked Polymeric Cyclic Polyethers

Numerous exchangers which are able to bind definite inorganic salts or organic compounds are obtained by condensation, substitution and copolymerization reactions with cyclic polyethers of different structure and ring size. They have various applications[20, 34−42].

Ewald Blasius and Klaus-Peter Janzen

2.3.1.3.1 Preparation of the Exchangers

The different syntheses of exchangers with cyclic polyethers as anchoring groups are listed in Table 3.

The condensation of dibenzo crown compounds with formaldehyde in formic acid results in cross-linkage by methylene groups. With monobenzo crown compounds and monobenzo cryptands, additional cross-linking agents, such as toluene, xylene, phenol, or resorcinol are used.

It is also possible to bind monobenzo and dibenzo crown compounds at the polystyrene surface via methylene bridges by condensation reactions with formaldehyde.

Polymerization reactions of monovinylbenzo crown compounds with divinyldibenzo crown compounds or divinylbenzene as cross-linking agents lead to polystyrene-like matrices.

Substitution reactions can be performed with chloromethylated polystyrene as well as with silica gel. Monobenzo crown compounds substituted by amino, hydroxymethyl or bromoalkyl groups are linked to the matrix by —C—NH—C, —C—O—C—, —C—C—C— or —Si—NH—C—, —Si—O—C— and —Si—C—C— bonds. Noncyclic ethers, too, can be bound by chloromethylated polystyrene in a similar way.

Table 3. Synthesis of exchangers containing cyclic polyethers as anchoring groups

Type of synthesis	Matrix	Starting materials	
		Anchor group	Cross-linking agent
Condensation	Methylene bridges	Dibenzo-crown compounds	Formaldehyde
	Methylene bridges cross-linking agent		Formaldehyde Toluene (Xylene, phenol, resorcinol)
	Methylene bridges Polystyrene		
Polymerization	Polystyrene		
Substitution	Polystyrene Amino bridges Polystyrene Ether bridges Silica gel, Methoxy bridges Silica gel Alkane bridges		Chlormethylated polystyrene Silica gel

2.3.1.3.2 Properties of Exchangers

The most notable properties of exchangers containing cyclic polyethers as anchoring groups, compared to those of commercially available exchangers, are listed in Table 4.

Table 4. Notable Properties of exchangers with cyclic polyethers as anchoring groups in comparison with commercially available exchangers

---

High resistance to chemicals, temperature and radiolysis
Neutral ligands as anchoring groups
Simultaneous uptake of cations and anions to maintain electroneutrality
Degradation or stripping of the solvent shell of the cation and anion
Binding of the ions is facilitated in solvents which are less polar than water, e.g. methanol
Stability of the polyether complexes depends on cation, anion, solvent, size of polyether ring and on the number, type and position of the heteroatoms (O, N, S)
Activation of anions
Salt uptake: only O as heteroatom: independent of pH
      O and N as heteroatoms: pH-dependent at pH < 3
Elution by pure solvents: no pollution of the eluate

Furthermore, the following factor is important:
Maximum capacities and stabilities of the polyether complexes show parallels. Thus, the acid and base theory of Pearson is valid. In methanol the maximum capacities are about 1—2 mmol/g whereas in water they are about half of it.

The selectivity orders in methanol and water follow those of the monomer complexes with a few exceptions.

With the anions the salt uptake generally increases in the order of their polarizability. In some cases hydroxide ions occupy an exceptional position.

The thermal stability of the polymers decreases with rising ring size. Condensation resins are thermally more resistant than copolymerization resins. In comparison with the —C—C— bonds in copolymers, the linkage of the anchoring group to the styrene matrix by —C—NH—C— or —C—O—C— bonds causes a decrease of the thermal stability.

Condensation resins are resistant to polar solvents such as acetone, dichloromethane, acetonitrile, ethanol, methanol, tetrahydrofuran as well as to 5M HCl, 5M HNO_3 and 5M NaOH. Likewise, oxidants such as dilute $KMnO_4$ or H_2O_2 solutions do not change the capacity whereas it decreases by 3 to 8% after extraction of the exchangers by benzene or cyclohexane.

On the other hand, exchangers with a styrene matrix and copolymers of vinyl-substituted crown compounds are also resistant to benzene and cyclohexane. With all exchangers containing nitrogen in the anchoring group, a complete protonation of the nitrogen atoms occurs at pH <2. Then, complex formation with salts does not take place any longer. Due to the quaternary amino groups formed, an anion exchanger results (Fig. 2).

Condensation resins are more resistant to radiolysis than polymerization resins. For example, no change in the capacity of an exchanger containing dibenzo[24]crown-8 as the anchoring group and prepared by condensation is observed after radiation of 10^9 rad. This stability is probably due to the presence of hydroxymethyl groups acting as radical scavengers.

The establishment of the equilibrium needs more time in water than in methanol. It is achieved sooner with condensation resins than with polymerization resins. The required time is sufficiently short to enable separation on silica gel modified by benzo[15]crown-5 or benzo[18]crown-6 to be carried out by means of high-performance liquid chromatography.

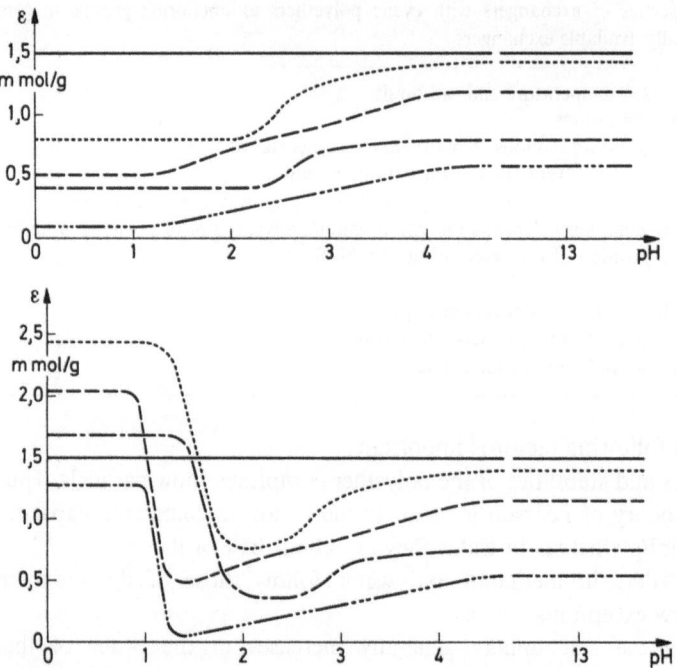

Fig. 2. Dependence of the maximum capacities for K^+ (above) and SCN^- (below) from KSCN on pH. ——— §-dibenzo[18]crown-6, ·········· §-dibenzo-monoaza[18]crown-6, --------- §-dibenzo-pyridino[18]crown-6, — · — §-benzomonoaza[15]crown-5

2.3.1.3.3 Selectivities and Applications of Exchangers

Cyclic polyether-containing exchangers are widely used in many fields of inorganic and organic chemistry. They are also helpful in solving special problems in radiochemistry.

Mention must be made of the use of cyclic polyethers bound to silica gel or polystyrene in ion chromatography. In contrast to the stationary phases used till now, these permit separation of cations and anions over the whole pH range in pure solvents, especially in water. A suppressor column is unnecessary. Therefore,

Table 5. Analytical applications of cross-linked polymeric polyethers

---

Separation of cations (e.g. alkaline earth metals, heavy and precious metals)
Separations of anions (e.g. halides, pseudohalides)
Separation of non-salt-like organic compounds
Trace enrichment (e.g. radionuclides)
Determination of water in inorganic and organic compounds

Column chromatography (low-pressure and high-pressure liquid chromatography, ion chromatography, column chromatography)
Thin-layer chromatography
Thin-layer electrophoresis

---

conductivity detectors can be applied without any problems. Numerous organic compounds can likewise be separated.

Thus, the anchoring group benzo[24]crown-8 is most suitable for separating Cs(I) from the other cations. Likewise, the order of elution Hg(II)/Cd(II)/Zn(II) and Zn(II)/Cd(II)/Hg(II) with benzo[15]crown-5 and dibenzo[24]crown-8, respectively, as anchoring groups can be predicted.

The selectivity orders also reveal the possibility of separations. In contrast to conventional strongly acidic and basic exchangers, they are not in accordance with the Hofmeister and Schulze-Hardy series. Rather, they show parallels with the capacities. This fact can be easily interpreted by the acid-base theory of Pearson.

The selectivity orders discussed in the following sections include ions of the same charge when based on the partition coefficients determined by batch tests and ions of different charge when based on the mass distribution ratios determined by column tests.

More than 30 polymers with cyclic polyethers as anchoring groups are known. The applications of some exchangers investigated in detail (Fig. 3) are described in the following section.

A simple rule helps in finding the suitable exchanger in order to solve a special separation problem. An exchanger prefers that cation for which the following relation is best fulfilled.

$$\frac{\text{diameter of the cation (according to Goldschmidt)}}{\text{diameter of the polyether ring}} = 0.80$$

In Table 6 the corresponding values for cations of the first main group of the periodic system as well as for Zn(II), Cd(II), Hg(II), Tl(I) and Pb(II) are given.

2.3.1.3.3.1 §-Benzo[15]crown-5

a) Selectivity series for this anchoring group
  Cations: K(I) > Rb(I) > Cs(I), Na(I) > Li(I)
      Ba(II) > Sr(II) > Zn(II) > Cd(II), Fe(II) > Ca(II) > Mn(II),
      Co(II), Ni(II), Mg(II), Hg(II)

Table 6. Quotient of ring diameter and diameter of cations

	Å	Li^+ 1.36	Na^+ 1.96	K^+ 2.66	Rb^+ 2.96	Cs^+ 3.34
B[15]C-5	1.95	0.69	1.00	1.36	1.52	1.71
DB[18]C-6	2.90	0.47	0.67	0.92	1.02	1.20
DB[21]C-7	3.79	0.36	0.52	0.71	0.79	0.89
DB[24]C-8	4.00	0.34	0.49	0.67	0.74	0.83

	Å	Zn^{2+} 1.36	Cd^{2+} 1.84	Hg^{2+} 1.86	Tl^+ 3.02	Pb^{2+} 1.34
B[15]C-5	1.95	0.71	0.94	0.95	1.58	0.69
DB[18]C-5	2.90	0.48	0.63	0.64	1.04	0.45
DB[21]C-7	3.79	0.37	0.49	0.50	0.81	0.36
DB[24]C-8	4.00	0.35	0.46	0.47	0.76	0.34

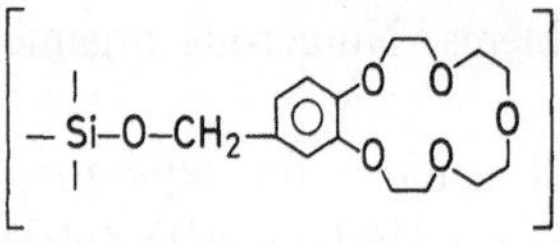

I: §-benzo [15] crown-5
Ø 1,7-2,2 Å
a   b

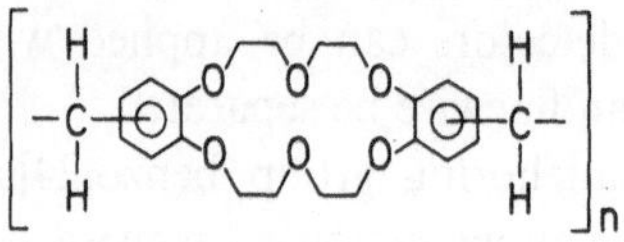

II: §-dibenzo [18] crown-6
Ø 2,6-3,2 Å
a   b

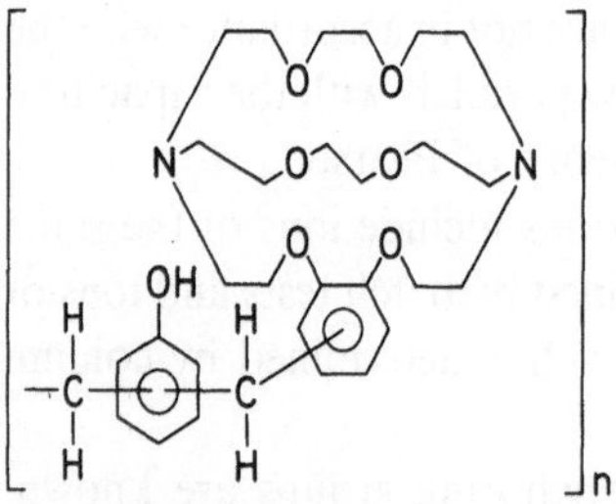

III: §-cryptand 2_benzo·2.2
Ø ~ 2,8 Å (estimated)

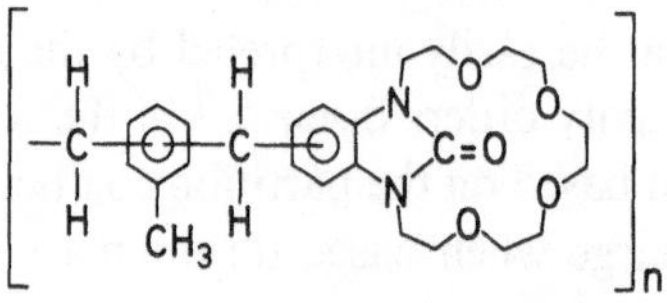

IV: §-benzimidazolono [18] crown-6
Ø 2,9 Å (estimated)

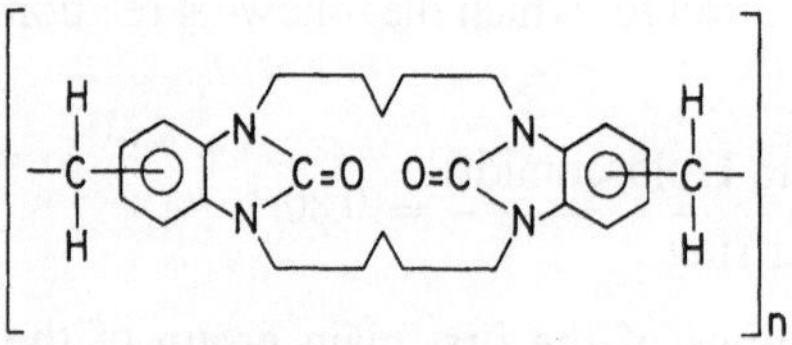

V: §-dibenzimidazolono-bispentamethylene

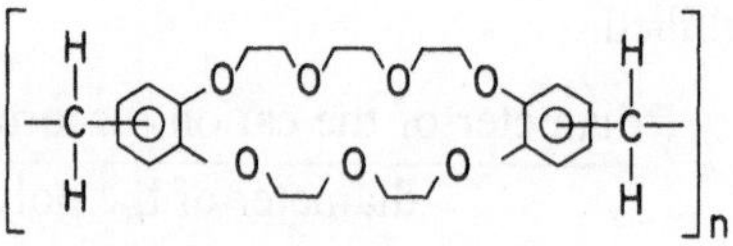

VI: §-dibenzo [21] crown-7
Ø 3,4-4,3 Å
a   b

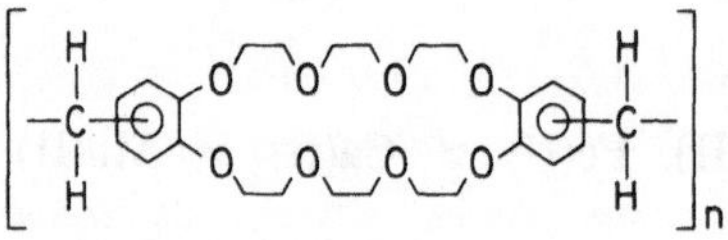

VII: §-dibenzo [24] crown-8
Ø > 4 Å (estimated)

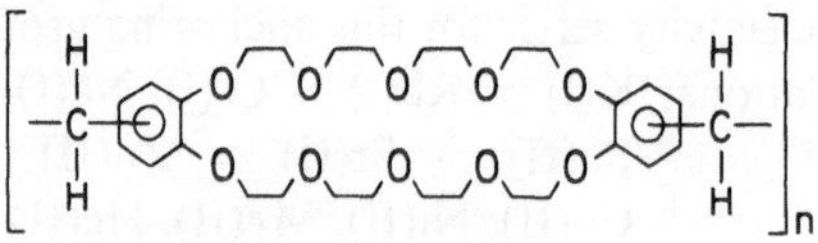

VIII: §-dibenzo [30] crown-10
Ø > 4 Å (estimated)

Fig. 3. Ring diameters according to the atom model of **a** Corey-Pauling-Koltun, **b** Fisher-Hirschfelder-Taylor

Anions: $SCN^- > I^-, PO_4^{3-} > NO_3^-, Br^- > Cl^- > F^-$

Alkylammonium salts: methyl- > ethyl-substituted

mono- > di- > tri- > tetra-substituted

K(I) is preferably bound within the group of alkali elements and Ba(II) among the bivalent cations.

b) Separations by §-Benzo[15]crown-5 (Matrix: silica gel)

Silica gel modified by B[15]C-5 is suitable for ion chromatography. The fast separation of alkali chlorides is shown in Fig. 4.

The eluent is methanol. At a pressure of 24 MPa the separation is completed within 18 minutes.

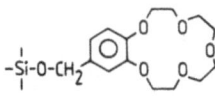

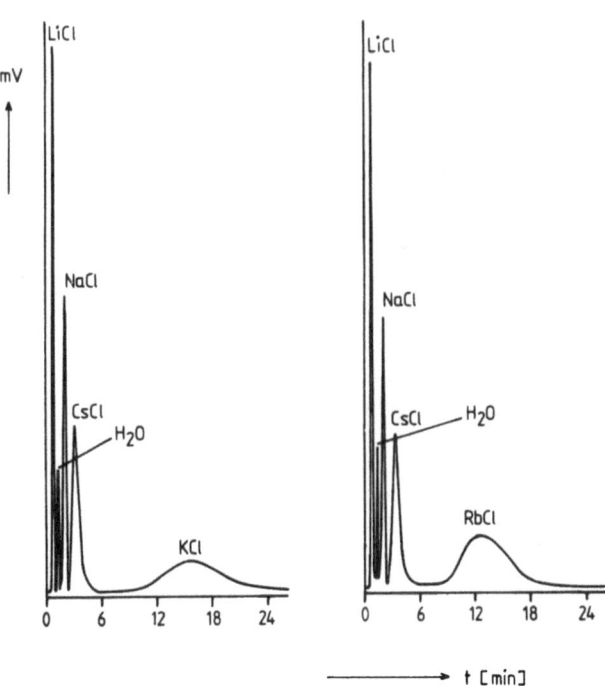

Fig. 4. Separation of alkali halides by elution with methanol

left: 0.020 mg LiCl
0.080 mg NaCl
0.330 mg KCl
0.276 mg CsCl
Elution rate: 5.0 cm³/min,

right: 0.020 mg LiCl
0.080 mg NaCl
0.423 mg RbCl
0.276 mg CsCl
Pressure: 24 MPa

2.3.1.3.3.2 §-Dibenzo[18]crown-6

a) Selectivity series for this anchoring group

Cations: K(I) > Rb(I) > Cs(I) > Na(I) > NH₄(I) > Li(I)
Ra(II) > Ba(II) > Eu(II) > Zn(II), Sr(II) > Cd(II) > Ca(II)
> Fe(II) > UO₂(II) > Cu(II), Co(II) > Ni(II) > Hg(II), Mg(II)
Fe(III) > Cr(III) > Pr(III) > Ce(III) > La(III)

Anions: SCN⁻ > I⁻ > NO₃⁻ > Br⁻ > Cl⁻ > PO₄³⁻ > OH⁻ > F⁻, SO₄²⁻

Alkylammonium salts: methyl- > ethyl- > n-propyl- > butyl-substituted

b) Separations by dibenzo[18]crown-6 (matrix: methylene bridges).

This is the most versatile exchanger. It preferably binds K(I) among the alkali elements and Ra(II) among the bivalent cations. With the latter it forms a particularly strong complex. Therefore, the exchanger is very suitable for the separation and enrichment of Ra(II) in the presence of Ba(II) [38].

Very small amounts of ⁹⁰Sr(II) can be separated from larger amounts of Ca(II), a problem that is important for the determination of ⁹⁰Sr(II) in the fall out [39].

201

Besides numerous separations of alkali, alkaline earth and transition metal salts, separation of ammonium chlorides is possible (Fig. 5).

With the same cation, anions are isolated according to their polarizability. The amount of the eluent required for elution increases with increasing polarizability. Separation time and efficiency also depend on the cation. Figure 6 shows the separation of the sodium halides.

The separation of Na_2SO_4 traces from concentrated NaCl solutions can be achieved up to a molar ratio of $1:10000$ [20].

c) Water determination by dibenzo[18]crown-6 (matrix: methylene bridges)

As an example for the quantitative application of elution chromatography on exchangers with cyclic polyethers as anchoring groups the determination of water in methanol as solvent must be mentioned. Methanol itself (Merck p.a.) contains at most 0.01% H_2O. However, this water content does not interfere with the measurement since a differential refractometer serves as detector.

Additional water in methanol or other organic solvents gives rise to a H_2O peak. Thus constitutional or adsorbed water in salts and organic compounds can be determined. Figure 7 shows the elution curves of some salts of different water contents.

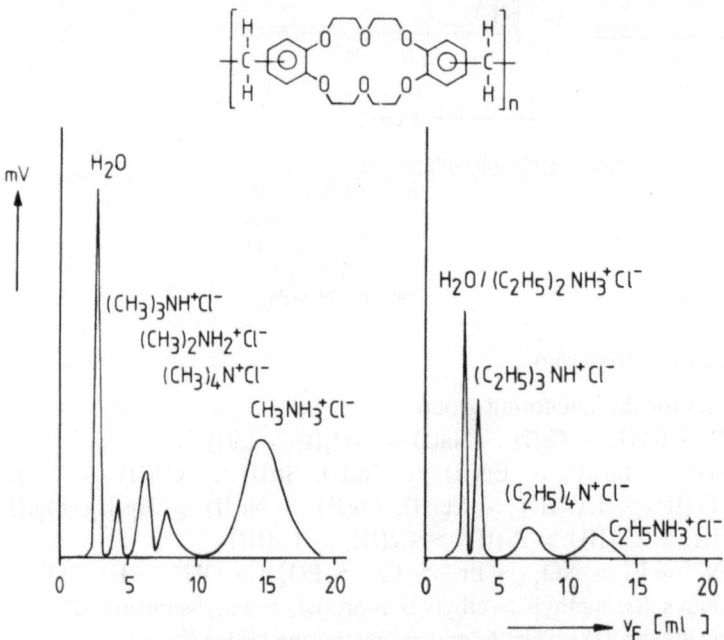

Fig. 5. Separation of alkyl ammonium chlorides by elution with methanol
left: 0.094 mg monomethyl ammonium chlorides   right: 0.065 mg monoethyl ammonium chloride
  0.049 mg dimethyl ammonium chlorides   0.088 mg diethyl ammonium chloride
  0.048 mg trimethyl ammonium chlorides   0.067 mg triethyl ammonium chloride
  0.081 mg tetramethyl ammonium chlorides   0.165 mg tetraethyl ammonium chloride

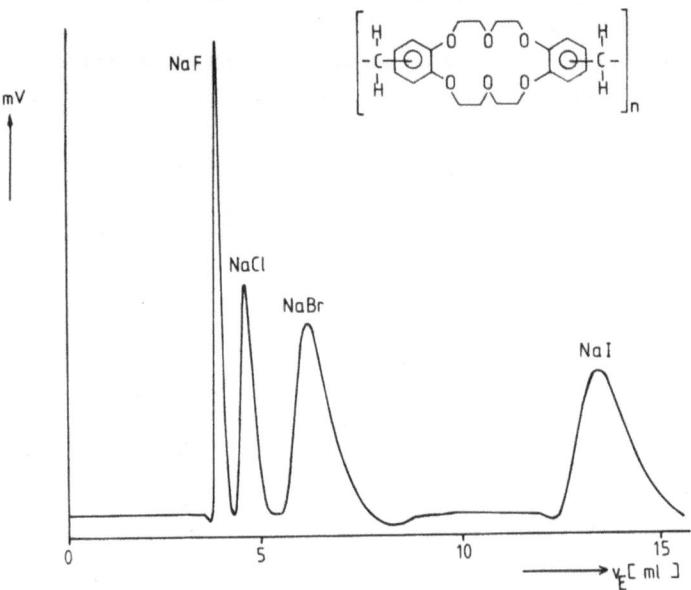

Fig. 6. Separation of sodium halides by elution with water 0.42 mg NaF, 0.35 mg NaCl, 0.62 mg NaBr and 2.25 mg NaI

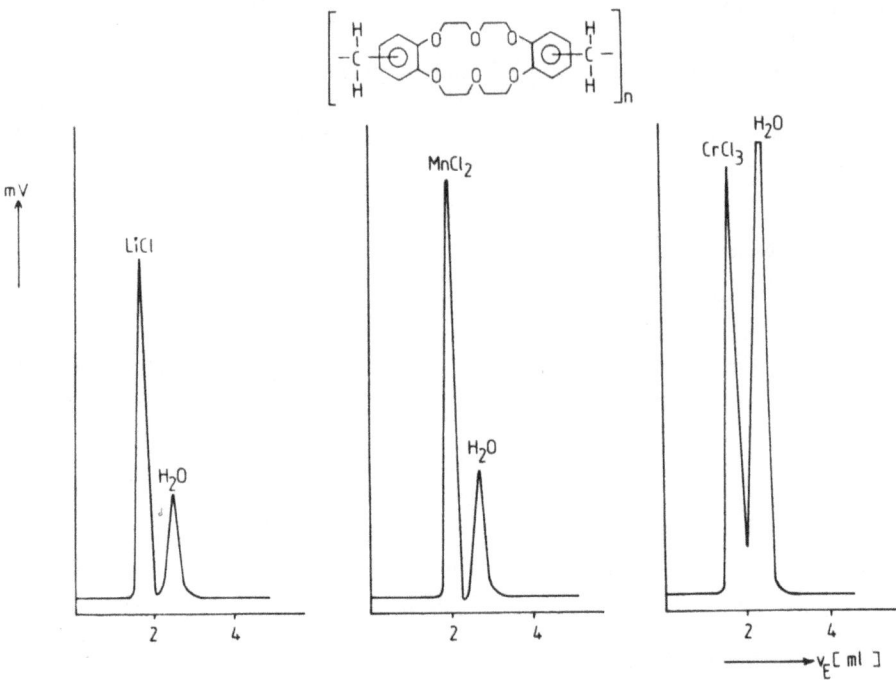

Fig. 7. Determination of water elution with methanol left: 0.02 mg LiCl $\cdot$ H_2O, middle: 0.09 mg $MnCl_2$ $\cdot$ 4 H_2O, right: 0.35 mg $CrCl_3$ $\cdot$ 6 H_2O

203

With benzoquinone one finds, in addition to the H_2O peak, the peak of an impurity, probably hydroquinone [40].

By quantitative evaluation of the peak areas the water content can be determined down to 1 ppm at the present time. This method is recommended if the water content of compounds can be ascertained only inaccurately or not at all by means of gas chromatography or other methods. For instance, Cu(II) salts and benzoquinone react with components of the Karl Fischer reagent.

The constitutional water of most salts is removed by the complex formation with cyclic polyethers.

2.3.1.3.3.3 §-Cryptand $2_{benzo}2.2$

a) Selectivity series for this anchoring group

Cations: K(I) > Na(I) Rb(I) > Cs(I) > Li(I)

for chlorides in water

K(I) > Na(I) > Rb(I) > Cs(I) > Li(I)

for chlorides in methanol

Anions: $SCN^- > I^- > Br^- > Cl^-$

for potassium salts in water

Polymers with cryptands as anchoring groups are commercially available under the tradenames "Kryptofix 221 B polymer" and "Kryptofix 222 B polymer". The $2.2.1_{benzo}$ cryptand preferably binds Na(I) whereas the $2.2.2_{benzo}$ cryptand preferably binds K(I). The $2_{benzo}2.2$ and $2_{benzo}2_{benzo}2$ cryptands can be cross-linked with formaldehyde or formaldehyde/phenol [36, 37]. The syntheses of the anchoring groups are laborious.

b) Separations by §-cryptand $2_{benzo}2.2$ (matrix: phenol/methylene bridges).

This exchanger has the best analytical properties of all polymers with cryptands as anchoring groups. In water as eluent both alkali ions and anions of the VII main group of the periodic system can be separated.

2.3.1.3.3.4 §-Benzimidazolono[18]crown-6

a) Selectivity series for this anchoring group

Cations: Rb(I), Cs(I) > Ra(I) > Mn(II) > K(I) > Ba(II) > Sr(II) > Ca(II), Fe(II), Fe(III), Co(II), Na(I), Cu(II), Cd(II), Zn(II), Hg(II), Cr(III), Li⟨I⟩, Mg(II)

for chlorides in methanol

Ba(II) > Sr(II) > K(I), Rb(I), Cs(I) > Na(I), NH_4(I), Ca(II), Pb(II), Tl(I), Cu(II) > Li(I), Mg(II)

for nitrates in methanol

no distinct selectivity for cations in water.

Anions: $SCN^- > I^- > Br^- > OH^- > Cl^- > F^-$, ClO_3^-, ClO_4^-, BrO_3^-, IO_3^-, NO_3^-, SO_4^{2-}, PO_4^{3-}

for Na(I) salts in methanol and water.

The exchanger binds Rb(I) and Cs(I) equally strongly. The stabilities of the complexes with alkaline earth cations grow with increasing atomic number of the elements. Ra(II) is bound appreciably more strongly than Ba(II).

b) Separations by §-benzimidazolono[18]crown-6 (matrix: toluene/methylene bridges).

With this exchanger many separations of cations and anions are possible. At pH < 3.5 protonation of the nitrogen atom takes place. In contrast to the polymers containing oxygen as heteroatom in the ring the application is therefore restricted to the pH range above 3.5. The exchanger is easily synthesized. The separation of all the alkaline earth ions is shown in Fig. 8.

Due to the protonation of the nitrogen atom, Ra(II) can be easily eluted by HCl. On the other hand, large eluent volumes are necessary with §-dibenzo[18]-crown-6 which only contains oxygen atoms in the ether ring [41].

2.3.1.3.3.5 §-Dibenzimidazolono-bispentamethylene

a) Selectivity series for this anchoring group

Cations: Hg(II) > Ca(II) > Cs(I) > Rb(I) > K(I) > Mn(II) > Ba(II) > Sr(II),
Co(II) > Fe(II), Fe(III), Cu(II), Ni(II), Zn(II), NH_4(I) > Li(I)
for chlorides in methanol

Cd(II) > Ba(II) > Mn(II) > Sr(II) > Ca(II) > K(I), Rb(I),
Cs(I) > Na(I), NH_4(I), Mg(II), Cu(II), Ni(II), Co(II), Fe(II), Fe(III),
Cr(III)
for chlorides in water

Pb(II) > Tl(I) > Ba(II), Sr(II), Ca(II) > Cs(I) > Mg(II), Tl(I),
K(I) > Cu(II), Na(I), Li(I), NH_4(I)
for nitrates in methanol

Pb(II), Cs(I) > Sr(II), Ba(II) > Na(I), Li(I), Mg(II), Ca(II) > NH_4(I),
Ag(I), Cu(II), Tl(I)
for nitrates in water

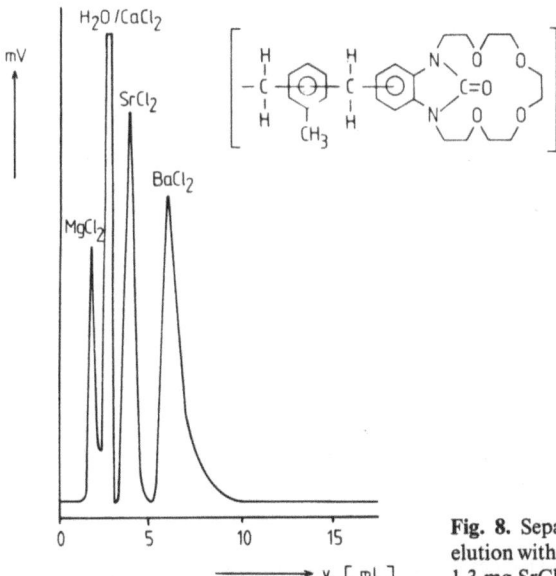

Fig. 8. Separation of alkaline-earth chlorides by elution with methanol 1.2 mg $MgCl_2$, 0.1 mg $CaCl_2$, 1.3 mg $SrCl_2$ and 1.9 mg $BaCl_2$

205

Anions: $ClO_4^- > SCN^- > I^- > Br^- > Cl^- > ClO_3^- > OH^- > IO_3^- > NO_3^-$
$> PO_4^{3-}, SO_4^{2-}, F^-, BrO_3^-$
for Na^+ salts in methanol and water.

In methanol the exchanger binds particularly heavy metal salts and, to a smaller extent, alkali metal salts. With water as eluent one can advantageously carry out separations of anions such as of the potassium halides, of sodium borate from boric acid and of sodium cyclamate from saccharin.

2.3.1.3.3.6 §-Dibenzo[21]crown-7

a) Selectivity series for this anchoring group
Cations: $Rb(I) > K(I) > Cs(I) > NH_4(I) > Na(I) > Li(I) > Ba(II) > Sr(II)$
$> Ca(II) > Mg(II)$
Anions: $SCN^- > I^- > Br^- > Cl^-$.

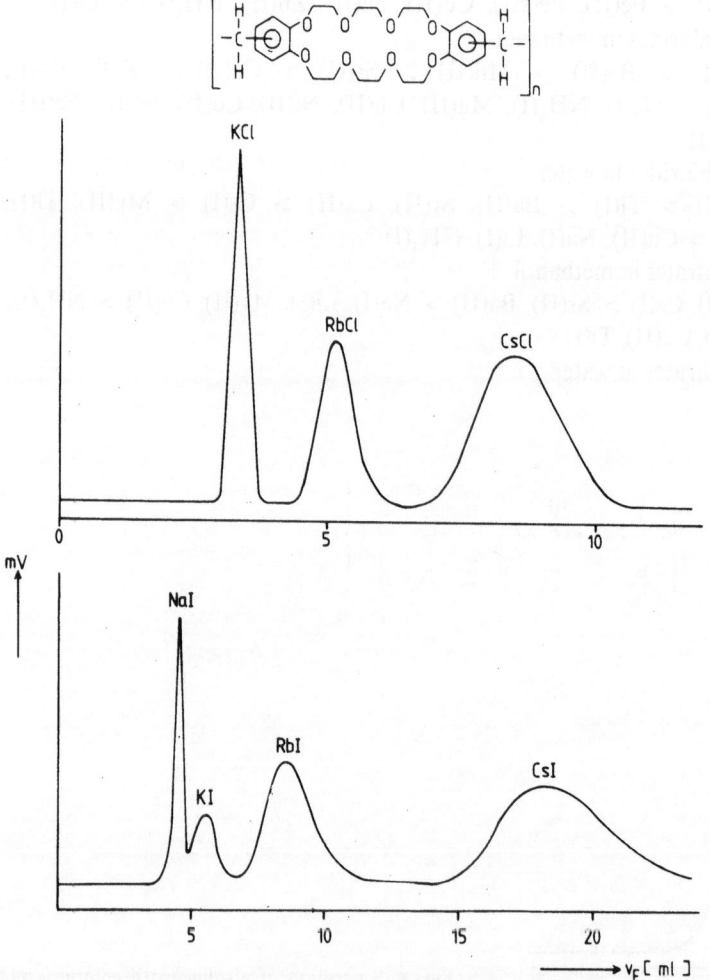

Fig. 9. Separation of alkali chlorides and iodides by elution with water above: 0.30 mg KCl, 0.42 mg RbCl and 0.54 mg CsCl, below: 0.12 mg NaI, 0.08 mg KI, 0.25 mg RbI and 0.36 mg CsI

With increasing ring size the maximum of salt uptake of the alkali elements shifts to the cations of higher atomic number. §-DB[18]C-6 and §-DB[21]C-7 preferably bind K(I) and Rb(I), respectively.

b) Separations by §-Dibenzo[21]crown-7 (matrix: methylene bridges).

This exchanger is well suited for the separation of alkali metal salts.

2.3.1.3.3.7 §-Dibenzo[24]crown-8

a) Selectivity series for this anchoring group

Cations: Cs(I) > Rb(I) > K(I) > Na(I) > Li(I) > Hg(II) > Pb(II) > Ba(II) > Cd(II) > Mn(II) > Ca(II) > Sr(II) > Mg(II) > Cu(II) > Co(II) > Ni(II) in water

Cs(I) > Rb(I) > K(I) > Hg(II) > Pb(II) > Na(I) > Ba(II) > Cd(II) > Mg(II) > Cu(II) > Fe(III) > Nb(IV) > Zr(IV) > Ce(III) > Sb(III) > Cr(III) > Li(I) > Mg(II) > Ni(II) > Zn(II) > Co(II) in 1 mol/l HNO_3.

The anchoring group is selective for Cs(I). The second selectivity series contains the components of a waste solution of medium activity.

b) Separations by §-Dibenzo[24]crown-8 (matrix: methylene bridges).

Fig. 9 shows the elution chromatogram of higher alkali metals chlorides separated in water within 80 minutes. With a more polarizable anion the separation takes longer.

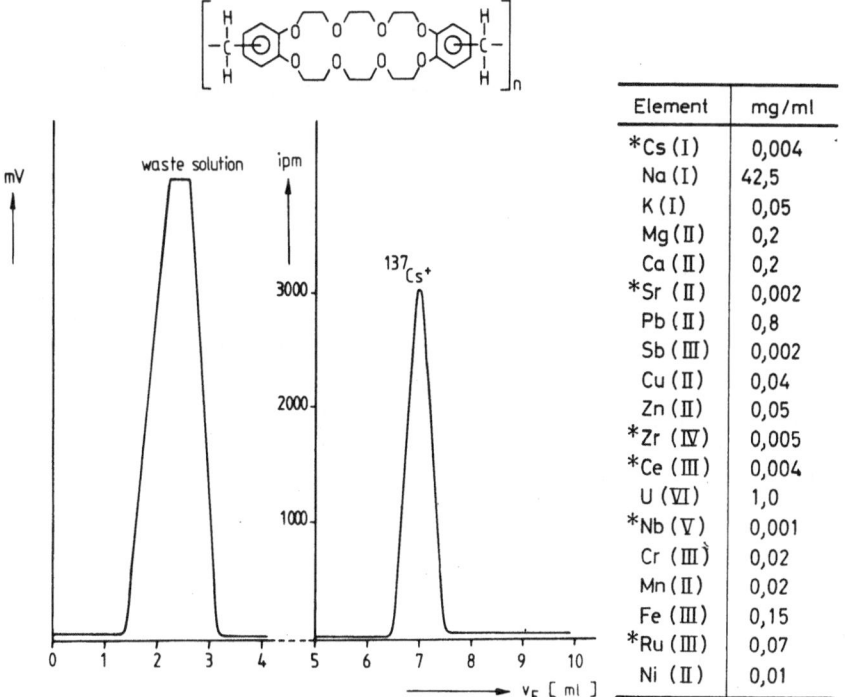

Element	mg/ml
*Cs (I)	0,004
Na (I)	42,5
K (I)	0,05
Mg (II)	0,2
Ca (II)	0,2
*Sr (II)	0,002
Pb (II)	0,8
Sb (III)	0,002
Cu (II)	0,04
Zn (II)	0,05
*Zr (IV)	0,005
*Ce (III)	0,004
U (VI)	1,0
*Nb (V)	0,001
Cr (III)	0,02
Mn (II)	0,02
Fe (III)	0,15
*Ru (III)	0,07
Ni (II)	0,01

Fig. 10. Separation of 0.0004 mg Cs+ (labelled by 5 µCi ^{137}Cs) from simulated nitric (1 mol/l) middle-active waste solution by elution with water

The high selectivity of the anchoring group DB[24]C-8 for Cs(I) permits the separation of Cs(I) from a mixture containing many other cations (Fig. 10).

2.3.1.3.3.8 §-Dibenzo[30]crown-10

a) Selectivity series for this anchoring group

The anchoring group exhibits no distinct selectivity for alkali and alkaline earth salts. However, for organic compounds which may interact with the anchoring group, a selectivity series may be established.

Aliphatic compounds: thiourea > urea

Aromatic compounds: $-CH_2Cl > -CH_3$

$-SH > -OH$

$-Br > -F > -Cl$

$-NO_2 > -NH_2 > -CH_3 > -SO_3H$

Heterocycles: pyrrole > thiophene > furan

triazine > pyrazine > pyridine.

b) Separations by §-Dibenzo[30]crown-10 (matrix: methylene bridges).

Compounds containing sulfur atoms are generally strongly bound by the exchanger. The separation of benzene from water and thiophene is possible. Diphenyl disulfide and thianthrene can also be separated. This was important for the elucidation of the reaction of aryl mercaptans with SF_4.

With increasing number of nitrogen atoms in the six-membered ring, the binding forces of the exchanger grow [40]. The eluent volumes required for these heterocycles increase in the order O, S, N [40].

A further example of the analysis and purity control of pharmaceutical products is the separation of the components of an analgesic [40].

In addition, numerous separations based on matrix effects are possible.

Table 7. Electrophoresis of neutral amino acids

solid support	paper [43]	paper [44]	paper [45,46]	§-DB[18]C-6
electrolyte buffer	pyridine acetate buffer	0.75 mol/l formic acid/ 1.0 mol/l acetic acid (1:1)	0.6 mol/l formic acid/ 2 mol/l acetic acid	1.09 mol/l formic acid/ 2.6 mol/l acetic acid
pH	3.6	2.25	1.9	1.64
voltage and	100 V/cm	55 V/cm	60 V/cm	40 V/cm
duration of electrophoresis	180 min	150 min	200 min	60 min
R_B-values				
Gly	133	143	140	97
αABS	107	—	106	135
Ser	68	102	102	156
Val	100	100	100	100
Ileu	—	97	94	33
Pro	51	81	—	135
Phe	79	79	—	20
Tyr	79	70	67	135
Try	88	59	—	3

2.3.2 Thin-Layer Chromatography and Thin-Layer Electrophoresis

The exchangers can also be applied as adsorbents in thin-layer chromatography and electrophoresis.

For this purpose, poly(ethylene terephthalate) sheets [42] are coated with suspensions of powdered exchangers in poly(vinyl alcohol) solutions. The exchanger layers are non-abrasive and resistant to fracture. Many organic solvents can be used, with the exception of chloroform, dichloromethane and dioxane.

a) Neutral amino acids

Table 7 shows a comparison of the results obtained by thin-layer electrophoresis on the exchanger layers with those obtained by paper electrophoresis.

Thus, it is possible to separate certain neutral amino acids, e.g. isoleucine, valine, serine and α-aminobutyric acid, which show non-resolved zones on paper, even when applying a higher voltage or longer separation times.

b) Benzoic acid, salicyclic acid and sorbic acid

The preservatives benzoic acid, salicyclic acid and sorbic acid can readily be separated by thin-layer chromatography on layers of the exchanger §-DB[18]C-6. This exchanger layer proved to be superior to the adsorbents usually used (Table 8).

Table 8a. Thin-layer chromatography of benzoic acid, salicyclic acid and sorbic acid on various layers

	cellulose [47]	silica gel/kieselguhr [47]	silica/gel/kieselguhr [47]
layer			
solvent	n-butanol/NH$_3$/H$_2$O (70:20:10 v/v)	hexane/acetic acid (96:4 v/v)	petroleum ether/ether (80:20:1 v/v)
development	5–6 h	20 cm in 1.5 h; if neccessary multiple d.	triple development
R$_B$-values			
salicyclic acid	1.00	1.00	1.00
sorbic acid	1.07	1.28	0.91
benzoic acid	0.91	1.54	1.11

Table 8b.

	kieselguhr [48]	§-dibenzo[18]crown-6
layer		
solvent	CHCl$_3$/acetic acid (90:10 v/v)	methanol
development		8–10 cm in 1 h
R$_B$-values		
salicyclic acid	1.00	1.00
sorbic acid	1.11	1.83
benzoic acid	1.08	1.40

3 Determination Methods by Means of Monomeric Cyclic Polyethers

The introduction of chromophoric groups into cyclic polyethers yields selective photometric reagents. The complexing properties of cyclic polyethers can be utilized for electrochemical determination methods.

3.1 Photometry

In principle, all picrate extractions by cyclic polyethers offer the possibility of direct photometric determination of the extracted salts (cf. Sect. 2.2.1). Another route involves intramolecular introduction of chromophoric substituents into cyclic polyethers.

4'-Picrylaminobenzo[15]crown-5 forms with K(I) an intensely colored ion pair [49]. The reddish-yellow powder of the crown compound is only moderately soluble in water but can be dissolved in many organic solvents. The saturated aqueous solution is reddish-yellow. In order to separate K(I) or Rb(I) from Na(I), the aqueous solution is extracted by chloroform containing 0.002 M 4'-Picrylamino-benzo[15]crown-5 and 1 M triethylamine. The latter is added to provide an excess of the anionic form of the crown compound.

From the difference between the adsorption of the chloroform solution at 560 nm before and after extraction, the content of K(I) or Rb(I) is determined. The determination of 10—400 ppm K(I) is possible in the presence of 2000 ppm Na(I).

3.2 Electrochemical Determination Methods

Potentiometric, conductometric, polarographic and voltammetric determinations are possible. Many studies on the application of cyclic polyethers as components of ion-sensitive electrodes have been reported.

3.2.1 Potentiometry

Na(I) and K(I) can be titrated with aqueous solutions of the 2.2.1 cryptand and 2.2.2 cryptand, respectively, using a cation-selective electrode [50]. The titration is performed in weakly alkaline solution the pH of which is adjusted to 9—10 by triethanolamine. Good results are achieved with alkali salt concentrations of 10^{-3} to 10^{-4} mol/l. The aqueous solution of the cryptands is stable for six weeks.

For Li(I) the 2.2.1 cryptand is used. However, the potential drop is small. The 2.2.1 cryptand ought to be more suitable for the determination of Ag(I) because the pK of the complex is 10.6 [51].

3.2.2 Conductometry

0.001 mol/l CsCl can be determined by conductometric titration in methanol/chloroform (90/10 v/v) with [18]crown-6 or polyvinyl[18]crown-6. The decrease of conductivity at the end point of the titration with polyvinyl[18]crown-6 is particularly high since the mobility of the Cs(I) complex with the linear polyether is less than that with the monomeric polyether [52, 53].

Na(I) can be titrated with dicyclohexano[18]crown-6. However, there is no sharp end point [54]. The conductivities of the single-protonated cryptands 2.1.1, 2.2.1, 2.2.2 and $2_B.2.2$ in water are practically equal, namely about 23.0 ± 0.5 cm^2 equiv$^{-1}\Omega$ at 25 °C [55]. Conductivity measurements of Na(I) and K(I) picrates as well as of the corresponding complexes with the 2.2.1 and 2.2.2 cryptands in tetrahydrofuran show that the dissociation constants of the solvent-separated ion pairs scarcely differ from those of the ligand-separated ion pairs [56].

3.2.3 Polarography

The stability constants of the complexes of dicyclohexano[18]crown-6 with sodium or potassium salts can be determined polarographically. The reduction at the dropping mercury electrode is reversible [57]. For the determination of the stability constants of other complexes, acetonitrile is recommended as solvent [58]. A stability constant of 1.4×10^3 is found in this solvent for the complex of [18]crown-6 and HgO [59]. The cations complexed by the crown compounds or cryptands are strongly adsorbed on the dropping mercury cathode [60].

The polarographic behavior of the complex of the 2.2.2 cryptand with Tl(I) in propylene carbonate has also been described [61]. The standard potential of the redox system Tl(I)/Tl(III) is +0.3 V, as measured by means of the saturated calomel electrode.

The diffusion coefficient is about 2.4×10^{-6} cm^2/s for Tl(I) and 1.3×10^{-6} cm^2/s for the corresponding complex. The reduction of the complex proceeds very slowly.

The first reduction product of the complex of the 2.2.2 cryptand with K(I) in propylene carbonate is K(0). The complex is then decomposed with the formation of amalgam. The limiting diffusion current is independent of the concentration of of the cryptand [62].

The following half-wave potentials have been determined for the alkali metal complexes with the 2.2.2 cryptand in propylene carbonate [63, 64]. As reference electrode a saturated calomel electrode has been used.

Li(I)	Na(I)	K(I)	Rb(I)	Cs(I)
−1.97	−1.84	−1.95	−1.97	−1.97

3.2.4 Ion-Sensitive Electrodes

The application of ion-sensitive electrodes with cyclic polyethers as carrier molecules has been described in numerous reviews [65–76].

The best K(I)-selective glass membrane electrodes show a selectivity of 20:1 for K(I)/Na(I). In contrast, electrodes based on valinomycin display selectivities of 4000:1 for K(I)/Na(I) and 100:1 for K(I)/NH$_4$(I) [77, 78].

The selectivities of cyclic polyethers as components of ion-sensitive electrodes are not influenced by the relative but by the absolute transfer velocity in the carrier molecules [79, 80]. In order to differentiate the relative transfer velocities, especially those multidentate ligands are suitable [81, 82] the coordination centers of which are so arranged that they are able to replace the hydrate shell of the cation to be determined and which are in addition sufficiently soluble in the membrane phase. Acyclic ligand matrices, e.g. some representatives of oxa-analogous azelaic or suberic diamids with long hydrocarbon chains or aromatic groups, fulfil these requirements [83, 86]. Depending on the number of coordination atoms and on the type of the chain ends, there are ion-sensitive carriers for Na(I) [86], Ca(II) and Ba(II) [84].

In Table 9 the carrier molecules and selectivity constants are listed.

The selectivity of carrier b for Ca(II), compared to those for Mg(II) and Zn(II) [88], is very striking and important for the differentiation of blood sera.

Table 9. Selectivity constants of carrier molecules in PVC matrices in acidic solution [87]

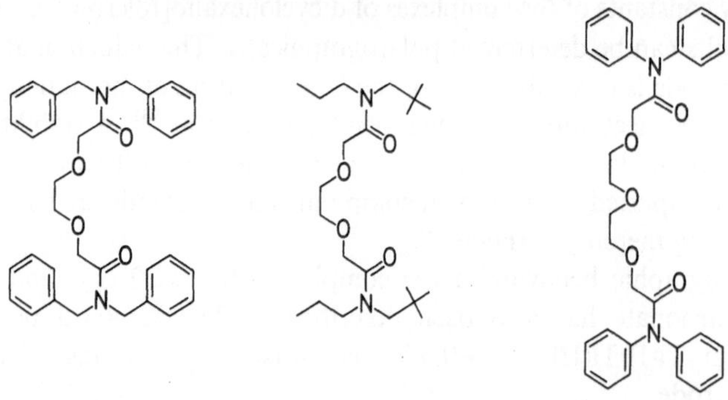

	a	**b**	**c**

M	M:Na(I)	M:Ca(II)	M:Ba(II)
Mg(II)	3×10^{-4}	3×10^{-4}	4×10^{-5}
Ca(II)	6×10^{-3}	1	8×10^{-4}
Ba(II)	4×10^{-2}	8×10^{-3}	1
Na(I)	1	2×10^{-2}	2×10^{-2}
K(I)	2×10^{-1}	2×10^{-2}	4×10^{-2}
NH$_4$(I)	4×10^{-3}	2×10^{-2}	1×10^{-2}
H(I)	1	10	9×10^{-2}
Zn(II)	—	10^{-3}	—

Particularly favorable is the fact that the membrane electrode does not interact with Na(I) and K(I) under physiological conditions. For similar reasons, carrier a which is selective for Na(I) in the presence of protons, is interesting. The Ba(II)-selective carrier c shows a good reproducibility of the e.m.f. of the measuring circuit and a long life. Another Na(I)-selective ligand [89] permits the intracellular determination of Na(I) in blood serum [90].

4 Conclusion

Cyclic polyethers are versatile in analytical and preparative chemistry.
a) Monomeric cyclic polyethers

A question of analytical interest is to which extent monomeric chiral polyethers are able to influence the selectivity of chiral membranes for enantiomers [91].

Several monographs on the applications of monomeric cyclic polyethers in phase transfer catalysis have been published [92, 93]. For technological applications, it has been suggested to use monomeric cyclic polyethers in non-aqueous electrochemical cells [94, 95] as corrosion inhibitors for metals [96] and for isotope enrichment by extraction [97-99].

b) Cross-linked polymeric cyclic polyethers

In the field of analytical chemistry the application in ion chromatography should be particularly mentioned. Using high elution rates, separations can be performed quickly and simply by polyethers bound to silica gel. A disadvantage is the low chemical resistance of this stationary phase to alkaline solutions. Better properties are shown by polystyrene beads to the surface of which polyethers are linked by methylene bridges [42].

In the field of preparative chemistry, salt conversion and anion activation are to be noted. For this purpose, copolymers of vinyl and divinyl crown compounds or other cross-linking agents are being prepared at present. They exhibit a spherical shape [42].

5 References

1. Stetter, H., Frank, W.: Angew. Chem. 88, 760 (1976), Int. Ed. Engl. 15, 686 (1976)
2. Eisenman, G., Ciani, S. M., Szabo, G.: J. Membr. Biol. 1, 294 (1969)
3. Jwachido, T., Sadakane, A., Toei, K.: Bull. Chem. Soc., Jpn. 51, 629 (1978)
4. Frensdorff, H. K.: Am. Chem. Soc. 93, 4685 (1971)
5. Sekine, T., Wakabajashi, H., Hasegawa, Y.: Bull. Chem. Soc. Jpn. 51, 645 (1978)
6. Sadakane, A., Jwachido, T., Toei, K.: ibid. 48, 60 (1975)
7. Kina, K., Shiraishi, K., Jshibashi, N.: Bunseki Kagaku 27, 291 (1978)
8. Kimura, T., Washima, I., Shimoro, T.: Chem. Lett. Jpn. 1977, 563
9. Takeda, Y., Goto, H.: Bull. Chem. Soc. Jpn. 52, 1920 (1979)
10. Takeda, Y., Goto, H.: ibid. 52, 2501 (1979)
11. Takeda, Y., Kato, H.: ibid. 52, 1027 (1979)
12. Mitchell, J. W., Shanks, D. L.: Anal. Chem. 47, 642 (1975)
13. Sumiyoshi, M., Nakahara, N.: Talanta 24, 763 (1977)
14. Marcus, Y., Asher, L. E.: J. Phys. Chem. 82, 1246 (1978)
15. Haynes, A. H., Pressman, B. C.: J. Membr. Biol. 18, 1 (1974)
16. McDowell, W. J., Shoun, R. R.: Report (1977), CONF 7090F-4; C. A. 88, 392 (1978)
17. Helgeson, R. C., Timko, J. M., Cram. D. J.: J. Amer. Chem. Soc. 95, 3023 (1973)
18. Kimuru, T., Iwashimo, T., Hamada, T.: Anal. Chem. 51, 1113 (1979)
19. Takeda, Y., Oshio, K., Segawa, Y.: Chem. Lett. 5, 601 (1979)
20. Blasius, E. et al.: Talanta 27, 127 (1980)
21. Yoshio, M. et al.: Anal. Lett. 11, 281 (1978)
22. Sevelie, D., Meider, H.: J. Inorg. Nucl. Chem. 39, 1403 (1977)
23. Gloe, K. et al.: Z. Chem. 19, 382 (1979)
24. Kopolov, S., Hogen-Esch, T. E., Smid, J.: Macromolecules 6, 133 (1973)
25. Smulek, W., Lada, W.: Radiochem. Radioanal. Lett. 30, 199 (1977)
26. Melson, F. et al.: J. Chromatogr. 20, 107 (1965)
27. Delphin, W. H., Horwitz, E. P.: Anal. Chem. 50, 843 (1978)
28. Koltloff, I. M.: ibid. 51, 1R (1979)
29. Helgeson, R. C. et al.: J. Amer. Chem. Soc. 96, 6762 (1974)
30. Sousa, R. L., Hoffmann, D. H., Kaplan, L., Cram, D. J.: J. Amer. Chem. Soc. 96, 7100 (1974)
31. Dietrich, B., Lehn, J. M., Simon, J.: Angew. Chem., Int. Ed. Engl. 13, 406 (1974)
32. Behr, J. P., Lehn, J. M., Vierling, P.: J. Chem. Soc., Chem. Comm. 1976, 621
33. Russel, G. A., Walraff, G., Gerlock, J. L.: J. Phys. Chem. 82, 1161 (1978)
34. Blasius, E., Adrian, W., Janzen, K.-P., Klautke, G.: J. Chromatogr. 96, 89 (1974)
35. Blasius, E., Janzen, K.-P.: Chem.-Ing.-Techn. 47, 594 (1975)
36. Blasius, E., Maurer, P. G.: J. Chromatogr. 125, 511 (1976)
37. Blasius, E., Maurer, P. G.: Makromol. Chem. 178, 649 (1977)
38. Blasius, E. et al.: Z. Anal. Chem. 284, 337 (1977)
39. Blasius, E., Janzen, K.-P., Neumann, W.: Mikrochim. Acta 1977, II 279
40. Blasius, E. et al.: J. Chromatogr. 167, 307 (1978)
41. Blasius, E. et al.: Talanta 27, 107 (1980)
42. Blasius, E. et al.: J. Chromatogr. 201, 147 (1980)

Ewald Blasius and Klaus-Peter Janzen

43. Clotten, R., Clotten, A.: Hochspannungselektrophorese, Stuttgart: Georg Thieme Verlag 1962, pp. 362
44. Visakorpi, J. K., Puranen, A. L.: Scand. J. Clin. Laborat. Invest. *10*, 196 (1958)
45. Werner, G., Westphal, O.: Angew. Chem. *67*, 251 (1955)
46. Wieland, T., Pfleiderer, G.: ibid. *69*, 199 (1957)
47. Copius-Peereboom, J. W., Beekes, H. W.: J. Chromatogr. *14*, 417 (1964)
48. Lemieszek-Chodorowska, K., Snycerski, A.: Ref. Anal. Abstracts *23*, 1937 (1972)
49. Tagaki, M., Nakamura, H., Ueno, K.: Anal. Lett. *10*, 1115 (1977)
50. Czerwenka, G., Scheubeck, L.: Z. Anal. Chem. *276*, 34 (1975)
51. Lehn, J. M., Sauvage, J. P.: J. Am. Chem. Soc. *97*, 6700 (1975)
52. Kopolow, S. et al.: J. Macromol. Sci. Part A *7*, 1015 (1973)
53. Hoegen-Esch, T. E., Kopolow, S., Smid, J.: Macromolecules, *6*, 133 (1973)
54. Pedersen, C. J., Frensdorff, H. J.: Angew. Chem. Int. Ed. Engl. *11*, 16 (1972)
55. Cox, B. G., Schneider, H., Stroka, J.: J. Am. Chem. Soc. *100*, 4746 und 6003 (1978)
56. Boileau, S., Hemery, P., Justice, J. C.: J. Solution Chem. *44*, 873 (1975)
57. Angostiano, A., Caselli, M., Della Monica, M.: J. Electroanal. Chem. *74*, 95 (1976)
58. Koryta, J., Mittal, L. M.: J. Electroanal. Chem. *16*, App. 14 (1972)
59. Pospisil, L., Mittal, L. M., Kuta, J., Koryta, J.: J. Electroanal. Chem. *46*. 203 und 251 (1973)
60. Britz, D., Knittel, D.: Electrochim. Acta *20*, 891 (1975)
61. Peter, F., Gysselbrecht, J. P., Gross, M.: J. Electroanal. *86*, 115 (1978)
62. Peter, F., Gross, M.: J. Electroanyl. Chem. *74*, 315 (1976)
63. Peter, F., Gross, M.: ibid. *61*, 245 (1975)
64. Peter, F., Gross, M.: ibid. *53*, 307 (1974)
65. Eisenman, G. et al.: Progr. Surface and Membrane Sci. (Danielli, J. F., Rosenberg, H. D., Capenhead, D. A., (ed.)), New York: Academic Press, Vol. 6 1973 pp. 140–242
66. Ovchinninkov, Y. A., Ivanov, V. T., Shlorob, A. M.: Membrane Active Complexones, Amsterdam: Elsevier 1974
67. Covington, A. K.: CRC Crit. Rev. Anal. Chem. *1974*, 355
68. Simon, W., Morf, W. E., Meier, P. Ch.: Struct. Bonding *16*, 115 (1973)
69. Morf, W. E., Wuhrman, P., Simon, W.: Anal. Chem. *48*, 1031 (1976)
70. Thomas, A. P. et al.: Anal. Chem. *49*, 1567 (1977)
71. Lindner, E. et al.: Anal. Chem. *50*, 1627 (1978)
72. Buck, R. P.: Anal. Chem. *50*, 17 R (1978)
73. Buck, R. P.: ibid. *48*, 26 R (1976)
74. Buck, R. P.: ibid. *46*, 28 R (1974)
75. Buck, R. P.: ibid. *44*, 270 R (1972)
76. Camman, K.: Das Arbeiten mit ionensensitiven Elektroden. Springer-Verlag, Heidelberg, New York, 1973
77. Stefanak, L., Simon, W.: Microchem. J. *12*, 125 (1967)
78. Pioda, L. A. R., Stankova, V., Simon, W.: Anal. Lett. *2*, 665 (1969)
79. Wuhrmann, P., Thoma, A. P., Simon, W.: Chimia *27*, 637 (1973)
80. Amman, D., Pretsch, E., Simon, W.: Tetrahedron Lett. *1972*, 2473
81. Morf, W. E., Simon, W.: Helv. Chim. Acta *54*, 2683 (1971)
82. Kirch, M., Lehn, J. M.: Angew. Chem. *87* (1975); Angew. Chem. Int. Ed. Engl. *14*, 555 (1975)
83. Thomas, R., Simon, W., Oehme, M.: Nature *285*, 754 (1975)
84. Amman, D., Pretsch, E., Simon, W.: Helv. Chim. Acta *56*, 1780 (1973)
85. Güggi, M., Pretsch, E., Simon, W.: Anal. Chim. Acta *91*, 107 (1977)
86. Amman, D., Pretsch, E., Simon, W.: Anal. Lett. *7*, 23 (1974)
87. Weber, E., Vögtle, F.: Kontakte (Merck) *2*, 16 (1978)
88. Fuchs, C. et al.: Klin. Wschr. *50*, 824 (1976)
89. Güggi, M. et al.: Helv. Chim. Acta *59*, 2417 (1976)
90. Le Blanc, O. H. et al.: A. Appl. Physiol. *40*, 644 (1976)
91. Newcomb, M., Helgeson, R. C., Cram, D. J.: J. Amer. Chem. Soc. *96*, 7367 (1974)
92. Weber, W. P., Gokel, B. W.: Phase Transfer Catalysis in Organic Synthesis Springer-Verlag, Berlin, Heidelberg, New York, 1977
93. Dehmlov, E. V., Dehmlov, S. S.: Phase Transfer Catalysis, Weinheim: Verlag Chemie 1980
94. U.S. Pat. 4, 132, 837, X Cl. 429-194, H 01 M6144 (1978'; C. A. 90: 1715295 (1978)

95. Jpn. Pat. 7901, 827; C Vl. H01 M617 (1979)
96. U.S. Pat. 3, 129, 521; Cl. 48–62, C. 23C (1975) C. A. *84*, 152334 (1975)
97. Jepson, B. E., DeWitt, R.: J. Inorg. Nucl. Chem. *38*, 1175 (1976)
98. Seminara, A., Siracusa, B.: Inorg. Chim. Acta *20*, 105 (1976)
99. Costen, R. M. et al.: Inorg. Nucl. Chem. Lett. II, 469 (1975)

Crown Compounds
as Alkali and Alkaline Earth Metal Ion
Selective Chromogenic Reagents

Makoto Takagi and Keihei Ueno

Department of Organic Synthesis, Faculty of Engineering, Kyushu University, Fukuoka 812, Japan

Table of Contents

This article deals with the recent development of crown ethers which have added chromophoric functional groups within the molecule (crown ether dyes). These compounds are designed to bring about specific color changes on the interaction with such metal cations as alkali and alkaline earth metals, thus being able to serve as probes or photometric reagents selective for these metal ions. The chromophoric groups can bear a dissociable proton (or protons) or can be nonionic. In the former, the ion exchange between the proton and appropriate metal cations causes the color change, while in the latter, the coordination of the metal ion to the chromophoric donor or acceptor of the dye molecule induces a change of the charge transfer band of the dye. The proton-dissociable crown ether dyes are especially suited for extraction photometric determination of alkali and alkaline earth metal ions. Similarly, the neutral crown ether dyes are potentially useful for such determinations in lipophilic homogeneous media.

1 Introduction

Since the dawn of inorganic analysis and metal complex chemistry, a wide variety of ion selective organic chromogenic reagents have been developed and come to practical use to solve problems in metal analysis [1,2]. Selective chromogenic reagents, alone or in combination with masking agents, have satisfactorily been used for colorimetric determination of most of the commonly encountered metal ions. Some of them are indispensable as metal indicators for standard chelatometric titrations.

While ion selective chromogenic reagents have become popular as early as 1950 for most of the metal ions, attempts to develop those selective to alkali metal ions remained unsuccessful. As to alkaline earth metal ions, the reagents with satisfactory selectivity are still rather rare, since they often suffer competition from transition metals. The discovery of crown ethers in 1967 [3] definitely gave a clue to cultivate this last field of organic reagent-aided metal analysis, but it still took ten years for such alkali and alkaline earth metal-selective chromogenic reagents to receive practical significance. In this review, recent research efforts are summarized for the development of crown ether-based chromogenic reagents which selectively respond to alkali and alkaline earth metal ions. Some critical comments are also added on the use of these reagents for analytical problems of these metal ions.

2 Monoprotonic Crown Ether Dyes

Designing alkali metal selective crown ether dyes is not much different in principle than that of classical chromogenic chelating agents; a monoprotonic chromophore is introduced into a crown ether skeleton in such a proximity to the ethereal function that the dissociation of the chromophoric proton is assisted by the complexation of the positively charged metal ion with the crown ether macrocycle [4]. Conceptually, the function of metal binding and the function of its detection may be separately carried by the crown ether and the protonic chromophore, respectively. However, the anionic chromophore can more or less contribute to the metal binding ability of the crown ether, and thus, in the actual molecule, the two functions can not be separated from each other.

2.1 Derivatives of Benzocrown Ethers

The first crown ether dyes designed according to the above principle are 4'-picryl-amino-substituted derivatives of benzo[15]crown-5 (B15C5)[1], 1–3, reported by Takagi, Nakamura, and Ueno [4,5]. The synthesis is readily attained by first nitrating and then reducing B15C5 leading to 4'-amino-B15C5, which is then reacted with picryl chloride under basic conditions to yield 1.

The crown ether dye 1 is almost insoluble in water, but is soluble in aqueous alkali

---

1 Notations frequently used in this report are listed and explained on p. XI

or in common organic solvent such as methanol, chloroform or benzene. The distribution between water and the organic solvent is strongly in favor of the latter, and under ordinary conditions, the amount of *1* dissolved in water which is in equilibrium with an equal volume of chloroform is negligible as compared with that in the organic phase. *1* is appreciably acidic and the colour of the solution turns from orange (λ_{max} = 390 nm; ε, 12,900 M^{-1} cm; 1 M = 1 mol dm^{-3}) to blood red (λ_{max} = 445 nm; ε, 20,000) on dissociation of the amino proton.

When a chloroform solution of *1* (HL) is brought into contact with an alkaline aqueous solution of alkali metal salts, K^+ and Rb^+ are strongly extracted into the chloroform solution and the color of the chloroform solution changes from orange to blood-red, indicating the formation of L^- species in the organic solution. Na^+ is extracted only slightly, while Li^+ not at all. The use of strong alkali such as sodium hydroxide and lithium hydroxide causes the coloration of the aqueous rather than the chloroform phase, indicating that the ML salt of these metal ions is poorly soluble in chloroform; the enforced dissociation of HL by the strong alkali results in the transfer of the dissociated crown ether species (L^-) from the organic to aqueous phase. Thus, the extent of coloration of the organic phase can be used as a measure for alkali metal (potassium and rubidium) present in the original aqueous solution.

In the typical extraction of alkali metal ions, equal volumes of an aqueous solution of alkali metal chloride (10^{-3}–1 M) and a chloroform solution at 2×10^{-3} M in *1* and 1 M in triethylamine are shaken at 25 °C, and the change in optical absorption of the organic layer is measured at 560 nm. The increase in molar absorptivity of HL upon formation of metal complex ML (or ML · HL, see below) is 4580 M^{-1} cm at this wavelength. The extraction is most satisfactorily explained by the reaction in Scheme 1. The extraction constants concerned are defined by Eqs. (1)–(4). The

$$(HL)_a \rightleftharpoons (L^-)_a + (H^+)_a$$

$$(M^+)_a \quad \text{aqueous phase}$$

_____

organic phase

$$(HL)_o \qquad (ML)_o \xrightarrow{(HL)_o} (ML \cdot HL)_o$$

parentheses with suffixes indicate the chemical species (suffix "a" for aqueous phase and "o" for organic phase), and the brackets stand for concentrations (M) in an ordinary sense.

$$(HL)_o + (M^+)_a \rightleftharpoons (ML)_o + (H^+)_a \tag{1}$$

$$K_{ex(ML)} = \frac{[(ML)_o] [(H^+)_a]}{[(HL)_o] [(M^+)_a]} \tag{2}$$

$$2 (HL)_o + (M^+)_a \rightleftharpoons (ML \cdot HL)_o + (H^+)_a \tag{3}$$

$$K_{ex(ML \cdot HL)} = \frac{[(ML \cdot HL)_o] [(H^+)_a]}{[(HL)_o]^2 [(M^+)_a]} \tag{4}$$

Makoto Takagi and Keihei Ueno

Table 1. Extraction constants (K_{ex}) of picrylamino-substituted benzocrown ether dyes (HL) for alkali metal ions

Crown ether	$pK_{HL}{}^a$		Li^+	Na^+	K^+	Rb^+	Cs^+
1^b	10.55	$pK_{ex(ML)}$	$—^c$	~ 13	—	—	—
		$pK_{ex(ML \cdot HL)}$	—	~ 10	7.55	8.5	10.4
2^d	8.63	$pK_{ex(ML \cdot HL)}$	—	~ 8.8	~ 7.1	~ 7.9	~ 8.3
4^b	10.58	$pK_{ex(ML \cdot HL)}$	—	12.49	10.39	10.70	—
		$pK_{ex(ML \cdot HL)}$	—	—	—	—	7.55

[a] $K_{HL} = [H^+][L^-]/[HL]$, water dioxane (9:1, v/v), 0.10 M LiCl, 25 °C; [b] Water chloroform (0.1 M triethylamine), pH 11.46, 25 °C; [c] Dashes indicate no measurable extraction according to this reaction; [d] Water chloroform (0.5 M triethanolamine), pH 10.5, 25 °C

Table 1 summarizes the extraction constants of *1*, *2*, and *4*[6] for alkali metal ions. As for 15C5-type crown ethers, *1* and *2* differ from each other in their acidity, the latter being higher by 2 logarithmic units due to the electron-withdrawing effect of the added nitro group in *2*. Since the extraction constants defined in Eqs. (2) and (4) are a kind of acid dissociation constants and thus implicitly involve the acid dissociation constants in aqueous solution[5], the extraction constants obtained for *2* in Table 1 are appreciably higher than those for *1*. This means that the extraction of alkali metals takes place at lower pH for *2* than for *1*. In agreement with this, the extraction study is made at lower pH for *2* by using triethanolamine rather than triethylamine. However, for practical use of these crown ether reagents for extraction colorimetric determination of alkali metal ions, it is best to use EDTA · Li (lithium salt of ethylenediaminetetraacetic acid) as a pH-buffering agent, since EDTA also serves as a potent masking agent for interfering polyvalent metal ions (see below). Li^+ has no measurable tendency to interact with the crown ethers.

It can be seen from Table 1 that picrylamino-substituted B15C5 ethers extract alkali metal ions predominantly in the form of 1:2 (metal:ligand) complexes (ML · HL). The extraction of the 1:1 complex (ML) is inferred only for Na^+. The 18C6-type dye *4*, on the other hand, extracts alkali metal ions predominantly through the formation of 1:1 complexes, except for Cs^+ which is extracted as a 1:2 complex (CsL · HL). A fitness-relationship between the cavity size of crown ethers and the ionic radius of metal ions has been well documented and discussed in reviews[7, 8]. It is widely known that 15C5-type crown ethers readily form 1:2 complexes with metal

1 n = 1 , X = H
2 1 , X = NO_2
3 1 , X = Br
4 2 , X = H
5 2 , X = NO_2

ions of larger ionic radius than Na$^+$ [9, 10]. An extensive study has been made on the complex formation and extraction equilibria in water-chloroform system for *1*, *2*, *4*, and *5* [5, 6].

It is to be noted that a general comparison of extraction efficiency or extraction selectivity of alkali metals between the 15C5- and 18C6-type crown ethers can not be simply made from the extraction data under particular conditions. This is because the mode of metal extraction differs. Thus, the fraction of the metal extracted, or of the crown ether HL converted to L$^-$ species in organic solution, is quite dependent on the concentration of crown ethers used. However, by using the constants in Table 1 it is easy to calculate these quantities as a function of particular extraction conditions.

An extraction photometric determination of 40–400 ppm and 4–40 ppm K$^+$ by *1* and *5*, respectively, has been described [5, 6]. Potassium in sea water was determined by using *1* in water (0.2 M Li · EDTA, pH 11.46) — chloroform (1 M triethylamine, 1×10^{-3} M *1*). The presence of Na$^+$ upto 200 fold molar excess of K$^+$ did not interfere the determination. Li · EDTA masked magnesium and calcium which otherwise interfered with the determination by forming hydroxide precipitates. K$^+$ in portland cement was similarly determined by using *5* in water (0.1 M Li · EDTA, pH 12.35) — chloroform (1×10^{-4} M *5*).

Under the above determination conditions, usually only serveral percent of K$^+$ in aqueous solution is transferred into organic solution, and correspondingly, the calibration lines are not entirely linear but often show a curvature. This is in contrast with the traditional concept of colorimetry of metals, where the complete conversion into colored metal complexes is taken as a prerequisite. The extraction constants of the crown ether dyes are much too small compared to the standard of such traditional extraction photometric reagents like dithizone and 8-hydroxyquinoline (for the determination of heavy metals). The low extraction constants necessitate the use of relatively high pH conditions for the effective extraction of alkali metals; a pH control (pH buffer) is also important in order to obtain reproducible results.

Pacey and coworkers described some new variations of picrylamino type benzocrown ethers in which one of the nitro groups of the picryl moiety was substituted by either a cyano or trifluoromethyl group [11, 12]. The crown ether dyes *6–8* have their general properties in common with dyes *1–5*, but improvement is attained in their spectral properties. Thus, while a considerable spectral overlap between HL and ML (or ML · HL) species is observed for *1–5*, such overlap is diminished for *6–8* (Fig. 1).

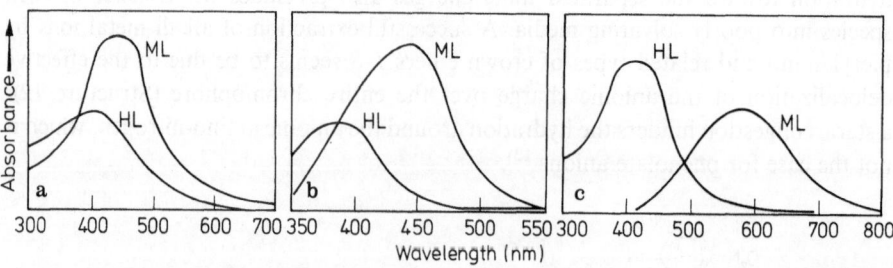

Fig. 1. Absorption spectra of HL and L$^-$ for crown ether dyes *1*, *7*, and *8*. 2×10^{-4} M crown ether dye. **a** *1* in water dioxane (1:1); **b** *8* in water acetonitrile (3:2); **c** *7* in water acetonitrile (3:2) (Reproduced with permission from: G. E. Pacey, Y. P. Wu and B. P. Bubnis, Analyst *106*, 636 (1981))

4'-(2",4"-Dinitro-6"-trifluoromethylphenyl)amino-B15C5 (*8*) showed the best extraction efficiency among these modified crown ether dyes, and it is claimed that its calibration graph gives a linear range of 5–700 ppm for K^+ in the presence of 3000 ppm of Na^+ [12].

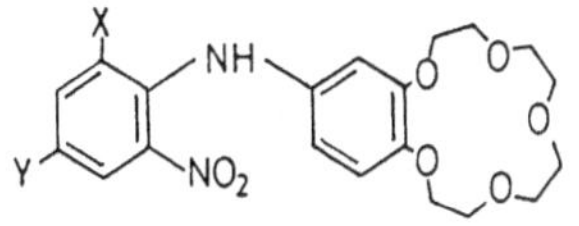

6	X = NO₂	Y = CN
7	NO₂	CF₃
8	CF₃	NO₂

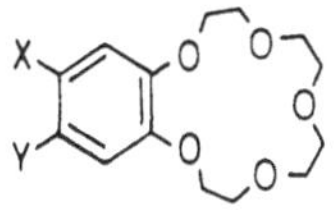

9	X =	(chromophore),	Y = H
10		−N=N−(Cl)(OH),	H
11		−N=N−(NO₂),	OH

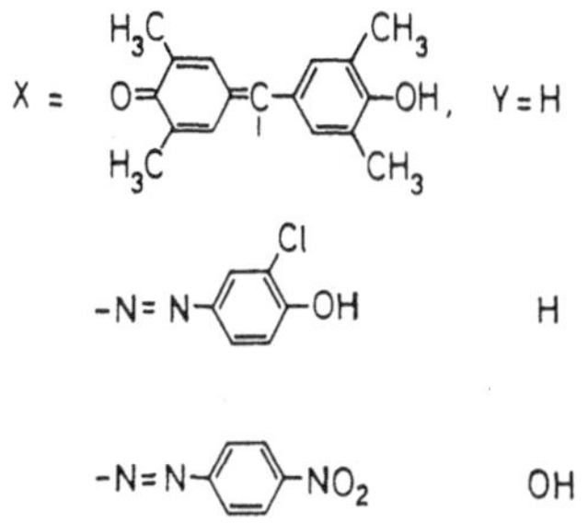

Various other monoprotonic chromogenic crown ethers are readily derived from benzocrown ethers. Dyes *9–11* are such examples. However, few proved to be useful as a reagent for extraction photometry of alkali metals; the metal extraction efficiency was too low [13]. The low extractability of these metal complexes are related to their "intramolecular" ion pair structure in the organic solution, where the alkali metal cation is located in the cavity of the crown ether and the anionic charge is located in the remote chromophoric portion of the molecule. The charge-separated species are not stabilized in non-polar organic solvents. A strong hydration toward the separated ionic charges also precludes the transfer of such species into poorly solvating media. A successful extraction of alkali metal ions by picrylamino and related types of crown ethers *1–8* seems to be due to the effective delocalization of the anionic charge over the entire chromophore (Structure *12*); a steric congestion hinders the hydration around the anionic amino-nitrogen, which is not the case for phenolate anions [13].

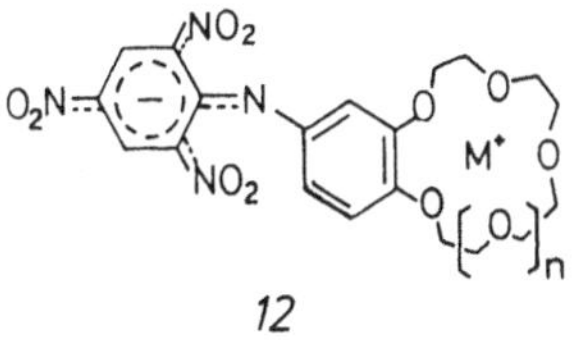

12

2.2 N-Side-armed Monoazacrown Ethers

To enhance the metal extraction efficiency or to increase the extraction constant [Eqs. (2) and (4)] one can introduce a chromophoric side arm into a crown ether skeleton in such a manner that the deprotonated anionic group can interact directly with or coordinate to the metal ion which is bound in the crown ether cavity. Synthetically, this is most readily attained by using azacrown ethers. Thus, p-nitrophenyl type and umbelliferone type azacrown ethers 13–16 (HL) are described by Nakamura, Takagi, Ueno, and coworkers [14, 15]. These are synthesized by reaction of 2-hydroxy-4-nitrobenzyl bromide (13, 14) or by the Mannich reaction of 4-methylumbelliferone (15, 16) with the appropriate monoazacrown ethers.

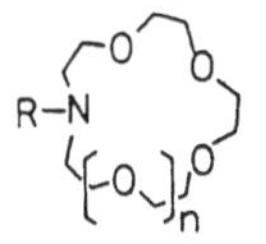

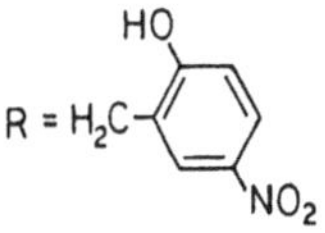

Like picrylamino-substituted benzocrown ethers, crown ethers 13–16 distribute predominantly into organic solution (water 1,2-dichloroethane system), and selectively extract certain alkali metal ions, causing the color (13, 14) or fluorescence (15, 16) change in the organic phase. In the fully protonated forms (H_2L^+), the dyes have two dissociable protons on the phenolic oxygen and the ammonium nitrogen. The two proton dissociation constants are fairly apart, however, and only the ammonium proton (weaker acid) takes part in the alkali metal extraction reactions which can be formulated in a similar manner to Scheme 1 and Eqs. (1) and (2). In contrast to picryl-amino-substituted benzocrown ethers, there was no tendency of 1:2 complex forma-

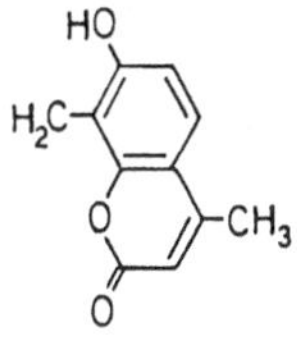

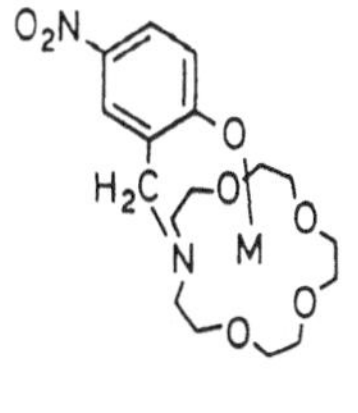

17

tion, suggesting that the residual coordinating site of the crown ether-bound metal ions is occupied from the axial direction by the anionic side arm of the crown ether (Structure *17*).

Table 2 summarizes some equilibrium data for the extraction of alkali metal ions by p-nitrophenyl type reagents *13* and *14*. Similar extraction behavior, especially with

Table 2. Extraction constants (K_{ex}) of N-side-armed monoazacrown ether dyes (HL) for alkali metal ions

Crown ether	pK_{H_2L}[a]	pK_{HL}[a]	$pK_{ex(ML)}$[b]				
			Li$^+$	Na$^+$	K$^+$	Rb$^+$	Cs$^+$
13	5.79	9.69	9.15	9.76	9.86	—[c]	—
14	5.77	9.59	10.29	9.46	8.93	9.63	10.62

[a] $K_{H_2L} = [H^+][HL]/[H_2L^+]$, $K_{HL} = [H^+][L^-]/[HL]$, 0.10 M $(CH_3)_4NCl$, 25 °C; [b] Water 1,2-dichloroethane, 25 °C; [c] Dashes indicate no extraction

respect to metal selectivity, is observed with the fluorescent reagents *15* and *16*. In accord with the lower proton dissociation constants (by 0.5–0.7 log units) of *15* and *16* as compared with *13* and *14*, metal extraction constants $K_{ex\ ML)}$ are also lower (by 0.1–1.1 log unit) for *15* and *16* than those for *13* and *14* [15]. It is interesting that while the neutral species of the crown ethers (HL) assume a zwitter-ionic (ammonium phenolate) structure in water, they exist predominantly in an uncharged, aminophenol form in 1,2-dichloroethane or other non-polar solvents. In accordance with this, the maximum absorption of dyes *13* and *14* in 1,2-dichloroethane displaces from 325 and 327 nm to 411 and 417 nm, respectively, on the extraction of alkali metals. Thus the spectral overlap between HL and ML species is very small.

Comparing Table 1 and Table 2, one observes a small increase of extraction constants for azacrown ethers, but the selectivity among the metals is diminished. However, a selectivity of *13* and *15* toward Li$^+$ is noteworthy. This is probably due to the contribution of the covalent interaction between the phenolate anion and Li$^+$. Extraction photometric and fluorimetric determinations of ppm level Li$^+$ are feasible by using *13* and *15*, respectively. As to the latter, the maximum of excitation spectra for the intense fluorescence at 440 nm shifts from 326 (HL) to 380 nm (ML) upon complex formation. Thus, for fluorimetry, the excitation is made at 380 nm, while fluorescence is measured at 440 nm. The 18-membered azacrown ethers *14* and *16* are selective for K$^+$. None of the crown ethers *13–16* is selective for Na$^+$.

2.3 Side-armed Crown Ethers

Four of this type of crown ether dyes (*18–21*, HL) have been described [16, 17]. These are prepared via three synthetic reactions starting from commercially available guaiacol glyceryl ether. Crown ethers *18* and *19* are similar in solubility — distribution behavior to those crown ethers discussed above, but *20* and *21* are somewhat different because of their enhanced acidity and water solubility.

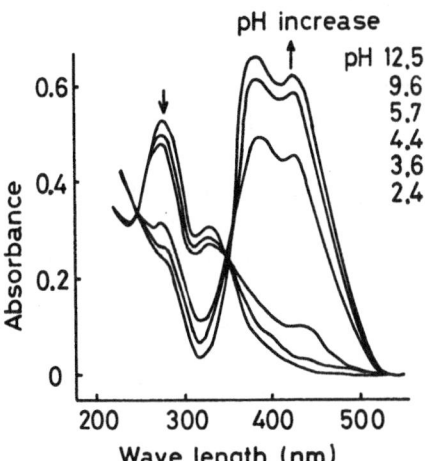

18 n = 1, X = H, Y = -N=N—⟨ ⟩—NO₂

19   2   H   "

20   1   NO₂   NO₂

21   2   NO₂   NO₂

When a 1,2-dichloroethane solution of *18–21* is brought into contact with an aqueous solution at various pH values, the distribution of the reagent species between the two phases is strongly dependent on the pH and the cation present in the aqueous solution. As to *18* and *19*, the distribution of the reagent species in the aqueous phase is negligible under ordinary pH conditions (pH ≤ 11). When the aqueous solution contains a salt of the metal whose ionic diameter fits that of the crown ether, the metal becomes extracted as the pH is raised, giving a violet coloration to the organic phase. When the metal does not fit the crown ether, the organic phase remains yellowish. On the other hand, the distribution of *20* and *21* gradually shifts to the aqueous phase as the pH is raised, until the distribution becomes constant at above pH 10. At this pH region, all the reagent species are dissociated to L^- both in the aqueous and organic phases. The distribution of L^- is determined solely by the distribution behavior of the metal salt of the reagent or of the complex (ML), between the aqueous and organic phases. Thus, when the aqueous solution contains the metal ion favorable for complex formation with *20* or *21*, the complex is extracted and the organic phase turns orange. Conversely, if the metal ion does not

pH increase

pH 12.5
9.6
5.7
4.4
3.6
2.4

Absorbance

0.6
0.4
0.2
0

200   300   400   500

Wave length (nm)

Fig. 2. Absorption spectra of *20* in 1,2-dichloroethane for the extraction of Na^+. $[(20)_0] = 1 \times 10^{-5}$ M and $[(Na^+)_a] = 2 \times 10^{-2}$ M before equilibration

225

fit to the reagent, the organic phase becomes colorless because the reagent stays only in the aqueous phase. Figure 2 shows the spectral behavior of *20* in the extraction of Na^+.

Tables 3 and 4 summarize some equilibrium constants and spectral properties. The extraction constants of *18–21*, especially of *20* and *21*, are considerably higher than those of reagents *1–8* and *13–16*; large proton dissociation constants of these reagents are obviously contributing to this outcome.

Table 3. Extraction constants (K_{ex}) of side-armed crown ether dyes (HL) for alkali metal ions

Crown	pK_{HL}[a]	$pK_{ex(ML)}$[b]				
		Li^+	Na^+	K^+	Rb^+	Cs^+
18	7.51	9.8	8.40	9.20	9.55	>10
19	7.54	8.8	9.32	7.11	7.75	>10
20	3.16	>5.6	3.60	4.15	4.76	5.6
21	3.27	4.8	4.01	2.22	2.68	3.76

[a] $K_{HL} = [H^+][L^-]/[HL]$, water dioxane (9:1, v/v), 0.10 M $(CH_3)_4NBr$, 25 °C;
[b] Water 1,2-dichloroethane, 25 °C

Table 4. Spectral properties of side-armed crown ether dyes (HL) and their alkali metal complexes

Crown ether	$\lambda_{max}(\varepsilon \cdot 10^{-3})$[a]			
	HL[b]	L^-[b]	NaL[c]	KL[c]
18	390(18)	518(30)	560(37)	567(37)
19	390(18)	518(30)	571(35)	575(35)
20	273(12)	385(13)	423(14)	426(13)
21	273(10)	385(11)	430(13)	433(12)

[a] λ (nm), ε (molar absorptivity, $M^{-1} cm^{-1}$); [b] Water dioxane (9:1, v/v), 0.10 M $(CH_3)_4NBr$, 25 °C; [c] 1,2-Dichloroethane

With regard to the selectivity between Na^+ and K^+, 15C5-type reagents prefer Na^+, whereas 18C6-type reagents prefer K^+. This is simply ascribed to the "size selectivity" of the crown ether macrocycles. However, a closer inspection of Table 3 reveals some interesting facts: Selectivity for K^+, as measured by the extraction constant (Table 3), is higher for *21* than for *19* (log K_{KL} − log K_{NaL} = 1.79 for *21* and 2.21 for *19*). Similarly, a comparison of selectivity between *20* and *18* reveals that *18* has a higher selectivity for Na^+ (log K_{NaL} − log K_{KL} = 0.80 for *18* and 0.55 for *20*). This means that for the reagents of the same crown ether skeleton, the more basic pendent anion (*18* and *19*) favors the extraction of Na^+, while the less basic pendent anion (*20* and *21*) favors K^+. This in turn indicates that coordination interaction between the phenolate and the metal is more important for Na^+ in the extraction process. In other words, the structure of the extracted K^+ complexes tends to be more of an intramolecular ion-pair rather than a chelate. In the same notion, sodium complexes have more of a chelate structure. In the light of this concept, Li^+ belongs to a sodium family, while Rb^+ and Cs^+ belong to a potassium family. The difference in the mode of metal — phenolate interaction is schematically illustrated in formulae 22 and 23.

In order to achieve a high selectivity among alkali metals in solvent extraction with monoprotonic crown ether reagents, two critical factors have to be considered. The first is the well known "size selectivity" by the crown ether macrocycle. The second is the coordination interaction (22) and the ion-pair (23) interaction of the anionic site with the crown ether-bound metal ion. A localized negative charge (high basicity) strengthens the coordination interaction and favors the extraction of lighter alkali metal ions. A delocalized negative charge (low basicity) favors the formation of ion-pair type complexes and leads to more efficient extraction of the heavier alkali metals.

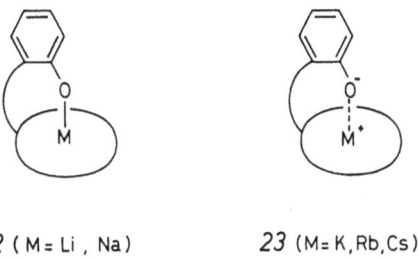

22 (M = Li, Na)   23 (M = K, Rb, Cs)

Reagent 20 is useful for the extraction photometric determination of Na^+ in human blood serum. A acid-deproteinated serum sample (0.1 ml) is mixed with an alkali metal-free tetramethylammonium borate solution (0.1 M, 4 ml) and diluted to 10 ml with distilled water (pH 10.3). The mixture is shaken with a 1,2-dichloroethane solution of 20 (2×10^{-4} M, 10 ml) for 10 min. After phase separation by centrifugation, the absorbance of the organic phase is measured at 423 nm against a reagent blank. A calibration line, which is prepared with aqueous solutions of sodium chloride is linear of Na^+ concentrations between 0 and 6 ppm. K^+ in the blood serum does not interfere with the Na^+ determination since the former concentration is less than one tenth of the latter. The determination of K^+ is similarly attained at ppm level by using 21. In this case, a presence of ten-fold excess Na^+ does not cause any serious interference.

One of the remarkable features of the dinitrophenol-type reagents 20 and 21 is, as mentioned before, that they do not distribute into the organic phase under photometric conditions unless metal ions of right size are present (pH $\geq$ 10). This stands in contrast to the reagents 18 and 19 as well as phenolic monoazacrown ethers and picrylamino-substituted benzocrown ethers. These latter reagents stay essentially in the organic phase; if one tries to achieve a higher metal determination-sensitivity by increasing the metal extraction efficiency through the use of higher reagent concentration, the reagent blanks (absorption due to HL species) inevitably increase. Though the spectral overlap between HL and ML is avoided for the phenolic reagents, the blank absorption is not negligible and its increase causes uncertainties when the accurate photometric measurements are required. The absence of blank absorption for 20 and 21 insures an accurate and sensitive determination as well as good reproducibility. Moreover, the determination is not sensitive to pH variation in the extraction procedure. These are the advantages of dinitrophenol-type reagents over the others from the practical point of view.

Extraction photometry is often conveniently replaced by photometry in homogeneous solution by using surfactant micelles as a pseudo-phase for extraction. Reagents such as *18* and *19* which do not dissociate and distribute into an aqueous phase should, in principle, be useful for such purposes. Unfortunately, the alkali metal complexes of the crown ether dyes are not quite lipophilic enough to be concentrated into the core of the micelle, but both HL and ML species seem to be concentrated on the surface. Consequently, the formation of ML and the ordinary proton dissociation of HL (to give M'L, where M' stands for cationic component of the pH-buffering agent) take place simultaneously at the micelle surface and can not be fully distinguished from each other spectrometrically [18]. The problems involved in homogeneous photometry are yet to be solved.

It is to be added that, again in the extraction with *20* and *21*, the fraction of metal transferred to the organic phase can amount only to 10–20% of the total metal originally present in the aqueous solution under ordinary conditions. Accordingly, the "apparent" molar absorptivity of the extracted metal complex is only 2200 as compared with the "true" value of 14,000 M^{-1} cm^{-1} shown in Table 4. This in turn means that the sensitivity of the determination by *20* or *21* may be increased even more by increasing the reagent concentration. However, due to contamination from glassware, the handling of Na^+ and K^+ below ppm levels requires special precautions [19], and such ultrasensitivity is practically not always worthwhile. With regard to contamination problems, it may also be pointed out on the side that proton-dissociable chromogenic crown ethers discussed above as well as those to be discussed below (extraction of alkaline earth metal ions) can be contaminated with alkali metals through storage in organic solution (1,2-dichloroethane) when kept in glass bottles more than a month. The contaminations, however, are readily removed by washing with aqueous boric acid before usage.

As to the (molecular) absorption photometry of alkali metals in general, another type of extraction photometry is also available. The method is based on ion-pair extraction of crown ether complexes of alkali metals by anionic dyes [19, 20]. Ordinary crown ethers as well as bicyclic crown ethers (cryptands) are used, while bromocresol green, resazurin and picrate are the typical of the pairing anions. These crown ethers and cryptands are not colored, but the experimental procedures are essentially similar to those described above for chromogenic crown ethers. Li^+, Na^+, and K^+ can be selectively determined at ppm levels.

2.4 "Crowned" Phenols

Kaneda and coworkers synthesized interesting crown ether dyes in which a phenolic hydroxyl group constitutes an integrated part of the crown ether skeleton (*24*, *25*) [21]. They described that upon addition of crystalline lithium chloride or perchlorate to a yellow solution of *24* in chloroform, a drastic color change to purple red took place when pyridine (800 times molar excess of *24*) was simultaneously added. Other alkali metal salts produced no such change. The coloration is very selective to Li^+, though

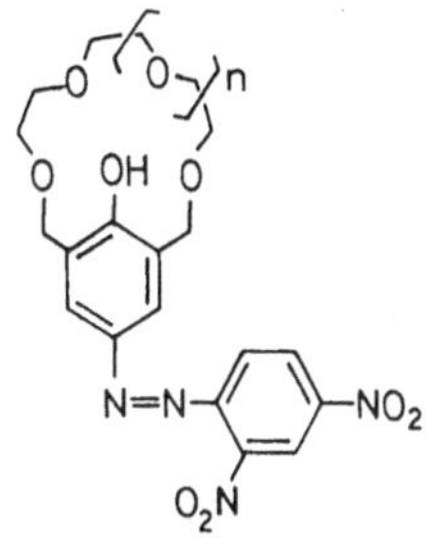

24   n = 1
25   n = 2

it is said that the coloration is sensitive to experimental conditions. Later, Nakashima and coworkers [22] further studied this system and described conditions under which 25 to 250 ppb Li^+ was determined photometrically in chloroform-dimethylsulfoxide triethylamine mixture (94.5 : 5 : 0.5; v/v). Sodium ions did not interfere, but K^+, Rb^+, Ca^{2+}, Sr^{2+}, Ba^{2+}, and Mg^{2+} did and were to be removed before the determination was performed. The coloration of 24 in organic media is no doubt related to the deprotonation of phenolic proton under combined influence of the proton-removing amine and the metal cation stabilizing the resultant phenolate anion, but the complexity of the system does not seem to allow a quantitative approach to the equilibria involved. The high sensitivity claimed for this method is remarkable, but presently a rather unusual medium utilized and strict precautions in sample preparation (exclusion of water) seem to limit its use for general purposes. It is said that reagents 24 and 25 work rather poorly in the liquid-liquid solvent extraction of alkali metals [23].

2.5 Benzocrown Ethers with Azo-linked Side Arm

Shinkai, Manabe and coworkers [24] described an interesting family of crown ether dyes, which responded to light and changed their complexation and extraction selectivity (for this topic see S. Shinkai, O. Manabe, this series 67). The dyes shown by a general structural formula 26 can be photochemically isomerized to 27 according to the E/Z-isomerization of the azo linkage. The transformation is reversible, the thermodynamically stable E-form (26) being recovered by a spontaneous,

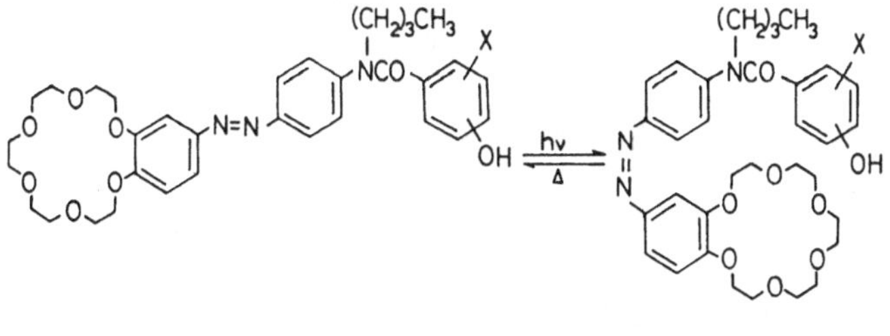

26   27

thermal process. Dyes *26* include three isomeric phenol dyes, *o*-, *m*-, and *p*-isomers according to the site of the phenolic hydroxyl with respect to the N-butylcarbamoyl substituent.

As is obvious from structural considerations, *E*-isomers bear no special characteristics other than the benzocrown ethers and phenols, since the phenol function is quite remote from the crown ether function and there can be no cooperation within the molecule for enhanced interaction with metal ions. On the other hand, isomerization to the *Z*-form brings the two functional portions close to each other, making it possible for the phenolate anion to interact directly with the crown ether-bound metal cation from axial directions. The mode of interaction is schematically shown by *28* for an *o*-phenol type dye.

The extraction of alkali metal ions into *o*-dichlorobenzene (with or without added 10% 1-butanol) was studied under strongly alkaline conditions (MOH $\leq$ 0.50 M). *E*-Isomers show a selectivity which simply follows the order of the size of alkali metal ions and differs from the unique stability of the metal complexes of 18C6. This suggests that alkali metal ions are extracted as a counterion of the phenolate anion and that the crown ether plays no significant role. On the other hand, the extraction with photo-isomerized *Z*-compounds shows varying selectivity among alkali metals according to the *o*-, *m*-, and *p*-isomeric phenols. This stands in agreement with the idea of the participation of such complex species like *28*, but in the lack of detailed equilibrium data, the extent of such structural contribution can not be rigorously assessed.

28

The major interest of the work of Shinkai, Manabe and coworkers was the photochemical control of metal transport through organic liquid membranes. The alkali metal transport is attained by consecutive processes of extraction and back-extraction (or stripping) on the high and low pH sides, respectively, of the membrane/ aqueous solution interface by using the monoprotonic crown ethers *26* as metal carriers (ionophores). By changing the extraction behavior of the crown ether reagents by photo-irradiation, an acceleration of some of the metal transport rates is realized. In fact, the study on *26* is one of the authors' extensive works on the photo-responsive crown ethers, and is dealt with in more detail in other chapters of this series of monographs.

Reagent *26* can also extract alkaline earth metal ions from their metal hydroxide solutions. A part of the phenolate anion in the organic phase can be replaced by other lipophilic monoanionic dyes such as 2,4-dichlorophenolindophenol (dissociated form, A⁻) added externally to the aqueous phase. Taking advantage of this reaction, a counter flow of proton and calcium 2,4-dichlorophenolindophenolate ($Ca^{2+} \cdot A^-$) through the liquid membrane is attained. It is often observed that cationic complexes (ML^+) produced from monoanionic crown ethers (L^-) and alkaline earth metal ions (M^{2+}) distribute to organic phase with the aid of added lipophilic anions by forming ion-pair type association complexes [18].

3 Diprotonic Crown Ether Dyes

The introduction of two of the proton-dissociable chromophores into the crown ether skeleton, according to the similar strategy outlined before, should lead to those dyes which are selective for divalent metal ions — above all to alkaline earth metals.

Dipicrylamino-substituted DB18C6 proved not to extract alkaline earth metal ions [13]. If the picture of intramolecular ion-pair type extraction by picrylamino-benzocrown ethers is correct, the dipicrylamino-substituted crown ether results in the extraction of alkaline earth metal complexes, which involves a charge separation

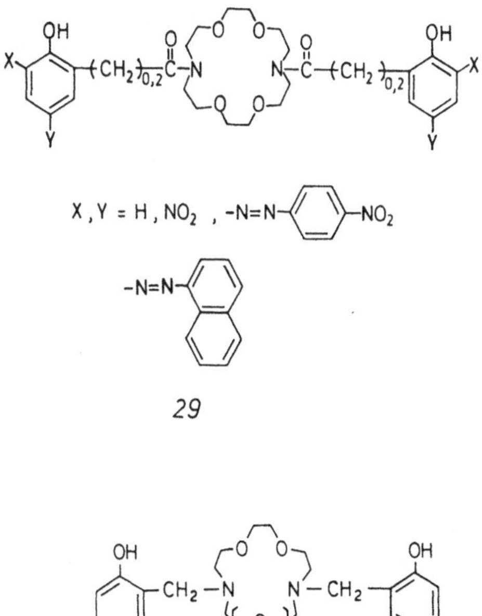

$X, Y = H, NO_2$, $-N=N-\langle \rangle-NO_2$

$-N=N-$ (naphthyl)

29

30	n = 1	,	X = NO_2
31	1		$-N=N-\langle \rangle-NO_2$
32	2		NO_2
33	2		$-N=N-\langle \rangle-NO_2$
34	2		$-N=N-\langle \rangle-NO_2$, O_2N

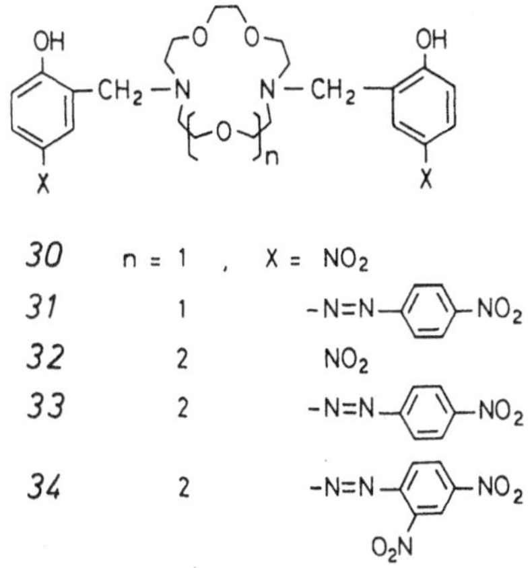

between two anionic single charges and a cationic double charge. Such structures should be highly unfavorable for extraction. Therefore, in order to achieve an efficient extraction of divalent metals, structures of crown ether dyes should be considered in which the two anionic chromophoric groups are capable of interacting directly with the metal ion from axial directions. This type of crown ethers are synthetically most easily obtained by using DA18C6 and other related diazacrown ethers. Some such synthetic products are shown in formulae *29–34*.

Crown ether dyes of structure *29* are poor extraction agents for alkaline earth metal ions, although some do extract these metals as well as Li^+ into chloroform at high metal concentrations (1 M) [13]. A low coordinating ability of the amide nitrogens in the crown ether skeleton is no doubt responsible for this behavior. On the other hand, crown ethers *30–34* unanimously show a strong affinity to alkaline earth metal ions and extract them efficiently into organic solvents such as 1,2-dichloroethane [25, 26].

The extraction of alkaline earth metal ions by reagent *30–34* can be formulated by Eqs. (5) and (6). The reagents do not distribute into aqueous phase under ordinary conditions; extraction constants $K_{ex(ML)}$ can be determined photometrically by standard procedures. For reagent *32*, the absorption maximum of the free reagent (H_2L) at 326 nm shifts to 406 nm upon extraction of Ca^{2+} through the formation of CaL species into 1,2-dichloroethane. Similarly, *33* shows a wave length shift from 392 to 518 nm, as illustrated in Fig. 3. The extraction according to Eqs. (5) and (6) is rigorously followed, and there is no indication of other types of complex formation.

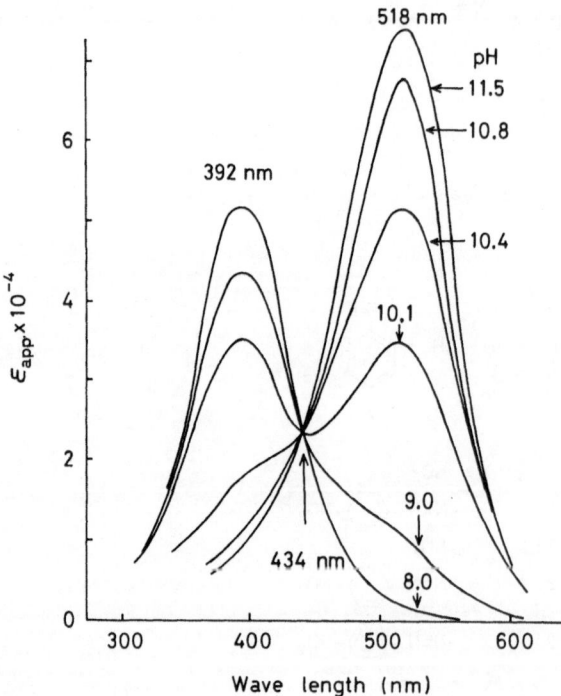

Fig. 3. Absorption spectra of *33* in 1,2-dichloroethane for the extraction of Ca^{2+}. $[(33)_0] = 6.1 \times 10^{-6}$ M and $[(Ca^{2+})_a] = 1 \times 10^{-3}$ M before equilibration

The proton dissociation and the extraction constants as well as some spectral properties are summarized in Tables 5 and 6.

$$(H_2L)_o + (M^{2+})_a \rightleftharpoons (ML)_o + 2 (H^+)_a \tag{5}$$

$$K_{ML}^{ex} = \frac{[(ML)_o] [(H^+)_a]^2}{[(H_2L)_o] [(M^{2+})_a]} \tag{6}$$

Table 5. Proton dissociation constants of crown ether dyes $(H_2L)^a$

Crown	pK_{H_4L}	pK_{H_3L}	pK_{H_2L}
32	4.03	6.56	9.8
33	5.52	9.00	10.9
34	5.01	8.72	10.3

[a] Water dioxane (9:1, v/v), 0.1 M $(CH_3)_4NBr$, 25 °C;
$K_{H_4L} = [H^+] [H_3L^+]/[H_4L^{2+}]$; $K_{H_3L} = [H^+] [H_2L]/[H_3L^+]$; $K_{H_2L} = [H^+] [HL^-]/[H_2L]$

Table 6. Extraction constants (K_{ex}) and spectral properties of crown ether dyes $(H_2L)^a$

Crown ether	H_2L		CaL		$pK_{ex(ML)}$		
	λ_{max}	$\varepsilon \times 10^{-3}$	λ_{max}	$\varepsilon \times 10^{-3}$	Ca	Sr	Ba
30	326	24	406	45	15.0	16.8	18.4
31	392	52	513	74	17.6	18.9	20.4
32	326	23	406	45^b	12.54	13.51	15.1
33	392	52	518	80	16.6	17.9	19.1
34	410	63	574	81	15.0	16.3	18.5

[a] Water 1,2-dichloroethane; spectral data are obtained in 1,2-dichloroethane saturated with water; λ (nm), ε (molar absorptivity, $M^{-1} cm^{-1}$), 25 °C; [b] Previously miss-typed as 55 [25)]

Metal extraction selectivity is: $Ca^{2+} > Sr^{2+} > Ba^{2+} \gg Mg^{2+}$ for both types of diazacrown ethers: 18C6- and 15C5-types. Mg^{2+} and alkali metals are extracted only to a minor extent. An especially high extraction constant is exhibited by the 18C6-type diazacrown ether with p-nitrophenol-chromophore (32). Unlike the extraction of alkali metal ions with monoprotonic crown ethers, a spectrophotometric concentration of 32 can extract Ca^{2+} quantitatively into 1,2-dichloroethane if the pH of the aqueous phase is adjusted high enough (pH $\geq$ 10). As seen in Table 5, 32 has much higher proton dissociation constants than 33 or 34. The constants K_{H_4L} and K_{H_3L} correspond to the dissociation of two phenolic protons, and K_{H_2L} to an ammonium proton. Thus, in aqueous solutions the crown ether dyes 32–34 as well as 30 and 31 exist in a zwitter ionic, ammonium phenolate form, while in 1,2-dichloroethane they are present in a neutral, amino-phenol structure.

An extraction photometric determination of Ca^{2+} in blood serum was successfully performed by using *32*. A serum sample (0.1 ml) is deproteinated by 1 ml of 0.33 M trichloroacetic acid, and a 0.5-ml aliquot of the supernatant liquid is diluted to 10 ml with tetramethylammonium borate buffer (pH 10.25). The solution is shaken with 10 ml of 2.5×10^{-4} M *32* in 1,2-dichloroethane. The absorbance of the organic solution is measured at 406 nm. The calibration line is linear at 0–0.8 ppm Ca^{2+} concentrations. As little as 0.1 ppm or 100 ppb Ca^{2+} is easily determined. By using *34*, the determination of Ca^{2+} below 300 ppb is feasible. The molar absorptivity 81,000 for the calcium complex is among the highest known for Ca^{2+} determinations. As to the generality and the convenience of Ca^{2+} determination, however, this type of crown ether reagent (extraction photometry) does not seem to offer particular advantages over many traditional metallochromic chelating agents such as Calmagite [1-(1-hydroxy-4-methyl-2-phenylazo)-2-naphthol-4-sulfonic acid], Chlorophosphon-azo-III [2,7-bis(4-chloro-2-phosphonophenylazo)chromotropic acid] and Eriochrome Black T [1-(1-hydroxy-2-naphthylazo)-6-nitro-2-naphthol-4-sulfonic acid], which are able to determine ppm to ppb level Ca^{2+} in aqueous media.

The extraction of other divalent metal ions have not been studied in detail, but Cu^{2+}, Ni^{2+}, Zn^{2+}, Cd^{2+}, and Pb^{2+} are extracted by *32* leading to errors in Ca^{2+} determination. These interferences can be removed by using NTA (nitrilotriacetic acid) as masking agent [26]. Complex formation constants of *32* are not yet determined in homogeneous solutions, but presumably the complexes of alkaline earth metal ions are less stable than those of other divalent metal ions. With ordinary chelating agents, transition metal and related heavy metal ions usually form much more stable complexes than alkaline earth metals do. However, the presence of the crown ether structure in *32* and related crown ether reagents (*33*, *34*) may well be expected to reduce the difference in the stabilities between the two families of metal ions. In fact, the complexation behavior of "crown complexanes" (*35*, H_2L) described by Takagi, Tazaki, Ueno and coworkers [27, 28] have partially supported this expectation.

Some of the complex formation constants of *35* with alkaline earth metal ions are quite close to those with divalent transition metal ions. Particularly, compound *35* with X = COOH, m = 1 and n = 2 forms complexes of almost same stability with Co^{2+}, Ni^{2+}, Ba^{2+}, Ca^{2+}, Sr^{2+}, and Zn^{2+} [28]. It may be pointed out on the side that the crystal structure of the Cu^{2+} complex of this ligand indicates that the two acetate groups coordinate in a *trans* configuration to the Cu atom bound in the

$$X-(CH_2)_m-N \qquad N-(CH_2)_m-X$$

X = COOH, PO_3H_2

35

distorted diazacrown ether ring [29]. Considering this fact as well as the size-fit of the Ca^{2+} ion and the crown ether cavity, it is very plausible that the calcium complexes of *32–34* involve the coordination of two phenolate anions from axial directions onto the metal ion held in the crown ether cavity, well completing an electrically neutral, lipophilic skin around the otherwise highly hydrophilic Ca^{2+}. Few synthetic reagents are known to extract Ca^{2+} with appreciable selectivity among common divalent metal ions (including transition and post-transition metals). In this sense, it is hoped that the crown ether dye of type *30–34* may find a use in the study of such biochemical processes as the transport of alkaline earth metal ions through a lipophilic barrier (cell membranes). The monoprotonic crown ether reagent discussed in the previous sections may be used for similar purposes for alkali metals. It can be readily shown that these reagents mediate the counter-flow of alkali metal ions (or alkali metals ions) and protons across lipophilic membranes (liquid membranes).

Fluorescent versions of the reagent of type *30–34* is easily obtained via a Mannich reaction of diazacrown ethers with 4-methylumbelliferone. Reagent *36* shows a similar metal extraction behavior as *32*, but the extraction constants are lower by 2–2.6 log units than *32*, reflecting reduced proton dissociation constants of *36*. Fluorimetry can, in principle, be more sensitive than absorption photometry. However, the lower extraction constant requires more basic conditions for efficient metal extraction, thus compensating the higher sensitivity of fluorimetry in practical applications for alkaline earth metal determination [15, 18].

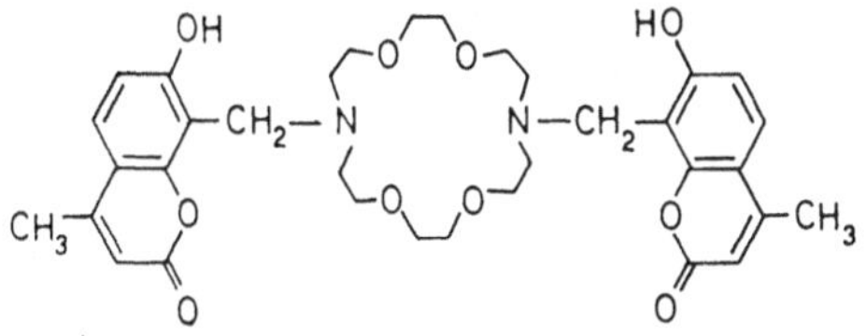

36

Diprotonic crown ethers can also be derived from DA24C8. Dye *37* offers a much larger cavity to accommodate metal ions, and it selectively extracts Ba^{2+} into 1,2-dichloroethane (Table 7) [18]. To our knowledge, *37* is the first photometric reagent

Table 7. Extraction constants of *37* (H_2L) for alkaline earth metal ions[a]

Metal ion	Mg^{2+}	Ca^{2+}	Sr^{2+}	Ba^{2+}
$pK_{ex(ML)}$	$\geqslant 18$	17.7	17.0	14.3

[a] Water 1,2-dichloroethane, 25 °C

selective for Ba^{2+} capable of determining the ion at ppm levels. Considering the size of the macrocycle, it would also be an interesting reagent for separation and concentration of Ra^{2+}.

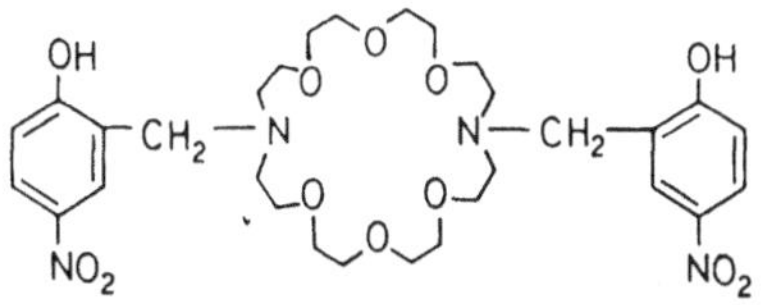

37

4 Uncharged, Neutral Crown Ether and Related Dyes

Shinkai, Manabe, and coworkers synthesized and studied a variety of bis(crown ether)s made by connecting two benzocrown ethers through azo linkage (*38*) [30]. This yields derivatives of azobenzene which, on photo-irradiation, isomerize reversibly from the (thermodynamically stable) *E*- to *Z*-configuration. The two isomers have different metal selectivities depending on the intramolecular cooperation of the two crown ether rings. However, spectral change caused by metal binding is, small as expected, and therefore this type of crown ethers and related bicyclic crown ethers [31,32] will not be discussed in further detail.

Takagi, Ueno, and coworkers described another type of azobenzene crown ethers in which the azo-linkage constitutes a part of the crown ether macrocycle (*39, 40*) [18,33]. A metallochromism is appreciable when the E-isomers are allowed to interact with excess alkali and alkaline earth metal ions in acetonitrile. However, this is not quite sufficient to be useful as a photometric aid for metal analyses. Quite recently, azobenzenophanes with a oligooxyethylene bridge at 4- and 4'-positions (instead of 2- and 2'-positions in *39* and *40*) have been successfully synthesized and studied [34].

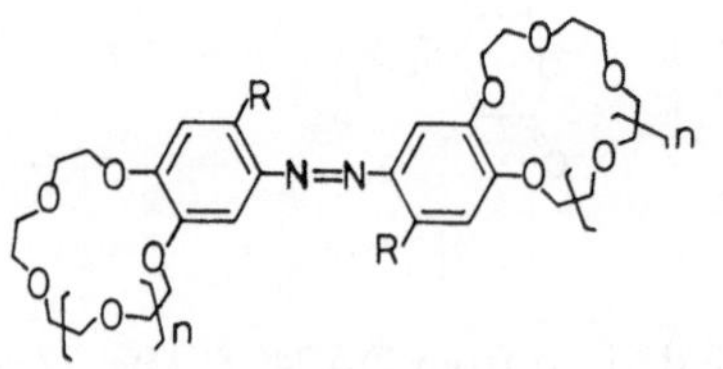

38

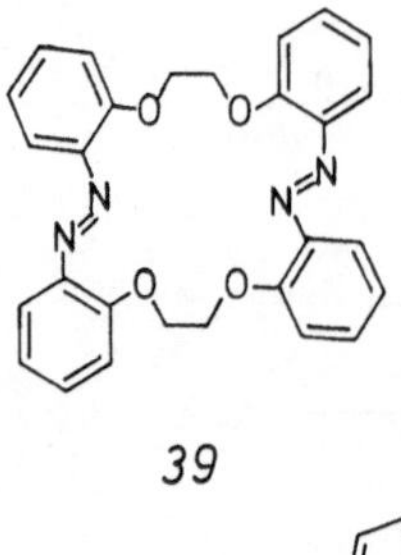

39

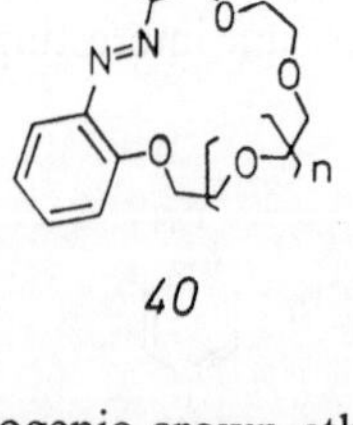

40

The most interesting chromogenic crown ethers have been described by Dix and Vögtle [35–37]. Some 40–50 crown ether dyes of various structures have been synthesized, the typical of which are shown in the structural formulae *41–50*. The leading principle of synthesis is the attachment of crown ethers of various ion selectivity to dyes in such a manner that the interaction with the alkali and alkaline earth metal ions causes color changes. Thus, the basic idea of molecular design is the same as that independently developed by Takagi and coworkers [4]. However, the main interest of Dix and Vögtle was for neutral, uncharged crown ether dyes, and therefore, the chromophores without dissociable protons were of specific concern in their study.

Dyes 41–50 have electron donor and acceptor sites within a molecule so that the charge transfer from the donor to acceptor according to electronic excitation gives rise to their strong visible light absorption. The interaction of metal ions with the dye molecule in a manner to inhibit the charge transfer (stabilization of the localized charge on the donor site) leads to the hypsochromic shift of the charge transfer band. On the contrary, the interaction to assist the charge transfer (stabilization of the charge on the acceptor site) results in a bathochromic shift.

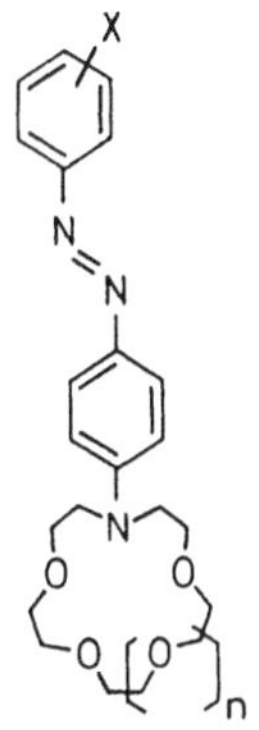

41

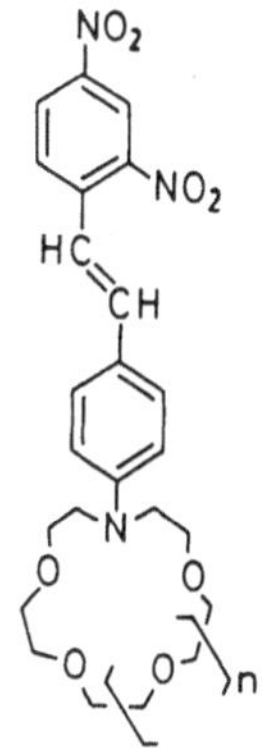

42 n = 1
43 n = 2

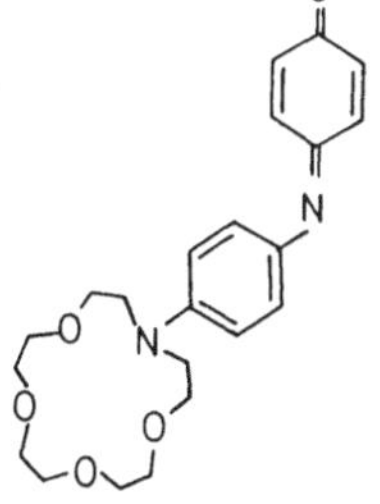

44

In dyes 41–44, the amino nitrogen which is an integrated part of the crown ether ring is simultaneously involved in the dye chromophore as a vital electron donor site. The two aromatic ethereal oxygens are in the same position in dyes 45 and 46. Thus, in these crown ether dyes, the interaction with metal cations tends to localize the electronic charge on the crown ether portion causing a hypsochromic shift of the charge transfer band which appears as the longest wave length absorption of these dyes. On the other hand, for dyes *48–50*, the metal cations interact with the acceptor site (carbonyl oxygen) of the dyes giving rise to a bathochromic shift of the charge transfer band (structures *51* and *52*). Figures 4 and 5 exemplify these relations.

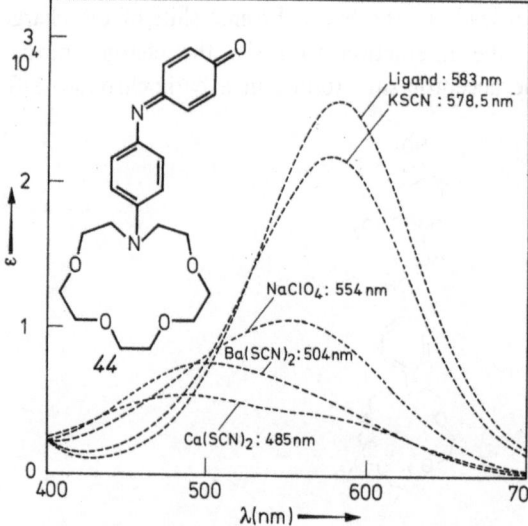

Fig. 4. Absorption spectra of *44* in the presence and the absence of metal salt: *44* > 100 in acetonitrile (Reproduced with permission from: J. P. Dix and F. Vögtle, Chem. Ber. *114*, 638 (1981))

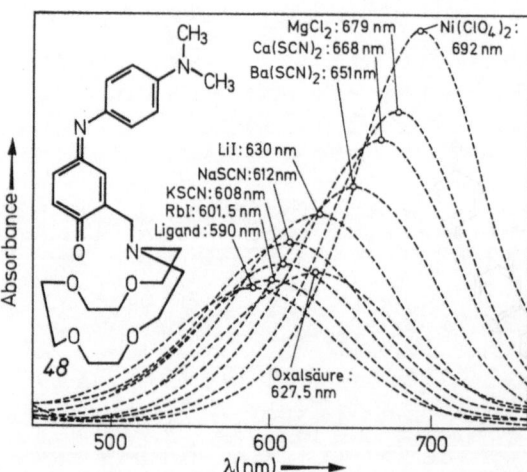

Fig. 5. Absorption spectra of *48* in the presence and the absence of metal salt. Metal salt: *48* > 100 in acetonitrile (Reproduced with permission from: J. P. Dix and F. Vögtle, Chem. Ber. *114*, 638 (1981))

Some of the typical spectral data are summarized in Table 8 [36, 37]. Dye *47* is somewhat peculiar in that both of the donor and acceptor sites of the dye chromophore can simultaneously interact with a metal cation. The dye shows two absorption maxima or shoulders, both of which are subjected to a regular shift on the addition of metal salts in acetonitrile. The long wavelength absorption (λ_1) corresponds to the charge tranfer band which involves the amino nitrogen. The short wavelength absorption (λ_2) corresponds to the benzenoid band [37].

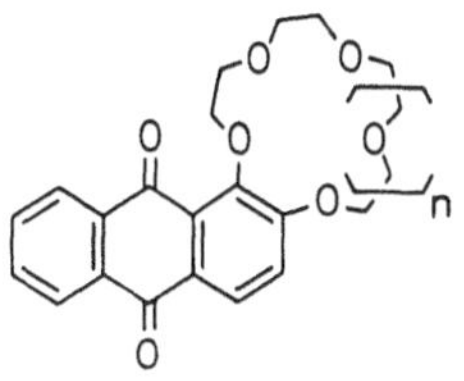

45 n= 1
46 n= 2

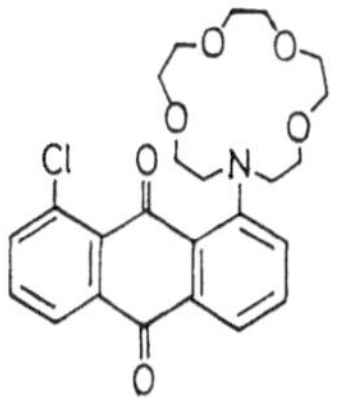

47

The extent of the absorption shift depends on
i   the fit between the size of crown ether and metal ion,
ii the surface charge density of metal cations, and
iii the solvent.

The point (i) is most clearly exemplified in Table 8 by 43. K^+ among alkali metals and Ba^{2+} among alkaline earth metals produce an almost exclusive absorption shift which is as larger as 100 nm in both acetonitrile and methanol. Point (ii) is easily seen in the behavior of 47 and 49 in acetonitrile. The spectral shift is most prominent for Li^+ among alkali metals. Moreover, if one compares alkali and alkaline earth metals, the latter obviously show larger spectral shifts. In this connection, it may be pointed out that a similar observation, though to a much smaller extent, can be made in the extraction of alkali and alkaline earth metal ions with proton-dissociable crown ethers. This is partially presented in Table 4 for the absorption maxima of NaL and KL species. Generally, for phenolic crown ether dyes, absorption maxima shift toward the longer wave lengths in the order, $Rb^+ = K^+ > Na^+ > Li^+$ ($\gg H^+$) and $Ba^{2+} \geq Sr^{2+} \geq Ca^{2+}$. As to point (iii), one notices the

Table 8. Absorption maxima (nm) of crown ether dyes and their variation in the presence of metal salts[a]

Metal salt	43	43[b]	45	47 λ_1[d]	47 λ_2[d]	49	49[c]
none	477	—[e]	373	493	~317; s	598	628
LiI	472(−5)[f]	—(0)	—	~430(−63); s	342(+25); s	633.5(+35.5)	628(0)
LiSCN	—	—	361(−12)	—	—	—	—
LiClO$_4$	—	—	—	—	—	—	—
NaClO$_4$	463.5(−13.5)	—(−5)	361(−12)	—	—	608.5(+10.5)	632(+4)
NaSCN	464(−13)	—(−7)	361(−12)	438(−55)	~330(+13); s	608(+10)	—
KSCN	382(−95)	—(−99.5)	368(−5)	487(−6)	~320(+3); s	598.5(+0.5)	634(+6)
RbI	468(−9)	—(−6)	371(−2)	—	—	598.5(+0.5)	633(+5)
MgCl$_2$	—	—	—	~480(−13)	344(+27)	681(+83)	629(+1)
Ca(SCN)$_2$ · 4 H$_2$O	381(−96)	—(−1.5)	~345(−28); s	~412(−81); s	349(+32)	676(+78)	667.5(+39.5)[g]
Ba(SCN)$_2$ · 2 H$_2$O	371(−106)	—(−99)	350(−23)	~415(−78); s	344(+27)	645(+47)	664.5(+36.5)

[a] Concentration of dyes: $10^{-4} - 10^{-5}$ M in acetonitrile, molar ratio metal : dye > 100; [b] Measured in methanol, metal : dye > 10; [c] Measured in methanol/water (1/1) with 1.5×10^{-2} M triethylamine; [d] An absorption maximum or a shoulder according to the conditions; [e] Dashes indicate that the value are not reported; [f] The figures in the parentheses indicate the spectral shift (nm) from the free dye; the negative values denote hypsochromic shifts, while the positive values bathochromic shifts; [g] CaCl$_2$ instead of Ca(SCN)$_2$ · 4 H$_2$O

enhanced metal selectivity of *43* on going from acetonitrile to methanol [reduction of spectral shift in methanol, particularly for Ca(SCN)$_2$]. However, it should also be pointed out that the shift widths are considerably diminished on going from acetonitrile to aqueous methanol for *49*; maximum shift is 83 nm in acetonitrile, while it amounts to only 39.5 nm in aqueous methanol.

Large spectral shifts which take place ion selectively suggest possible applications to the photometry of some alkali and alkaline earth metal ions in organic solution in a similar manner to that performed for crown ether dye *24*. Unfortunately, the spectral behaviors of dyes *41–50* have been studied only in the presence of excess metal salts, and their response to low concentrations of metal salts has not been reported. For practical application a right choice of solvents has to be made, since most of these dyes are expected to show considerable solvatochromism. It is pointed out that the absorption maximum of dye *49* moves from 598 nm in acetonitrile to 628 nm in aqueous methanol (Table 8). The effect of the anionic component of the salt as well as the interferences from acids and bases need to be carefully assessed.

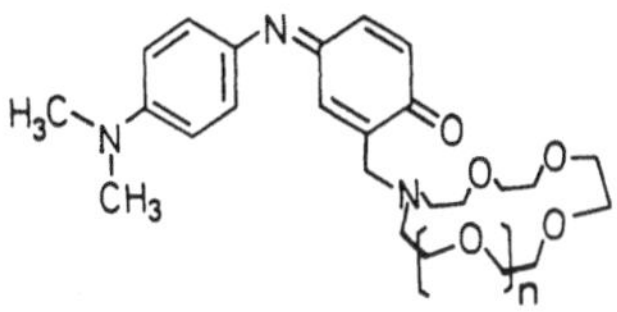

48 n = 1

49 n = 2

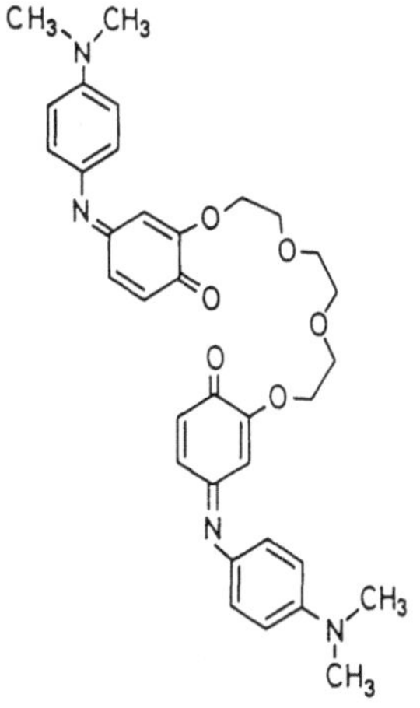

50

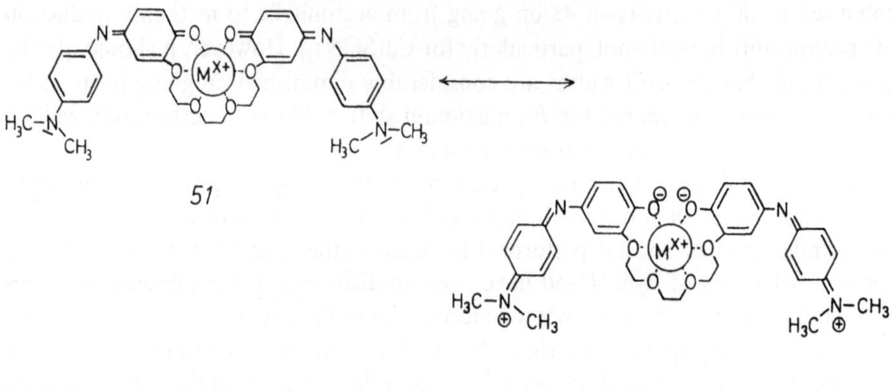

51

52

5 Conclusion

Crown ether dyes as chromogenic reagents for alkali and alkaline earth metal ions are still in the early stage of development. However, the basic idea of molecular design for such reagents has been fully materialized by a substantial number and variety of proton-dissociable as well as neutral crown ether dyes hitherto synthesized. Some of them are practically usable for extraction photometry of Na^+ and K^+. Molecular absorption photometry is especially suited for automatic flow analysis because of the operational simplicity and stability. The use of chromogenic crown ether reagents, however, may not only be limited to such elemental analysis. As pH indicators are used to check and monitor the acidity of aqueous solutions or even some organic solutions, the crown ether-based metallochromic indicators may eventually be used for the similar purpose on alkali and alkaline earth metal ions. Such indicators may also find a use as photometric probes in biochemical studies such as alkali and alkaline earth metal transport in cell membranes. The present state of arts is yet far from this goal, and a further study in this field would certainly be fruitful and rewarding.

6 References

1. Sandell, E. B.: Colorimetric Determination of Traces of Metals, New York—London, Interscience 1959
2. Cheng, K. L., Ueno, K., Imamura, T.: Handbook of Organic Analytical Reagents, Boca Raton, CRC Press 1982
3. Pedersen, C. J.: J. Am. Chem. Soc. 89, 7017 (1967)
4. Takagi, M. et al.: Anal. Lett. 10, 1115 (1977)
5. Nakamura, H. et al.: Talanta 26, 921 (1979)
6. Nakamura, H. et al.: Anal. Chem. 52, 1668 (1980)
7. Lamb, J. D. et al.: in: Coordination Chemistry of Macrocyclic Compounds; Melson, G. A. (Ed.), p. 145, New York, Plenum Press 1979
8. Dalley, N. K.: in: Synthetic Multidentate Macrocyclic Compounds, Izatt, R. M., Christensen, J. J. (Eds.), p. 207, New York, Academic Press 1978
9. Mallinson, P. R., Truter, M. R.: J. Chem. Soc., Perkin Trans. II 1818 (1972)

10. Ishizu, K. et al.: Chem. Lett. 227 (1978)
11. Pacey, G. E., Bubnis, B. P.: Anal. Lett. *13*, 1085 (1980)
12. Pacey, G. E. et al.: Analyst *106*, 636 (1981)
13. Yamashita, T. et al.: Bull. Chem. Soc. Jpn. *53*, 1550 (1980)
14. Nakamura, H. et al.: Chem. Lett. 1305 (1981)
15. Nishida, H. et al.: ibid. 1853 (1982)
16. Nakamura, H. et al.: Bunseki Kagaku *31*, E131 (1982)
17. Nakamura, H. et al.: Anal. Chim. Acta *139*, 219 (1982)
18. Takagi, M. et al.: unpublished work
19. Takagi, M. et al.: Anal. Chim. Acta *126*, 185 (1981)
20. Sumiyoshi, H. et al.: Talanta *24*, 763 (1977)
21. Kaneda, T. et al.: Tetrahedron Lett. *22*, 4407 (1981)
22. Nakashima, K. et al.: Chem. Lett. 1781 (1982)
23. Nakashima, K., Akiyama, S.: private communication
24. Shinkai, S. et al.: J. Am. Chem. Soc. *104*, 1967 (1982)
25. Nishida, H. et al.: Mikrochim. Acta I, 281 (1981)
26. Shiga, M. et al.: Bunseki Kagaku, *32*, E293 (1983)
27. Takagi, M. et al.: Chem. Lett. 1179 (1978)
28. Tazaki, M. et al.: ibid. 571 (1982)
29. Uechi, T. et al.: Acta Crystallogr. *B38*, 433 (1982)
30. Shinkai, S. et al.: J. Am. Chem. Soc. *104*, 1960 (1982) and references cited therein
31. Shinkai, S. et al.: Chem. Lett. 499 (1982)
32. Shinkai, S. et al.: Tetrahedron Lett. *20*, 4569 (1979)
33. Shiga, M. et al.: Chem. Lett. 1021 (1980)
34. Shinkai, S., Manabe, O.: J. Am. Chem. Soc. *105*, 1851 (1983)
35. Dix, J. P. Vögtle, F.: Angew. Chem. *90*, 893 (1978); Angew. Chem. Int. Ed. Engl. *17*, 857 (1978)
36. Dix, J. P., Vögtle, F.: Chem. Ber. *113*, 457 (1980)
37. Dix, J. P., Vögtle, F.: ibid. *114*, 638 (1981)

Photocontrol of Ion Extraction
and Ion Transport by Photofunctional Crown Ethers

Seiji Shinkai and Osamu Manabe

Department of Industrial Chemistry, Faculty of Engineering, Nagasaki University,
Nagasaki 852, Japan

Table of Contents

1 Introduction

Photoresponsive systems are ubiquitously seen in nature: photosynthesis, vision, phototropism, and phototaxis are typical examples. In these systems, light is used as a trigger to cause subsequent life processes. We have been interested in the application of the phenomena to biomimetic systems. To imitate the fundamental functions of such photoresponsive systems, one has to combine within a molecule a photoantenna to capture a photon with a functional group to mediate a subsequent event. Host molecules have drawn attention as simplified enzyme model systems. We assumed that the photocontrol of host molecules, as well as that of enzymes [1-3], might provide a somewhat new field of chemistry.

We selected a crown ether moiety as a functional group because it has the ability to associate with charged and uncharged species and its conformation is easily changeable owing to the flexible nature of the macrocyclic polyether. In other words, one may expect that, provided that the conformation of the crown ether ring is easily changed, both the ion-binding ability and the ion-selectivity would be largely affected.

In the artificial photoresponsive systems, chemical substances with photoinduced structural changes are candidates for the photoantenna. Chromophores offering possibilities as a photoantenna are listed in Fig. 1. Prerequisites for a good photoantenna behaviour are high quantum yield, high reversibility, and large structural change. Among these candidates, (E)-(Z) isomerism of azobenzene (a), spiropyran-merocyanineequilibrium (d), and dimerization of anthracene (f) have frequently been employed.

The last problem is how to transmit the photoinduced structural changes occurring in the chromophore to the crown ether. The most expeditious method is to connect the crown ether with the chromophore through a covalent bond. In fact, most of the artificial photoresponsive systems used so far follow this concept. However, the utilization of secondary valence forces, ion flux, conformational change of polypeptides, etc., as the natural transducers, may be also interesting.

2 Classification of Photoresponsive Crown Ethers

2.1 Capped Crown Ethers

The most direct and expeditious method to transmit the photo-induced configurational change, occurring at the chromophore moiety, to the crown ether subunit is to construct a "chromophore bridge" at the crown ring. Considerations on CPK models showed that the distance between N(7) and N(16) of the diazacrown ether *1* is almost equal to that between the two carbonyl groups of 3,3'-bis(chlorocarbonyl)-azobenzene *2*. Thus, the first photoresponsive crown ether *3* was synthesized from *1* and *2* in 23 % yield using high-dilution conditions [4]. When 4,4'-bis(chlorocarbonyl)-azobenzene was used instead of the 3,3'-isomer, the reaction system resulted in either polymers (e.g. *11–13*) or cylindrical ionophores (section 1.3).

The conformational change of the crown ether moiety of *3* which expectedly occurs in conjugation with (Z)-(E) isomerism of the azobenzene cap was inferred from

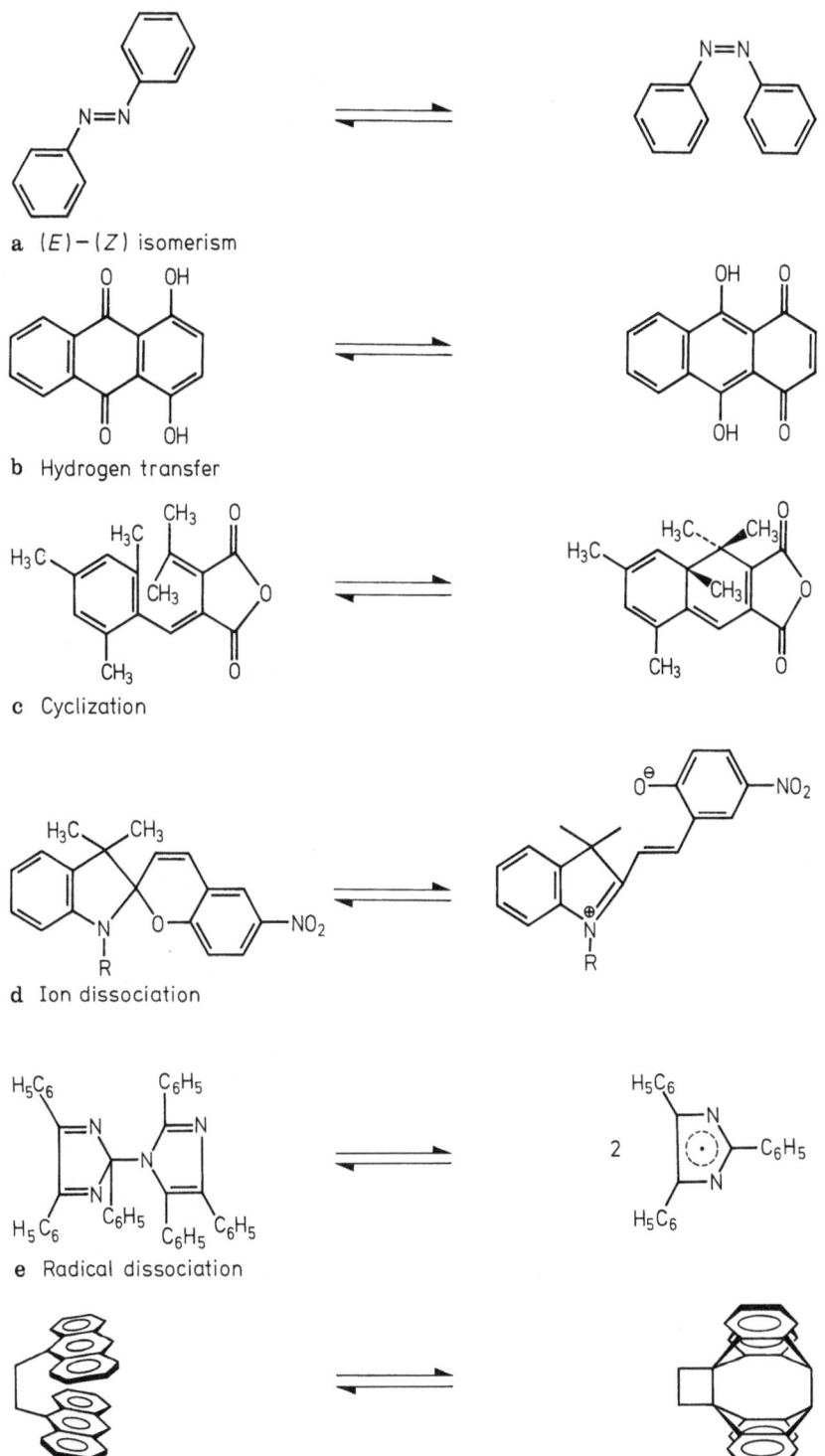

a $(E)-(Z)$ isomerism

b Hydrogen transfer

c Cyclization

d Ion dissociation

e Radical dissociation

f Dimerization

Fig. 1. Candidate chromophores for photoantennas

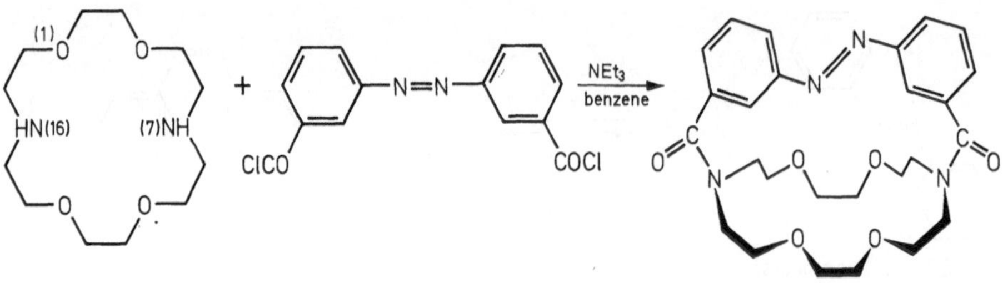

1   2   (E)-3

the CPK model studies. They show that the azobenzene moiety of (E)-3 projects vertically over the crown ether plane, whereas that of (Z)-3 is almost parallel to the crown ether plane. The structure was later confirmed by an X-ray crystallographic study [5]. The CPK models also suggest that in (E)-3 the *ortho* hydrogen of the azo linkage penetrates partially into the plane of the crown ether. This would probably interfere with deep complexation of alkali metal cations.

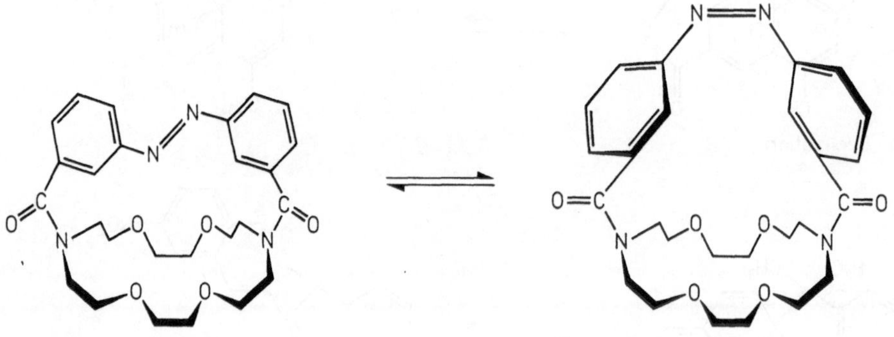

(E)-3   (Z)-3

Under UV-light irradiation, the absorption band of (E)-3 decreased rapidly and reached a photostationary equilibrium where the ratio of (E)-3/(Z)-3 was 40/60. The solvent extraction of alkali metal salts of Methyl Orange from water to benzene indicated that (i) the extractability of (E)-3 changes in the order $K^+ > Na^+ > Li^+ > Rb^+, Cs^+$, whereas that of (Z)-3 is in the order $K^+ > Na^+ > Rb^+ > Li^+, Cs^+$, (ii) large alkali metal cations such as Rb^+ and Cs^+ are hardly extracted by (E)-3, (iii) (E)-3 extracts Li^+ which cannot be extracted by (Z)-3, whereas (E)-3 cannot extract Rb^+ which is extracted by (Z)-3, and (iv) Na^+ is better extractable by (E)-3 than by (Z)-3 but the reverse is true for K^+. These findings consitantly suggest that the crown ether ring of (Z)-3 is somewhat expanded by the (E)-to-(Z) isomerization of the azobenzene-cap.

The azo-linkage allows the coordination of heavy-metal ions, however such a coordination is generally weak [6,7]. For 2,2'-azopyridine, one may expect the complexation with the azo-linkage to be stabilized by neighboring pyridine nitrogens. In fact, Shinkai et al. [8] have found that (E)-2,2'-azopyridine acts as a much better ligand than photo-isomerized (Z)-2,2'-azopyridine. Based on CPK models (and the X-ray crystallographic study) [5,8], one may predict that pyridine nitrogens of (E)-4 are directed toward the crown ether plane and coordinate metal ions in the crown ether ring, whereas those of (Z)-4 have not such a coordination ability due to the distorted configuration. Thus, this may be an example of a photoresponsive crown-cryptand transformation.

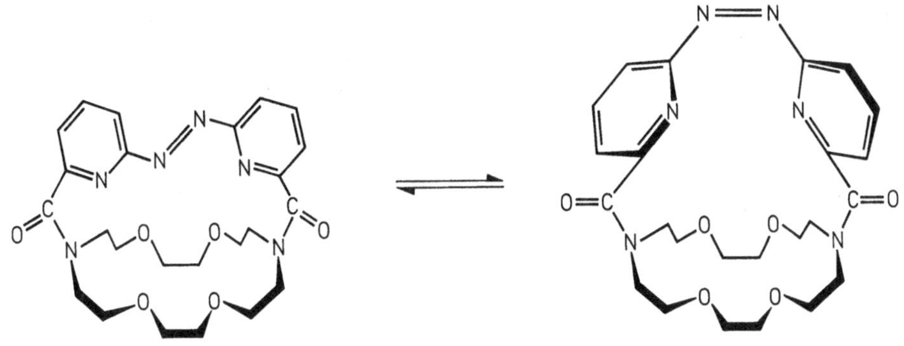

$(E)-4$ $(Z)-4$

The result of solvent extraction with 4 indicated that the photoisomerization of the 2,2'-azopyridine bridge affects the extractability of alkali metal cations only to a smaller extent. In contrast, (E)-4 extracts considerable amounts of heavy metal ions (Cu^{2+}, Ni^{2+}, Co^{2+}, and Hg^{2+}), whereas photoisomerized (Z)-4 scarcely extracts these metal ions [8]. Such a difference in the extractability was not observed between (E)-3 and photoisomerized (Z)-3. Thus, the affinity of 4 toward heavy metal ions is primarily governed by the configuration of the 2,2'-azopyridine bridge. Shinkai et al. [9] further synthesized the following capped thiacrown ethers, 5 (X=CH) and 6 (X=N).

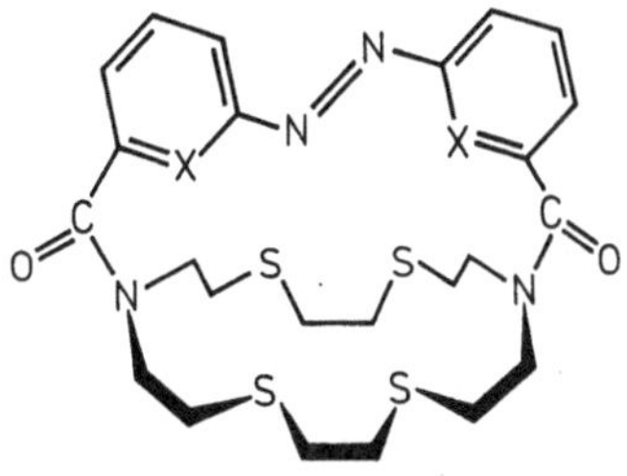

$(E)-5$ (X=CH) and $(E)-6$ (X=N)

The thermal (Z)-to-(E) isomerization of azobenzene has been controversial: the reaction may proceed either via a rotation about the N=N bond or flip-flop inversion of one of the nitrogen atoms (Fig. 2) [10]. The difference between the two mechanisms is that the rotational mechanism proceeds via a dipolar transition state accompanying a large volume change.

Nowadays, it is believed that the thermal isomerization of "usual" azobenzene derivatives proceeds through an inversion mechanism: for example, the lack of major solvent or pressure effect on the rate of isomerization may be cited as an evidence against rotation [10–13]. Whitten et al. [10,14] showed, however, that sizable kinetic solvent effects are observed for azobenzene with push-pull substituents (e.g., dimethylamino and nitro) stabilizing the dipolar transition state of the rotational mechanism. Since (Z)-3 has an azobenzene covalently bridged to an azacrown ether and the rotation of the benzene rings is sterically restricted, this molecule provides the "standard" thermodynamic parameters for the inversion isomerism [15]. Thus, it was demonstrated that the thermal isomerization of most (Z)-azobenzenes occurs via an inversion mechanism, the activation parameters ($\Delta H^{\ne}$ and $\Delta S^{\ne}$) including those of (Z)-3 being subject to a good $\Delta H^{\ne} - \Delta S^{\ne}$ compensation relationship. The pressure effect also supported the inversion mechanism because the $\Delta V^{\ne}$ was too small (ca. -0.4 to $-0.7\,\text{ml mol}^{-1}$) to consider the rotational mechanism [16]. Interestingly, simple azobenzenes have small negative $\Delta V^{\ne}$ values, whereas only (Z)-3 has a small positive $\Delta V^{\ne}$ value ($2.0\,\text{ml mol}^{-1}$).

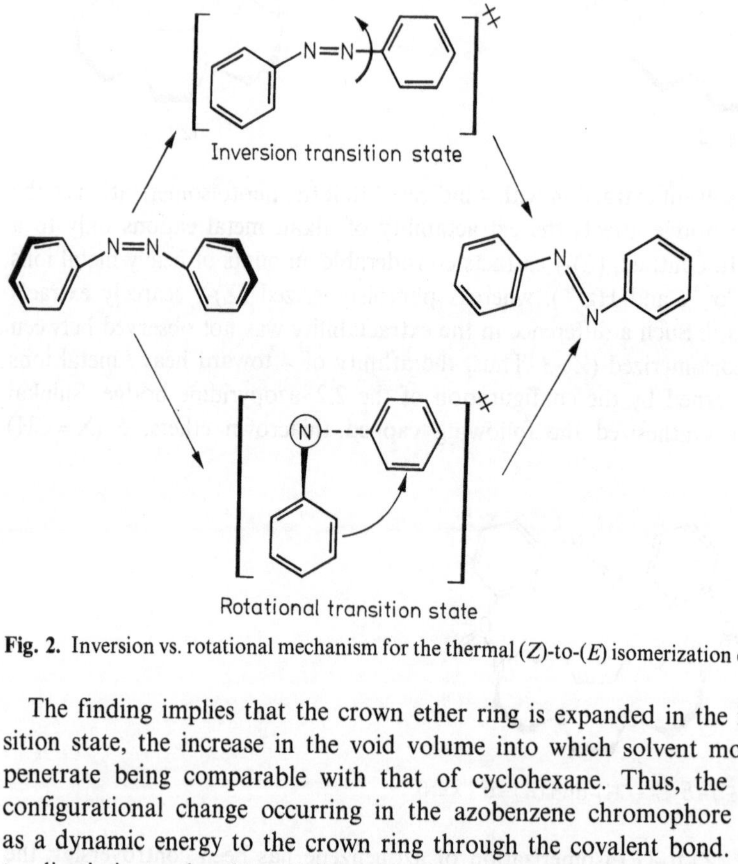

Fig. 2. Inversion vs. rotational mechanism for the thermal (Z)-to-(E) isomerization of (Z)-azobenzene

The finding implies that the crown ether ring is expanded in the inversion transition state, the increase in the void volume into which solvent molecules cannot penetrate being comparable with that of cyclohexane. Thus, the photo-induced configurational change occurring in the azobenzene chromophore is transmitted as a dynamic energy to the crown ring through the covalent bond. A similar idea to discriminate the thermal isomerization mechanisms on the basis of the azobenzenophane molecules such as 7 and 8 has been reported by Rau and Lüddecke [17].

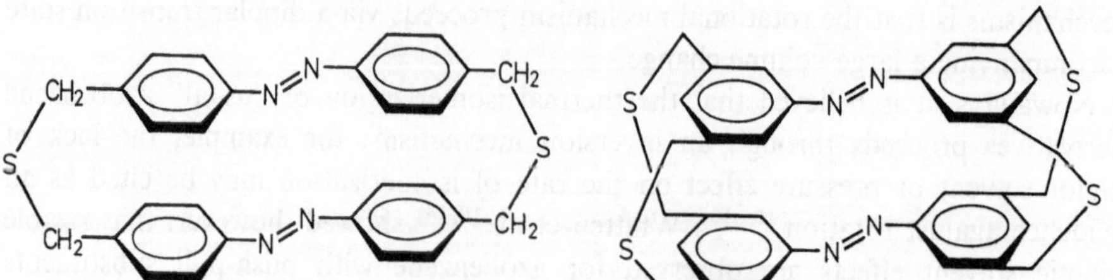

Shinkai et al. [4] have found that the rate of the thermal isomerization of (Z)-3 is suppressed by K^+ and ammonium ions. The crown ether ring of (Z)-3 must expand once in the transition state [16], so that the interaction between the crown ether and the complexed cation must be disrupted (or weakened) temporarily. This requires an additional free-energy of activation for the isomerization process. Probably, the transition state species of 3 exhibits the ion-affinity and the ion selectivity quite different from those in the ground state.

Cyclodextrin is another interesting host-molecule. Ueno et al. [18] synthesized azobenzene-capped β-cyclodextrin 9 to regulate the binding ability of β-cyclodextrin (β-CD). They found that 9 presents an induced circular dichroism band in the azobenzene π—π* region (355 nm) before irradiation, whereas it shows another band in the azobenzene n—π* region (445 nm) after irradiation. Since both circular dichroism bands nearly disappeared on the addition of guest molecules in large excess, the association constants could be determined from the change in the band intensity.

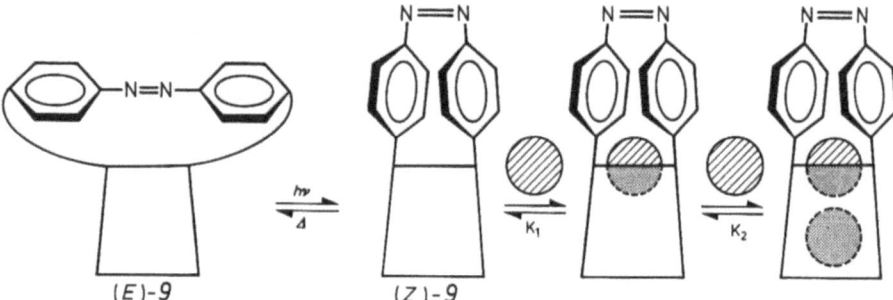

The examination of the association constants revealed that (i) (E)-9 has the association constants generally smaller than β-CD, whereas (Z)-9 has the greater binding ability than β-CD and (ii) (Z)-9 tends to include two guest molecules. These results allow to consider the poor binding ability of (E)-9 to be associated with the "shallow floor" concept suggested by Emert and Breslow [19]. On the other hand, the improved binding ability of (Z)-9 stems from the expansion of the cavity due to the capping with (Z)-azobenzene. The circular dichroism spectrum of 9 shows in some cases an abrupt change on the addition of guest molecules with ionic nature [20]. The non stoichiometric behavior was regarded as evidence for a conformational transition of the cyclodextrin moiety caused by outside binding of the guest molecules.

The rate of hydrolysis of p-nitrophenyl acetate (PNPA) catalyzed by 9 according to Eq. (1) is accelerated by photoirradiation (eq. 1) [21].

The reaction proceeded according to the Michaelis-Menten-type kinetics:

$$9 + \text{PNPA} \underset{k_{-1}}{\overset{k_1}{\rightleftharpoons}} 9 \cdot \text{PNPA} \overset{k_2}{\rightarrow} \text{products}.$$ The maximum rate constant k_2 for (Z)-9 was smaller than that for (E)-9, indicating the unfavorable geometry of the (Z)-9 · PNPA complex to hold the substrate molecule in the correct catalytic position. Probably, PNPA is included too deeply in the expanded cavity of (Z)-9 to attain a favorable attack of the alkoxide ion of the cyclodextrin moiety at the carbonyl group of PNPA. On the other hand, (Z)-9 has a considerably smaller K_m value $(= k_{-1}/k_1)$ than (E)-9, indicating that the substrate is included deeply in the expanded hydro-

phobic cavity of (Z)-9. These situation leads to an increased overall hydrolysis rate (k_2/K_m) for (Z)-9; it was about five times greater than that for (E)-9.

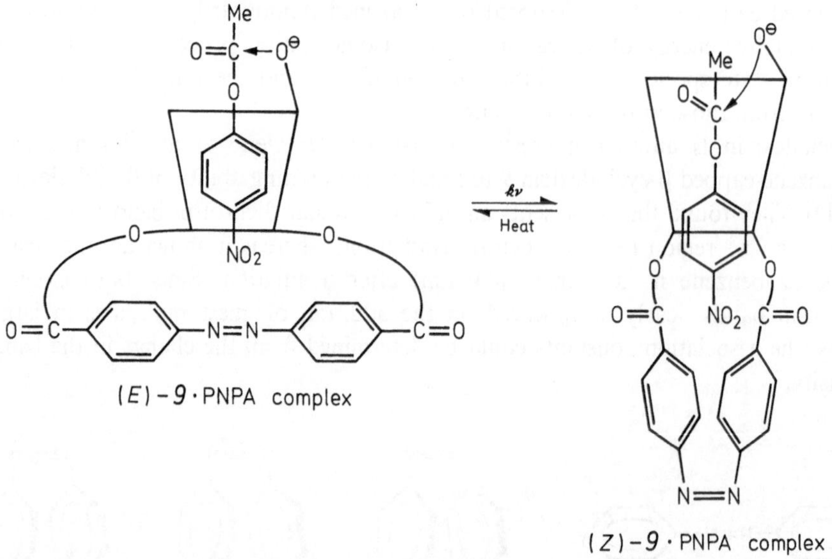

$$\text{(E)}-9 \cdot \text{PNPA complex} \xrightarrow[\text{Heat}]{h\nu} \text{(Z)}-9 \cdot \text{PNPA complex} \tag{1}$$

2.2 Polymeric Photoresponsive Crown Ethers

Photoregulation of the conformation of synthetic polymers has been attempted in several systems to elucidate the function of photobiological transducers and also to construct photoreceptor analogs. The basic idea for regulating the conformation of synthetic polymer chains is to prepare a polymer system containing a chromophore capable of transforming the light energy into a conformational change of the polymer chain. Thus, a number of polymers containing either azobenzene or spiropyran were synthesized (e.g., 10–12)[22–26]. With these photoresponsive polymers, the solution viscosity, pH change, polarity change of film surface, etc. was photocontrolled. These phenomena are all due to the photoinduced changes of the chromophoric groups.

The combination of these photoresponsive behaviors of synthetic polymers with a crown ether group may lead to novel photofunctional systems. Shinkai et al.[4] synthesized 13 by the polycondensation of 1 and 4,4'-bis(chlorocarbonyl)azobenzene. Conformational behavior of 13 was conveniently assessed by viscosity measurements. The intrinsic viscosity of (E)-13 (0.057 dl/g) in o-dichlorobenzene decreases to 0.047 dl/g under UV-light irradiation[4]. The result implies that the (Z)-azobenzene unit produced by photoirradiation changes the polymer conformation to a more compact one.

The solvent treatment with (E)-13 showed the extractability of $Na^+ > K^+ > Cs^+ > Li^+, Rb^+$[4], indicating that (E)-13 also belongs to a class of polymers with crown ethers in the main chain, because the crown polymers with a crown moiety in the side chain show a higher affinity for alkali metal cations with large ion radii. The crown polymers with the crown group in the main chain also exhibit a cation affinity analogous to that of the monomeric analogs[27–30]. As expected, the extract-

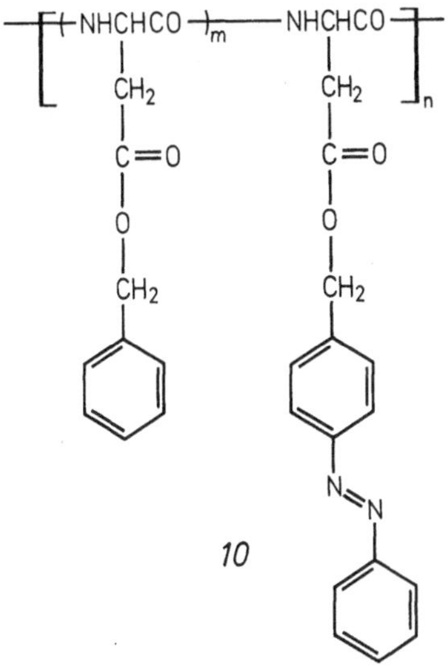

10

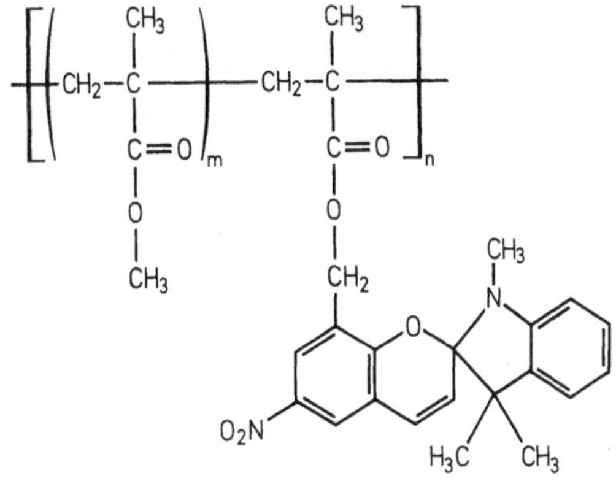

11

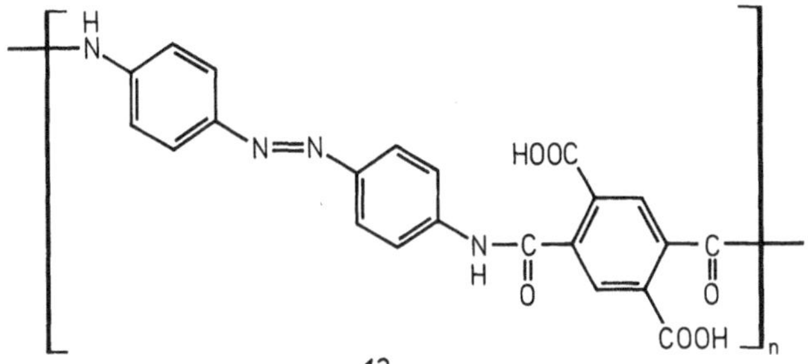

12

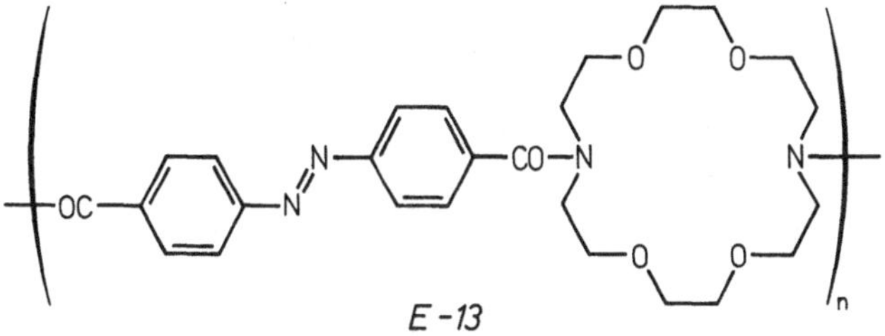

E-13

ability of *13* was affected by UV-light irradiation: under photoirradiation the extractability of K^+ was somewhat enhanced but that of Na^+ was markedly reduced. It was also found that (*E*)-*13* extracts dipotassium salt of phthalic acid in preference to those of isophthalic acid and terephthalic acid and the extractability decreases under photoirradiation [4].

Polymer supports carrying various functional groups have attracted much attention in recent years. In particular, immobilization of crown ethers in polymer supports is one of the most noteworthy achievements demonstrating the usefulness of polymer supports. In the photoresponsive system, polymer supports may be useful as a "fixed point" to induce the conformational change of crown ethers by light. The distance between the 4- and 4'-positions of (*E*)-azobenzene is 9.0 Å, while that of (*Z*)-azobenzene is 5.5 Å. When a molecule containing both a crown ether and an azobenzene is immobilized in a polymer support, the contraction of the azobenzene moiety due to the photo-induced (E)-to-(Z) isomerization is compensated by the elongation of the "flexible" crown ether moiety. *14* immobilized in the cross-linked polystyrene beads shows this behaviour [31].

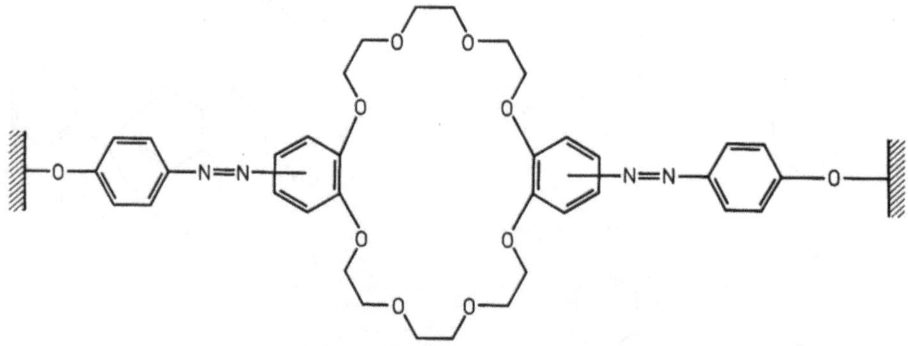

14

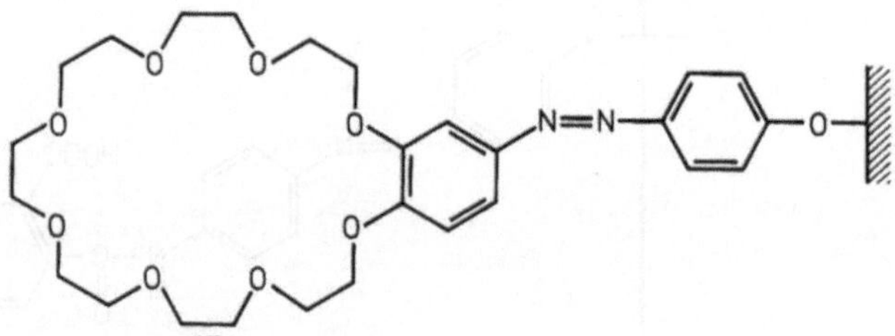

15

When *14* was mixed with the DMF solution of cesium *p*-nitrobenzoate in the dark, the concentration of Cs^+ in the solution decreased rapidly and reached an equilibrium within 20 min. When this solution was subjected to photoirradiation by a UV-lamp inducing the (E)-to-(Z) isomerization of the azobenzene moiety, the concentration of Cs^+ in the solution gradually increased. On the other hand, such a photoresponsive ion-adsorption was hardly observed for *15* in which only one terminal azophenoxide group was immobilized in the polymer support [32]. These results consistently suggest that, as shown in Fig. 3, the photoinduced (E)-to-(Z) isomerization of the azobenzene moiety is capable of changing the conformation of the crown ether into a more stetched one which has a poor ion-binding ability relative to the "normal" crown ether.

Several exciting facts were found by measuring the rate of the thermal (Z)-to-(E) isomerization occurring in the polymer beads [31, 32]. In the 3.2% crosslinked polystyrene beads bearing *14*, the rate of the ion-adsorption to the polymer beads was almost comparable with that of the thermal (Z)-to-(E) isomerization of the azobenzene unit. In the 9.0% cross-linked polystyrene beads, however, the former was much slower than the latter. Probably, diffusion of Cs^+ ion in the polymer matrix is involved in the rate-limiting factor in the highly crosslinked polystyrene beads. It is also interesting that the thermal isomerization in the 9.0% crosslinked polystyrene beads is faster by about one order of magnitude

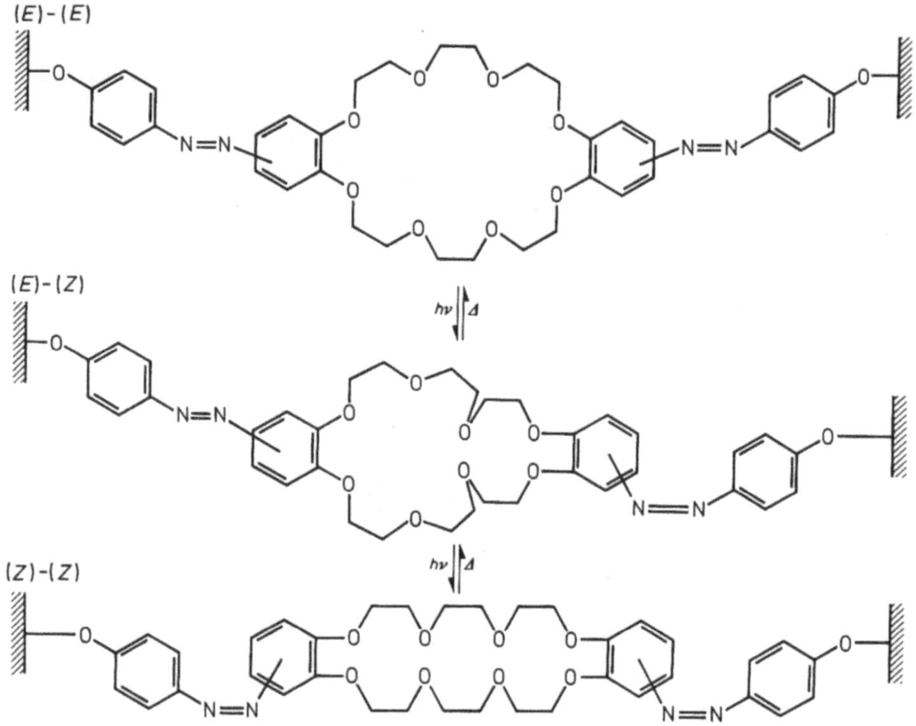

(E)-(E)

(E)-(Z)

$h\nu \upharpoonleft\downharpoonright \Delta$

(Z)-(Z)

$h\nu \upharpoonleft\downharpoonright \Delta$

Fig. 3. Photoisomerism of the azobenzene-crown-azobenzene bridge immobilized in the polymer support

than that in the 3.2% crosslinked polystyrene beads. The most fascinating rationale is that the photoinduced dynamic tension being stronger in the highly cross-linked polymer facilitates the thermal isomerization (i.e., the relaxation). If so, this may be an interesting example for the conversion of dynamic into chemical energy.

Similar phenomena were observed for the adsorption of K^+ on the immobilized [18]crown-6 compounds _16_ [32].

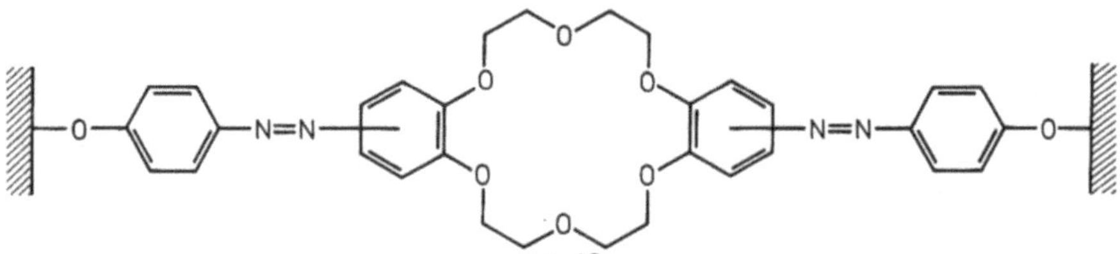

(E)-16

2.3 Cylindrical Ionophores

A cylindrical ionophore, in which two macrocyclic ligands are linked by two (or more than two) pillars, represents an interesting type of receptor molecule which allows to bind diammonium ions and more than one metal ion. Since the metal affinity of cylindrical ionophores is crucically governed either by the pillar length or by the ring size of ligands, one may expect that the binding affinity and the binding selectivity is readily controlled by changing the length of the two-pillars. Thus, ionophores with azobenzene pillars act as "elastic" cyclindrical ionophores in response to photo-induced (E)-(Z) isomerization of the azobenzene pillars.

The first photoresponsive cylindrical ionophore synthesized by Shinkai et al. [33] is shown by formula *17*. As described before, the reaction of *1* with 3,3'-bis(chlorocarbonyl)azobenzene under high dilution conditions yielded *3* (i.e., 1:1 adduct), whereas the reaction of *1* with 4,4'-bis(chlorocarbonyl)azobenzene gave *17* (i.e., 2:2 adduct). The difference can be rationalized in terms of the steric factor that the distance between N(*7*) and N(*16*) of *1* is almost equal to that between two carbonyls of the 3,3'-isomer but significantly shorter than that between two carbonyls of the 4,4'-isomer. Similarly, *18* and *19* were obtained by condensation under high-dilution conditions.

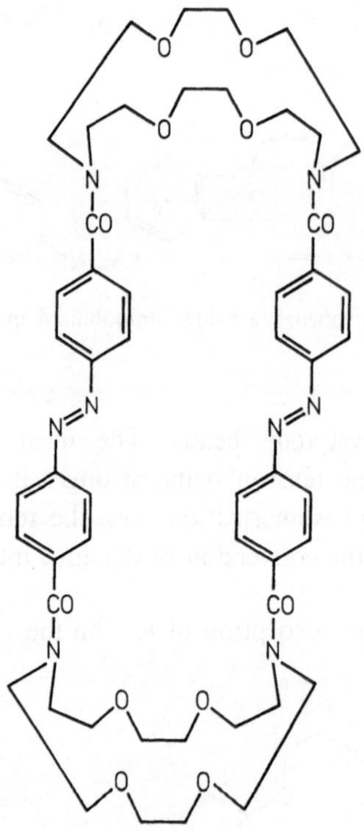

(E)-*17*

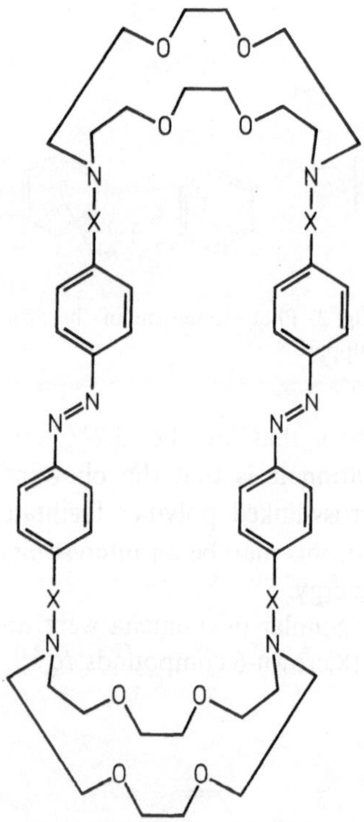

(E)-*18* $(X = OCH_2CO)$
(E)-*19* $(X = CON(CH_3)CH_2CH_2)$

From CPK model considerations it was speculated that, assuming that the polymethylene chain adopts an extended conformation, the distances between the two crown rings to (E)-*17* and (Z)-*17* (two azobenzenes are (Z)-forms) are almost equal to those between the two terminal ammonium groups of decamethylenediamine and hexamethylenediamine, respectively. It was shown that (E)-*17* extracts decamethylene and dodecamethylene diammonium salts efficiently but hardly extracts tetramethylene and hexamethylene diammonium salts, whereas hexamethylene diammonium salt is most extractable under UV-light irradiation [33]. The results are basically compatible with the prediction from CPK model studies. Similarly, *18* allows to extract diammonium salts with long methylene units but scarcely extracts those with short methylene units [34].

(*E*)-*19* having four tertiary amine functionalities strongly binds alkali metal ions, but UV-light irradiation has almost no effect on the binding ability. This is probably due to the too long distance between the two crown ether rings. Instead, *19* acts as a photoresponsive host molecule in aqueous solution [35]. (*E*)-*19* is very soluble in water at pH 5 by protonating the azacrown nitrogens and strongly associates hydrophobic guest molecules such as Methyl Orange. When the two azobenzene pillars were photo-isomerized to the (*Z*)-form, the complex was dissociated and the increase in the free Methyl Orange was apparently correlated to the concentration of (*Z*)-*19*. The result indicates that the association ability of the host molecule is crucially subject to the molecular shape.

2.4 Phane-Type Crown Ethers

In order to design a photoresponsive crown ether exhibiting a large change in the ion-binding ability, one should choose a large geometrical change without steric hindrance. In fact, however, these two requirements are quite opposing factors. Crown ethers containing a photoresponsive chromophore like a phane unit might satisfy the two opposing requirements.

Desvergne et al. [36] utilized the photo-dimerization of anthracene to design a "switched-on" crown ether *21*. In the presence of a lithium salt photoirradiation of *20* gave the photocyclo-isomer *21*, which was fairly stable with the lithium salt as in Eq. (2). When the Li$^+$-*21* complex was shaked in a polar solvent such as acetonitrile, *21* really reverted to the open form *20* by competing solvation of Li$^+$ with a polar solvent. The remarkably enhanced stability of *21* in the presence of Li$^+$ ion was rationalized in terms of a "cation-lock".

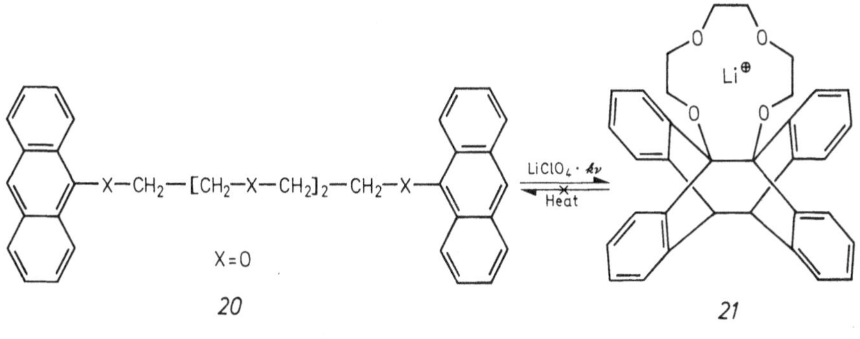

$$\text{(2)}$$

Yamashita et al. [37] synthesized *22* in which the intermolecular photodimerization of anthracene, which may possibly take place in *20*, is completely suppressed. In order to assess the effect of metal ions on photochromic process, half-life time of *22* in the photo-reaction and of *23* in the thermal reaction (3) were measured with and without metal ions. The half-lifetime of the dark reaction $23 \rightarrow 22$ was prolonged by metal ions compared to that of ion-free case, and the effect on the half-life time was in the order $Na^+ > K^+ > Li^+$. On the other hand, the photo-reaction $22 \rightarrow 23$ is not dependent on metal ions. This may be attributed to much larger energy of the photo-transition than the stabilization energy due to the ion complexation.

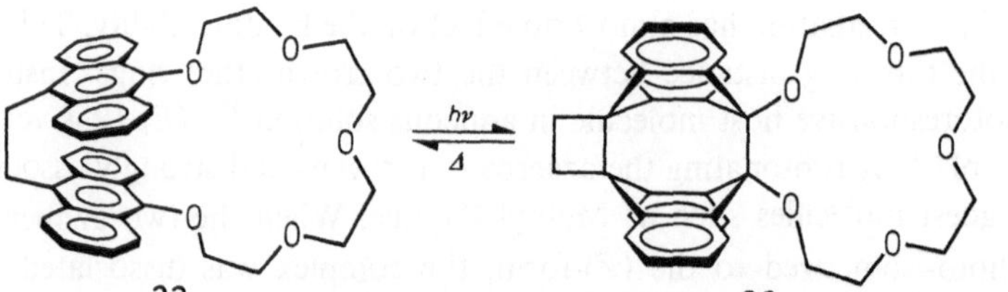

$$(3)$$

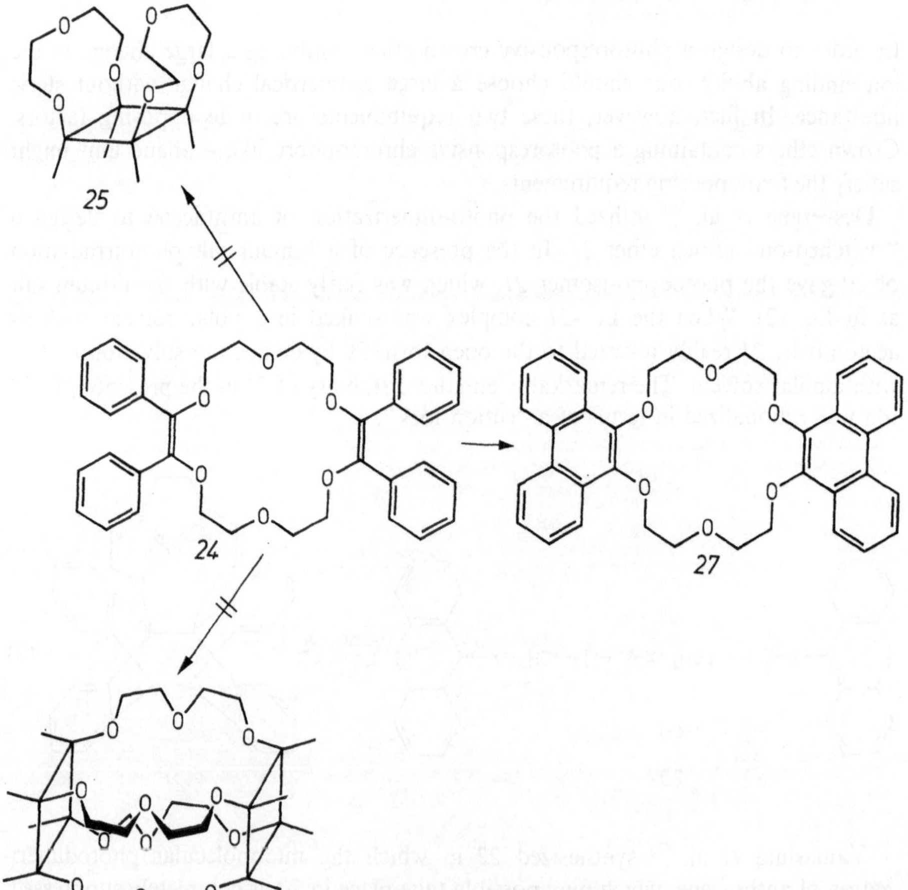

The foregoing successful results suggest that the photochemistry of unsaturated crown ethers appears to as a novel application. Eichner and Merz [38] attempted to synthesize a basket crown 25 or a cage crown 26 by photo-reaction with 24 (4). However, when *24* was irradiated in benzene with catalytic amounts of iodine and oxygen bubbling through the solution, *27* was obtained as the only reaction product in 50–60% yield. Thus, the photo-oxidation to phenanthrene exceeded the photo-dimerization to cyclobutane.

Shinkai et al. [34] synthesized (*E*)-*28* by condensation of the corresponding diamine with 4,4'-bis(chlorocarbonyl)azobenzene under high-dilution conditions. The CPK model of *28* suggested that the polyoxyethylene chain is almost linearly extended when the azobenzene segment adopts the (*E*)-configuration, whereas that of *28* with the (*Z*)-configuration forms a crown-like loop. The result of the solvent extraction with *28* revealed that (*E*)-*28* completely lacks the affinity with alkali metal cations, while photoisomerized (*Z*)-*28* is capable of extracting large alkali metal cations such as Rb^+ and Cs^+. Thus, an ionophoric loop appears in response to photoirradiation; the loop size is roughly comparable with that of [21]crown-7 (21C7)[1].

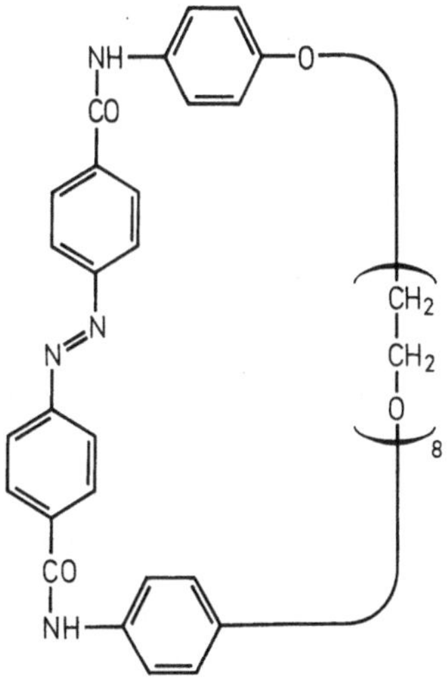

(*E*)-*28*

A large change in the ionophoric nature of the crown-like loop is effected if one can incorporate the azobenzene unit directly into the ring structure (i.e., as azo-benzenenophane). Shiga et al. [39] synthesized the azobenzenophanes (*E*)-*29* and (*E*)-*30*. (*E*)-*29* exhibited the affinity to alkali metal cations such as Na^+ and K^+ similar to "regular" crown ethers. In contrast, photoisomerized (*Z*)-*29* completely lacked the affinity to the metal ions. As speculated from the CPK model, the con-formation of the crown ring of (*Z*)-*29* is highly strained and unfavorable for the interaction with metal ions. However, both the photo- and thermal isomerization of *29* and *30* are extremely slow and the attempt to (*Z*)-isomerize more than 20% leads to an irreversible spectral change. This is probably due to the large steric crowding expected for the transition state.

In the example above, the photoresponsive "all-or-nothing" change was attained but the good reversibility was not. To attain both at the same time, one has to utilize

---

1 Notations frequently used in this report are listed and explained on p. XI

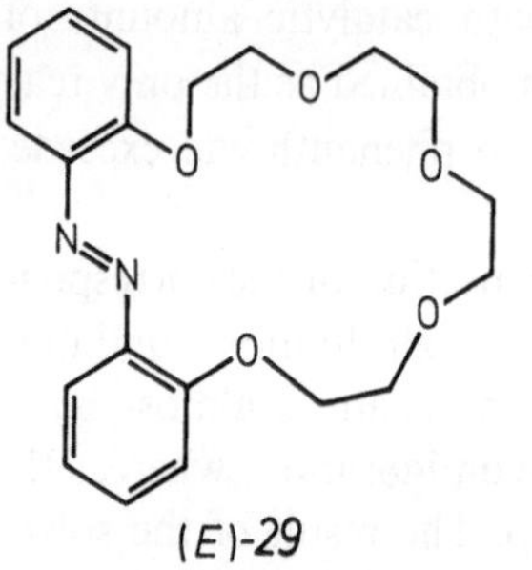

(E)-29 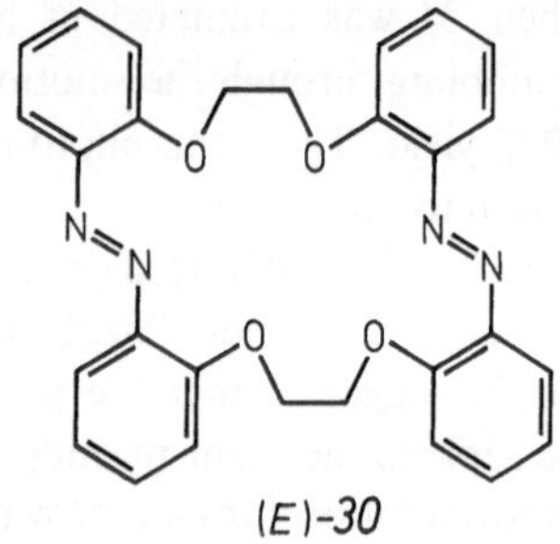   (E)-30

a large geometrical change without inducing any steric strain. Shinkai et al. [40] synthesized azobenzenophane-type crown ethers *31* (n = 1, 2, 3), with 4,4′-positions of azobenzene linked by a polyoxyethylene chain.

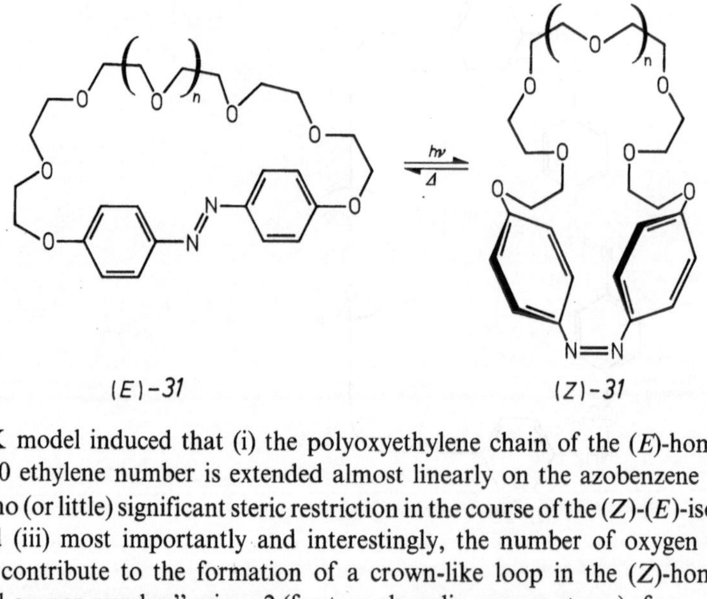

(E)-31   (Z)-31

The CPK model induced that (i) the polyoxyethylene chain of the (E)-homologs with 6 to 10 ethylene number is extended almost linearly on the azobenzene plane, (ii) there is no (or little) significant steric restriction in the course of the (Z)-(E)-isomerization, and (iii) most importantly and interestingly, the number of oxygen atoms which can contribute to the formation of a crown-like loop in the (Z)-homologs is the "total oxygen number" minus 2 (for two phenolic oxygen atoms): for example, (Z)-*31* (n = 1) forms a polyoxyethylene loop analogous to [15]crown-5 (15C5). The (Z)-(E) isomerism (5) was reversible.

The solvent extraction of alkali metal cations with *31* indicated that the predictions based on CPK models are reasonable. The (E)-isomers totally lacked the affinity for metal ions, whereas, the (Z)-isomers extracted considerable amounts of alkali metal cations to the organic phase. The maximum extractability was observed for Na^+ in (Z)-*31* (n = 1), K^+ in (Z)-*31* (n = 2), and Rb^+ in (Z)-*31* (n = 3). The result implies that *31* satisfies the reversible "all-or-nothing" change in the ion-binding ability, and the spheric recognition patterns, typical of crown ethers in solution. The ion selectivity supports that the ring size of (Z)-*31* (n = 1), (Z)-*31* (n − 2), and (Z)-*31* (n = 3) correspond to those of 15C5, 18C6, 21C7, respectively. The rate of the thermal (Z)-to-(E) isomerization is also suppressed by the complexed metal ions. The association constants of the (Z)-homologs were thus estimated by analyzing the plots of the rate vs. the metal concentration.

2.5 Azobis (Crown Ethers)

In the preceding excamples, the photocontrol of the crown ether function is achieved on the basic idea that the conformational change in the crown ether ring is induced by the photoresponsive configurational change in the intramolecular chromophore. Here, the idea is that the spacial distance between two crown ether rings changes in response to photoirradiation. It has been established that alkali metal cations exactly fitting the size of the crown ether form a 1:1 complex, whereas those which have larger ion radii form a 1:2 complex. This view was substantiated clearly by using bis(crown ethers) [41, 42], polymeric crown ethers [29, 30] and was also shown by the crystal structure of crown-alkali metal cation complexes [43]. For example, 15C5 and its analogs form a 1:1 cation/crown complex with Na^+, whereas a 1:2 cation/crown complex is formed with K^+. Kimura et al. [41] reported that the maleate derivative ((Z) form) of benzo[15]crown-5 B15C5 extracts K^+ from the aqueous phase 14 times more efficiently than the fumarate derivative ((E) form) owing to the formation of the intramolecular 1:2 complex. The result suggests that the bis(crown ether) in which the C=C double bond is replaced by the azo-linkage exhibits a photoresponsive ion-extraction behavior.

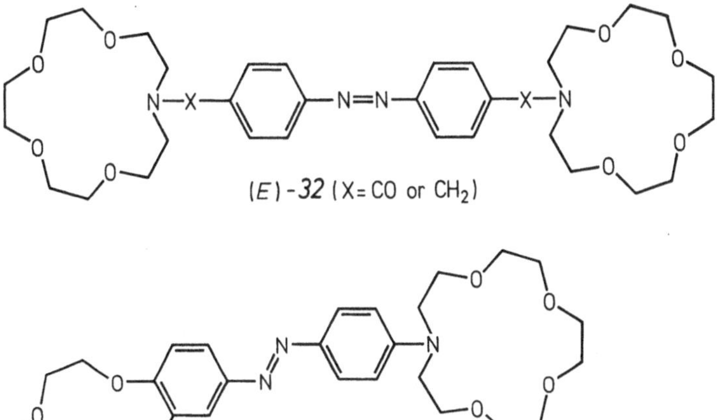

(E)-32 (X= CO or CH$_2$)

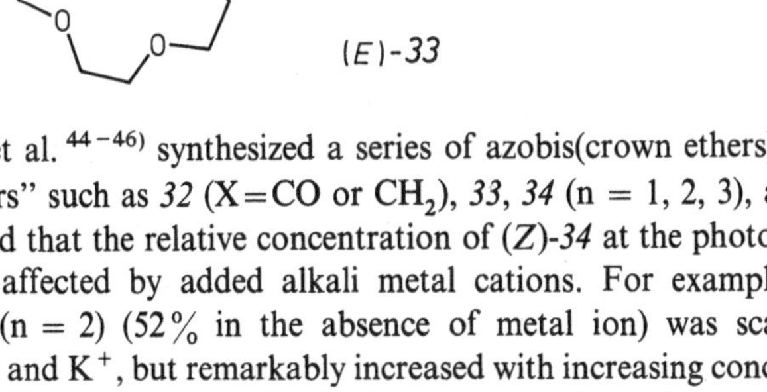

(E)-33

Shinkai et al. [44−46] synthesized a series of azobis(crown ethers) called "butterfly crown ethers" such as 32 (X=CO or CH$_2$), 33, 34 (n = 1, 2, 3), and 35 (n = 2, 3). It was found that the relative concentration of (Z)-34 at the photostationary state is sensitively affected by added alkali metal cations. For example, the content of (Z) in 34 (n = 2) (52% in the absence of metal ion) was scarcely affected by added Na^+ and K^+, but remarkably increased with increasing concentrations of Rb^+ and Cs^+ [45, 46]. In particular, the (Z)-isomer was enhanced up to 98% at high Rb^+ concentrations. Similarly, the rate of the thermal (Z)-34 (n = 2) to (E)-34 (n = 2) isomerization was markedly suppressed by Rb^+ and Cs^+, Rb^+ exhibiting the greatest inhibitory effect. The photostationary state and the thermal isomerization rate of 34 (n = 3) were also affected by Rb^+ and Cs^+, but the Cs^+ provided on influences greater than Rb^+ [45,46]. These novel findings were attributed to the "tying effect" of two crown ethers by one alkali metal cation, which is expected only for the sandwich-type 1:2 cation/crown complexes (e.g., *36*).

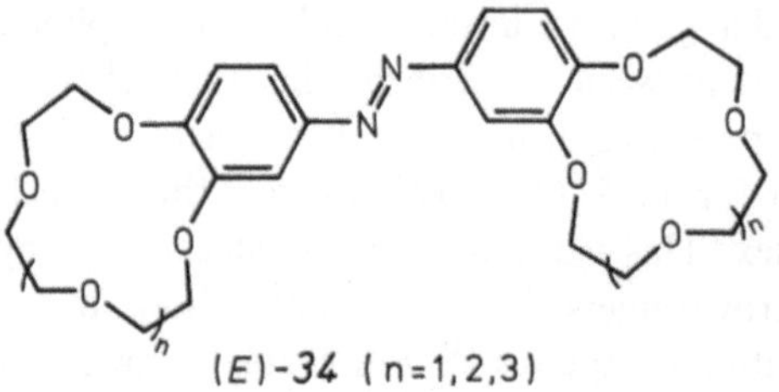

(E)-*34* $(n=1,2,3)$

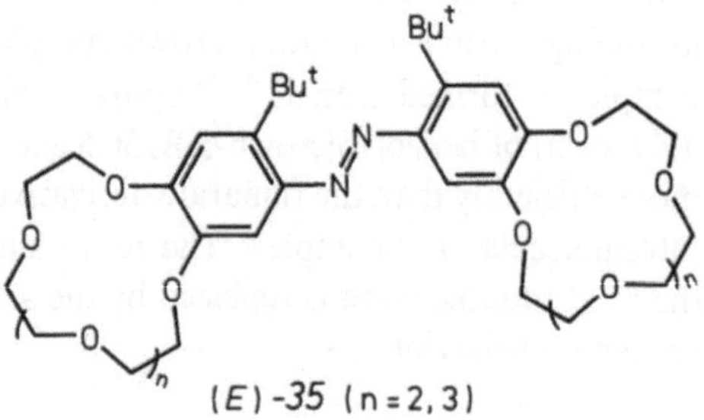

(E)-*35* $(n=2,3)$

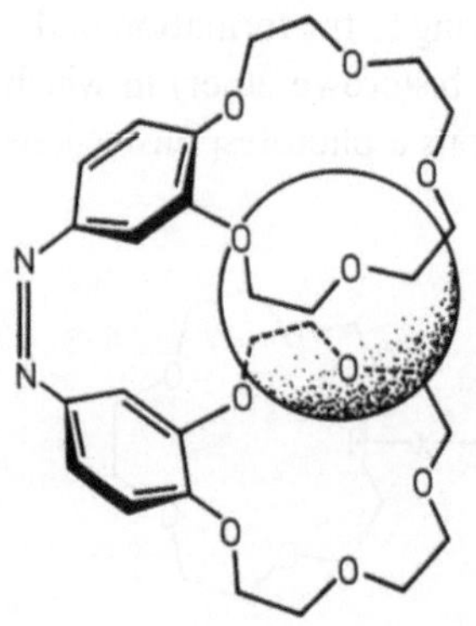

36

The influence of added alkali metal cations on the photostationary state and the isomerization rate were also observed for *35* (n = 2, 3); but when comparing crown series *35* with *34*, the changes of *35* occurred with smaller alkali metal cations: for example, K^+ increases the $(Z)\%$ of *35* (n = 2), whereas the $(Z)\%$ of *34* (n = 2) was hardly affected [46]. The result suggests that the inner cavity between two crown ether rings of (Z)-*35* might be somewhat smaller than that of (Z)-*34*. The difference between crown series *34* and *35* is ascribed to the buttressing effect of the *tert*-butyl groups. (Z)-Azobenzenes are sterically crowded; the introduction of *tert*-butyl groups into the ortho positions of the azo linkage would further increase the steric crowding of Z-*35*.

The influence of the photoirradiation on the ion-binding ability of *34* and *35* was quite remarkable: for example, (E)-*34* (n = 2) extracted Na^+ 5.6 times more efficiently than (Z)-*34* (n = 2), whereas (Z)-*34* (n = 2) extracted K^+ 42.5 times more efficiently than (E)-*34* (n = 2) [45]. The selectivity is expressed by the ratio $((E)/(Z)$ of extractability for Na^+ against that for K^+ and was found 238-fold. Conceivably, Na^+ was extracted as a 1:1 cation/crown complex, while K^+, Rb^+ and Cs^+ (the ion radii of which are somewhat greater than the size of B15C5) were extracted as 1:2 cation/crown sandwich-type complexes as *36*. A systematic investigation of the crown series *34* and *35* [45,46] established that the extractability of the (Z)-forms follows the same order as the inhibitory effect on the thermal isomerization rate and the (Z) percentage at the photostationary state. Conceivably, both orders are associated with the stability of the sandwich-type complex with the Z-isomers.

Shinkai et al. [44] also found that (Z)-32 has a higher extractability for K$^+$ and Rb$^+$ than (E)-32. This was also rationalized by the formation of a sandwich-type complex with (Z)-32. Being different from the crown series 34 and 35, however, neither the photostationary state nor the thermal isomerization of 32 was affected by added alkali metal cations. The contrasting behavior may be related to the difference in the "rotational freedom" of the crown ether groups.

Compounds 34 and 35 employ benzocrown ethers as ion-binding groups, so that the geometrical change induced by the isomerization of the N=N double bond is readily trasmitted to the crown ethers. On the other hand, 32 has two free-rotating single bonds between the crown ether and the azobenzene, which may act as an absorber of the geometrical change occuring at the N=N bond. As a result, the isomerization of the N=N bond may start without disruption of the interaction between the cation and crown ethers. In contrast, the extractability of 33 was scarcely changed by photoirradiation, indicating that the two crown ethers act quite independently [44]. Of course, neither the photostationary state nor the thermal isomierzation was affected at all by added alkali metal cations [44]. Probably, these effects are associated not only with the rotational freedom but also with the molecular symmetry. As shown in 37, (Z)-33 has an unsymmetrical cavity between disordered crown caps and thus becomes disadvantageous to catch "round" metal cations.

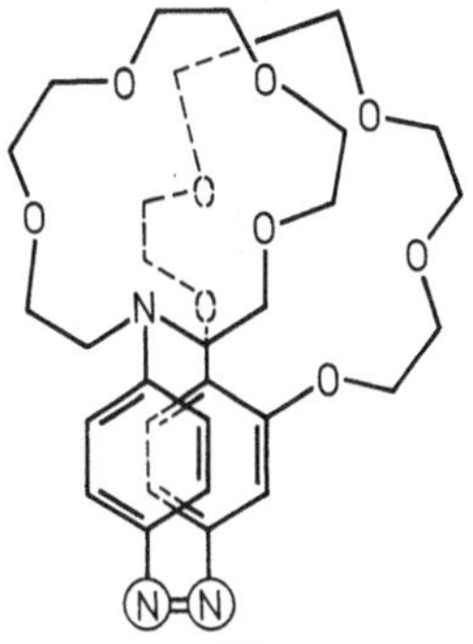

37

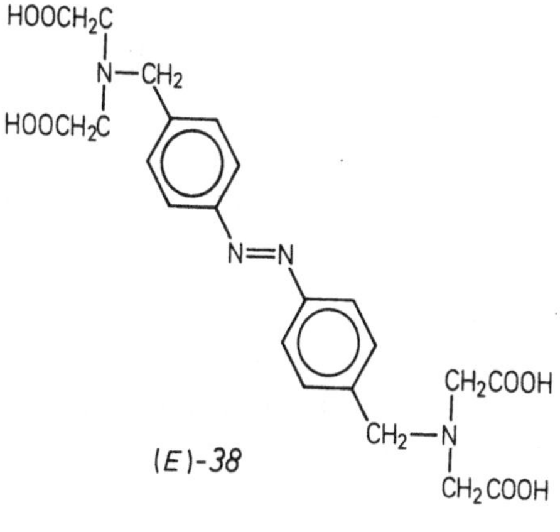

(E)-38

Blank et al. [47] reported on an EDTA-like photoresponsive ligand *38*. The planar (*E*)-*38* could not bind Zn^{2+}. On exposure to light of 320 nm, (*E*)-*38* was converted to a nonplanar chelating agent (*Z*)-*38*, which, because of two cooperative iminodiacetic acid groups, bound Zn^{2+} with a binding constant estimated to be 1.1×10^5 M^{-1}. The interconversion between (*E*)- and (*Z*)-*38* is reversible, suggesting possible application to photo-driven ion pumps.

2.6 Anion-Capped Crown Ethers

In a biochemical field, the more often employed ionophores for alkaline earth metal cations are the polyether antibiotics such as monensin and nigericin representing monobasic, carrier-type ionophores. They feature coupled-transport of cations and protons [48]. It is known that coupled-transport can be imitated in an artificial system using anion-capped crown ethers such as *39*, *40*, and *41*; their ion-binding ability is associated with the correctly placed anionic group on the crown ether ring [49, 50]. These results imply that an ionophore having a crown ring and an anionic cap in a suitable geometrical position acts as an efficient receptor for alkaline earth metal cations.

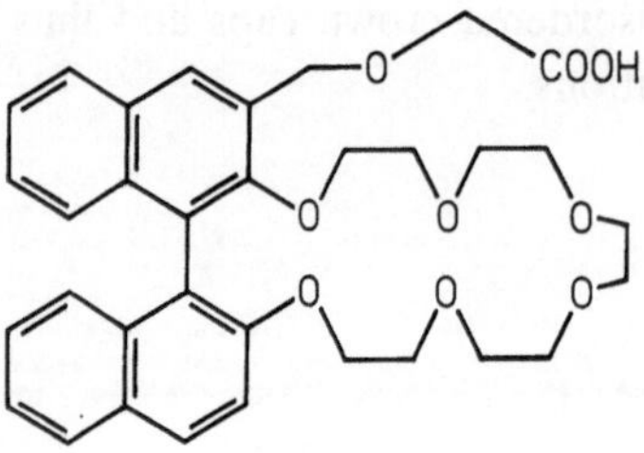

39

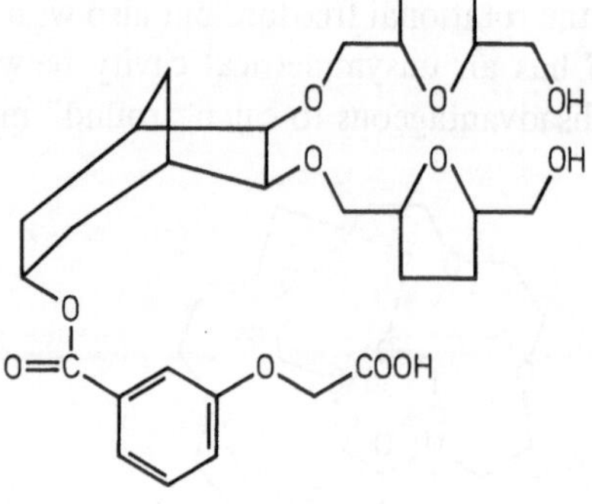

40

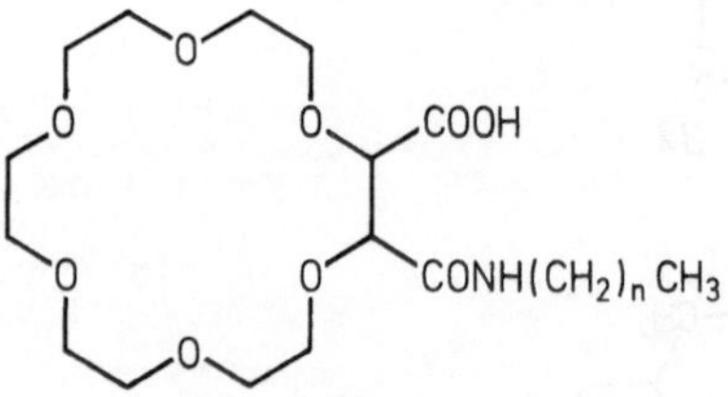

41

The above concept is directly applicable to the design of photoresponsive anion-capped crown ethers. If one synthesizes a crown ether putting on and off an anion-cap in response to photoirradiation, it would lead to the photocontrol of the binding of alkaline earth metal cations. Structure *42* was the first synthetic attempt to design such a molecule [51]. It was found that photo-isomerized (*Z*)-*42* binds K^+ more effectively than (*E*)-*42*. This was attributed to the effect of the carboxylate group capped on the crown ether ring. However, the extractabity was generally low owing to the hydrophilic nature of *42*.

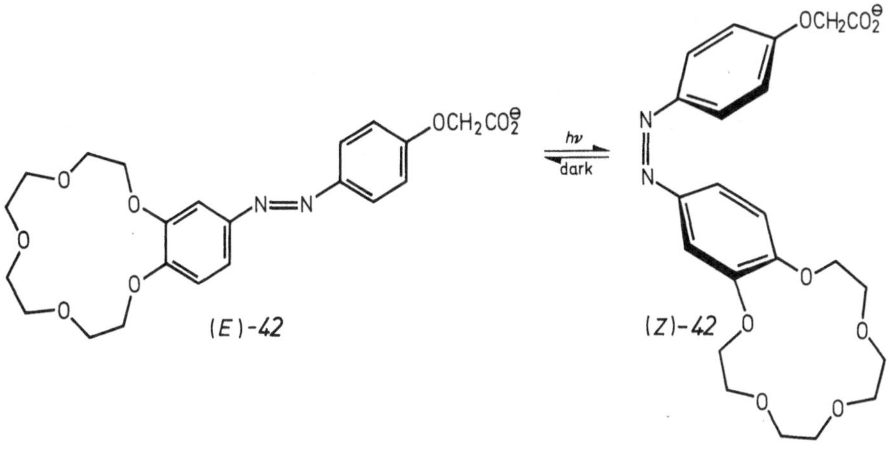

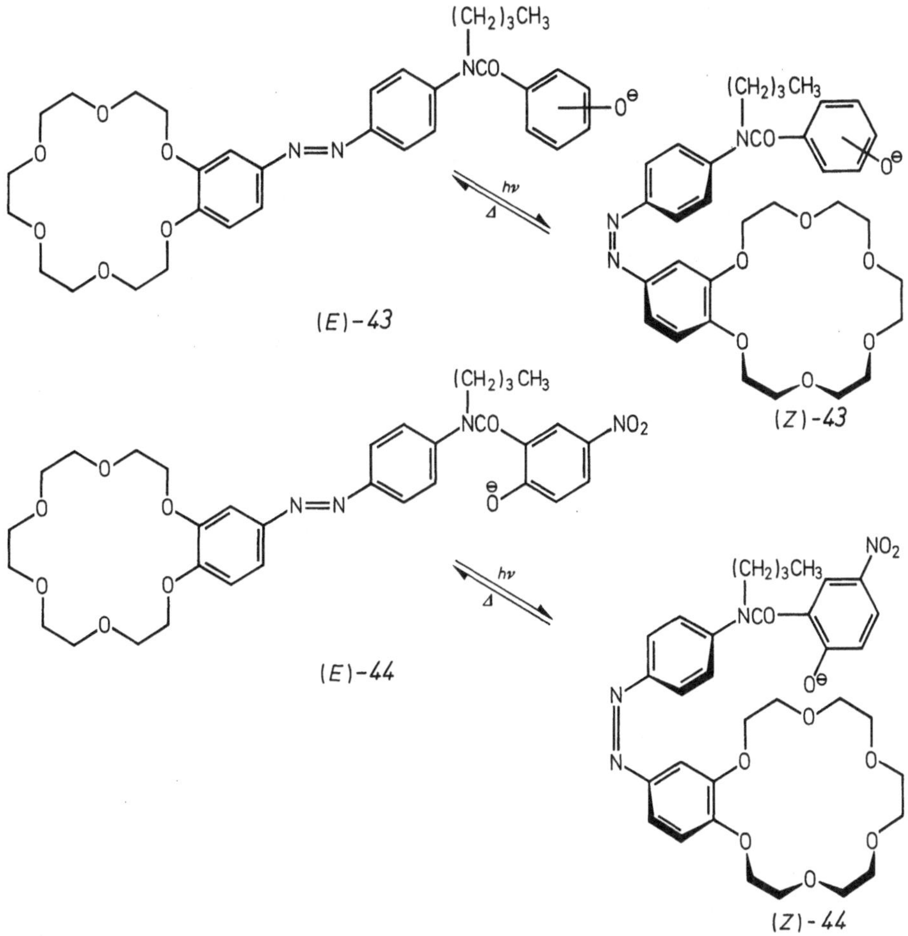

Shinkai et al. [52] subsequently synthesized *43* (*m*-OH and *p*-OH) and *44*. In order to enhance the hydrophobic nature, the carboxylate group was substituted with the phenolate group and the *n*-butyl group was newly introduced. The 5-nitro group in *44* was introduced to lower the pK_a of the phenol group so that (*Z*)-*44* may extract Ca^{2+} ion at neutral pH. Examinations of CPK models revealed that the formation of the complex sandwiched between the crown ring and the phenoxide cap is possible in case of (Z)-43 (m-OH) and (Z)-44 but not with (Z)-43 (p-OH), because in (Z)-43 (m-OH) and (Z)-44 the phenoxide exactly sits on the top of the crown ring, while that of (Z)-43 (p-OH) is located parallel to the crown ether plane.

The extraction with (Z)-43 gave the maximum selectivity for Rb^+ and (Z)-44 efficiently extracted K^+, Rb^+, Cs^+, and Ca^{2+}. The extractability was improved in most cases by UV-light irradiation, indicating that the photoisomerized anionic group contributes to the ion-binding. Unlike 34 and 35, the sharp ion-selectivity was not observed in anion-capped systems.

2.7 Crown Ethers with Other Responsive Functions

The preceding chapters surveyed photoresponsive crown ethers "excited" by a light trigger. In that class of the crown ethers, the light energy is transferred in the order, light→chromophore→covalent bond→crown ether. One may consider other trigger energies or other transfer routes which may collaborate in "responsive crown ethers".

The oldest example is a class of *pH-responsive crown ethers* which change their ion-binding ability in response to medium pH. Anion-capped crown ethers such as 39-41 belong to this group [49, 50], and 42-44 are pH-responsive as well as photoresponsive [52]. As an attempt to imitate the functions of the polyether antibiotics such as monensin and nigericin, a number of acyclic polyethers bearing both a hydroxyl and a carboxylic group at the chain ends have been synthesized. Typicals examples are 45-48 [53-57]. Similarly, a polyether 49 with a quinolyl group instead of a hydroxyl group acts as a pH-responsive receptor for Na^+ [58].

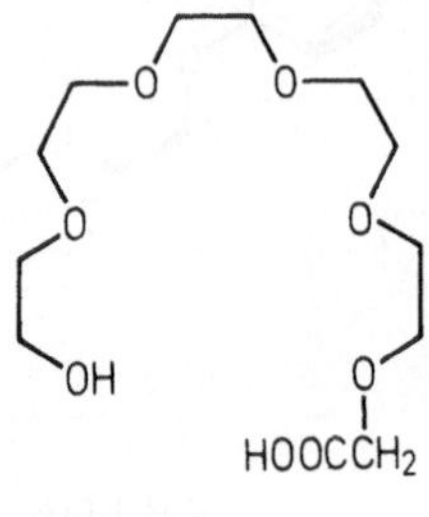

45

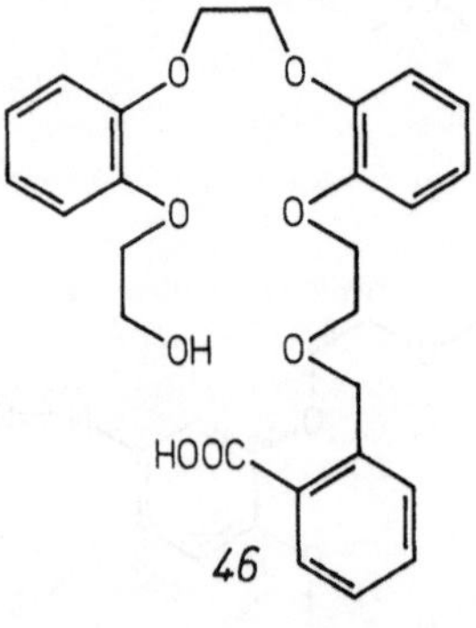

46

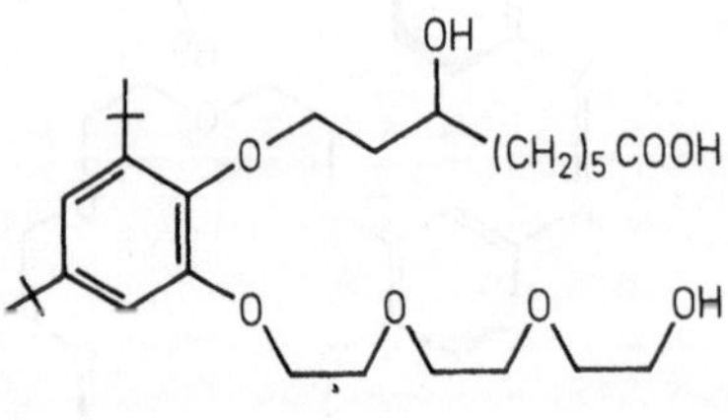

47

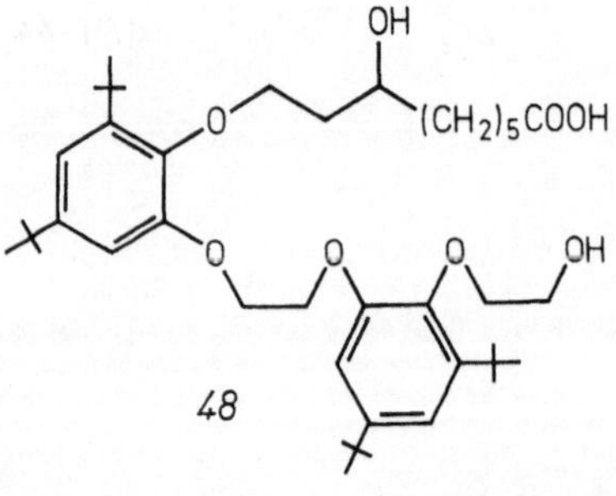

48

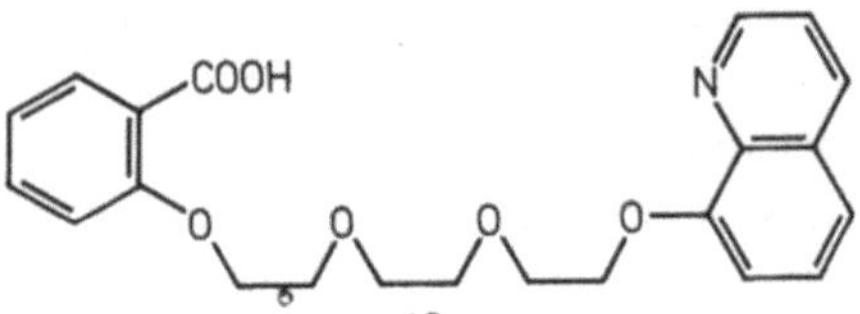

49

In *50* two binding sites are present: the crown ether to bind alkali metal and ammonium ions and the 2,2'-bipyridyl function to bind other metal nuclei [59]. These sites, though separated, are not expected to behave independently because chelation of metals at the bipyridyl function forces the aromatic nuclei toward coplanarity; this leads to a restricted conformational freedom of the crown ring. Thus, the ion-binding ability of the crown site of *50* is "controlled" by the complexation of the 2,2'-bipyridyl site [59]. This relation reminds us of the allosteric effects which frequently appear in enzyme chemistry.

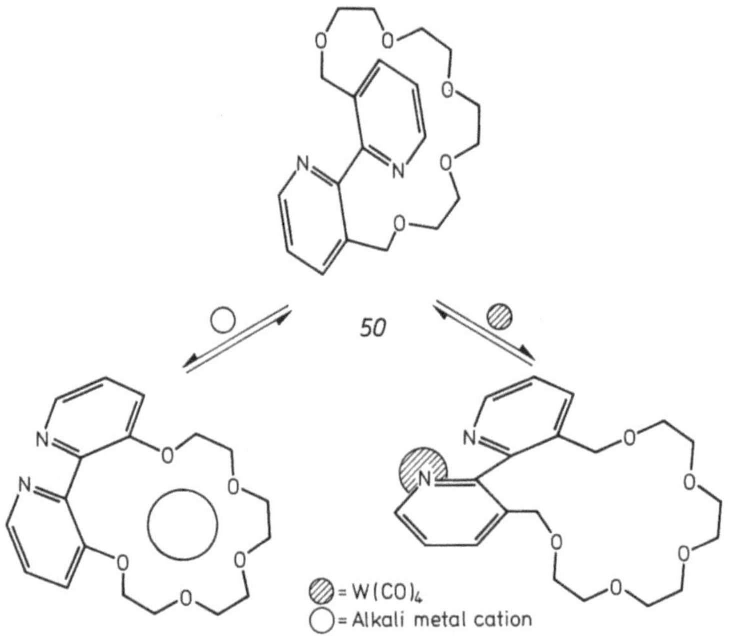

50

⊘ = W(CO)₄
○ = Alkali metal cation

It was found from NMR measurement, association constants, and ion-transport that *50* itself binds K^+ in preference to Na^+, whereas Na^+ is bound more efficiently in the presence of $W(CO)_4$. The rapid racemization of the structure forces the benzylic oxygens into conformations in which only one can participate in the formation of a crown-like cavity.

Other iteresting examples are *redox-responsive crown ethers*. The redox functions in the enzyme-chemical field are frequently mediated by prosthetic groups. Since most prosthetic groups are capable of catalyzing, although weakly, the enzyme-mediated reactions even in the absence of apoenzymes, they have attracted much attention to bioorganic chemists. In contrast to holoenzymes having both the catalytic site and the recognition site, however, they consist only of the catalytic site. One may thus expect that the prosthetic molecules bearing the recognition site within a molecule behave as a more attractive enzyme model.

The ability of crown ethers to associate with charged and uncharged substrates resembles to early reaction steps in enzyme-mediated reactions. Therefore, crown ethers are useful as a potential recognition site for this purpose. Compounds *51–54* represent such a type of molecular assembly containing within the same molecule both the prosthetic group and the crown ether ring. Among them *51* and *52* may act as crown ether NADH mimics [60–62]. E.g. substrates (sulfonium salts or pyridinium salts containing an extra ammonium group, etc.) in the initial step complex with the crown ether moiety and are then reduced by the dihydropyridine function through a pseudo-intramolecular process.

Crown Compound *53* belogs to the flavins (vitamin B$_2$ family). The redox properties of *53* are sensitively affected by complexed alkali metal cations [63]. In *54* the B15C5 moieties are appended to a porphyrin skeleton at the methine posi-

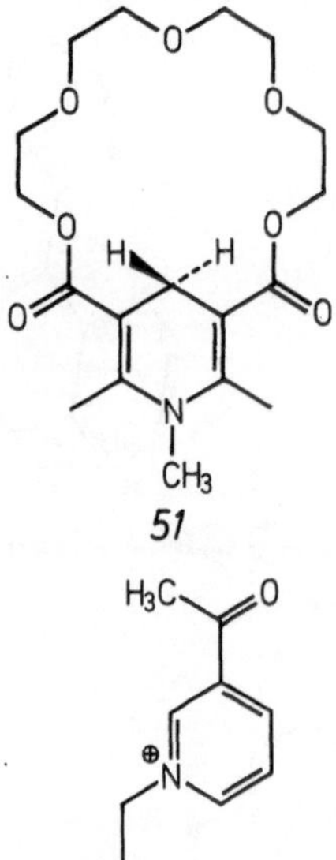

51

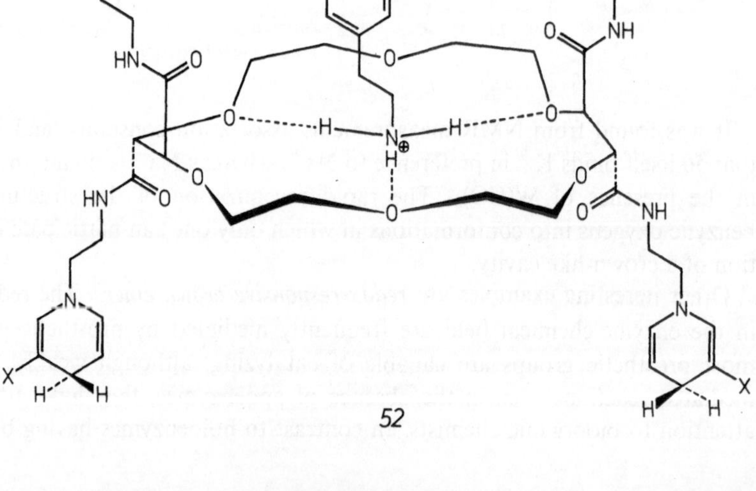

52

tions [64]. It was found that the cations (K^+, Ba^{2+}, and NH_4^+) capable of forming 1:2 cation/crown complexes efficiently quench the fluorescence of the porphyrin. These examples show the crown ether moieries to play a role of a recognition site of these prostetic molecules. It may be said paradoxically, however, that the ion-binding ability of the crown ether moieties is "controlled" by the redox state of the prosthetic molecules. However, there exist few examples for the application of prosthetic molecules to control the crown functions in which the crown ether moieties play a primary role.

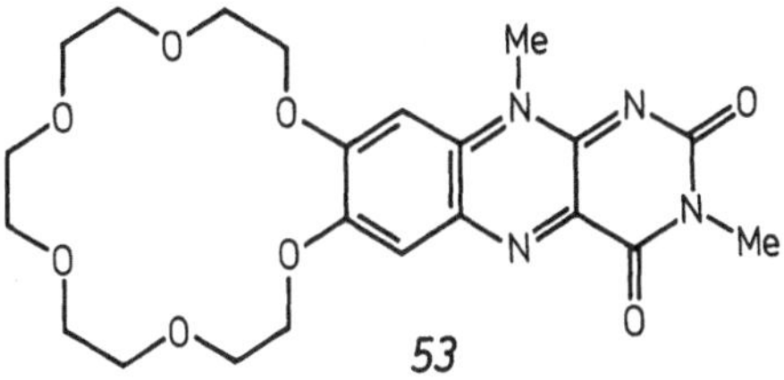

53

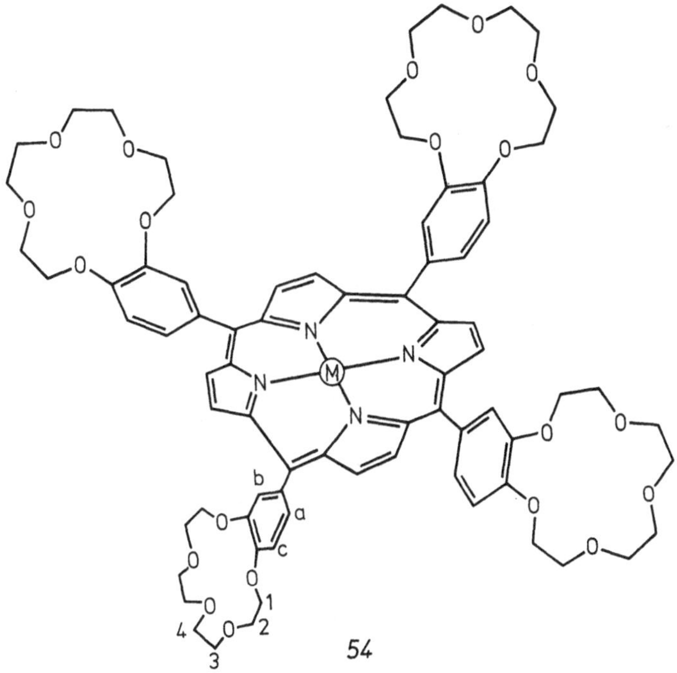

54

Crown ethers bearing thiol functions have been synthesized [65–67]. The object of these investigations is again to imitate the function of thiol-dependent enzymes and to utilize the crown ethers as a recognition site. Recently, Shinkai et al. [68, 69] attempted to control the ion-binding ability by the redox function of thiol groups.

In *55*, the interconversion between monocrown and biscrown was thus achieved by redox reaction, and the oxidized form with a methylene spacer (X=CH$_2$) formed a 1:2 cation/crown sandwich-type complex with K^+ and Rb^+. On the other hand, the interconversion between lariat crown and cryptand may be effected in *56* by redox reaction of the thiol caps.

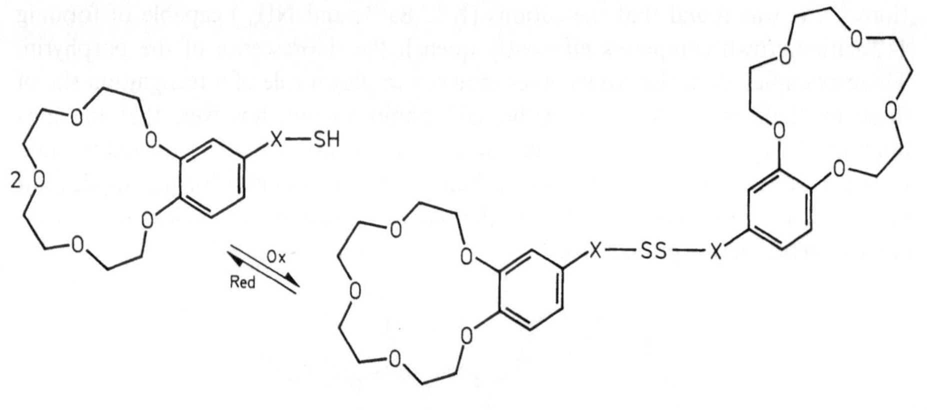

55

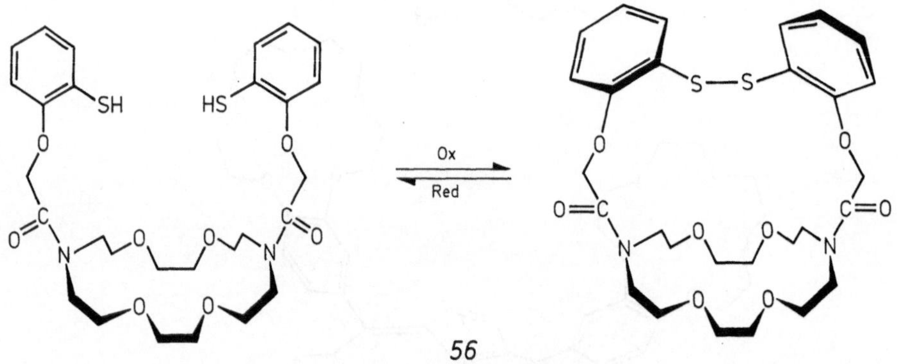

56

Recently, a temperature-responsive crown ether has been exploited by Warshawsky and Kahana [70]. $\Delta H°$ and $\Delta S°$ values for complexation between cation and crown ether are usually negative and small. Consequently, the sign and value of the free energy, $\Delta G°$, may depend on the absolute temperatures. In homogeneous systems

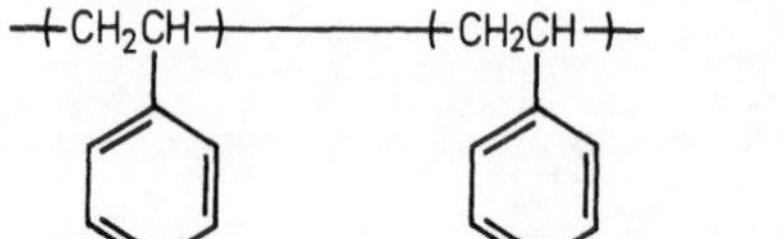

57

this has little significance. In heterogeneous systems such as equilibria between in-soluble polymer and solution, however, it is reasonable to assume that the opposing effects of $\Delta H°$ and $\Delta S°$ could be exploited to induce temperature-regulated release of salts from their insoluble polymeric crown complexes.

A polymeric crown unit *57* in crosslinked polystyrene beads [70] in a column could be saturated with KCl; a sudden heating to 40 °C caused a spontaneous elution by "thermal shock", and a 3-fold increase in the eluant of the original ion concentration. The authors plan the application to control the temperature of phase transfer catalysis and to thermoregulate polymeric delivery systems of Na^+/K^+. The latter, however, is more easily exploited by responsive crown ethers.

Also, the functionalization of a "rope-skipping" crown ether *58* [71] and a "breath-ing" crown ether *59* [72] may lead to an application of novel responsive crown ethers.

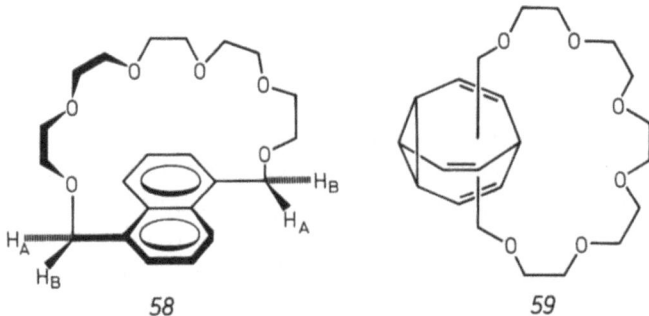

58   59

3 Light-Driven Ion Transport

3.1 Ion Transport Mediated by Photoresponsive Crown Ethers

Cations are known to be transported through membranes by synthetic macrocyclic polyethers as well as by antibiotics. In particular, some polyether antibiotics feature the interconversion between the cyclic and acyclic form in the membrane phase, a feature which is believed to lead to the high functionality of these antibiotics as ion carriers [48, 49, 73, 74]. For example, the reduction of the ion-affinity through the interconversion to the acyclic form is responsible for the rapid release of ions from the ion-carrier complex in the membrane phase to the second (OUT) aqueous phase. Kobuke et al. [75] and Kirch and Lehn [76] independently established by examination of synthetic macrocyclic polyethers that the best carrier for the ion transport is a ligand giving a moderately stable rather than a very stable complex. This general rule was further corroborated by Lamb et al. [77]. The rule clearly indicates a dilemma occurring in the artificial transport system: the very stable complex, which may rapidly extract ions into the membrane phase, cannot release the ion efficiently from the complex. As a result, the plots of K (association constant) vs. transport rate provide a maximum at a moderate K value. In other words, the rule implies that the ion transport with simple synthetic ion carriers cannot exceed a maximal.

ceiling velocity. Thus, the method of the reversible interconversion of the ionophore structure by which the natural ionophores skillfully break the dilemma becomes very attractive. The idea thus arises that, provided that the binding ability of the ion carrier can be changed by light, it would lead to acceleration or control of the ion-transport rate by an on-off light switch.

Shinkai et al. [44] found in the transport of alkali (K^+, Rb^+, and Cs^+) picrates with *32* ($X = CO$ or CH_2) across a liquid (*o*-dichlorobenzene) membrane in a U-tube that the rate is significantly accelerated by UV-irradiation which isomerizes the (*E*)-forms to the (*Z*)-forms. When UV and visible irradiation were conducted alternately, the interconversion between (*Z*)- and (*E*)-forms occurred rapidly. They independently confirmed that the (*Z*)-forms of *32* ($X = CO$ or CH_2) are rapidly isomerized to the (*E*)-forms by visible light [44]. However, the rates of the ion transport were smaller than those under UV irradiation. These results imply that the ion-release to the second (OUT) aqueous phase is not involved in the rate-limiting step and the light-driven ion transport is solely attributed to the enhanced binding ability of the (*Z*)-forms which extracts cations from the first (IN) aqueous phase to the liquid membrane phase. In other words, the rate enhancement observed for *32* ($X = CO$ or CH_2) is simply correlated to the enhancement of the (*Z*) concentration in the liquid membrane phase, whereas the reversible interconversion between the (*Z*)- and the (*E*)-forms has little significance. As described above, the complexes with (*Z*)-*32* ($X = CO$ or CH_2) are not stable enough to render ion-release to the rate-limiting step [44]. One may say, therefore, that in order to utilize visible light as well as UV light as a source of the ion transport accelerator a bis(crown ether) has to be employed, the (*Z*)-form of which forms a stable complex with metal cations.

The (*Z*)-forms of *34* form relatively stable 1:2 cation/crown sandwich-type complexes with large alkali metal cations [45, 46]. At the K^+ transport with *34* (n = 2), Shinkai et al. [45, 78] found that the rate with picrate as counter-anion is rather retarded by UV irradiation, whereas that with less hydrophobic 2-nitro-diphenyl-amine-4-sulfonate as counter-anion is significantly increased. The result suggests, as shown in Fig. 4, that when a less hydrophobic counter-anion is used, the ion extraction from the IN aqueous phase to the membrane phase is rate-limiting and UV irradiation facilitates the ion extraction to enhance the overall transport velocity. On the other hand, when a fully hydrophobic counter-anion (e.g., picrate) is used, the ion extraction is no longer the rate-limiting step and the relatively slow K^+-release to the OUT aqueous phase from the complex with (*Z*)-*34* (n = 2) becomes rate-limiting. In other words, the thermal (*Z*)-to-(*E*) isomerization is not fast enough to eliminate ion-release from the rate-limiting step. As a result, the rate of the ion transport is retarded under UV irradiation.

One may expect that if the conversion of (*Z*)-*34* (n = 2) to (*E*)-*34* (n = 2) is efficiently accelerated by visible light, the ion-release step would no longer be rate-limiting and the overall rate of the K^+ transport would be significantly enhanced even when a hydrophobic counter-anion is used. It was found that the (*Z*)-to-(*E*) isomerization of *34* is effectively accelerated by visible light, but the rates are, like the thermal isomerization [45, 46], suppressed by added alkali metal ions; for example, the inhibitory effect on the photoisomerization of (*Z*)-*34* (n = 2) is in the order of $Rb^+ > K^+ > Na^+$ [79]. As expected, alternate irradiation of the membrane by UV and visible light significantly accelerated the rate of K^+ and Rb^+ transport with

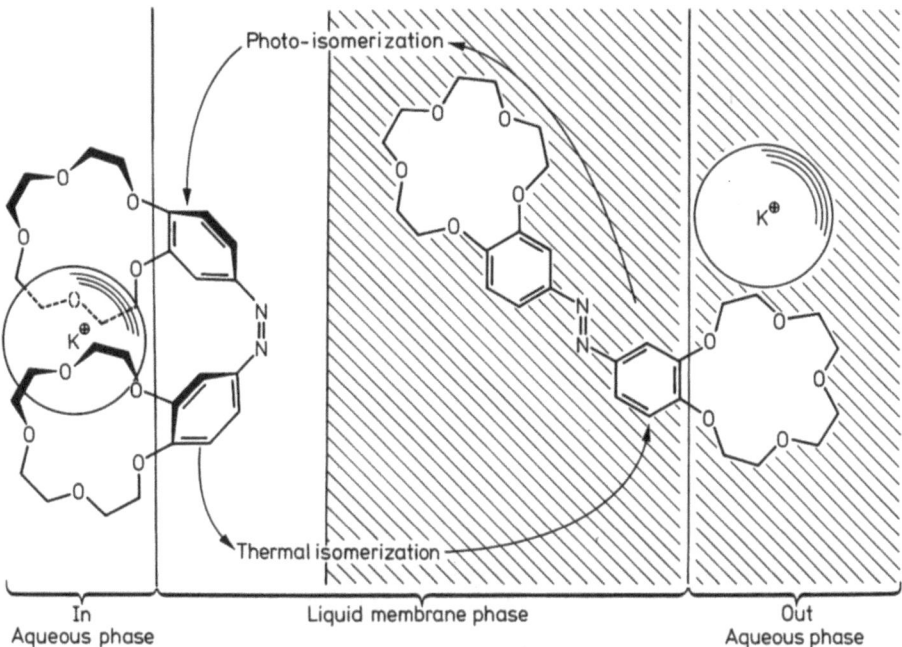

Fig. 4. Schematic representation of light-driven ion transport with *34* (n = 2)

picrate as counteranion, that of K^+ transport being accelerated more efficiently than that of Rb^+ transport [79]. The transport phenomenon may be illustrated schematically as Fig. 5.

In contrast, the rate of Na^+ transport was accelerated by UV irradiation and was retarded by alternate irradiation. The trend is similar to the transport with *32* (X=CO or CH_2) in which the ion extraction with the (Z)-forms is exclusively involved in the rate-limiting step.

As described above, the best carrier for ion transport is a ligand that gives a moderately stable rather than a very stable complex. It is of interest that the light-driven ion-transport system provided a similar conclusion. Thus, for less-stable complexes such as $Na^+[(Z)-34$ (n = 2)], the rate is proportional to the concentration of the (Z)-formed carrier, irradiation by visible light which efficiently mediates (Z)-to-(E) isomerization retarding the transport rate. In contrast, for very stable complexes such as $Rb^+[(Z)-34$ (n = 2)], the extraction is markedly facilitated by UV light which mediates (E)-to-(Z) isomerization, but the stable metal-(Z)-form complexes cannot be isomerized rapidly by visible light. Hence, the most efficient light-driven ion-transport is realized in moderately stable complexes such as $K^+[(Z)-34$ (n = 2)].

Niwa et al. [80] prepared a polymer membrane of poly(vinyl chloride) containing *34* (n = 2) and a liquid crystal compound *60* and examined the photoresponsive K^+ transport across this membrane. Since *60* has the strong absorption band at UV region, the (E)-to-(Z) photoisomerization of *34* (n = 2) takes place at the surface of the polymer membrane. On the other hand, the membrane is transparent in the visible region, so that the visible-light-mediated (Z)-to-(E) isomerization would be possible

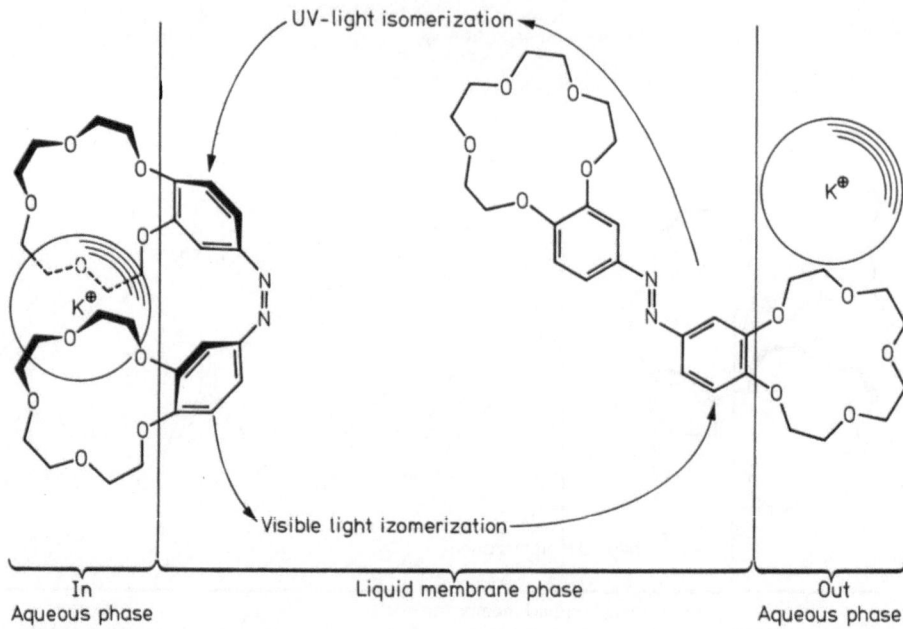

Fig. 5. Schematic reprsentation of ion transport accelerated by alternate irradiation of UV and visible light

anywhere in the membrane phase. In the dark, KCl slowly permeated through the membrane from the KCl-containing IN aqueous phase to the OUT aqueous phase. When UV light and visible light were irradiated alternately from the side of the IN aqueous phase and from the side of the OUT aqueous phase, respectively, the KCl flux rapidly increased (about two fold). When a hydrophobic salt (potassium *p*-toluene sulfonate) was used, the permeation in the dark was almost negligible. The permeation of this salt through membrane was readily induced by alternate irradiation of two different lights. When the lights were switched off, the flux again became negligibly small. In a separate study, they confirmed spectrophotometrically that the photoinduced (*E*)-(*Z*) intercoversion occurred in the polymer membrane. The result shows that the permeation through the polymer membrane is photocontrolled using *34* as a photofunctional valve.

$$CH_3CH_2O-\!\!\!\bigcirc\!\!\!-CH=N-\!\!\!\bigcirc\!\!\!-(CH_2)_3CH_3$$

60

The crown ether *44* with its photoresponsive anionic cap carries alkali metal cations without counter-anion and alkaline earth metal cations with a counter-anion. Furthermore, the counter-flow of protons plays an important role. In a U-tube transport apparatus containing a liquid (90 vol% o-dichlorobenzene + 10 vol% *n*-butyl alcohol) membrane, the rates of Na$^+$ and K$^+$ transport were acclerated by UV irradiation [52]. Similarly, UV irradiation was effective for the Ca^{2+} transport with

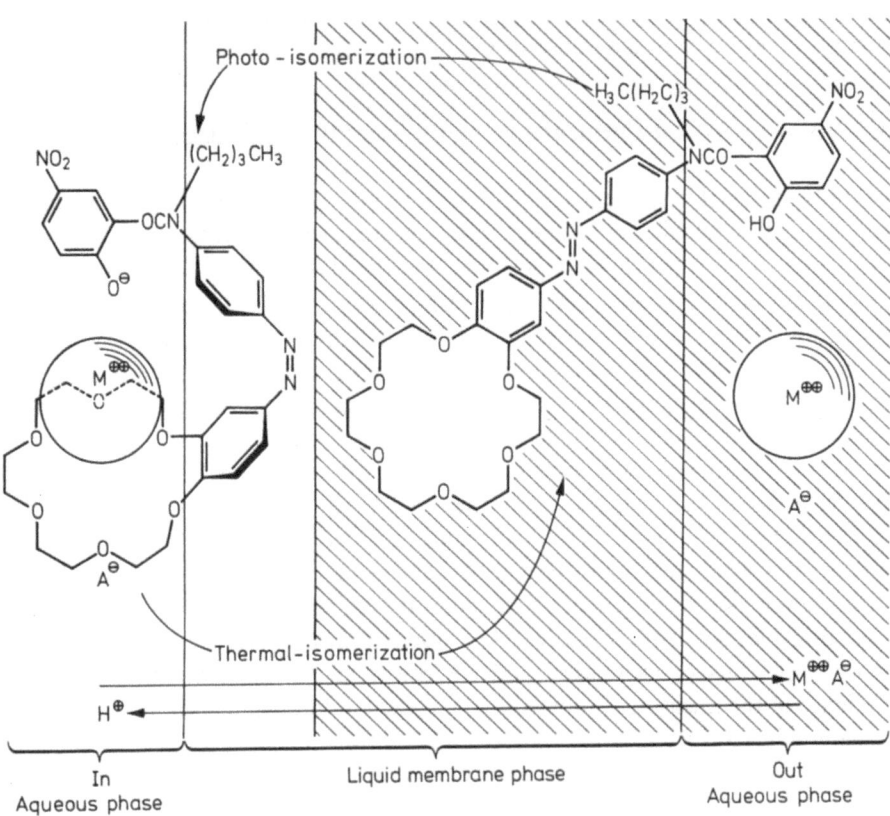

Fig. 6. Schematic representation of membrane transport of alkaline earth metal cations mediated by UV light and pH gradient

picrate as a counter-anion. When the OUT aqueous phase was water, Na$^+$ was transported by *44* whereas Ca^{2+} was scarcely transported in spite of the increasing Ca^{2+} concentration in the membrane phase. When the OUT aqueous phase was substituted by an 0.01 N HCl aqueous solution, the rates of Na$^+$ and Ca^{2+} were enhanced by 4.0-fold and 276-fold, respectively. The difference in the magnitude of the rate acceleration suggests that neutralization of the phenoxide group of *44* is important for the release of Ca^{2+} from the Ca^{2+}-(Z)-*44* complex. Therefore, the transport of Ca^{2+} is mediated both by UV irradiation and by the counter-flow of protons, the rate increase achieved by these two effects being 760-fold. Summarizing these observations, one may illustrate the Ca^{2+} transport phenomenon as Fig. 6.

(E)-*4* is capable of transporting Cu^{2+} across a liquid membrane [8]. When the liquid membrane phase was subjected to UV light irradiation for 1.5 h, the rate of Cu^{2+} transport was suppressed gradually and a new rate of Cu^{2+} transport was obtained. The magnitude of the rate decrease was approximately equal to the concentration of photoisomerized (Z)-*4*. When the liquid membrane phase was subjected to visible light irradiation which mediates the (Z)-to-(E) isomerization, the transport rate was enhanced again. The results clearly demonstrate that the rate of Cu^{2+} transport is reversibly controlled by an on-off light switch and (Z)-4 in the membrane phase scarcely contributes to the transport of Cu^{2+}.

3.2 Photocontrolled Permeability of Liposomal Membranes

In biological systems, there are several pigments which receive light signals to control the physiological phenomena. The typical example is a visual excitation which is initiated upon the light absorption by the 11-(Z)-retinal chromophore of rhodopsin (photoresponsive compound) which isomerizes to the all-(E) form. This photo-induced configurational change is followed by a conformational change of opsin (protein), Ca^{2+} permeation across membrane, and a conductance change in retinal receptor membrane [81]. Rhodopsin-phospholipid membrane vesicles were studied to localize initial transduction events in a reconstituted system [82-85]. It was found that unexposed rhodopsin-containing phospholipid vesicles are sealed to ion movement while they become permeable after brief exposure to blue-green light. Selected ions (Ca^{2+}, Mn^{2+}, Co^{2+}, Ni^{2+}, and Mg^{2+}) were thus photo-released from the interior of loaded membrane vesicles [84]. The number of ions released/rhodopsin bleached was dependent on the light intensity, and high yields (40–160) fo Ca^{2+}/ rhodopsin bleached were observed. The results indicate that rhodopsin spans the phospholipid bilayer membrane, and are consistent with an increase in the permeability of the membrane initiated by light excitation of rhodopsin [84, 85].

Light-induced release from phospholipid versicles was achieved in a more simplified system using an azobenzene-containing surfactant 61 [86].

$$Br^{\ominus}(CH_3)_3\,N^{\oplus}\!\!-\!\!(CH_2)_n\,O\!-\!\!\bigcirc\!\!-\!N\!=\!N\!-\!\!\bigcirc\!\!-\!OC_{12}H_{25}$$

(E)-61 (n = 2,4)

Embedding (E)-61 in phospholipid vesicles provided the enhancement of the liposome shrinkage. This was attributed to the perturbation of the membrane structure to yield small channels for the water permeation. The photoinduced configurational change of (E)-61 to (Z)-61 enhanced the osmotic shrinkage of the embedded liposomes. The result shows that (Z)-61 further enlarges the channel. The photoresponsive change in the channel size was further demonstrated by the data on the release of bromothymol blue (BTB) [86]: the rates of the BTB release from the (Z)-61 loading liposomes were markedly larger than those from the (E)-61 loading liposomes. Conceivable, linear-shaped (E)-61 disturbs the membrane structure to a smaller extent relative to distorted (Z)-61.

The transport of amino acids across biological membranes is assumed nowadays to be coupled with Na^+ transport [87]. The artificial transport of amino acids has been investigated by serveral groups [88-91], but they mostly employed protected amino acids instead of biologically relevant "free" amino acids.

An efficient transport system of freee amino acids was reported by Behr and Lehn [88]; they demonstrated that a cationic surfactant is useful for the transport from the basic IN aqueous phase to the acidic OUT aqueous phase, while an anionic surfactant is useful for the transport from the acidic IN aqueous phase to the basic OUT aqueous phase. The difficulty of the amino acid transport stems from the poor lipophilicity of their zwitterionic structure.

Sunamoto et al. [92] utilized photoinduced interconversion between spiropyran *62* and merocyanine *63* to mediate the transliposomal amino acid release. Compound *62* showed a normal photochromism in bilayer membranes (i.e., the photocoloration and thermobleachihg cycle). UV irradiation for 20 min followed by visible light irradiation for 5 min on the liposome suspension brought about significant transfer of phenyl-alanine (Phe) from the interior of the liposomes to the exterior. In the dark under the same conditions no Phe was spontaneously released. Thus, as shown in Fig. 7, *63* is capable of extracting zwitterionic amino acids into the membrane phase by neutralizing the charges each other.

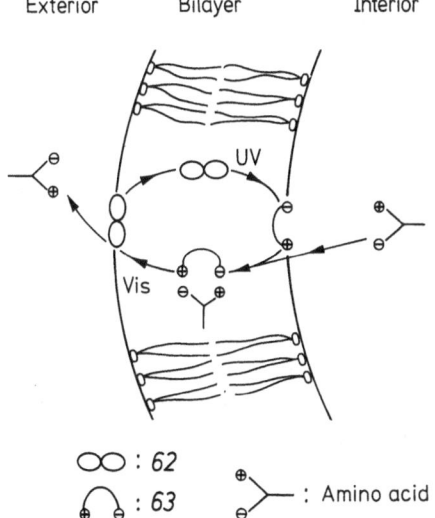

62   U.V. / Vis light or heating   *63*

Exterior   Bilayer   Interior

∞ : *62*
⌒ : *63*
: Amino acid

Fig. 7. Transport of amino acids across liposomal membranes with *62*

3.3 Other Ion Transport Systems with Responsive Functions

Here, we wish to survey the ion transport systems switched by energies other than light energy. The most widely investigated energy source is the acid-base neutralization, but this will be omitted from this article as the method has frequently been reviewed. Rebek and Wattley [93] used the remote-controlled crown ether *50* with the bipyridyl unit as a second metal binding site. The ion transport across a liquid (chloroform) membrane indicated that *50* carries K^+ 3.8 times more efficiently than Na^+ while the *50*-$W(CO)_4$ complex carries Na^+ 1.3 times more efficiently than K^+. The remarkable inversion of the K^+/Na^+ transport preference demonstrates that these variables are subject to control within a single carrier by remote binding forces. Therefore, the binding-induced conformational change in the crown ether ring is responsible for the regulation of the ion transport rate.

Another interesting and useful energy source is a redox system. Shinbo et al. [94] designed a liquid membrane system consisting of $K_3Fe(CN)_6$ (aqueous phase)/ N,N,N',N'-tetramethyl-*p*-phenylenediamine (TMPD) (dichloroethane)/ascorbic acid (aqueous phase). As the redox reaction in the interfaces $TMPD \rightleftarrows TMPD_{ox}^+$, picrate anion was carried from the oxidative aqueous phase to the reductive aqueous phase against its concentration gradient. They also demonstrated that in a liquid membrane system consisting of $K_3Fe(CN)_6$ (aqueous phase)/dibutylferrocene (dichloroethane)/ $K_4Fe(CN)_6$ (aqueous system) the redox reactions enforce a coupled counter-flow of protons and anions (perchlorate ions) [94].

The above examples imply that in order to demonstrate coupled cation transport by redox energy, for example, it is necessary to choose a carrier which is negative when reduced and neutral when oxidized. Grimaldi and Lehn [95] satisfied the prerequisite with a nickel bisthiolene carrier *64*. In the presence of dicyclohexano[18] crown-6 (DCH18C6), *64* carried a cation (K^+) and an electron from the reductive ($S_2O_4^{2-}$) to the oxidative ($K_3Fe(CN)_6$) aqeous phase. Similarly, Matsuno et al. [96] succeeded in the coupled transport of Cu^{2+}, an electron, and an anion from the reductive (H_2NOH) to the oxidative (O_2) aqueous phase using bathocuproine *65*.

64   *65*

Redox-driven ion transport was also achieved by using *55* ($X = CH_2$) [69]. Since the oxidized bis(crown ether) has a large affinity relative to the reduced mono(crown ether), K^+ was rapidly extracted from the oxidative aqueous phase and released to the reductive aqueous phase. It was demontrated that the interconversion between the oxidized and the reduced form of *55* can be mediated by an electrochemical method [69]. Recently, an ion gate membrane in which redox-responsive polymers such as poly-(vinylferrocene), poly(tetrathiafulvalene), and (poly(pyrrole) are coated on a gold minigrid sheet was developed by Burgmayer and Murray [97]. The impedance of the membrane changes in response to the voltage, and the permeability of anion (Cl^-) can be controlled by an electrochemical energy source.

A pH difference may be utilized in an indirect manner for the pH-induced structure change in carrier molecules. Shimidzu et al. [98] synthesized a polymer membrane *66* containing a butyrolactone unit. The membrane exhibited active and selective transport of alkali metal cations [99]. This is due to the pH-dependent cyclic→acyclic interconversion of the lactone ring (Fig. 8). Thus, the active transport was effectively promoted by the pH gradient.

Similar pH-dependent active transports were attained in the polymer membrane sustaining a phenolphthalein analogue [100]. Active transports of anions was also

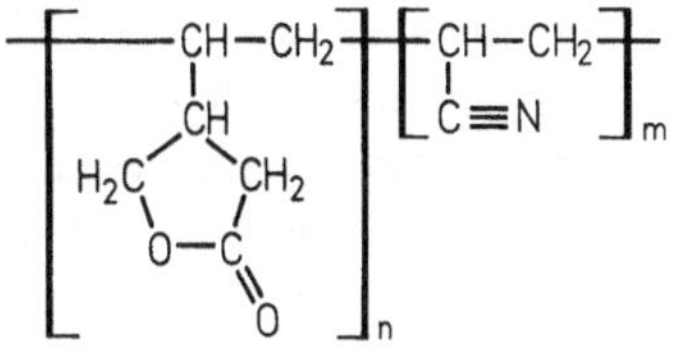

66

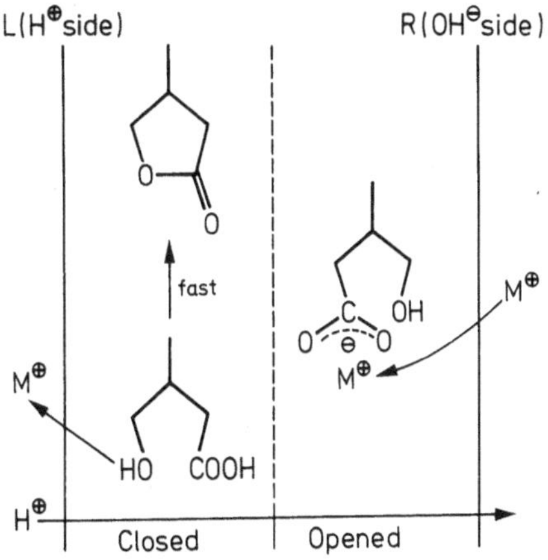

Fig. 8. Active transport achieved in the lactone-containing polymer membrane

demonstrated using the pH-induced structural change. It is known that N-hydroxy-ethyl amide group *67* exhibits tautomersim and transforms very rapidly into amino-ethyl ester *68* in an acid solution. When a polymer having pendant N-hydroxyl groups is synthesized to form a membrane and the membrane is fixed in a cell as a partition film, in which one side solution is acidic and the other side solution is basic, it is expected that an anion may be carried out through the membrane owing to the tautomerism. Thus, the active transport of Cl^- was achieved from the acidic to the basic aqueous solution across the polymer membrane [101].

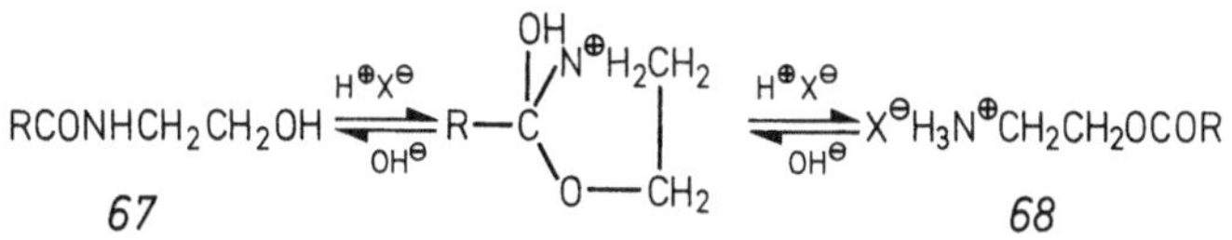

67   *68*

Electron transport systems with switch functions are another fascinating topic in relation to the objects of the present article. It seems impossible, however, to cover this widely extended field here.

4 Conclusions

Since the unexpected discovery of dibenzo[18]crown-6 by Pedersen in 1967, the chemistry of crown ethers has rapidly been established as a new field. In the initial stage of the crown chemistry the syntheses of the crown compounds and the

279

characterizations of their metal complexes were the main objects of the investigation, the functional facet of crown ethers has been left less exploited for a long time. To let the crown ether work — this is the basic idea for the design of functional crown ethers. Natural ionophores actually "work" in biological systems. The basic idea of the photoresponsive crown ethers is originating of the responsive action of polyether antibiotics. The novel phenomena were attained because they combine within the same molecule a photoantenna which acts as a photoresponsive trigger and a functional group (crown ether) which causes a subsequent event. The trigger may be substituted by a component sensitive to pH, redox potential, electron current, spectral pattern, etc. The functional group may be substituted by, for example, cyclodextrin, micelles, membranes, polypeptides, etc. Furthermore, there exist many methods other than the covalent bond to transduce the energy in the antenna to the functional group. We expect that chemistry of responsive systems, including that of photoresponsive crown ethers, will be further advanced as an engineering application to switch-functioned systems.

5 References

1. Bieth, J. et al.: Proc. Natl. Acad. Sci. U.S.A. *66*, 850 (1970)
2. Berezin, I. V. et al.: FEBS Lett. *30*, 329 (1974)
3. Karube. I.. Nakamoto. Y.. Suzuki, S.: Biochim. Biophys. Acta *445*, 774 (1976)
4. (a) Shinkai, S., Ogawa, T., Nakaji, T., Kusano, Y., Manabe, O.: Tetrahedron Lett. *1979*, 4569,
   (b) Shinkai, S. et al.: J. Am. Chem. Soc. *102*, 5860 (1980)
5. Ammon, L., Shinkai, S.: to be submitted
6. Fisher, D. P. et al.: J. Am. Chem. Soc. *99*, 2811 (1977)
7. Nakamura, A., Aotake, M., Otsuka, S.: ibid. *96*, 3456 (1974)
8. (a) Shinkai, S., Minami, T., Kouno, T., Kusano, Y., Manabe, O.: Chem. Lett. *1982*, 499,
   (b) Shinkai, S. et al.: J. Chem. Soc. Perkin 2 *1982*, 2741
9. Shinkai, S., Shigematsu, K., Honda, Y. Manabe, O.: to be submitted
10. Wildes, P. D., Pacifici, J. G., Irick, G., Whitten, D. G.: J. Am. Chem. Soc. *93*, 2004 (1971) and references cited therein
11. Harberfield, P. et al.: J. Am. Chem. Soc. *97*, 5804 (1975)
12. (a) Asano, T.: J. Am. Chem. Soc. *102*, 1205 (1980),
    (b) Asano, T. et al.: ibid. *104*, 4900 (1982)
13. Nishimura, N. et al.: Bull. Chem. Soc. Jpn. *49*, 1381 (1976)
14. Schanze, K. S., Mattox, T. F., Whitten, D. G.: J. Am. Chem. Soc. *104*, 1733 (1982)
15. Shinkai, S. et al.: Chem. Lett. *1980*, 1303
16. Asano, T. et al.: J. Am. Chem. Soc. *103*, 5161 (1981)
17. Rau, H., Lüddecke, E.: ibid. *104*, 1616 (1982)
18. (a) Ueno, A. et al.: ibid. *101*, 2779 (1979),
    (b) Ueno, A. et al.: Chem. Lett. *1979*, 841
19. Emert, J., Breslow, R.: J. Am. Chem. Soc. *97*, 670 (1975)
20. Ueno, A. et al.: Chem. Lett. *1979*, 1007
21. Ueno, A. et al.: J. Chem. Soc. Chem. Commun. *1981*, 94
22. Irie, M., Hayashi, K.: J. Macromol. Sci.-Chem. **A13**, 511 (1979)
23. Irie, M. et al.: Macromolecules *14*, 262 (1981)
24. Blair, H. S., et al.: Polymer *21*, 1195 (1980)
25. Irie, M. et al.: J. Polym. Sci. Polym. Lett. Ed. *17*, 29 (1979)
26. Negishi, N..et al.: Kobunshi Ronbun Shu *37*, 293 (1980)
27. Blasius, E., Maurer, P.-G.: Makromol. Chem. *178*, 649 (1977)
28. Bormann, S. et al.: Tetrahedron *31*, 2791 (1975)
29. Smid, J. et al.: J. Am. Chem. Soc. *97*, 5932 (1975)

30. Wong, L., Smid, J.: ibid. *99*, 5637 (1977)
31. Shinkai, S., Kinda, H., Manabe, O.: ibid. *104*, 2933 (1982)
32. Shinkai, S. et al.: unpublished result
33. Shinkai, S., Honda, Y., Kusano, Y., Manabe, O.: J. Chem. Soc. Chem. Commun. *1982*, 848
34. Shinkai, S. et al.: Bull. Chem. Soc. Jpn. *56*, 1700 (1983)
35. Shinkai, S., Ueda, K., Honda, Y., Manabe, O.: to be submitted
36. (a) Desvergne, J.-P., Bouas-Laurent, H.: J. Chem. Soc. Chem. Commun. *1978*, 403,
    (b) Bouas-Laurent, H. et al.: Pure & Appl. Chem. *52*, 2633 (1980)
37. Yamashita, I. et al.: Tetrahedron Lett. *1980*, 541
38. Eichner, M., Merz, A.: ibid. *22*, 1315 (1981)
39. Shiga, M., Takagi, M., Ueno, K.: Chem. Lett. *1980*, 1021
40. (a) Shinkai, S., Minami, T., Kusano, Y., Manabe, O.: Tetrahedron Lett. *23*, 2581 (1982),
    (b) Shinkai, S. et al.: J. Am. Chem. Soc. *105*, 1851 (1983)
41. Kimura, K. et al.: Chem. Lett. *1979*, 611
42. Maeda, T., Kimura, K., Shono, T.: Bull. Chem. Soc. Japan *55*, 3506 (1982)
43. Mollison, P. R., Turter, M. R.: J. Chem. Soc. Perkin 2 *1982*, 1818
44. Shinkai, S. et al.: J. Chem. Soc. Perkin 1 *1981*, 3279
45. (a) Shinkai, S., Ogawa, T., Kusano, Y., Manabe, O.: Chem. Lett. *1980*, 283,
    (b) Shinkai, S. et al.: J. Am. Chem. Soc. *103*, 111 (1981)
46. Shinkai, S. et al.: J. Am. Chem. Soc. *104*, 1960 (1982)
47. Blank, M. et al.: Science *214*, 70 (1981)
48. (a) Choy, E. M. et al.: J. Am. Chem. Soc. *96*, 7085 (1974),
    (b) Cussler, E. L.: Am. Inst. Chem. Eng. J. *17*, 1300 (1971)
49. Wierenga, W. et al.: J. Am. Chem. Soc. *101*, 1334 (1979)
50. Frederick, L. A. et al.: J. Chem. Soc. Chem. Commun. *1980*, 1211
51. Shinkai, S. et al.: Tetrahedron Lett. *21*, 4463 (1980)
52. Shinkai, S., Minami, T., Kusano, Y., Manabe, O.: J. Am. Chem. Soc. *104*, 1967 (1982)
53. Yamazaki, N. et al.: Tetrahedron Lett. *1978*, 2429
54. Kubokawa, H. et al.: Chem. Lett. *1982*, 1937
55. Yamazaki, N. et al.: J. Macromol. Sci.-Chem. **A13**, 321 (1079)
56. Gardner, J. O., Beard, C. C.: J. Med. Chem. *21*, 357 (1978)
57. Hiratani, K. et al.: Chem. Lett. *1980*, 477
58. Hiratani, K.: ibid. *1981*, 21
59. Rebek, J., Jr. et al.: J. Am. Chem. Soc. *101*, 4333 (1979)
60. Van Bergen, T. J., Kellogg, R. M.: ibid. *99*, 3882 (1977)
61. De Vries, J. G., Kellogg, R. M.: ibid. *101*, 2759 (1979)
62. Behr, J.-P., Lehn, J.-M.: J. Chem. Commun. *1978*, 143
63. Shinkai, S. et al.: Tetrahedron Lett. *24*, 1539 (1983)
64. Thanabal, V., Krishnan, V.: J. Am. Chem. Soc. *104*, 3643 (1982)
65. Chao, Y., Cram, D. J.: ibid. *98*, 1015 (1976)
66. Matsui, T., Koga, K.: Tetrahedron Lett. *1978*, 1115
67. Lehn, J.-M., Sirlin, C.: J. Chem. Soc. Chem. Commun *1978*, 949
68. Minami, T., Shinkai, S., Manabe, O.: Tetrahedron Lett. *1982*, 5167
69. Shinkai, S. et al.: Chem. Lett. *1983*, 747 and unpublished results
70. Warshawsky, A., Kahana, N.: J. Am. Chem. Soc. *104*, 2663 (1982)
71. Brown, H. S. et al.: J. Org. Chem. *45*, 1682 (1980)
72. Schroder, G., Witt, W.: Angew. Chem. Int. Ed. Engl. *18*, 311 (1979)
73. Pressman, B. C.: Ann. Rev. Biochem. *45*, 501 (1976)
74. Westly, J. W.: Ann. Rep. Med. Chem. *10*, 246 (1975)
75. Kobuke, Y. et al.: J. Am. Chem. Soc. *98*, 7414 (1976)
76. Kirch, M., Lehn, J.-M.: Angew. Chem. Int. Ed. Engl. *14*, 555 (1975)
77. Lamb, J. D. et al.: J. Am. Chem. Soc. *102*, 6820 (1980)
78. Shinkai, S. et al.: J. Chem. Soc. Chem. Commun. *1980*, 375
79. Shinkai, S., Shigematsu, K., Sato, M., Manabe, O.: J. Chem. Soc. Perkin 1 *1982*, 2735
80. Kumano, A. et al.: Chem. Lett. *1983*, 1327
81. Ebrey, T. G., Honig, B.: Quat. Rev. Biophys. *8*, 129 (1975)
82. Montal, M.: Biochim. Biophys. Acta *559*, 231 (1979)

83. Kaupp, U. B. et al.: ibid. *552*, 390 (1979)
84. O'Brien, D. F.: Photochem. Photobiol. *29*, 679 (1979)
85. Tyminski, P. N. et al.: Biochemistry *21*, 1197 (1982)
86. (a) Kano, K. et al.: Chem. Lett. *1980*, 421,
    (b) Kano, K. et al.: Photochem. Photobiol. *34*, 322 (1981)
87. Ring, K.: Angew. Chem. Int. Ed. Engl.*9*,345(1970)
88. Behr,J.-P.,Lehn,J.-M.:J.Am.Chem.Soc.*95*,6108(1973)
89. Newcomb,M.etal.:ibid.*101*,4941(1979)
90. Maruyama, K. et al.: Tetrahedron Lett. *22*, 2001 (1981)
91. Tsukube, H.: ibid. *22*, 3981 (1981)
92. Sunamoto, J. et al.: J. Am. Chem. Soc. *104*, 5502 (1982)
93. Rebek, J., Jr., Wattley, R.: ibid. *102*, 4853 (1980)
94. (a) Shinbo, T. et al.: Nature *270*, 277 (1977),
    (b) Shinbo, T. et al.: Chem. Lett. *1979*, 1177
95. Grimaldi, J. J., Lehn, J.-M.: J. Am. Chem. Soc. *101*, 1333 (1979)
96. Matsuo, S. et al.: Chem. Lett. *1981*, 1543
97. Burgmayer, P., Murray, R. W.: J. Am. Chem. Soc. *104*, 6139 (1982)
98. Shimidzu, T. et al.: Makromol. Chem. *178*, 1923 (1977)
99. Shimidzu, T. et al.: Macromolecules *14*, 170 (1981)
100. Shimidzu, T. et al.: Polym. J. *12*, 363 (1980)
101. Ogata, N. et al.: J. Polym. Lett. Ed. *25*, 1419 (1980)

Bioorganic Modelling
Stereoselective Reactions with Chiral Neutral Ligand
Complexes as Model Systems for Enzyme Catalysis

Richard M. Kellogg

Department of Organic Chemistry, University of Groningen, Nijenborgh 16,
9747 AG Groningen/Netherlands

Table of Contents

1 Introduction

In 1926 Leopold Ruzicka, on the occasion of his inaugural address as newly appointed professor of organic chemistry at the University of Utrecht, observed

„Es ist also das Endziel der organisch-chemischen Forschung einen bestimmten Teil des Materials zu liefern für die Beantwortung einer die Menschheit tief bewegenden Frage: welches sind die naturwissenschaftlich klar faßbaren Grundlagen der Lebensvorgänge? Gar mannigfaltig sind die Wege und Möglichkeiten, die der chemischen Forschung zur Erreichung dieses Zieles zur Verfügung stehen. Es sind nicht nur jene wichtig, die direkt an das Hauptziel führen; auch scheinbar nebensächliche und fernliegende Einzelheiten können manchmal in unerwarteter Weise die große letzte Aufgabe fördern"[1].

The words remain true. The years since 1926 have brought us, however, to a significant new stage. The point has now been reached that, armed with remarkable amounts of information on the structure and mechanism of action of large biomolecules, steps can now be contemplated to blend that knowledge with good understanding of synthetic and mechanistic chemistry into man-made compounds designed to carry out specific functions.

There are many directions in which this can proceed. This review, however, will be for reasons of subject matter and space restricted to a "state-of-the-art" discussion of only one aspect of such work. This aspect will be the bioorganic modelling of certain enzymic processes with fairly small molecules, naturally occurring or synthetic, which have the ability to complex the substrate in a rapid pre-equilibrium, just as in an enzyme. Because of the subject content of this book these compounds will be in almost all cases macrocycles and they will usually have also the capacity for the recognition of enantiomers of a potential substrate.

Most effort so far has been concentrated on the design of synthetic systems capable of mimicking the selectivity and/or speed of enzymic processes. A portion of such efforts will also be the subject of this chapter. However, a few generalizations should be made at this point. First, if the path to the design and synthesis of "artificial enzymes" is to be followed, blind and unthinking imitation of Nature should be avoided. Drastic but creative innovations dictated by synthetic access are in order. For example, it is not possible to synthesize all too readily a long polypeptide with a defined order of amino acids and having specified properties. Other approaches must be followed. At the current stage of synthetic knowledge the use of molecules far smaller than enzymes is virtually mandatory. Synthetic strategies with a degree of flexibility can often be developed for the preparation of covalently bonded organic molecules with molecular weights 10^{-3} to 10^{-4} those of typical enzymes. This means the loss of enormous amounts of structure as found in enzymes; it is a moot point, however, whether this structure represents redundant or necessary information in regard to the chemistry of the reaction to be catalyzed.

A further simplification often necessary is the use of organic solvents rather than water. However, the tendency that many enzymes have to develop active sites in hydrophobic pockets from which water is fairly well excluded provides some excuse for this pragmatic course of action [2].

The consequences of using synthetic molecules far smaller than normal enzymes, assuming that the desired catalytic aspects can and have been built into the small molecule, will doubtlessly be felt both in speed and selectivity of action. However, one can usually tolerate great losses in reaction speed relative to an enzyme and still have a system that is chemically interesting and useful. Absolute selectivity with regard to substrate is also usually not wanted. Most often reaction of a *family* of structurally related compounds is desired and hence some tolerance in the "fit" is usually needed. What is required is a good recognition of the site at which a reaction should occur, if there is a choice of sites, and good chiral discrimination between enantiomeric substrates. The difficulties should not be underestimated of achieving a balance between these desirable aspects and what in reality is possible.

To achieve some of the desirable features mentioned above one piece of enzymic chemistry should be embodied, namely that the surface on which the catalyzed reaction occurs should be complementary (properly positioned hydrogen bonds, optimal orientation of charges, relief of strain, etc.) to the *transition state* of the reaction to be catalyzed. [3, 4] This is doubtlessly one of the things available in the large amount of structure present in an enzyme perfected through evolution; optimal complementarity will be harder to achieve in a smaller synthetic structure. This problem of complementarity will be a major barrier in achieving rate enhancements approaching those of enzymes.

A final general point has to do with evolution. Many biochemically catalyzed processes have an important aspect of spontaneity in that a good part, if not all, of the chemistry will also proceed non-enzymatically. The remarkably spontaneous aspects of the cyclization of squalene epoxide to lanesterol come to mind [5] as does the stereochemically less complex but just as spectacular tropinone (1) synthesis of Robinson (eq. 1) [6].

$$HC(CH_2)_2CH \quad + \quad {}^{\ominus}O_2CCH_2CCH_2CO_2^{\ominus} \cdot Ca^{2\oplus} \quad + \quad CH_3NH_2 \quad \longrightarrow \quad \mathbf{1} \quad + \quad CO_2 \tag{1}$$

1

The point is that some enzymic processes may operate with active site components that by themselves are fairly capable for carrying out the desired reaction. These components have become incorporated into a peptide chain forming an enzyme that has become very efficient in the course of evolution. A detailed discussion of the association of chemistry and evolution has been given by Visser. [7] Such considerations are especially important with coenzymes, these being particularly loved subjects for model study (and will be discussed again later in this review). The point to be drawn from this is that bioorganic modelling of enzymic chemistry is easier if the chemistry tends to proceed spontaneously. Although this sounds like a statement of the obvious, confusion leading to poor design of experiments has occurred with certain coenzymes, for example biotin, which has consistently been assigned an activating role that it does not fulfil. [7a,b,8] These evolutionary points will be discussed where appropriate in the text.

2 Enantioselective Reactions

A major consideration of this review will be how to achieve enantioselective reactions using macrocyclic systems. Macrocycles are attractive for this purpose because of the possibility of defining well the structure of the complex leading to reaction. However, there is no magic in macrocycles; highly enantioselective syntheses using well-designed non-macrocyclic systems have been developed in recent years. Four pertinent examples are shown in eqs. 2–5. The success of these reactions lies undoubtedly for a good part in chelation, which enforces rigidity during the transition state for reaction. The most important point to be learned is that even with small molecules that good to excellent enantioselectivities can be achieved in reactions ranging from

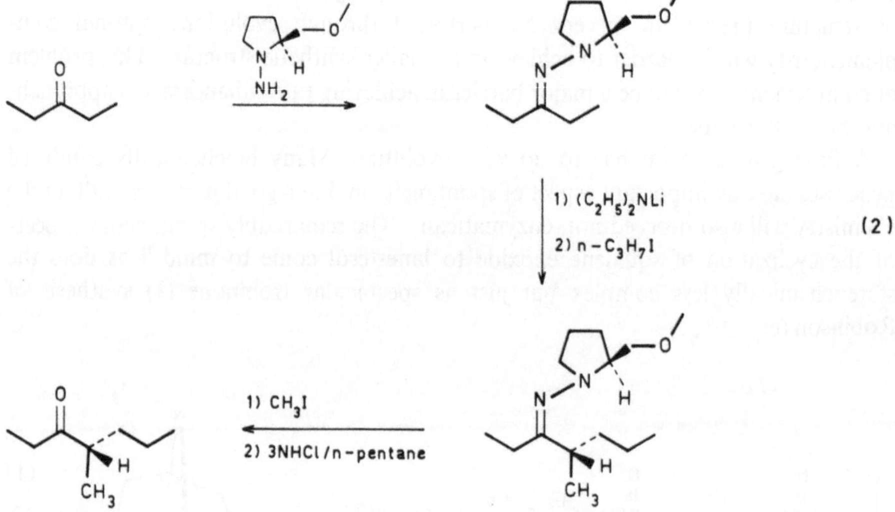

$$(2)$$

$$2$$

> 99% enantiomeric excess

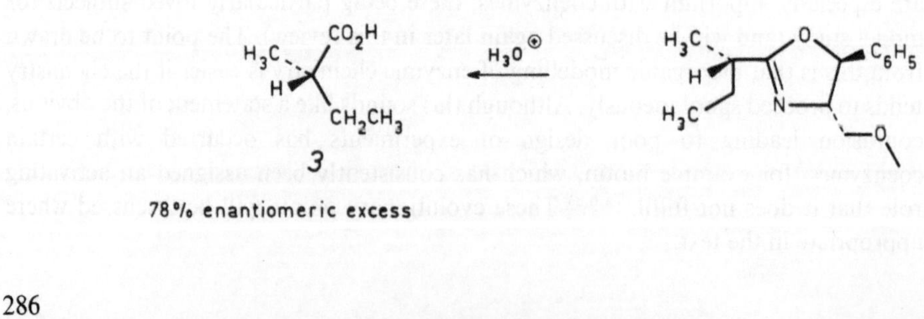

$$(3)$$

$$3$$

78% enantiomeric excess

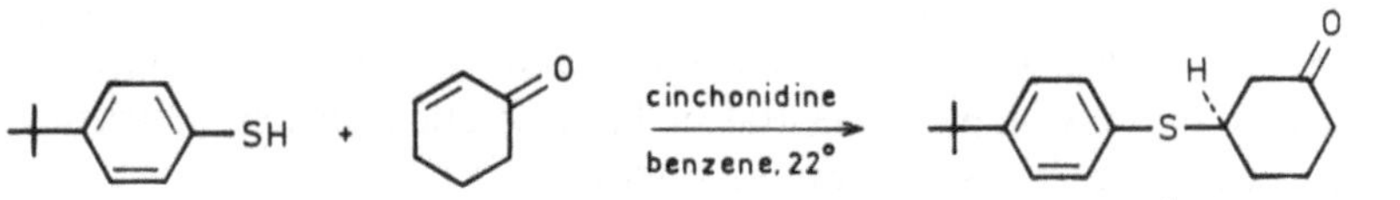

$$\text{(4)}$$

62% enantiomeric excess

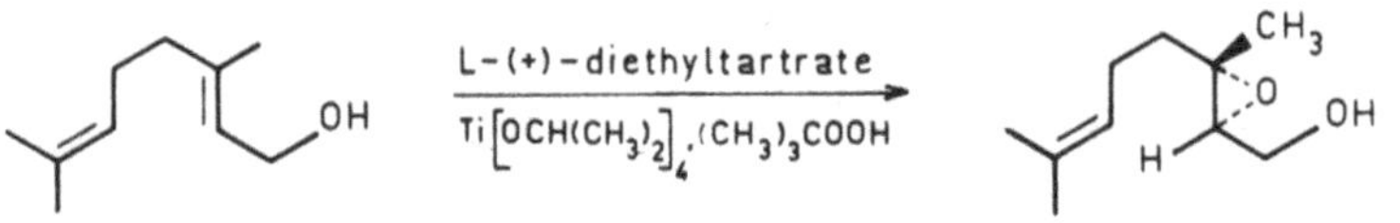

$$\text{(5)}$$

95% enantiomeric excess

carbon-carbon bond formation (2, 3) [9, 10] to carbon-sulfur bond formation (4) [11] to epoxidation (5) [12] of allylic alcohols by making use of only relatively small groups without extremely large steric demands. The free energy differences between the diastereometric transition states are not large — about 2 kcal/mol for an 80% enantiomeric excess [10a] — which illustrates how much can be accomplished with the cumulative effects of different small interactions.

3 Cyclodextrins

Cyclodextrins are cyclic glucose oligomers (6a) having the shape roughly depicted in (6b), as a cylindrical form with the primary hydroxyl group at the more restricted end of the cylinder (most of the hydroxyl groups have been omitted from the drawing).

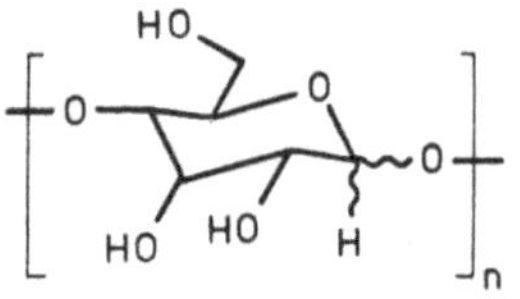

6a 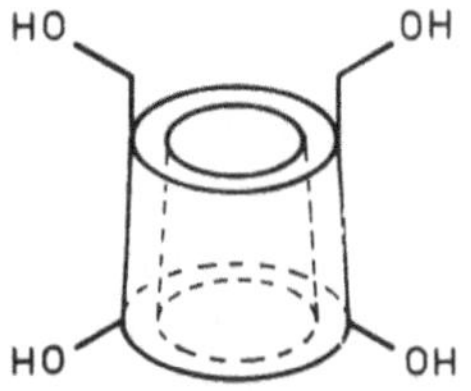   6b

The interior is relatively apolar relative to water and is large enough (internal diameter about 4.5 Å and depth about 6.7 Å for α-cyclodextrin, which has six α-linked glucose units and about 7.5 Å internal diameter for β-cyclodextrin, typesetting error which is a β-linked glucose heptamer [13] to accommodate apolar guests of appropriate size such as benzene derivatives.

This forms an ideal basis for the synthesis of "enzyme models" since naturally occurring material with built-in complexing ability can be used. Surprisingly effective methods have been developed for the reasonably selective functionalization of the primary hydroxyl groups allowing the incorporation of various catalytically active groups. In some cases recognition of enantiomers or enantioselective syntheses have been involved. [14] All this work has been competently and extensively reviewed [15] and hence need not be rediscussed here.

There are definite limitations to the cyclodextrins. The presence of many hydroxyl groups makes the problem of selective functionalization difficult. Moreover, the glycoside linkages are only stable in neutral or basic solution, thereby restricting somewhat the chemical reactions than can be studied.

4 Peptides

Although not macrocyclic structures, small peptides, not enzymes, used to catalyze organic reactions, should be briefly mentioned in the context of this chapter. Interest has centered for the greater part on hydrolytic reactions related to those catalyzed by, for example, chymotrypsin. [16] Such studies usually involve the use of activated esters, usually p-nitrophenolates (7), which hydrolyze at readily followable rates under weakly basic conditions or in the presence of various nucleophiles.

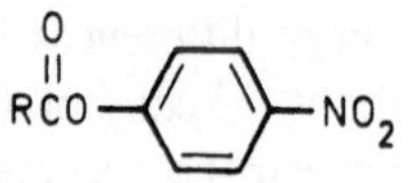

7

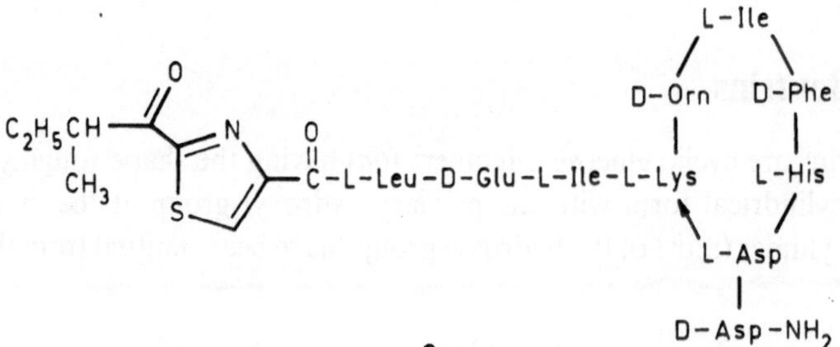

8

In an early example, commercial bacitracin (8), a cyclic antibiotic [17] (the structure of bacitracin F is given) was found to hydrolyze L-(9) about twice as fast as the D-enantiomer. [18] No selectivity was found for the corresponding alanine enantiomers. Most likely an imidazole group from histidine is involved in the hydrolysis, but owing to the structural complexity of the peptide it is not possible to determine the origin of the enantioselectivity.

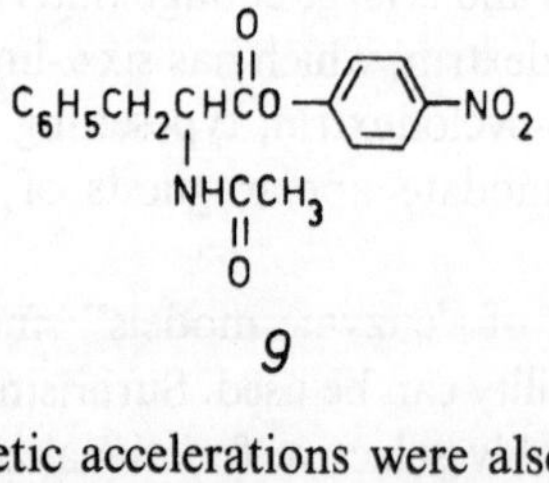

9

Mild kinetic accelerations were also found with the pentapeptides (10) and (11), which contain both histidine and serine, both of which are involved at the active site in β-chymotrypsin catalyzed hydrolyses. The catalytic constant for hydrolysis of p-nitrophenyl acetate (12) catalyzed by (10) is almost 16 times that of imidazole and for (11) the catalytic constant is 24.5 times greater. [19] However, the modestness of the catalytic effects is brought home on realizing that the catalytic constant for hydrolysis of (12) by α-chymotrypsin is 10^4 that of imidazole. With (11) fairly good chiral recognition was found for hydrolysis of the enantiomers of (13), the catalytic constants being 26 times greater than that for histidine for the D-enantiomer of (13) and 19 times greater for the L-enantiomer. Again, however, it is not possible to pinpoint the structural basis for these differences in hydrolysis rates.

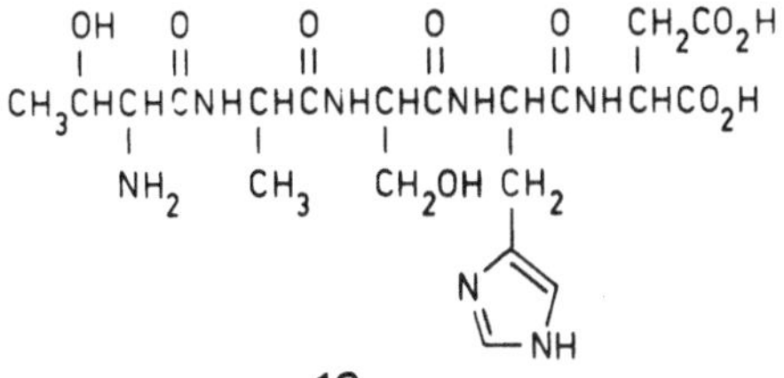

10

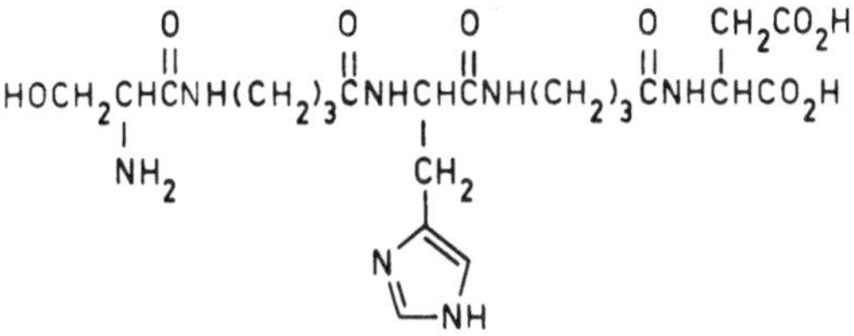

11

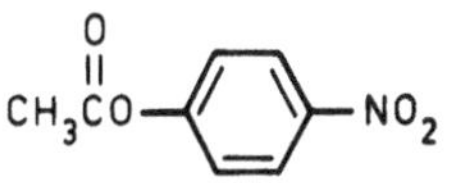

12

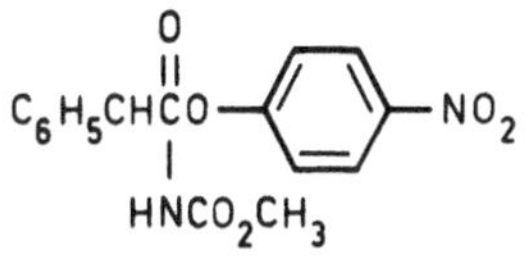

13

Naturally occurring peptides can also be used to catalyze reactions other than hydrolytic. Although the present survey is not exhaustive one can note, for example, that in the epoxidation of α,β-unsaturated ketones by basic hydrogen peroxide (eq. 6) that the polypeptide (14) acts as an efficient catalyst producing enantiomeric excesses of (15) of up to 93% [20].

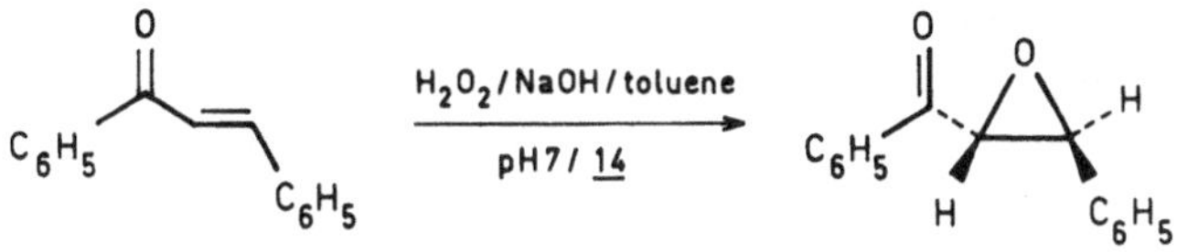

(6)

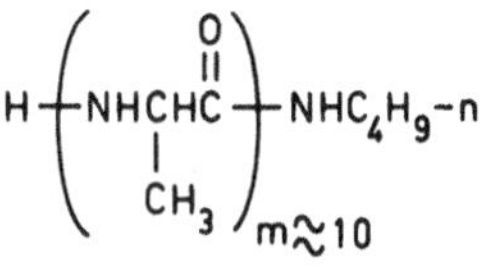

14

With a larger peptide, bovine serum albumin, molecular weight 66,000, which has three binding sites per molecule, both reductions and oxidations have been catalyzed. In the presence of stoichiometric amounts of bovine serum albumin acetophenone is reduced by sodium borohydride to 1-phenylethanol in up to 78 % enantiomeric excess (eq. 7). [21] The acetophenone is firmly complexed to the peptide during reaction.

$$C_6H_5\overset{O}{\overset{\|}{C}}CH_3 \ + \ NaBH_4 \ \xrightarrow[\text{albumin}]{\text{bovine serum}} \ C_6H_5\overset{OH}{\underset{H}{C}}CH_3 \qquad (7)$$

In a similar fashion sulfide (16) is oxidized to the sulfoxide (17) by $NaIO_4$ again in the presence of a stoichiometric amount of bovine serum albumin (eq. 8) in an enantiomeric excess of 81 %. [22] Somewhat similar experiments have been reported

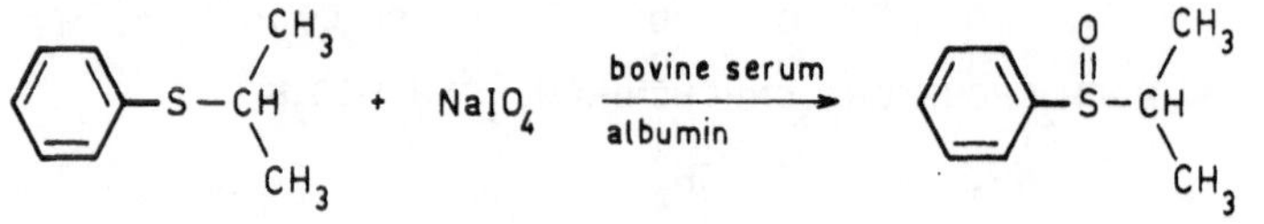

$$\underset{16}{} \qquad\qquad\qquad \underset{17}{} \qquad (8)$$

using kinetic resolution in the oxidation of a sulfoxide to an (achiral) sulfone. [22] In these experiments with bovine serum albumin there is again no structural basis for understanding the enantioselectivities observed. Nor are the reactions very general, at least at this stage of development.

5 Catalytically Active Synthetic Macrocycles

5.1 Crown Ethers — General Design Points

Following the development of effective synthetic routes to macrocyclic crown ethers as exemplified in eq. 9 for the synthesis of [18]crown-6 (18), extensive studies were initiated on the factors affecting their complexing ability. [23] The success of this work or, better said, the opportunity to carry it out depended on the availability of synthetic routes chiefly along the "template" lines depicted. [24] The synthetic opportunity to introduce significant structural modifications in a rational way is the aspect that lends the greatest advantage to the crown ethers. This synthetic flexibility is more difficult to

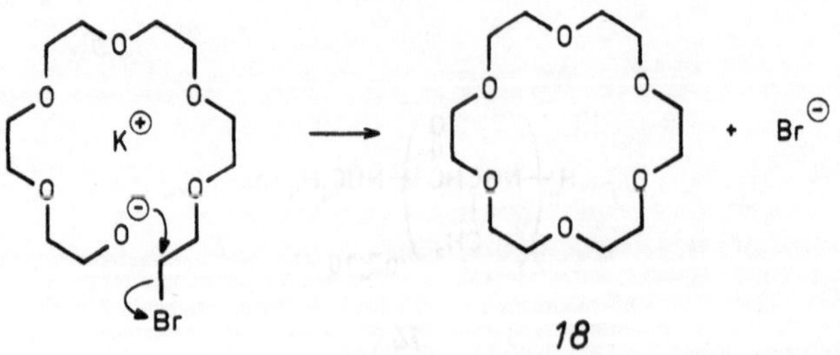

$$ + \ Br^{\ominus} \qquad (9)$$

$$\underset{18}{}$$

achieve with the cyclodextrins or, for example, with macrolide antibiotics, which have excellent complexing properties but usually so many functional groups that selective transformations on one particular substituent are not in any practical manner feasible (coupled often with scarcity of materials).

The classical model for binding of an *ion* in a crown ether involves three point bonding as illustrated for the ammonium ion complex with [18]crown-6 in *18-NH$_4^\oplus$a*, more easily drawn and visualized in *18-NH$_4^\oplus$b*, viewed from the top. [25] The most logical method of designing crown ethers capable of mimicking enzymic chemistry

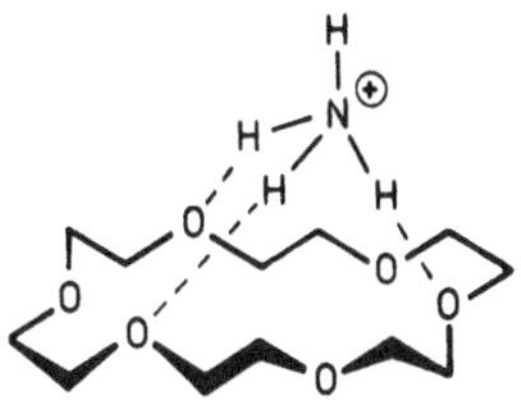

$18a$-NH$_4^\oplus$ 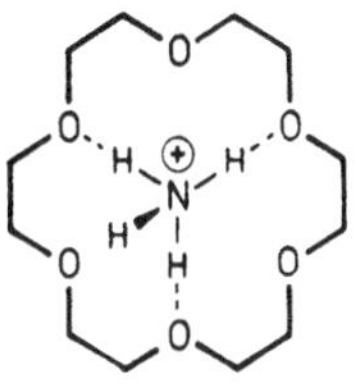 $18b$-NH$_4^\oplus$

is to assume the validity of the three point bonding model, use as guest an ammonium ion with a group R on which a reaction is to be carried out and then modify as desired the periphery of the crown ether.

The last aspect, modification of the periphery of the crown ether will be considered first. To the present time three main routes have been followed with as goal the synthesis of a *chiral* crown ether having also catalytically active groups in the periphery. Cram has brilliantly used bis-β-naphthol (*19*), which can be readily resolved into its enantiomers, as chiral component in the synthesis of crown ethers, an example being *20* (*a* and *b* being different projections) prepared using the potassium salt of the bis-phenolate of *19* and the bis-tosylate of pentaethyleneglycol. [26] The twisted nature of the bisnaphthyl linkage imparts a strong chiral bias to the system. A moderate price in complexing ability is paid for the disruption of the classical crown

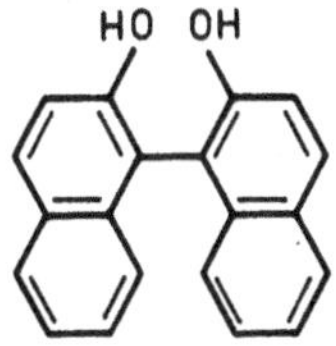

19

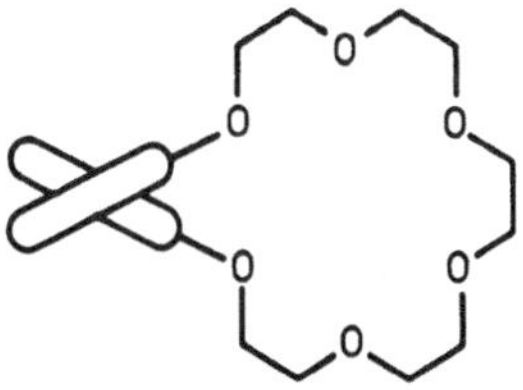

s-*20a* 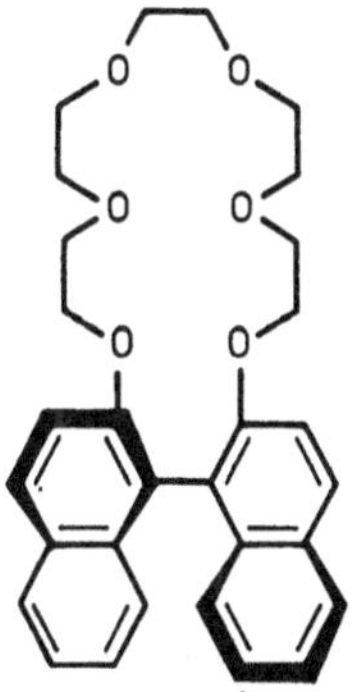   s-*20b*

ether system as well as the introduction of two *phenolate* oxygens of lowered basicity. The synthetic key to introduction of extra "arms", which can bear catalytically active groups, is a double Mannich reaction of (*19*) followed by acetylation and hydrolysis with LiAlH$_4$ to afford the *bis* (hydromethyl) derivative (*21*, eq. 10) which can be resolved easily [27].

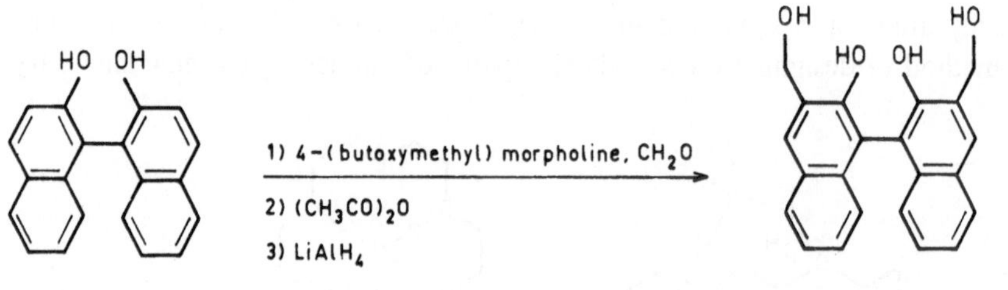

$$\text{19} \quad \xrightarrow[\substack{2)\ (CH_3CO)_2O \\ 3)\ LiAlH_4}]{1)\ 4-(butoxymethyl)\ morpholine,\ CH_2O} \quad \text{21} \qquad (10)$$

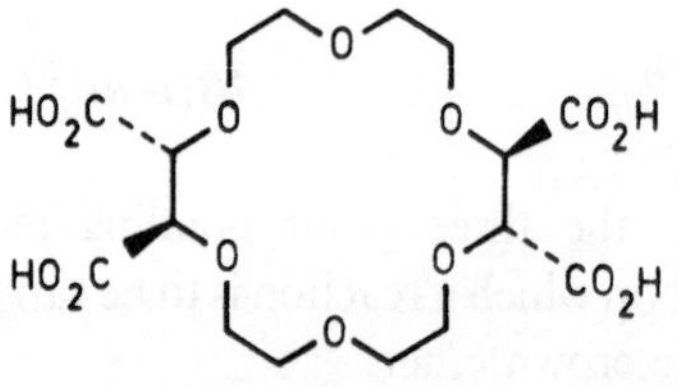

22

Another approach developed by the group of Lehn [28] requires a less drastic distortion of the periphery of the crown ether. For example in (*22*) derived from L-tartaric acid an [18]crown-6 periphery is still intact and four sites for attachment of

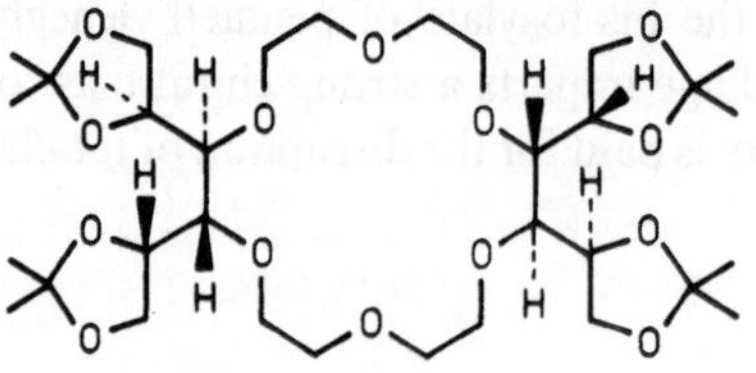

23

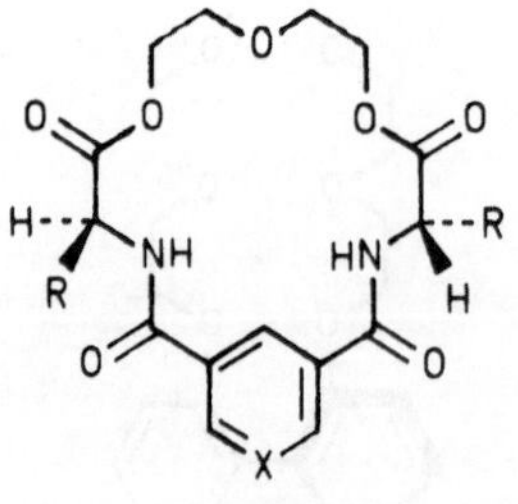

24

a) X=CH

b) X=N

functional groups are present. The D-enantiomer of tartaric acid is also available extending the synthetic flexibility derived by manipulation of the substituents [29].

Still another approach to chiral crown ethers that have potential applications for catalytic reactions is through incorporation of sugar residues as pioneered chiefly by Stoddart's group. [30] An example of one of the many compounds of this sort that has been synthesized (again by application of "templated" Williamson reactions) is (23) obtained from D-mannitol.

In our work we have developed methods to build heavily modified crown-like systems in which the chiral components are amino acids. [31] The general structural type is illustrated by (24a, b). In this case the macrocyclic crown ether system has been badly broken by extra substituents and heteroatoms (amide nitrogen, ester ether oxygen) of lowered basicity.

The chemical applications in catalytic reactions of these and related systems will be discussed in the following sections. One point of design that should be emphasized before proceeding on to the chemistry is that (20) and (22–24) have C_2 symmetry axes, which makes the faces of the system homotropic. Although this design feature is not mandatory, interpretations of complexing and reactions are greatly simplified if both faces of the reactive macrocycle are identical.

5.2 Crown Ether Derived Compounds That Imitate Proteases

Proteases like papain [32] have a cysteine sulfur as active nucleophile [33] and bind the substrate in a cleft. By using a crown ether ring instead of a cleft for binding, taking as substrate an ammonium salt that complexes via three-point bonding to oxygens of the crown ether, and introducing sulfur nucleophiles at the proper position to attack an activated carbonyl group in the ammonium substrate (ammonium salts of activated amino acid esters (25) — as illustrated in eq. 11 — are ideal) one has the potential for creating a synthetic protease, or at least a synthetic transacylase.

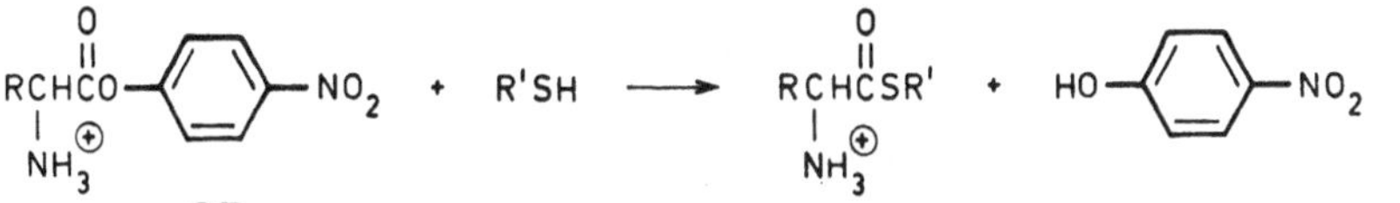

$$RCHCO-\langle\!\!\!\!\!\bigcirc\!\!\!\!\!\rangle-NO_2 \; + \; R'SH \; \longrightarrow \; RCHCSR' \; + \; HO-\langle\!\!\!\!\!\bigcirc\!\!\!\!\!\rangle-NO_2 \qquad (11)$$

25

The best examined system of this type is (26a, S-enantiomer illustrated) prepared and studied by Cram and coworkers. [26] The dimethyl derivatives (26b) had already been

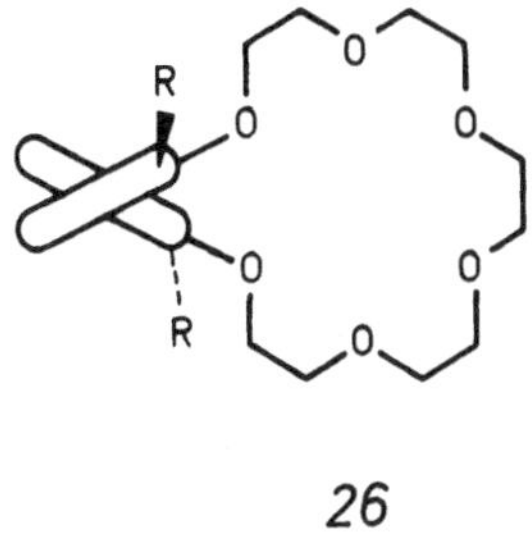

26

a) $R=CH_2SH$

b) $R=CH_3$

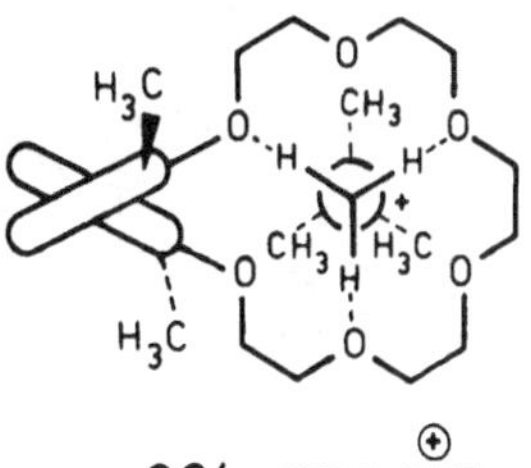

$26b-(CH_3)_3CNH_3^{\oplus}$

293

established to bind $(CH_3)_3CNH_3^\oplus$, picrate$^\ominus$ in $CHCl_3$ at 25° with $\Delta G° = -6.4$ kcal/mol by means of the anticipated three-point binding ($26b$-$(CH_3)_3CNH_3^\oplus$, picrate$^\ominus$). Similar binding is expected with $26a$ with CH_2SH side arms.

In separate experiments with $HOCH_2CH_2SH$ as the thiol it was established that, at least at low concentrations, the thiolate anion, $HOCH_2CH_2S^\ominus$, is the only kinetically significant nucleophile in the thiolysis of 25 and that there are no appreciable buffer effects in the organic solvent mixtures used. The measured rates, determined by the appearance of p-nitrophenol or -phenolate depending on the pH are, however, those of transacylation (eq. 11) rather than subsequent (slow) solvolysis (eq. 12).

$$\underset{\underset{\overset{\oplus}{NH_3}}{\overset{|}{R}CHCSR'}}{\overset{O}{\overset{||}{}}} + \text{H}-\text{Osol} \longrightarrow \underset{\overset{O}{\overset{||}{}}}{RCO}-\text{sol} + R'SH \qquad (12)$$

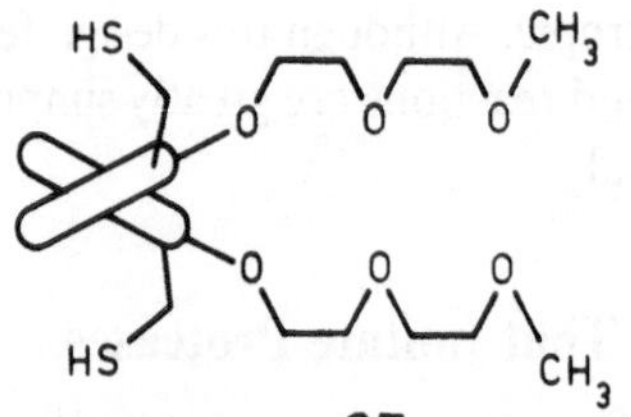

27

The effect of incorporating a cyclic structure is best evaluated by comparison with the transacylation rate constants for $26a$ with those for "open" analog (27) (S-enantiomer shown). For the transacylations of p-nitrophenolate esters of amino acid salts in 20% C_2H_5OH/CH_2Cl_2 at 25°, assuming no effective difference in pK's between ($26a$) and (27), cyclic $26a$ reacted consistently faster then "open" (27). For L-$25a$, $R = (CH_3)_2CHCH_2$—, $k_{26a}/k_{27} = 1170$, and for ($25b$), $R = (CH_3)_2CH$—, $k_{26a}/k_{27} = 160$, and for ($25a$), $R = C_6H_5CH_2$—, $k_{26a}/k_{27} = 490$ as typical values for the observed rate differences.

Not only is there a rate acceleration to be derived from the cyclic structure of ($26a$) but there is also structural recognition for the transacylation rates fall off with increasing size of the R group in (25). The rate for ($25c$, $R = (CH_3)_2CH$) is consistently lower than for ($25d$, $R = CH_3$), rate differences of 30–300 being observed depending on the series investigated. These effects probably arise from the fact that the tetrahedral intermediate from (S-$26a$) with (R-25) (or R-$26a$ with D-25) is more

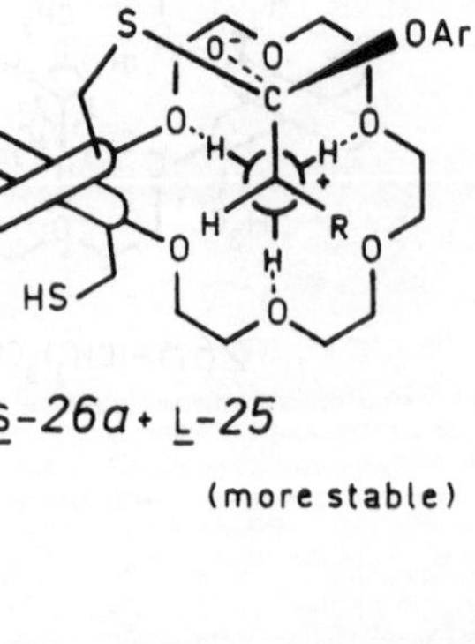

S-$26a$ + L-25 S-$26a$ + D-25

(more stable)   (less stable)

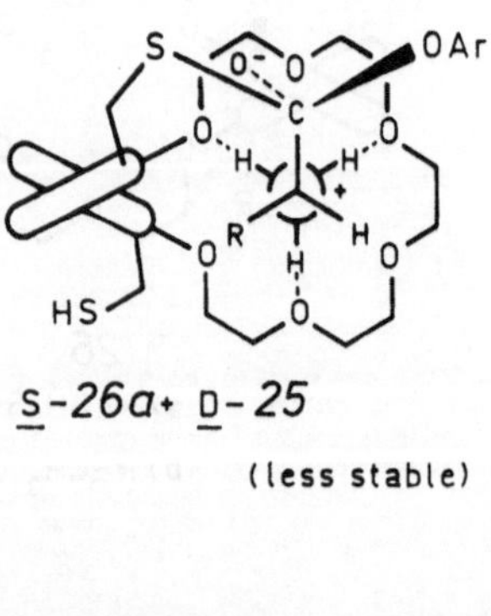

complementary to the macrocyclic structure than the tetrahedral intermediate formed from (R-*26a*) and *D-25* (S-*26a* with L-25). Note, of course, in the drawings that only the sterically most likely diastereomers of the tetrahedral intermediates are indicated. From CPK models the fit of the tetrahedral intermediates to the topology of the macrocycle appears to be reasonable but certainly not perfect.

From this work one can derive high hopes for the future because by *rational* design of complexes and before the experiment analyses of steric interactions (using CPK models) a synthetic catalyst was designed in a rational manner. The rate accelerations must of course to be made better and the problem of catalytic turnover has not yet been solved. In the reactions described the "catalyst" is actually used in about 50-fold excess. Solutions for these problems are challenges for the future.

Crown ethers based on tartaric acid (see *22*) have been used for similar transacylations. [34] The *tetra*-cysteine derivative (*28*) serves as a papain model. Because (*28*) has an undistorted [18]crown-6 periphery quite good complexation of

28 (L-cysteine derivative)

29

ammonium ions is expected. Transacylations of p-nitrophenol esters of amino acid hydrobromides are indeed observed. Owing to the length of the cystinyl substituents the activated ester group of the substrate must be well separated from the ammonium ion center. This is achieved readily with dipeptides as substrates, an example being the p-nitrophenolate ester of glycinyl-L-phenylalanyl hydrobromide (*29*), which prior to thiolysis is probably bound to (*28*) as shown. In $CH_2Cl_2/CH_3OH/H_2O$ (97.9/2/0.1 ratio) the L-enantiomer of (*29*) undergoes acyl transfer to (*28*) about 50 times more rapidly than the D-enantiomer of (*29*). (L-*29*) appears to fit better in the cavity of (*28*). A detailed analysis of steric effects is difficult, however. Thiolysis definitely occurs in and not outside of the complex as established by the inhibitory effect of KBr, which competes for a binding site in (*28*). Benzylation of the thiol residues removes the catalytic activity of (*28*) entirely.

In (*28*), as with bis-naphthyl system (*26*), there is no turnover because the "catalyst" is used in excess in order to obtain rate data. Also both for (*26*) and (*28*) it is unfortunate that the thiol groups in the complexes are not all useable. In (*28*) the conformational flexibility of the cysteinyl arm is likely detrimental to the efficiency of catalysis

295

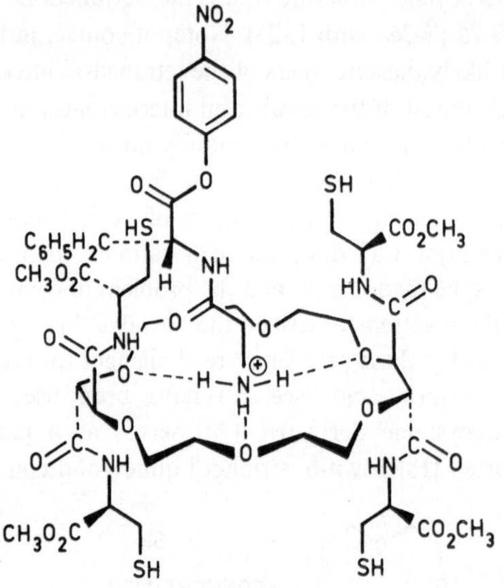

$$28 - 29^{\oplus}$$

although the good complexing ability and high amino acid content does give it "enzyme-like" character.

Thiol catalyzed transacylations using the principles discussed above have also been investigated using crown ether systems (30a–c) (31) and (32). [35, 36] The latter two compounds have particularly large chiral groups built into the periphery. The kinetic results with these compounds parallel those for (26a) and (28) with reasonable kinetic accelerations being found for transacylations. Chiral discriminations are modest, not exceeding a factor of 2 for D- and L-forms of (25). Very likely too much conformational flexibility is still present in these compounds for highly efficient chiral recognition.

We have developed synthetic routes to resorcinol-based crown ethers like (33) [37]. These have been established to complex ammonium salts and to undergo conformational changes wherein the phenyl ring acts as a "hinge" moving the catalytically

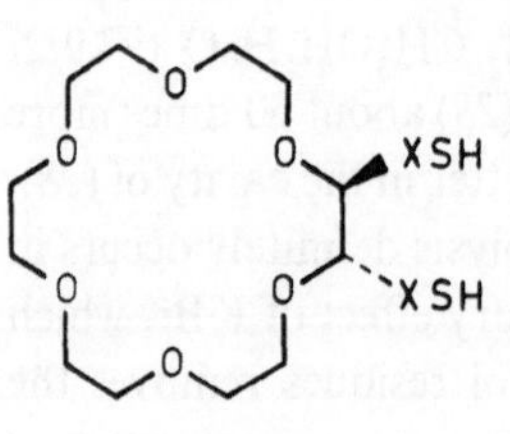

30
a) X = -CH₂-
b) X = -(CH₂)₃-
c) X = -CH₂O(CH₂)₂-

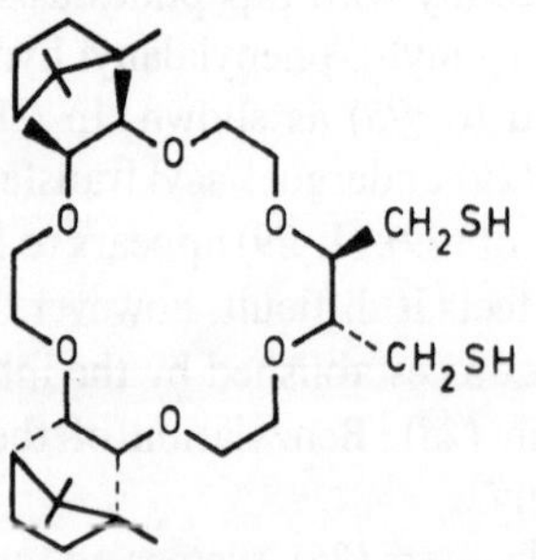

31

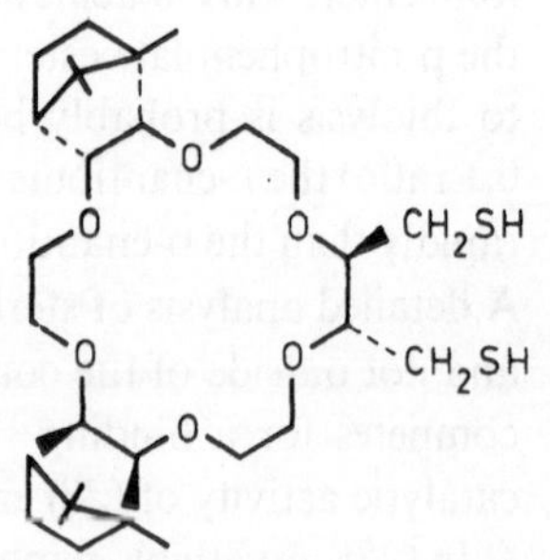

32

296

active thiol group back and forth relative to the macrocyclic ring. Transacylations with these compounds are being investigated.

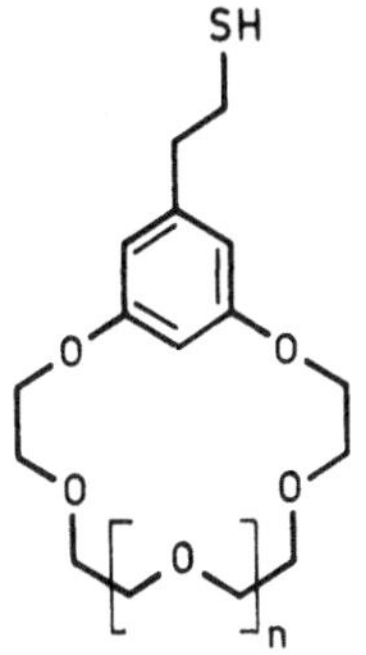

33

a) n = 2

b) n = 3

5.3 Cyclic Hydrocarbon Systems that Imitate Proteases

Crown ether derived systems have as attractive feature a good binding cavity for *ions*; the structural mixture of polar heteroatoms and connecting apolar hydrocarbon backbone also usually ensures a good solubility of the crown ether and its complexes in polar solvents.

There are, however, other binding forces that might be used to hold a, for example, *nonionic* substrate in a cavity. One such possibility that has been investigated is the use of "hydrophobic interactions". [38] This can be illustrated with an example. The macrocycles (*34*) and (*35*) [39], obtainable by means of acyloin condensation, followed by conversion to the oxime, possess functionalities that are potential catalysts for the hydrolysis of activated esters. [16] If hydrophobic interactions are important, an activated ester of a long chain fatty (i.e. hydrophobic) acid, for example, the p-nitro-phenolate ester of lauric acid, should associate *in water* with the hydrophobic

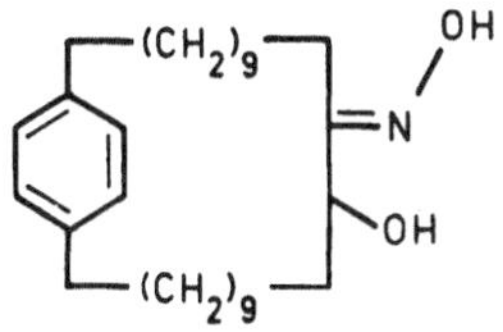

34 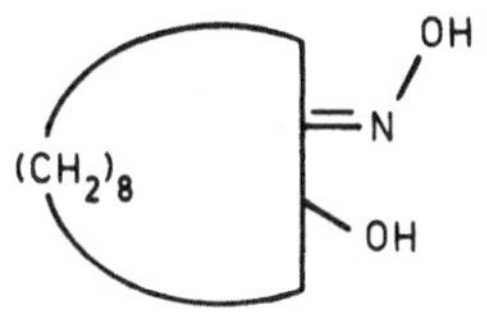   35

macrocycle. Although there is no control over the geometry of association, the increased local concentrations could produce rate enhancements. Indeed (*34*), but not (*35*), induces mild rate enhancements and the rate is inhibited by $Cu^{2\oplus}$ ions, which compete for the α-hydroxyoxime ligand.

A complication with such work, however, is the pronounced and expected tendency of such poorly soluble compounds to form micelles in water. [40] This tendency was underestimated in earlier work [41] on synthetic systems suggested to be models for esterases. To avoid micelle formation very low concentrations, usually $<10^{-6}$ M, must be used. The limits severely the range of practical applications.

A structurally more sophisticated hydrocarbon derived macrocycle is (36),[42] which also exhibits hydrolytic activity. At low concentrations (36), relative to imidazole alone, produces enhanced rates of hydrolysis of p-nitrophenyl esters of long chain acids. The suggestion that binding of substrate to catalyst occurs is supported by the fact that (37), which bears no catalytically active groups, *inhibits* the hydrolysis rate of p-nitrophenyl acetate by binding the substrate in the hydrophic cavity [43].

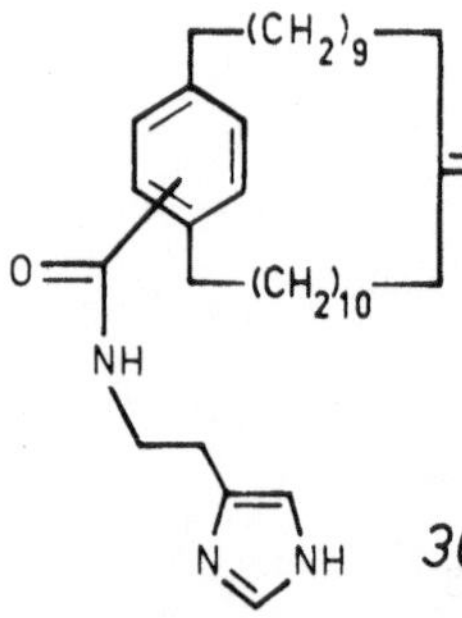

5.4 Macrocycles with Hydrogenase Activity

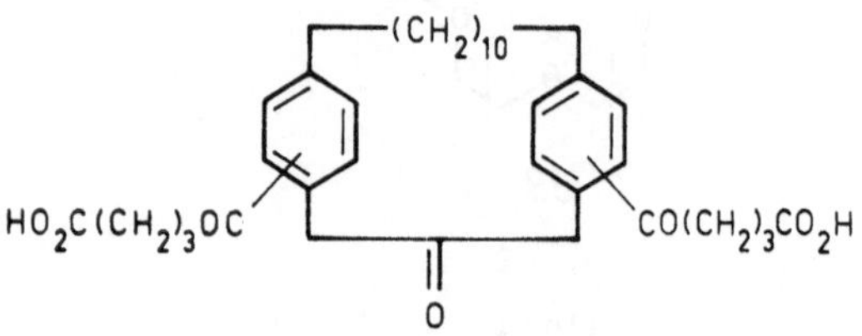

Excluding hydrolytic enzymes, about 70% of known enzymes require a cofactor and this figure becomes even higher if metal ions are included as cofactors. [44] An obvious chemical extrapolation from this information is to use the small coenzyme, or a similar molecule with the same chemistry, and to replace the accompanying protein (apoenzyme) and its function with synthetic material. To do this it is necessary that the chemical transformations desired be intrinsic to the coenzyme itself, the apoenzyme providing a binding site and carrying out an activating and regulatory role. As already mentioned coenzymes that are evolutionarily old are most likely to have the intrinsic chemistry desired.

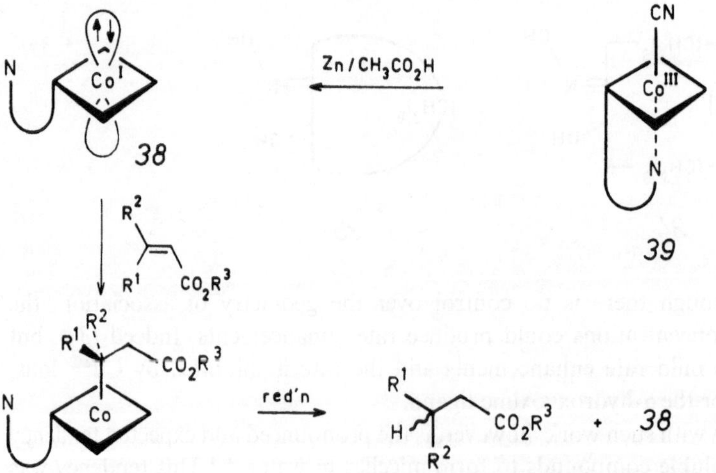

(13)

A case that may be mentioned briefly in the present context is that of cob(I)alamin (*38*) generated chemically from cyanocob(III)alamin (*39*). In extensive studies Fischli [45-49] has demonstrated that good degrees of enantioselectivity in *nonenzymic* reactions can be achieved as illustrated for a generalized hydrogenation scheme in eq. 13, in which the schematic drawings are intended to represent the corrin ring (perspective four-ring) and the attached benzimidazole in the "off" position.

This intrinsic reactivity needed for spontaneous reaction is certainly not present in every coenzyme, however. An example is biotine (*40*), which in its carboxylated form (*41*, eq. 14) is a form of *deactivated* rather than "activated CO_2". Activation by the accompanying apoenzyme is essential for CO_2 transfer. [8] This process has been discussed in terms of evolution [7a].

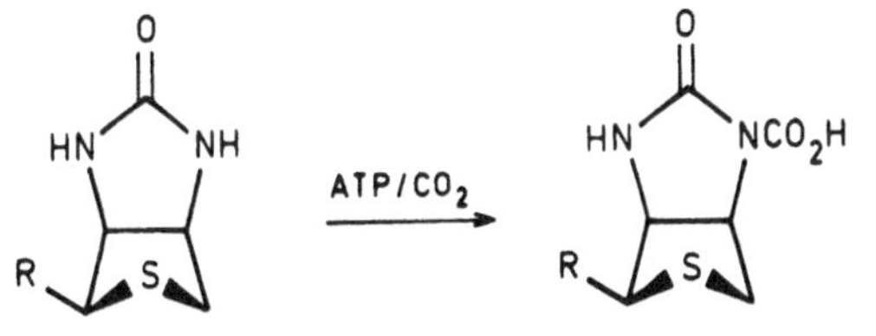

$$\text{40} \qquad \xrightarrow{\text{ATP}/\text{CO}_2} \qquad \text{41} \tag{14}$$

Nicotinamide adenine dinucleotide (*42*) in the 1,4-dihydro form is a coenzyme that does lend itself well for use in model systems. The reactive portion is the 1,4-dihydronicotinamide ring (*43*), which is capable of acting as "hydride" donor towards many substrates; the pyridinium salt (*44*) acts in turn as "hydride" acceptor (eq. 15). The sugar nucleotide tail can be deleted without fatal consequences for the reactivity although the 1,4-dihydropyridine form (*44*) by itself has little tendency to reduce anything except the most reactive of potential substrates. [50] The problem is not

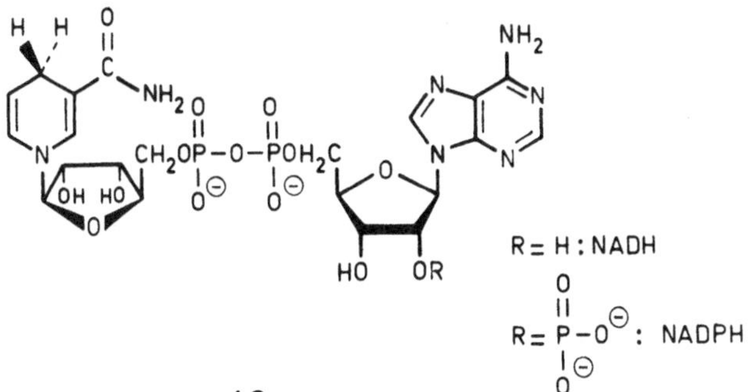

R = H : NADH

$$R = \overset{O}{\underset{\underset{O}{|}}{\overset{||}{P}}} - O^{\ominus} \; : \; \text{NADPH}$$

42

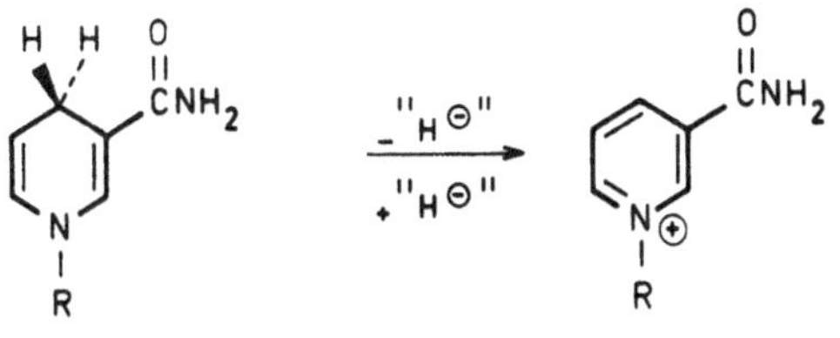

$$\text{43} \qquad \overset{-\,"H^{\ominus}"}{\underset{+\,"H^{\ominus}"}{\longrightarrow}} \qquad \text{44} \tag{15}$$

intrinsic lack of reduction potential of synthetic 1,4-dihydropyridines (N-benzyl-1,4-dihydronicotinamide (*43*, $R = CH_2C_6H_5$) has $E_0 = -361$ mV in nonaqueous solution compared to $E_0 = -315$ mV for NADH in aqueous solution) [51].

For reduction of carbonyl compounds *catalysis* is necessary. One of the major contributing effects to *enzymic* catalysis is polarization of the carbonyl substrate by an electrophile (eq. 16), which is usually imidazole or a zinc ion embedded in the peptide chain [52].

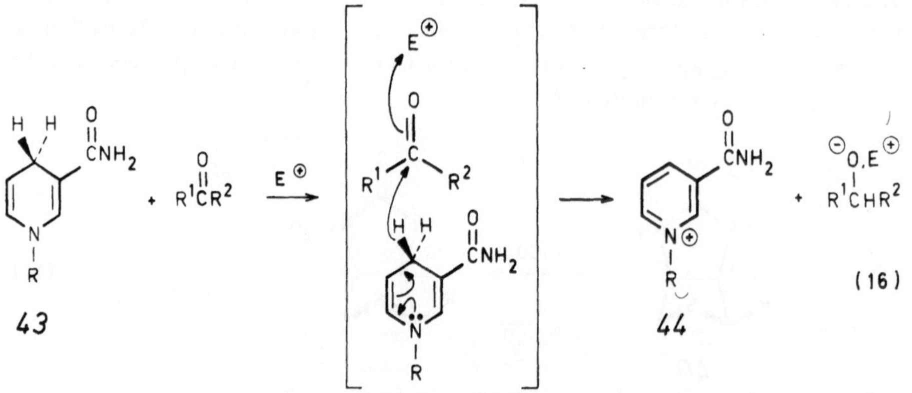

$$(16)$$

In *nonenzymic* reductions with synthetic dihydropyridines magnesium ions, usually as the perchlorate, $Mg(ClO_4)_2$, are the best electrophilic catalysts so far found. [53] The $Mg^{2\oplus}$ ion may in fact do more than just polarize the carbonyl group although this remains a point of contention. Other catalysts have been used but in general with less success than $Mg^{2\oplus}$ [54].

Two basically different strategies have been followed in combining dihydropyridines with crown ethers to obtain catalytically active systems. One approach is to attach one or more derivatives of (*43*) to or in the periphery of a crown ether, which has good complexing properties for metal ions. Those metal ions should be the electrophilic catalyst for coordination and activation of a carbonyl substrate (eq. 17). The success

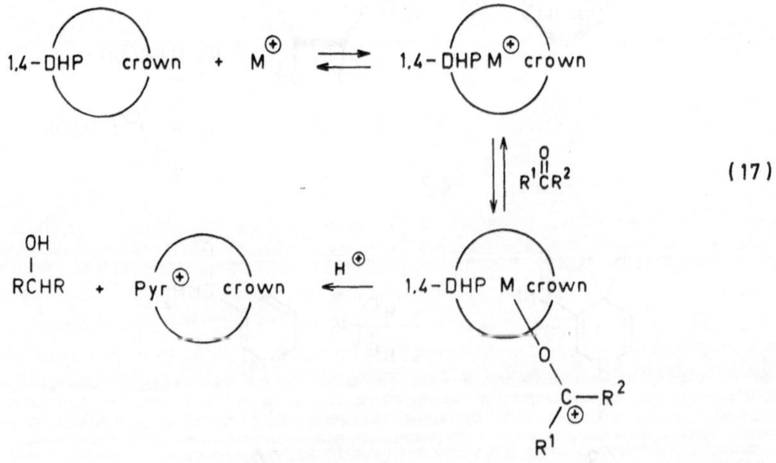

$$(17)$$

of this approach depends on a proper union of proximity effects and orientation of the dihydropyridine, electrophilic catalyst and carbonyl substrate.

A second strategy, more analogous with the syntheses of crown ethers with trans-acylation capabilities, is to use primary ammonium ions with the potential substrate in the substituent R of the ammonium ion; the dihydropyridine is attached to the crown ether (eq. 18). This approach poses restrictions on the types of reactions that can be carried out because there is no built-in possibility for electrophilic catalysis within the complex.

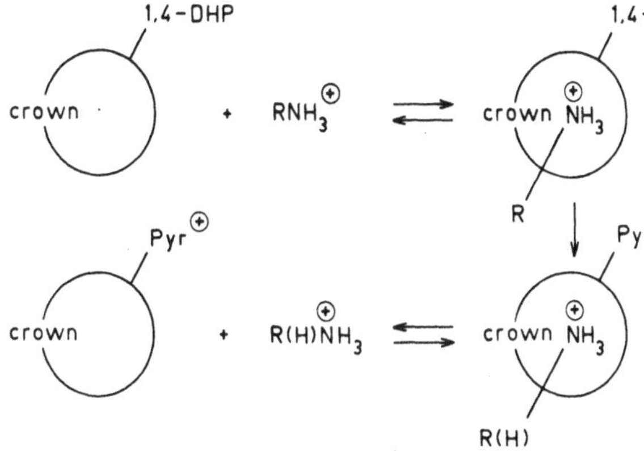

$$(18)$$

Before proceding to examples of these strategies the reader is reminded of some occasionally overlooked synthetic aspects of dihydropyridine/pyridinium salt chemistry. The pyridinium salt to 1,4-dihydropyridine conversion (*44* to *43*, eq. 15) is carried out virtually quantitatively and completely regioselectively with sodium dithionite, [55] and pyridinium salts are conveniently obtained from alkylation of pyridines with primary alkyl halides or similar reagents as illustrated in eq. 19. Sodium dithionite is the only reagent that carries out the pyridinium salt to 1,4-dihydropyridine reduction regioselectively. All other functionalities present in a synthetic system must be compatible with this (mild) reducing agent. [56] The dihydropyridines themselves must bear at least one electron withdrawing group

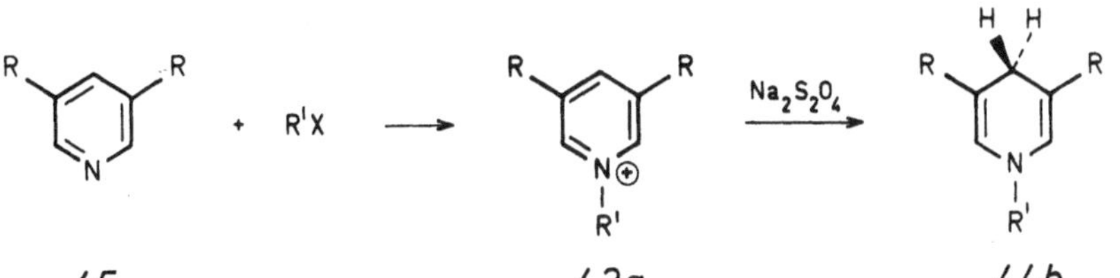

$$(19)$$

at the 3-position (for example 1,4-dihydronicotinamide *43*) or two at the 3- and 5-positions (see further) to have reasonable stability. As a final point, 1,4-dihydropyridines not alkylated at nitrogen are available (for example *46*, eq. 20) [57] but on hydride transfer to substrate S a pyridine remains and there are no regioselective ways of reducing a *pyridine* to a 1,4-dihydropyridine in contrast the reversibility of the pyridinium salt/N-alkyl-1,4-dihydropyridine link, which is one of its main attractions.

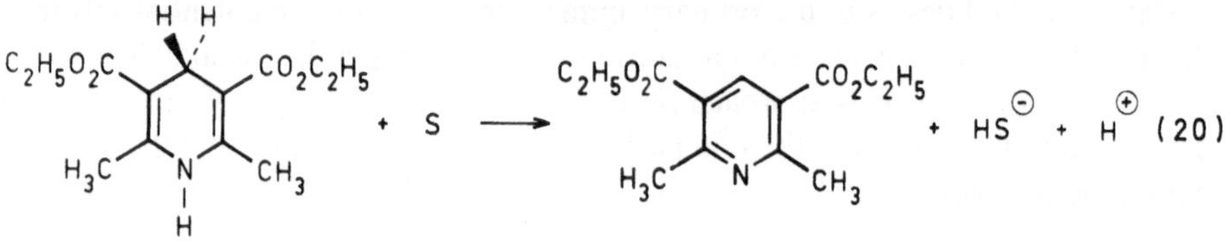

$$46 \qquad (20)$$

The approach of eq. 18 has been followed by Lehn's group. [58] The tartrate derived [18]crown-6 derivative (22) has been connected via the pyridine nitrogens to four nicotinamide derivatives to form (47). A potential substrate can be attached to an ammonium salt to give, for example, (48), which complexes in the crown ether (eq. 21). The only reaction so far investigated using this approach is, as can be seen from eq. 21, not a carbonyl group reduction but rather hydride transfer to a pyridinium salt.

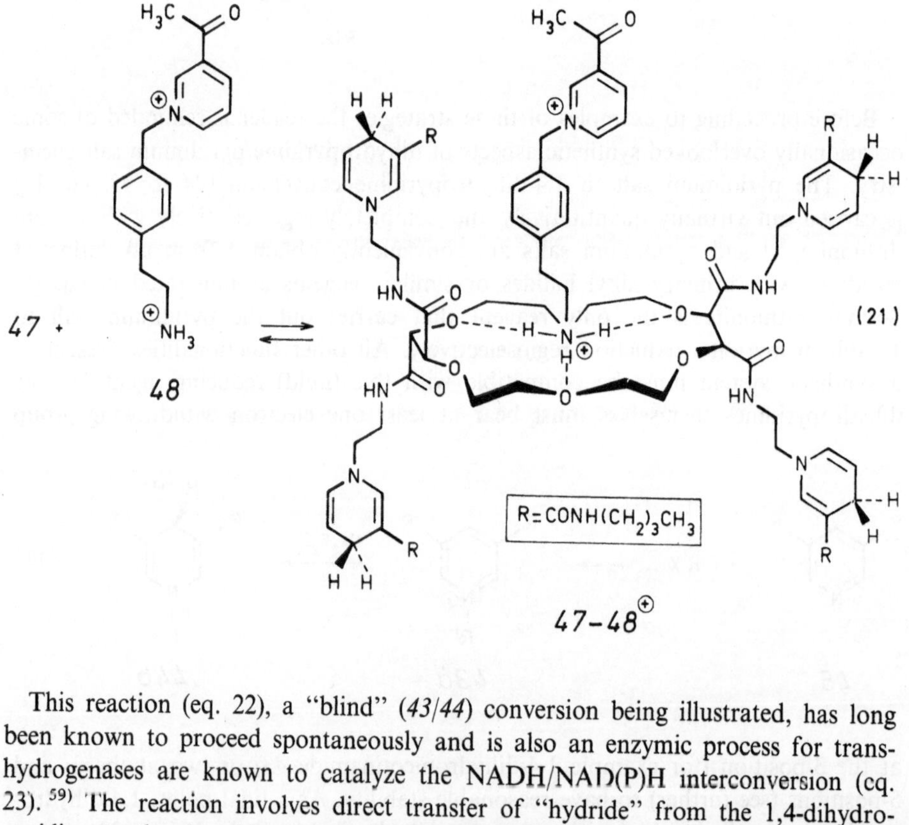

$$R = CONH(CH_2)_3CH_3$$

$$47 + 48 \qquad \rightleftharpoons \qquad 47\text{-}48^{\oplus} \qquad (21)$$

This reaction (eq. 22), a "blind" (43/44) conversion being illustrated, has long been known to proceed spontaneously and is also an enzymic process for trans-hydrogenases are known to catalyze the NADH/NAD(P)H interconversion (cq. 23). [59] The reaction involves direct transfer of "hydride" from the 1,4-dihydro-pyridine to the 4-position of the pyridinium salt both in the enzymic and non-enzymic examples. The position (eq. 22) of the equilibrium for 1,4-dihydropyridines and pyridinium salts with nonidentical substituents depends on the difference in reduction potentials of the two possible 1,4-dihydropyridines. [60] A considerable complication can be that the 2(6)-position of the pyridinium salt also acts as hydride accepter site, leading to isomeric dihydropyridines; this phenomenon has been studied in detail [59g,h].

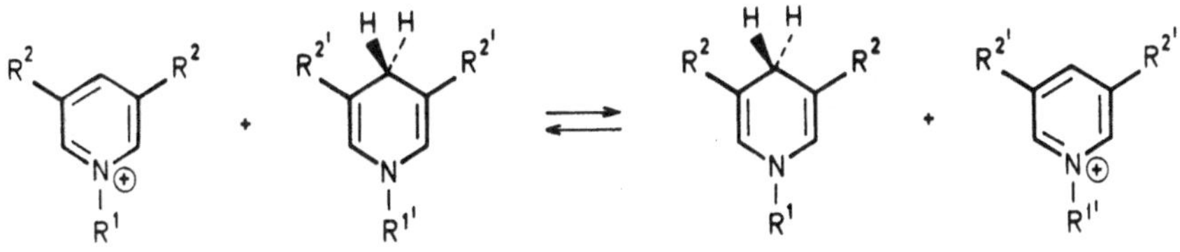

$$(22)$$

$$\text{NADPH} + \text{NAD}^{\oplus} \underset{}{\overset{enzyme}{\rightleftharpoons}} \text{NADP}^{\oplus} + \text{NADH} \qquad (23)$$

When ammonium salt (48) is allowed to combine with (47) (eq. 21) a complex (47–48⊕) having presumably the indicated structure is formed. In this connection one notes also that (48) (as well structurally related ions) is bound to, for example, (49) to which tryptophane methyl ester units have been attached. [61] This leads to development of a charge-transfer absorption in the complex (49–48⊕). That indole-nicotinamidium interactions can provide a charge-transfer absorption is well-known [62] and has been suggested to the responsible for the color of the NAD⊕/3-phosphoglyceraldehyde dehydrogenase complex [63].

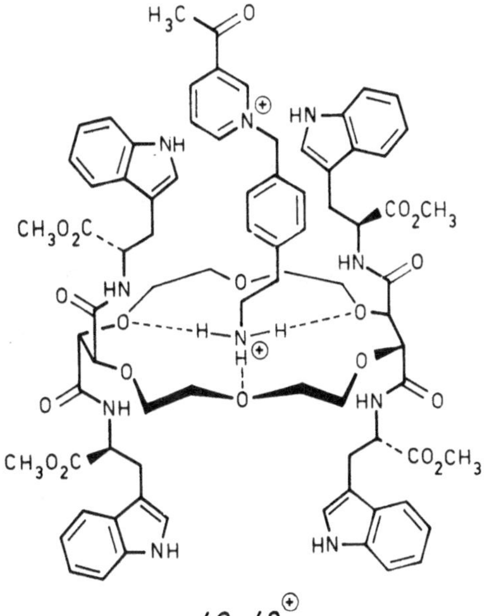

$$49\text{-}48^{\oplus}$$

Such extra bonding interactions, which might also be present in the 47–48⊕ complex, add extra stability and provide a nice framework for studying the geometrical requirements of intermolecular interactions between groups.

The rate of hydride transfer is indeed mildly enhanced (assignment of an exact number does not seem justified) in the (47–48⊕) complex; hydride transfer is also slowed by adding $K^\oplus$ ions, which displace (48) from its binding site. A disadvantage of (47) as ligand is that the catalytically active groups probably have too much conformational freedom and do not reach an optimal geometry for hydride transfer.

Our own approach to the combination of crown ether and dihydropyridine chemistry has involved constructing the dihydropyridine as an integral portion of the macrocyclic crown ether ring (see 24b, for example). The first synthetic approach involved ring-closure of an alicyclic precursor by means of the Hantzsch 1,4-dihydropyridine synthesis as illustrated for the preparation of (50, eq. 24) [64,65]. Such "Hantzsch esters" (general type 46) are attractive in that the acid functionalities at the 3,5-positions can be used as "handles" for attaching the (dihydro)pyridine

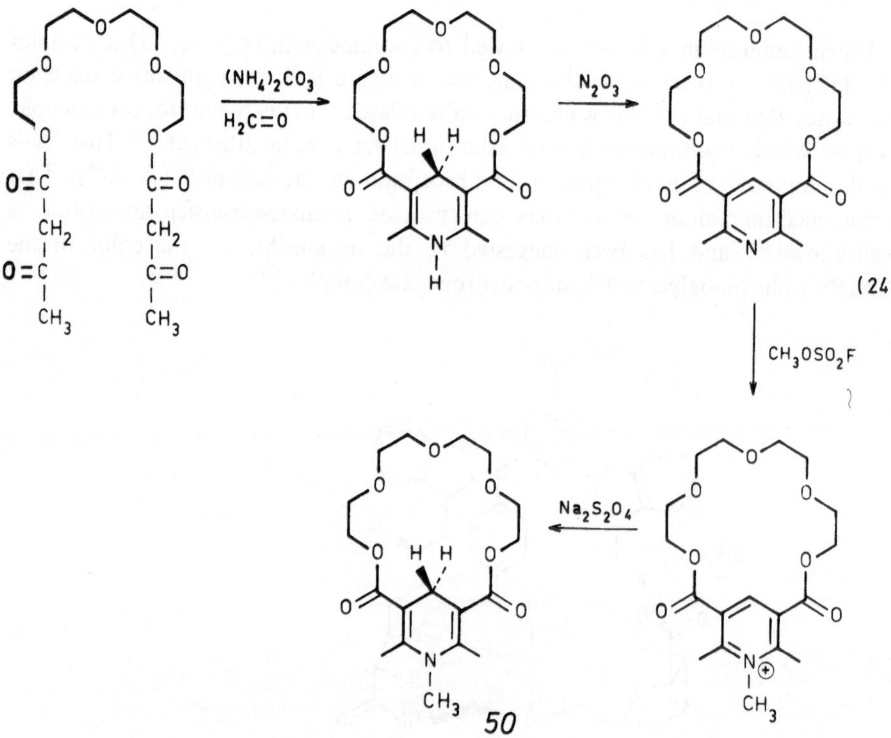

(24)

into the macrocycle ring. The success of the synthesis (20–25% yields) of eq. 24 probably lies in the use of $(NH_4)_2CO_3$, the $NH_4^\oplus$ ions acting both as ammonia source [66] as well as template [24].

The 3,5-bridged 1,4-dihydropyridines like 50 are not true crown ethers owing to the drastic structural changes in the ring. The stability constants for complexation by (50) and related structures have not yet been measured but there are sufficient indications that complexing power still remains. Compound (50) complexes positive ions in solution [67] and stable complexes with $NaClO_4$ have been isolated and the structure of one complex has been determined. [68] Unfortunately (50) and related compounds in the presence of various metal ions did not give reproducible reductions of alcohols as had been hoped (eq. 18). [69] That enhanced reactivity was nevertheless present in (50) was established in an unusual reaction. Phenacyl sulfonium salts (51) can be reduced by (52) as shown in eq. 25 [70]. The reaction involves a reductive cleavage of a carbon-sulfur bond rather than reduction of the

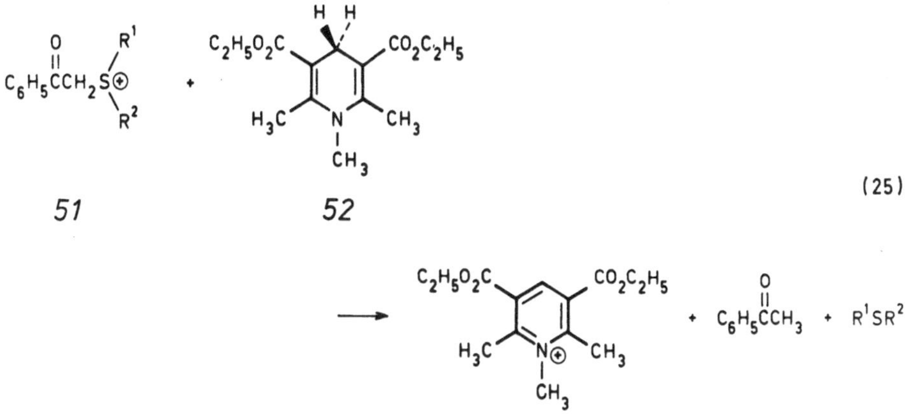

$$(25)$$

51   52

carbonyl group of (51); a chain electron transfer mechanism explains the results. [70] The bridged compound (50) carries out the same transformation of (51) ($R^1 = C_6H_5$, $R^2 = CH_3$) but, extrapolated to 75°, the reduction by (50) proceeds $2.7 \cdot 10^3$ times more rapidly than that by nonbridged (52) [69]. This accelerated reduction was ascribed to formation in a rapid preequilibrium step of a complex with the large sulfonium salt (51) perched on (50) as illustrated. Although the complexation constant with large (51) must be small, this explanation appears to be correct especially in view of the strongly inhibiting effect of $NaClO_4$ on the reduction; $Na^\oplus$ ions compete better than (51) for the binding sites in (50).

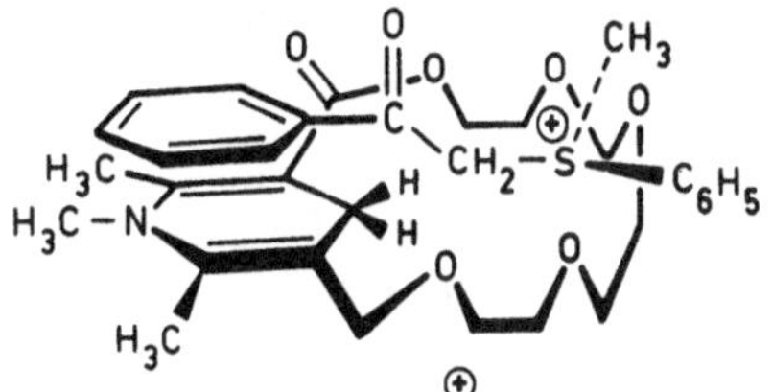

50-51$^\oplus$

Although encouraged by these results we realized that considerable redesign was in order. Moreover, difficulties in alkylating the pyridine nitrogen [65] made it necessary to remove the methyl groups at the 2,6-positions of the pyridine ring; this meant abandoning the Hantzsch approach (eq. 24).

A direction that redesign of the system could take was suggested by experiments with 1,4-dihydronicotinamide derivatives (55) with an optically active amine attached to the side chain; such compounds in the presence of $Mg^{2\oplus}$ are capable of reducing activated carbonyl compounds like ethyl phenylglyoxalate (53) to ethyl mandelate (54) with modest transfer of chirality (eq. 26) [71]. This approach has subsequently been developed by Ohno [72] into an extremely efficient method for the transfer of chirality in some reductions. We felt that the selectivity

305

and efficiency of these derivatives of (52) could be improved by restricting the conformational flexibility by bridging as shown in (56), which with properly substituted bridges would have aspects of a crown ether. Our anticipation was that

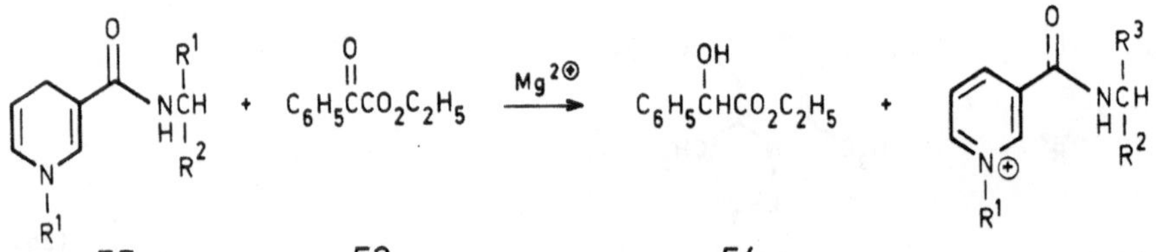

(26)

three point binding of an electrophilic catalyst would be possible using the two amide groups and a heteroatom in the bridge. If complexation of a carbonyl group occurred as illustrated in (56a) predictable enantioselectivity should result.

The chief synthetic problem in an approach to (56) would be the incorporation of the bridge thereby creating a macrocyclic ring. Numerous approaches to bridged pyridines have been described [73] but none met our needs. We therefore developed an independent route. We had been struck by the report that cesium-carboxylates in *DMF* are powerfully nucleophilic. [74] We were well rewarded on applying this report to commercially available pyridine-3,5-dicarboxylic acid (57), which could be converted into, for example, (58) in 90% yield (eq. 27) [75].

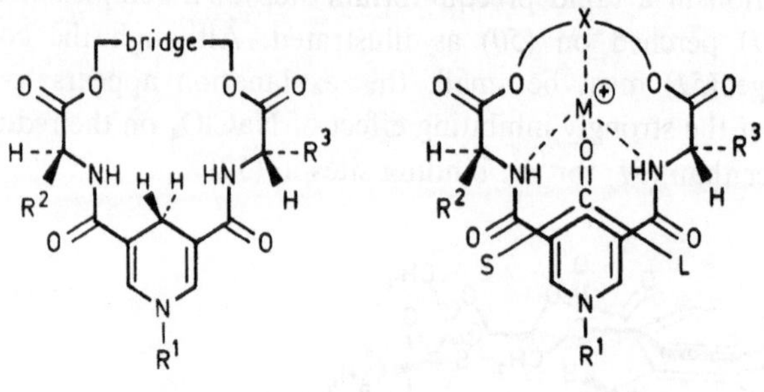

56

56-M-SCL

The use of cesium as anion activator in the formation of macrocycles has not

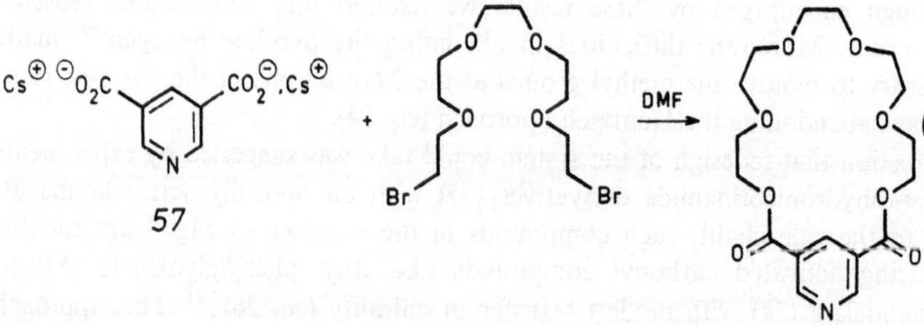

(27)

57

58

been reported previously [76]. This approach using cesium salts has subsequently been developed by us into a general method for the synthesis of macrolides, [77] macrocyclic bis-phenols, [78] macrocyclic sulfides, [79] and other applications [80].

Using the cesium salt approach the synthesis of compounds of general structure (56) could be accomplished in good yield. The synthetic approach is presented in eq. 28 for the synthesis of a specific compound (56a) [R = $(CH_3)_2CH$, bridge = $-(CH_2)_2O(CH_2)_2-$] [80]. It has been possible to prepare a series of compounds (56) in which the structural parameters of R group and length and shape of bridge have been varied. [31, 82] A considerable improvement in the synthetic scheme (eq. 28) has been the development of a direct coupling method for connecting the amino acids to the acid (eq. 29), which now allows multigram syntheses to be carried out.

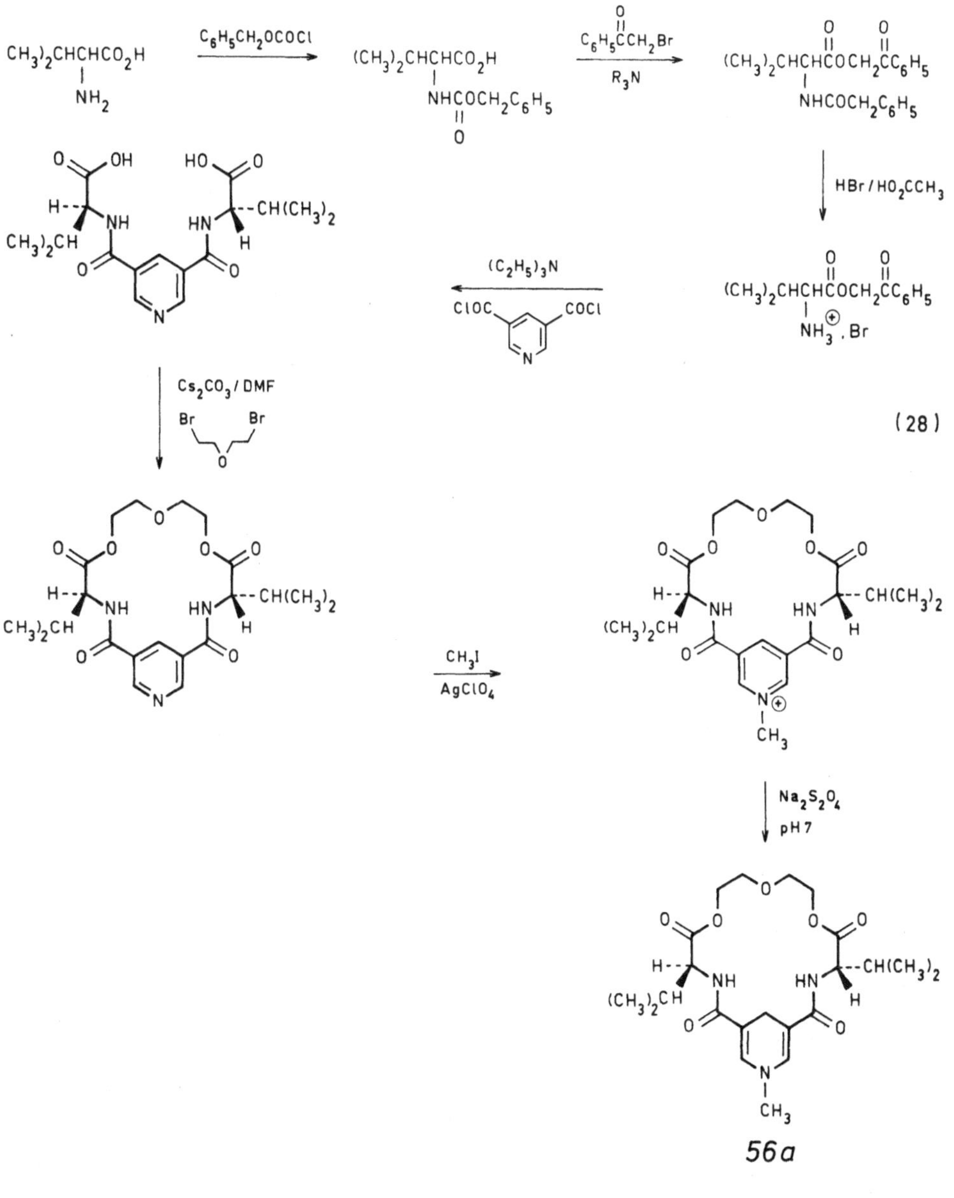

$$CH_3)_2CHCHCO_2H \quad \xrightarrow{C_6H_5CH_2OCOCl} \quad (CH_3)_2CHCHCO_2H \quad \xrightarrow[R_3N]{C_6H_5CCH_2Br} \quad (CH_3)_2CHCHCOCH_2CC_6H_5$$

with NH_2 → $NHCOCH_2C_6H_5$ → $NHCOCH_2C_6H_5$

$$\downarrow HBr/HO_2CCH_3$$

$$(CH_3)_2CHCHCOCH_2CC_6H_5$$
$$NH_3^{\oplus}.Br^-$$

$$\xleftarrow[ClOC \quad COCl]{(C_2H_5)_3N}$$

(28)

$$\downarrow Cs_2CO_3/DMF \quad Br\!-\!\!-\!O\!-\!\!-\!Br$$

56a

$$\downarrow CH_3I \quad AgClO_4$$

$$\downarrow Na_2S_2O_4 \quad pH\,7$$

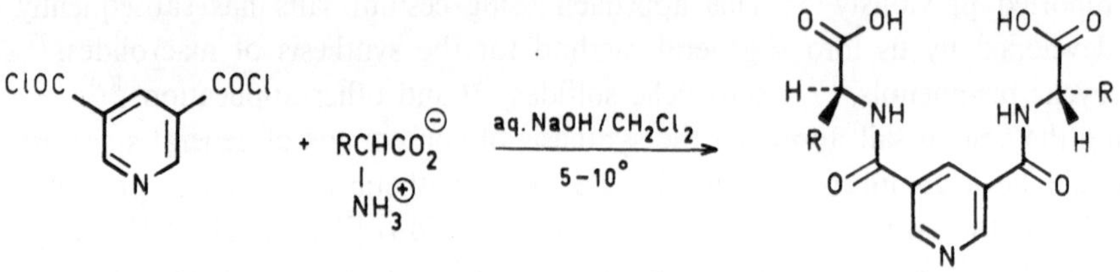

$$(29)$$

The compounds (56) are indeed capable of reducing activated carbonyl derivatives. The results of numerous experiments using valine, phenylalanine, and proline derived crown ethers are given in the Table [81,82]. The overall reaction is that shown in eq. 30.

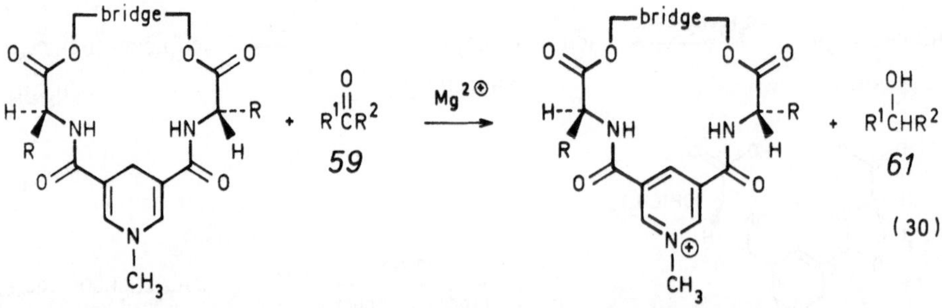

$$(30)$$

For comparison purposes noncyclic compounds 62a, b were also examined.

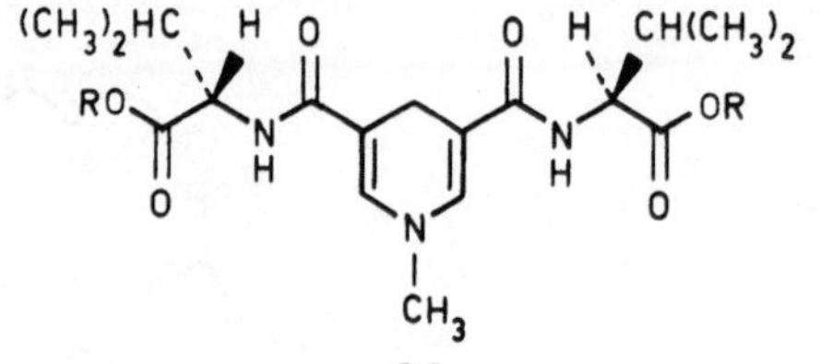

62

a) R = CH$_3$

b) R = CH$_3$O(CH$_2$)$_2$–

These reduce ketone (59) [R^3 = C$_6$H$_5$, R^2 = CO$_2$C$_2$H$_5$] cleanly to the corresponding alcohol (eq. 30) but in enantiomeric excesses of respectively 10% and 18% of the R-configuration.

The results reported in the Table are extremely promising. In contrast to the poor enantiomeric excesses obtained with (62a, b) all cyclic derivatives (56) except for proline (entry 19) give good to excellent transfer of chirality, in many cases high transfer. The results are completely stereochemically consistent in that (56) derived from L-amino-acids always produces S alcohols (the relative group priorities in all the reduced alcohols are identical allowing direct comparison). This points to similar stereochemical factors affecting the transition states for each reduction. These stereochemical factors can lead to good transfer of chirality although they

Table 1. Reduction of Activated Ketones (60) by 1,4-Dihydropyridines (56)

Entry	Amino Acid	Bridge	Major Enantiomer	Enantiomeric Excess	Substrate
1	L-valine	$-(CH_2)_2O(CH_2)-$	S	86%	$R^1=C_6H_5$; $R^2=CO_2C_2H_5$
2	L-valine	$-(CH_2)_2[O(CH_2)_2]_2-$	S	43	$R^1=C_6H_5$; $R^2=CO_2C_2H_5$
3	L-valine	$-(CH_2)_2[O(CH_2)_2]_3-$	S	54	$R^1=C_6H_5$; $R^2=CO_2C_2H_5$
4	L-valine	$-(CH_2)_4-$	S	±65	$R^1=C_6H_5$; $R^2=CO_2C_2H_5$
5	L-valine	$-(CH_2)_5-$	S	90	$R^1=C_6H_5$; $R^2=CO_2C_2H_5$
6	L-valine	$-(CH_2)_6-$	S	88	$R^1=C_6H_5$; $R^2=CO_2C_2H_5$
7	L-valine	$-(CH_2)_8-$	S	83	$R^1=C_6H_5$; $R^2=CO_2C_2H_5$
8	L-valine	$-(CH_2)_{10}-$	S	53	$R^1=C_6H_5$; $R^2=CO_2C_2H_5$
9	L-valine	$-(CH_2)_{12}-$	S	42	$R^1=C_6H_5$; $R^2=CO_2C_2H_5$
10	L-valine	$m-CH_2C_6H_4CH_2-$	S	86	$R^1=C_6H_5$; $R^2=CO_2C_2H_5$
11	D-valine	$-(CH_2)_6-$	R	85	$R^1=C_6H_5$; $R^2=CF_3$
12	L-valine	$-(CH_2)_2O(CH_2)_2-$	S	68	$R^1=C_6H_5$; $R^2=CONH_2$
13	L-valine	$-(CH_2)_2O(CH_2)_2-$	S	64	$R^1=C_6H_5$; $R^2=CONHC_2H_5$
14	L-valine	$-(CH_2)_2O(CH_2)_2-$	S	78	$R^1=C_6H_5$; $R^2=CO_2C_2H_5$
15	L-phenylalanine	$-(CH_2)_2O(CH_2)_2-$	S	87	$R^1=C_6H_5$; $R^2=CONHC_2H_5$
16	L-phenylalanine	$-(CH_2)_2O(CH_2)_2-$	S	84	$R^1=C_6H_5$; $R^2=CO_2C_2H_5$
17	L-phenylalanine	$-(CH_2)_2O(CH_2)_2-$	S	60	$R^1=m-C_6H_4OC_6H_5$; $R^2=CO_2CH_3$
18	L-phenylalanine	$-(CH_2)_2O(CH_2)_2-$	S	20	$R^1=m-C_6H_4OC_6H_5$; $R^2=CONH_2$
19	L-proline	$-(CH_2)_2O(CH_2)_2-$	—	0	$R^1=C_6H_5$; $R^2=CO_2C_2H_5$

do not produce in any of these systems significant rate accelerations (quantitative measurements have, however, not been carried out). All of the reductions are quite clean and give the alcohols (*61*) in usually about 60–80% yield after work-up and isolation (yields determined by ^{1}H-NMR are usually 85–90%).

Two of the most important conclusions to be drawn from these results are: (a) the original model (*56a*) used for the design of (*56*) must be incorrect. Comparison of entries 1–3 with entries 4–10 reveals that the enantiomeric excesses are not significantly influenced by the presence of a heteroatom as coordinating site in the bridge. (b) For bridges not longer than eight atoms (entries 1, 4–7, 10, 12–17) almost without exception irregardless of shape (entry 10) or content of the bridge high enantiomeric excesses are found.

Apparently the compounds (*56*) deviate so strongly from classical crown ether structures that other complexation factors — leading nevertheless to high enantioselectivity — become important. Because only weak complexes are formed in solution interpretation of the results remains speculative. We ascribe, however, the formation of S-alcohols from L-amino acid derivative (*56*) to the formation of a ternary complex having the structure crudely represented in (*63*). The main stereochemical

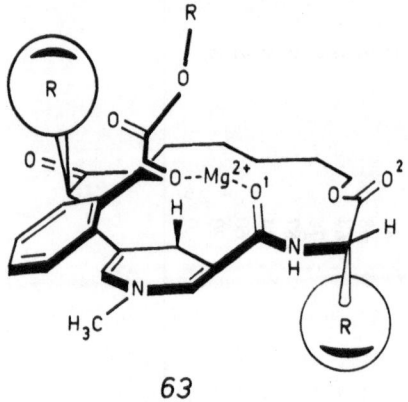

63

features are the amino acid substituents (no significant differences are seen between the isopropyl of valine compared to the benzyl of phenylalanine, both of which provide steric hindrance on *one* side of the molecule). From ^{13}C studies of the complexation of (*56*) with $Mg^{2\oplus}$ we conclude that 0–1 binds with the $Mg^{2\oplus}$ ion. [83] It is sterically attractive to allow 0–2 also to coordinate to $Mg^{2\oplus}$ although there is no direct spectroscopic evidence for this. This complexation of $Mg^{2\oplus}$ is *sterically directed*, i.e. it occurs specifically from the *least hindered* side of 0–1 and introduces a conformational change as the carbonyl groups move out of plane to optimal coordinating positions. The positioning of the substrate with the carbonyl carbon over the dihydropyridine "hydride" to be transferred is dictated by the steric bulk of the amino acid substituent. Examination of CPK models reveals that the phenyl group of, say, ethyl phenylglyoxalate (*53*) fits well only when it is placed over the dihydropyridine ring. This leads to transfer of hydride to the *re*-face of the carbonyl group producing the S-enantiomer of the alcohol in accord with experimental observation.

The drop in enantiomeric excesses of alcohol that occurs when the length of the bridge exceeds eight atoms is probably a consequence of increased conformational mobility of the macrocyclic ring, which allows the amino acid substituent to adopt conformations with it lying in the plane of the dihydropyridine ring. These conformational movements change the morphology of (56) with a consequent loss of steric discrimination for complexing of the carbonyl substrate. A similar effect is probably also involved for the proline derivative (entry 19, Table), which, owing to the 5-membered ring, adopts a heavily flattened conformation.

5.5 Macrocycles Capable of Catalyzing Other Bond-Forming or Bond-Breaking Reactions

Despite the obvious potential relatively few examples have been described of the catalysis of other reactions by macrocyclic compounds bearing reactive functional groups. One example is the promotion of macrolide formation (i.e. 65→66) [84] by functionalized crown ether (64) as shown in eq. 31. [85] Reasonable yields of macrolides are obtained for n of the order of 10 and m = 2 in (64). The intention of the approach of eq. 31 is to hold the nucleophilic alkoxide close to the carbonyl group to be attacked by chelation to $K^{\oplus}$ in the crown ether ring. Overcoming the entropy effect involved in *intramolecular* cyclization is the major problem of macrolide synthesis. [86] To accomplish this with (64) it is necessary to choose m and n so that optimal coordination can be achieved for the (presumed) tetrahedral

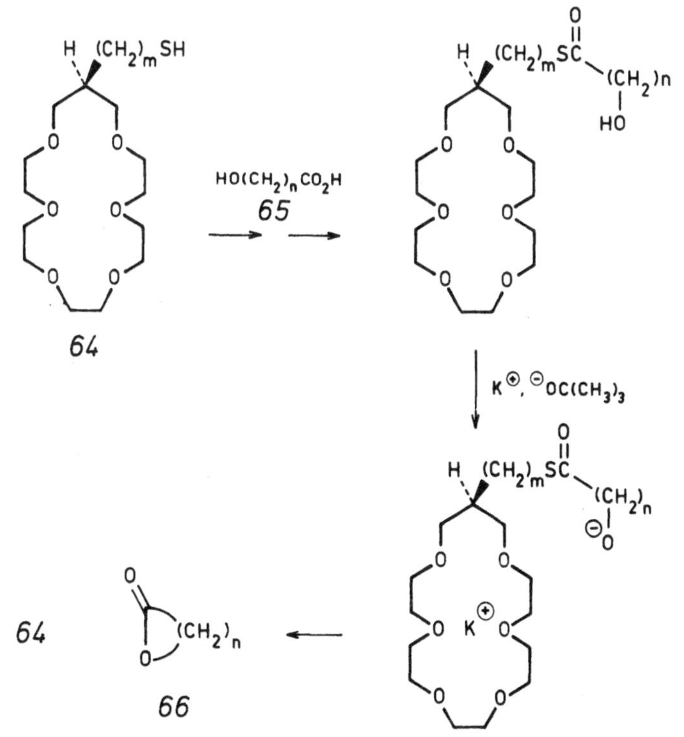

(31)

311

intermediate formed on attack of carbonyl by alkoxide as depicted schematically in eq. 32. This amounts to complementarity in the transition state for the reaction.

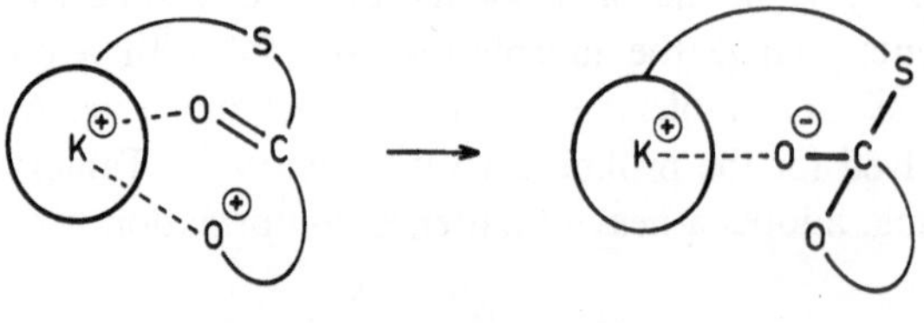

(32)

The potential for the catalysis by bioorganic modelling with macrocycles is enormous and is limited chiefly by ingenuity. For example, interesting possibilities for the future are held in sugar derived crown ethers like (23). [30] Likewise the light driven conformational changes seen in crown ethers (67) or cryptates (68) as shown in eqs. 33 and 34 offer an exciting basis for the development of catalytic systems [87,88].

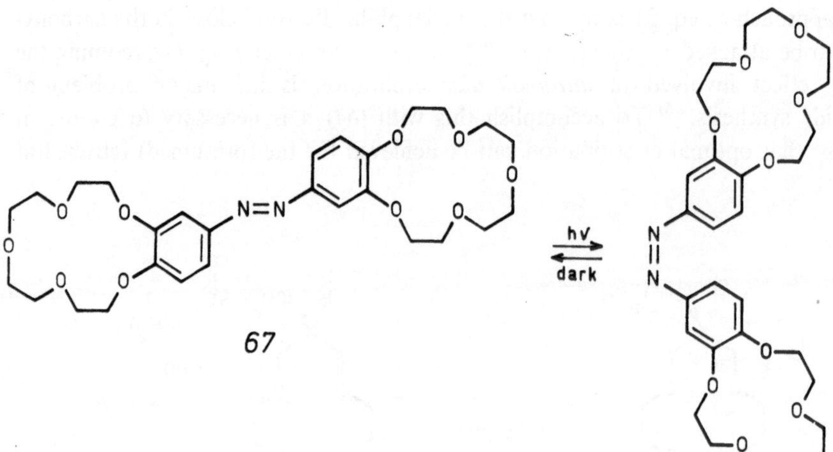

67

(33)

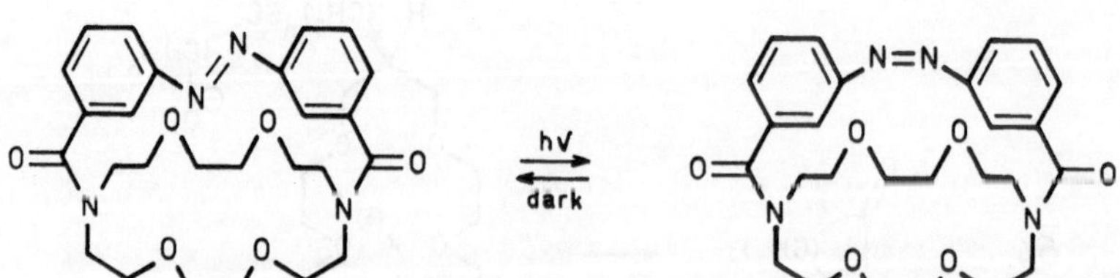

68

(34)

Other prospects are opened by binuclear complexes like (69), which is capable of complexing two copper ions. [89] The cavity between the two copper ions should

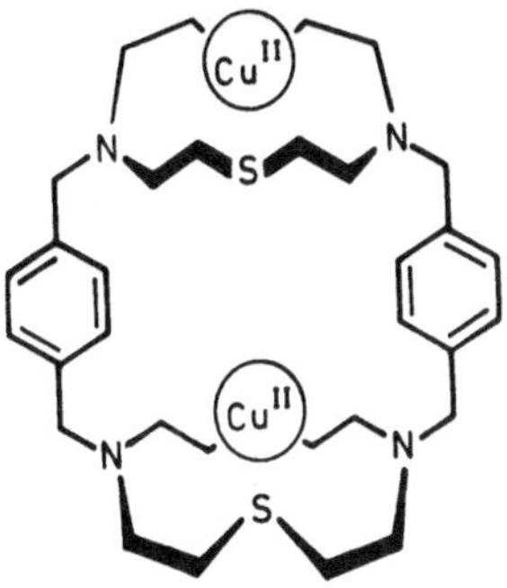

69

be large enough to accept a substrate offering thereby the opportunity to model oxygenase activity [90].

Finally, positively charged complexes linke (70) [91] are mentioned for their possibilities in the complexing and activation of anions. The development of general

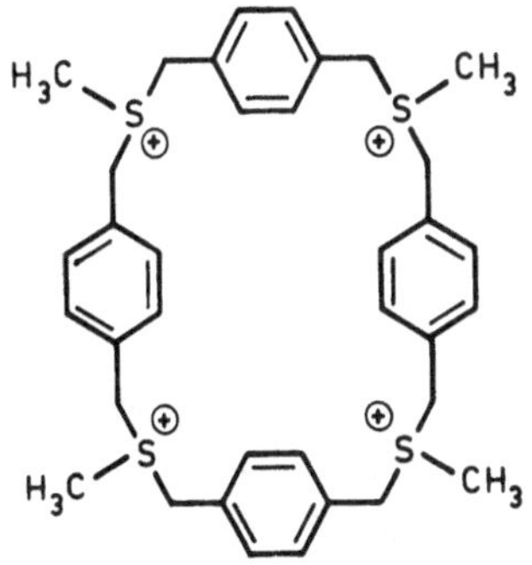

70

synthetic routes to macrocyclic sulfides with conformationally flexible chains [92] offers now the opportunity to develop new anion activating systems.

6 Summary

In terms of the reporting of accomplished chemistry this review can do no more than give an indication of the rapid progress in the branch of bioorganic modelling based on the use of macrocyclic compounds that (usually) act as complexing agents. What remains to be done, however, is to point out problems that have not been satisfactorily solved and to suggest other profitable areas of investigation.

From the material accumulated in this review one can draw the conclusion that especially crown (or cryptate) systems offer special advantages in bioorganic modelling because such compounds can — enzyme like — complex a potential substrate. On the basis of quite simple binding considerations, coupled with an analysis of steric interactions, accurate predictions of the stereochemistry of the complex can be made. The inclusion of catalytic groups in the crown (or cryptate) system and reactive functional groups in the substrate is then done in such a fashion that the stereoelectronic arrangement is compatible with the predicted geometry of the complex. However, the good complexing ability of the ligand is paradoxically often its greatest failing in terms of developing a system in which the functionalized ligand acts truly as a catalyst. As seen from much of the chemistry discussed in this review the ligand is incapable of the double task of complexing substrate but *releasing* product in an enzymic fashion, i.e. that *turnover* occurs.

How is this problem to be solved? Induced conformational changes are an obvious approach although the design of proper systems remains a challenge for which few suggestions outside of unlimited ingenuity can be given. Much of the solution to such problems will lie also in a much better understanding than we now have of *non-covalent* interactions and the stereochemistry of such interactions. [93] The assembly and disassembly of large molecular aggregates by the making and dissolution of non-covalent bonds is an art at which chemists are still relative amateurs.

A better understanding of non-covalent interactions may also provide the key to achieving also the twin goals of *both* speed and selectivity in bioorganic modelling. As far as enantioselectivity is concerned it is clear that this can be achieved fairly effectively by the use of relatively small, but appropriately placed, groups that force the substrate to complex in an enantioselective step with the ligand. In other words, the problem of enantioselectivity can be solved at the stage of *complex forming*, which is kinetically rapid. The problem of rate enhancement lies in the complex itself and can be solved effectively only by achieving optimal complementarity with the transition state of the reaction being catalyzed. Again the achievement of this goal lies in ingenuity of design.

Potential areas of applications of chiral crown ether (or cryptate) ligand systems in bioorganic modelling lie in, for example, the formation of carbon-carbon bonds, development of oxidative processes (i.e. $NAD^{\oplus}$-like mediated dehydrogenations of alcohols), synthesis of metal (perhaps binuclear) complexes with oxygenase activity, and the development of compounds with group transfer abilities. This short list includes only a few of the exciting challenges for the future that with ingenuity and imagination can be solved.

7 Acknowledgement

The work from my group has been carried out chiefly by T. J. van Bergen, J. Buter, P. Jouin, B. J. van Keulen, W. H. Kruizinga, P. Piepers, C. B. Troostwijk, and J. G. de Vries. To these coworkers and others more indirectly connected with the research reported here I owe a debt of gratitude for their enthusiasm, hard work, ideas, and good fellowship.

8 References

1. Ruzicka, L.: Über den Bau der Organischen Materie, publisher I. van Druten, Utrecht, 1926. I thank Prof. V. Prelog, E. T. H. Zürich, for making me aware of this inaugural address and Dr. A. Fürst, Hoffmann-La Roche Basel, for providing a copy

2. Lehninger, A. L.: Biochemistry, 2nd ed., Worth Publishers, Inc., New York, N.Y., (1975)
3. Pauling, L.: American Scientist *36*, 58 (1948)
4. Lienhard, G. E.: Science *180*, 149 (1973)
5. For example: (a) van Tamelen, E. E.: Acc. Chem. Res. *5*, 152 (1975), (b) Johnson, W. S.: ibid. *1*, 1 (1968)
6. (a) Robinson, R.: J. Chem. Soc. *111*, 762 (1917), (b) Robinson, R.: J. Chem. Soc. *1953*, 999
7. (a) Visser, C. M., Kellogg, R. M.: J. Mol. Evol. *11*, 163 (1978), (b) Visser, C. M., Kellogg, R. M.: ibid. *11*, 171 (1978), (c) Visser, C. M.: Naturwissenschaften *67*, 549 (1980), (d) Visser, C. M.: Bioorgan. Chem. *9*, 261 (1980)
8. Visser, C. M., Kellogg, R. M.: Bioorg. Chem. *6*, 79 (1977)
9. Enders, D., Eichenauer, H.: Angew. Chem., Int. Ed., Engl. *18*, 397 (1979)
10. (a) Meyers, A. I.: Acc. Chem. Res. *11*, 375 (1978), (b) Meyers, A. I., Knaus, G., Kamata, K., Ford, M. E.: J. Am. Chem. Soc. *98*, 567 (1976)
11. Hiemstra, H., Wynberg, H.: ibid. *103*, 417 (1981)
12. Katsuki, T., Sharpless, K. B.: ibid. *102*, 5974 (1980)
13. (a) Cramer, F., Saenger, W., Spatz, H.: ibid. *89*, 14 (1967), (b) James, W. J., French, D., Rundle, R. E.: Acta Crystal. *12*, 385 (1959)
14. For early work, see: (a) Cramer, F.: Angew. Chem. *64*, 136 (1952), (b) Cramer, F., Dietsche, W.: Chem. Ber. *92*, 1739 (1959), (c) Flohr, K., Paton, R. M., Kaiser, E. T.: J. Chem. Soc. Chem. Commun. 1971, 1621
15. (a) Bender, M. L., Komiyama, M.: Cyclodextrin Chemistry, in Reactivity and Structure in Organic Chemistry, vol. 6, New York, 1978, (b) Breslow, R.: Acc. Chem. Res. *13*, 170 (1980)
16. See, for example: Jencks, W. P.: Catalysis in Chemistry and Enzymology, McGraw Hill, New Yok, N.Y. (1969)
17. For example: Storm, D. R., Strominger, J. L.: J. Biol. Chem. *248*, 3940 (1973)
18. Elmore, D. T., Smith, J. J.: Biochem. J. *94*, 563 (1965)
19. (a) Sheehan, J. C., Bennett, G. B., Schneider, J. A.: J. Am. Chem. Soc. *88*, 3455 (1966), (b) Cruickshank, P. A., Sheehan, J. C.: ibid. *86*, 2070 (1964)
20. Julia, S., Masana, J., Vega, J. C.: Angew. Chem. *92*, 968 (1980)
21. Sugimoto, T., Matsumura, Y., Tanimoto, S., Okano, M.: J. Chem. Soc. Chem. Commun. *1978*, 926
22. Sugimoto, T., Kokubo, T., Miyazaki, J., Tanimoto, S., Okano, M.: J. Chem. Soc., Chem. Commun. *1979*, 402
23. For example: (a) Cram, D. J. Cram, J. M.: Science *183*, 803 (1974), (b) Cram, D. J., Cram, J. M.: Acc. Chem. Res. *11*, 8 (1978)
24. For example: (a) Busch, D. N.: Rec. Chem. Prog. *25*, 107 (1964), (b) Eschenmosher, A.: Pure Appl. Chem. *20*, 1 (1969)
25. Nagano, O., Kobayashi, A., Sasaki, Y.: Bull. Chem. Soc. Jpn. *51*, 790 (1978)
26. Chao, Y., Weisman, G. R., Sogah, G. D. Y., Cram, D. J.: J. Am. Chem. Soc. *101*, 4948 (1979) and many previous papers cited therein
27. Cram, D. J. et al.: J. Org. Chem. *43*, 1930 (1978)
28. (a) Girodeau, J. M., Lehn, J. M., Sauvage, J. P.: Angew. Chem. *87*, 813 (1975), (b) Behr, J. P., Lehn, J. M., Vierling, P.: J. Chem. Soc., Chem. Commun. *1976*, 621
29. See, for example: Seebach, D., Hungerbühler, E. in: Modern Synthetic Methods 1980 (ed. Scheffold, R.), Otto Selle Verlag, Frankfurt, 1980, p. 91
30. Review, Stoddart, J. F.: Chem. Soc. Rev. *8*, 85 (1979)
31. Jouin, P., Troostwijk, C. B., Kellogg, R. M.: J. Am. Chem. Soc. *103*, 2091 (1981)
32. (a) Drenth, J. et al.: Nature *218*, 929, (b) op. cit., Phil. Trans. Roy. Soc. Ser. B. *2579*, 231 (1970)
33. See, for example: Walsh, C.: Enzymatic Reaction Mechanisms, W. H. Freeman and Co., San Francisco, 1979
34. (a) Behr, J. P., Lehn, J. M., Vierling, P.: J. Chem. Soc. Chem. Commun. *1976*, 621, (b) Lehn, J. M.: Pure and Appl. Chem. *50*, 871 (1978), (c) Behr, J. P. et al.: Helv. chim. Acta *63*, 2096 (1980)
35. Matsui, T., Koga, K.: Tetrahedron Lett. *1978*, 1115
36. Sasaki, S., Koga, K.: Heterocycles *12*, 1305 (1979)
37. Van Keulen, B. J., Kellogg, R. M.: unpublished observations

38. Ben-Naim, A.: Hydrophobic Interactions, Plenum Press, New York, N.Y., 1980
39. Murakami, Y., Sunamoto, J., Kano, K.: Chem. Lett. *1973*, 223
40. Guthrie, J. P.: J. Chem. Soc., Chem. Commun. *1972*, 897
41. See, for example: (a) Blyth, C. A., Knowles, J. R.: J. Am. Chem. Soc. *93*, 3017, 3021 (1971), (b) Hershfield, R., Bender, M. L.: ibid. *94*, 1376 (1972)
42. Murikami, Y. et al.: Bull. Chem. Soc. Japan *50*, 3365 (1977)
43. Murakami, Y. et al.: J. Chem. Soc., Perkin I *1979*, 1560
44. White, H. B., III: Pyridine Nucleotide Coenzymes (ed. J. Everse, B. M. Anderson, K. S. You) in press
45. For example: Fischli, A., Süss, D.: Helv. chim. Acta *62*, 48 (1979)
46. Fischli, A.: ibid. *62*, 882 (1979)
47. Fischli, A., Süss, D.: ibid. *62*, 2361 (1979)
48. Fischli, A., Müller, P. M.: ibid. *63*, 529 (1980)
49. (a) Fischli, A., Müller, P. M.: ibid. *63*, 1619 (1980), (b) Fischli, A., Daly, J. J.: ibid. *63*, 1628 (1980) and references cited therein
50. For example: (a) Karrer, P. et al.: Helv. chim. Acta *20*, 55 (1937), (b) Rafter, G. W., Colowich, S. P.: J. Biol. Chem. *209*, 773 (1954), (c) Abeles, R. H., Hutton, R. F., Westheimer, F. H.: J. Am. Chem. Soc. *79*, 712 (1957), (d) Mauzerall, D., Westheimer, F. H.: ibid. *77*, 2261 (1955), (e) review of dihydropyridines: Eisner, U., Kuthan, J.: Chem. Rev. *72*, 1 (1972)
51. (a) Jones, J. B., Taylor, K. E.: Can. J. Chem. *54*, 2974, 2969 (1976), (b) Taylor, K. E., Jones, J. B.: J. Am. Chem. Soc. *98*, 5687 (1976)
52. For discussions of enzyme action, see for example: (a) Eklund, H. et al.: FEBS Lett. *44*, 200 (1974), (b) Adams, M. J. et al.: Proc. Natl. Acad. Sci. USA *1974*, 1968, (c) Hill, E. et al.: J. Mol. Biol. *72*, 577 (1972), (d) Lazdunski, M.: in Prog. Bioogr. Chem. *3*, 81 (1974)
53. For example: (a) Ohno, A. et al.: J. Chem. Soc., Chem. Commun. *1978*, 328, (b) Gase, R. A., Pandit, U. K.: J. Am. Chem. Soc. *101*, 1979 (1979)
54. For example: Baba, N., Oda, J., Inouye, Y.: J. Chem. Soc., Chem. Commun. *1980*, 815 and references cited therein
55. (a) Yarmolinsky, M. B., Colowick, S. P.: Biochim. Biophys. Acta *20*, 177 (1956), (b) Biellmann, J. F., Callot, H. J.: Bull. Soc. Chim. Fr. *1968*, 1154
56. de Vries, J. G., Kellogg, R. M.: J. Org. Chem. *45*, 4126 (1980)
57. For some applications, see: Pandit, U. K. et al.: J. Chem. Soc., Chem. Commun. *1974*, 627 and references cited therein
58. Behr, J. P., Lehn, J. M.: J. Chem. Soc., Chem. Commun. *1978*, 143
59. For example: (a) Colowick, S. P. et al.: J. Biol. Chem. *195*, 95 (1952), (b) Kaplan, N. O., Colowick, S. P., Neufeld, E. F.: ibid. *195*, 107 (1952), (c) Kaplan, N. O., Colowick, S. P., Neufeld, E. F. ibid., *205*, 1 (1953), (d) Kaplan, N. O., Colowick, S. P., Neufeld, E. F.: Fed Proc. Fed. Am. Soc. Exp. Biol. *11*, 238 (1952), (e) Cilento, G.: Arch. Biochem. Biophys. *88*, 352 (1960), (f) Spiegel, M. J., Drysdale, G. R.: J. Biol. Chem. *235*, 2498 (1960), (g) Ludowieg, J., Levy, A.: Biochemistry *3*, 373 (1964), (h) van Bergen, T. J., Mulder, T., Kellogg, R. M.: J. Am. Chem. Soc. *98*, 1960 (1976), (i) Bergen, T. J., Mulder, T., van der Veen, R. A., Kellogg, R. M.: Tetrahedron *34*, 2377 (1978)
60. These hydride exchange processes can also be used to establish redox potentials for 1,4-dihydropyridines: Piepers, O., Kellogg, R. M.: publication in preparation
61. Behr, J. P., Lehn, J. M.: Helv. chim. Acta *63*, 2112 (1980)
62. Ash, R. P., Herriott, J. R., Deranleau, D. A.: J. Am. Chem. Soc. *99*, 4471 (1977)
63. Kosower, E. M.: ibid. *78*, 3497 (1956). But see also: Piepers, O., Kellogg, R. M.: J. Chem. Soc., Chem. Commun. *1980*, 1154
64. Hantzsch, A.: Liebigs Ann. Chem. *215*, 1 (1882)
65. Kellogg, R. M. et al.: J. Org. Chem. *45*, 2854 (1980)
66. Haley, C. A. C., Maitland, P.: J. Chem. Soc. *1951*, 3155
67. van Bergen, T. J., Kellogg, R. M.: J. Am. Chem. Soc. *99*, 3882 (1977)
68. van der Veen, R. H., Kellogg, R. M., Vos, A., van Bergen, T. J.: J. Chem. Soc., Chem. Commun. *1978*, 923
69. van Bergen, T. J., Kellogg, R. M.: unpublished
70. van Bergen, T. J., Hedstrand, D. M., Kruizinga, W. H., Kellogg, R. M.: J. Org. Chem. *44*, 4953 (1979)

71. (a) Pandit, U. K., Mas Cabré, F. R.: J. Chem. Soc., Chem. Commun. *1971*, 552, (b) Ohnishi, Y., Kagami, M., Ohno, A.: J. Am. Chem. Soc. *97*, 4766 (1975), (c) Endo, T., Hayashi, Y., Okawara, M.: Chem. Lett. *1977*, 391, (d) Endo, T., Kawasaki, H., Okawara, M.: Tetrahedron Lett. *1979*, 23, (e) Ohnishi, Y., Numakunai, T., Ohno, A.: ibid. *1975*, 3813, (f) van Ramesdonk, H. J., Verhoeven, J. W., Pandit, U. K., de Boer, T. J.: Rec. trav. Chim. Pays-Bas *97*, 195 (1978)
72. (a) Ohno, A., Ikeguchi, M., Kimura, T., Oka, S.: J. Chem. Soc., Chem. Commun. *1978*, 328, (b) Ohno, A., Ikeguchi, M., Kimura, T., Oka, S.: J. Am. Chem. Soc. *101*, 7036 (1979)
73. For examples and literature citations, see: (a) Newkome, G. R., Kawato, T.: *44*, 2693 (1979), (b) Newkome, G. R., Kawato, T., Nayak, A.: ibid. *44*, 2697 (1979), (c) Heimann, U., Vögtle, F.: Angew. Chem. *90*, 211 (1978), (d) Izatt, R. M. et al.: J. Am. Chem. Soc. *99*, 2365 (1977), (e) Newkome, G. R. et al.: Chem. Rev. *77*, 513 (1977), (f) Bradshaw, J. S., Stott, P. E.: Tetrahedron *36*, 461 (1980)
74. Wang, S. S. et al.: J. Org. Chem. *42*, 1286 (1977)
75. Piepers, O., Kellogg, R. M.: J. Chem. Soc., Chem. Commun. *1978*, 383
76. But see the use of CsF in the synthesis of crown ethers: Reinhoudt, D. N., de Jong, F., Tomassen, H. P. M.: Tetrahedron Lett. *1979*, 2067
77. Kruizinga, W. H., Kellogg, R. M.: J. Chem. Soc., Chem. Commun. *1979*, 286, J. Am. Chem. Soc. in press
78. van Keulen, B. J., Kellogg, R. M., Piepers, O.: J. Chem. Soc., Chem. Commun. *1979*, 235
79. Buter, J., Kellogg, R. M.: ibid. *1980*, 466
80. Strijtveen, B., Vriesma, B., Kellogg, R. M.: unpublished
81. de Vries, J. G., Kellogg, R. M.: J. Am. Chem. Soc. *101*, 2759 (1978)
82. Jouin, P., Troostwijk, C. B., Weremus Buning, G., Talma, A. G.: publication in preparation
83. Jouin, P., de Vries, J. G., Kellogg, R. M.: unpublished
84. Interest in macrolide formations is intense. For recent reviews, see: (a) Nicolau, K. C.: Tetrahedron *23*, 683 (1977), (b) Masamune, S., Bates, G. S., Corcoran, J. W.: Angew. Chem. Int. Ed. *16*, 585 (1977), (c) Back, T. G.: Tetrahedron *33*, 3041 (1977)
85. (a) Rastetter, W. H., Phillion, D. P.: Tetrahedron Lett. *1979*, 1469, (b) Rastetter, W. H., Phillion, D. P.: J. Org. Chem. *45*, 1535 (1980)
86. See, for another approach ref. 77
87. (a) Shinkai, S., Ogawa, T., Kusano, Y., Manabe, O.: Chem. Lett. *1980*, 283, (b) Shinkai, S. et al.: Tetrahedron Lett. *1979*, 4569, (c) Shinkai, S. et al.: J. Chem. Soc., Chem. Commun. *1980*, 375, (d) Shinkai, S. et al.: Tetrahedron Lett. *1980*, 4463, (e) Shinkai, S. et al.: J. Am. Chem. Soc. *102*, 5860 (1980)
88. For a somewhat related application in cyclodextrin chemistry, see: Ueno, A., Takahashi, K., Osa, T.: J. Chem. Soc., Chem. Commun. *1980*, 837. A somewhat different application is described by: Kojima, M., Toda, F., Hattori, K.: Tetrahedron Lett. *1980*, 2721
89. Kahn, O. et al.: J. Am. Chem. Soc. *102*, 5935 (1980)
90. See, for example: Vanneste, W. H., Zuberbühler, A.: in Molecular Mechanisms of Oxygen Activation (ed. O. Hayaishi), Academic Press, New York, N.Y., 1974, pp. 371—404
91. Tabushi, I., Sasaki, H., Kuroda, Y.: J. Am. Chem. Soc. *98*, 5727 (1976)
92. Buter, J., Kellogg, R. M.: J. Chem. Soc., Chem. Commun. *1980*, 466
93. See, for example, the interview with Lord Todd: Chem. Eng. News. *58*, 28, Oct. 6 (1980), and also Lehn, J. M.: Structure and Bonding *16*, 1 (1973)

Complexation of Uncharged Molecules and Anions by Crown-Type Host Molecules

Fritz Vögtle, Heinz Sieger, Walter Manfred Müller

Institut für Organische Chemie und Biochemie der Universität Bonn,
Gerhard-Domagk-Str. 1, D-5300 Bonn 1, FRG

Table of Contents

Fritz Vögtle, Heinz Sieger and Walter Manfred Müller

1 Introduction: Significance and Problems of Host-Guest Complexation Involving Uncharged Species

Definite complexes between various uncharged molecules are still relatively unexplored [1] although this type of complexation plays a fundamental role in biochemical processes, e.g. the base-pairing of nucleic acids [2], enzyme/substrate or pharmacon/receptor interactions. "*The molecular biology is a striking example of the fact that molecules which fit into one another sterically and functionally, lead together to fundamentally new functional possibilities*" [1d]. A suitable approach in this respect is "*the planning and synthesis of arrangements of various types of molecules, which are all constructed such that they fit into one another and thus by themselves and in an organized way build up to functional systems where the functional groups are held fast in the desired arrangement by the intermolecular forces*" [1d].

Though many data on hydrogen bond interactions have been gathered, it is still difficult today to suggest specific *hosts* (receptors) for the complexation of neutral (uncharged) *guest* molecules. In spite of the fact that much experience on other types of complex binding between organic neutral molecules has been gained, e.g. on the intermolecular charge-transfer complexation [3], there is a lack of knowledge about concave host substances which can clasp or enclose convex neutral molecules with the formation of a cavity or pseudocavity.

This survey mainly intends to give a description of the experimental and theoretical work hitherto undertaken to enclose small uncharged *guest molecules* in *host cavities* or -pseudocavities [4]. It should be noted that for a few complexes it is difficult to decide or has not been clarified whether the guest molecule is enclosed in an *intramolecular cavity* of a single host molecule or located in *intermolecular cavities* of the host lattice. This can usually only be clarified by X-ray structural analysis.

The emphasis in this report lies on new possibilities of inclusion with the smaller unimolecular cavities of *coronands* [5] (crown compounds), *cryptands* [5], *catapinands* and *podands* [4a, 6] (open chain crown-type compounds). The *neutral molecule complexes* of these compounds [7] are to be understood as adducts with small molecular particles; these adducts are not held together primarily by electrostatic (ionic) interactions. Because of their integral, simple stoichiometry they seem not to belong to the urea host lattice-type inclusions [8] or to the dianin compound type containing *intermolecular cavities* [9]. Thus, the numerous inclusion cavities or channels of urea [8] and thourea [8, 9], dianin [10], perhydrotriphenylene [11], tri-o-thymotide [12], deoxycholeic acids [13], β-hydroquinone [14], guayacanin [15], triphenylmethane [16], tris(o-phenylene-dioxy)cyclotriphosphazane [17] etc. [1b, c, f, g, j, l] are not described here. The inclusion compounds of α-, β- and γ-cyclodextrins and their derivatives [18] representing readily available large *intra*molecular cavities that can be used for enzyme-analogous stereoselective reactions, are also not reported.

The inclusion compounds, which are formed with two-dimensionally cross-linked *host lattices* such as those of clay minerals, graphite and mica, have been reported elsewhere [1c, f, k, l].

Fritz Vögtle, Heinz Sieger and Walter Manfred Müller

2 Monocyclic Crown Compounds as Hosts for Uncharged Guest Molecules

2.1 Thiourea, Urea and Related Polar NH-Bond-Containing Guest Molecules

Pedersen first reported crystalline complexes between uncharged molecules such as *thiourea* and crown ethers [19]. The stoichiometriy of the isolated crown complexes of thiourea, N-phenylthiourea, 1-phenylsemicarbazide, 1-phenylthiosemicarbazide, 4-phenylsemicarbazide, 2-thiazolinedione, thiobenzamide, and other neutral guest molecules (cf. Table 1) lies between one molecule of polyether and one to six molecules of the other components; a rule for these varying ratios could not be found. Pedersen

Table 1. Crystalline urea, thiourea and related complexes of the coronands *1–15*

Host	Guest	Stoichiometry Host:Guest	M.p. [°C]	Ref.
2a	thiourea	1:4	152–164	[19]
	thioacetamide	1:2	84–87	[33]
	thiobenzamide	1:2	81–82	[19]
2b	thiourea	1:1	127	[19]
3	thiourea	1:1	115–117	[19]
4	thiourea	3:1	123–124	[19]
5a	thiourea	1:1	165–166	[19]
5b	thiourea	2:7	105–106	[19]
6	thiourea	1:6	178–180	[19]
7	thiourea	1:6	175–180	[19]
		1:5	167–174	[19]
8	thiourea	1:6	168–173	[19]
	thiourea	1:6	175–176	[19]
	thiourea	1:6	189–193	[19]
8 (isomer A)	thiourea	1:6	168–172	[19]
8 (isomer B)	thiourea	1:6	197–198	[19]
8	thiobenzamide	1:2	154–156	[19]
	N-phenylthiourea	1:2	179–180	[19]
	1-phenylsemicarbazide	1:2	150–158	[19]
	1-phenylthiosemicarbazide	1:1	155–165	[19]
	4-phenylthiosemicarbazide	1:2	144–145	[19]
	2-thiazolidinethione	1:2	125–127	[19]
9c	formamide	1:2	92–97	[28a]
	acetamide	1:2	55–58	[33]
	thioacetamide	1:2	111, 121–123	[33]
	benzenesulfonamide	1:2	95	[21]
	urea	1:4	145–148	[20]
	thiourea	1:4	168–170	[20]
	dithiooxamide	1:1	149–154	[20]
10b	thiourea	1:4	145–148	[20]
12a	formamide	1:1	130, 139–141	[28a]
15	formamide	3:2	140, 152–159	[28a]

assumed that the size of the guest molecule (6–7 Å) and the stoichiometries exclude the possibility of a traditional inclusion in the urea lattice [8b, c] and that the exact nature of these complexes is unclear.

Of interest is that both *urea* and *thiourea* raise the solubility of *dibenzo[18]crown-6* (*5a*) in solution, too. Nevertheless, "*it is interesting that, in spite of the effect of urea ...*" crystalline complexes of *urea* with polyethers were not obtained when urea was handled in exactly the same way as thiourea [19]. When the crown ether ring, as in dibenzo[14]-crown-4 (*1*), is too small for the potential "guest" [19], a crystalline complex is not obtained either.

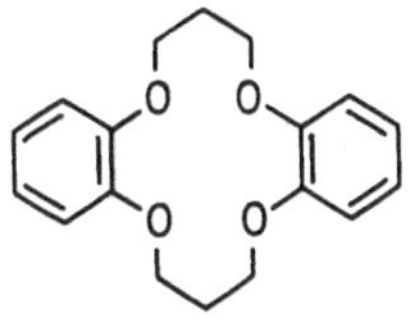

1

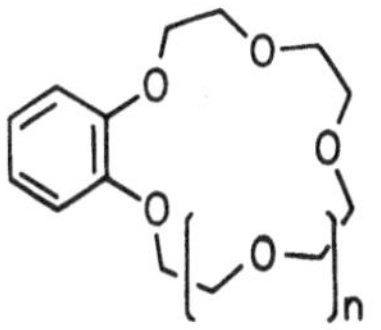

2a: n=1
2b: n=2

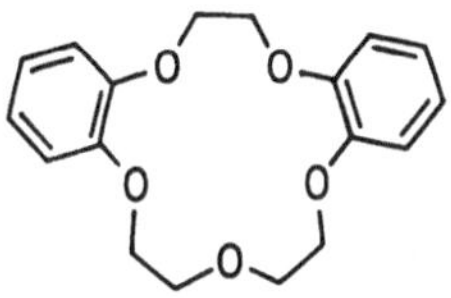

3

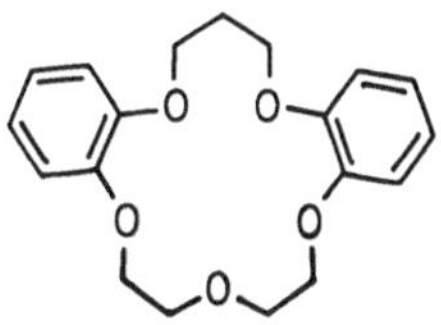

4

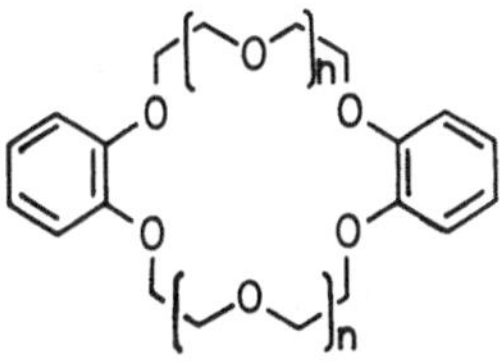

5a: n=1
5b: n=2

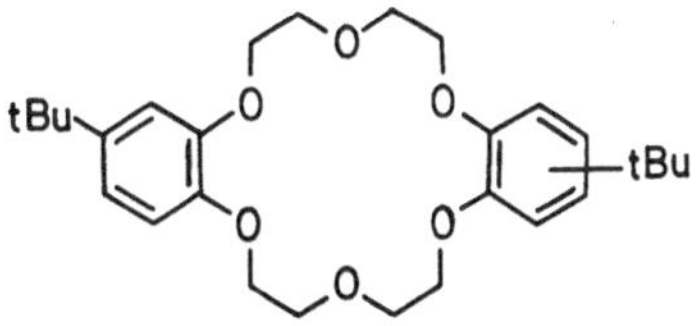

6

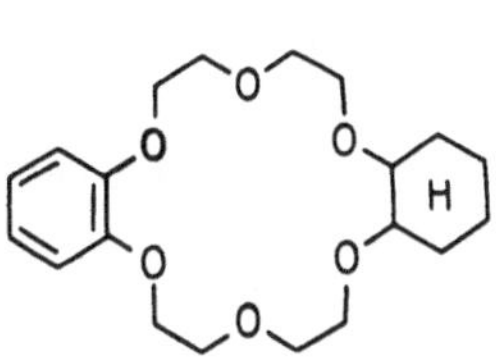

7

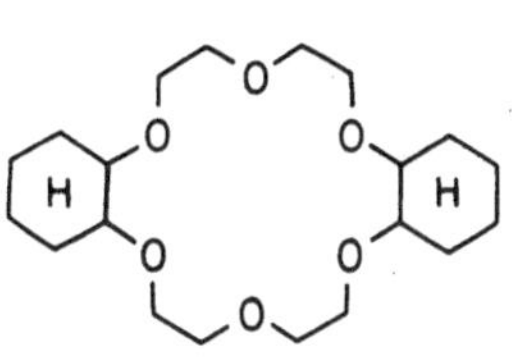

8

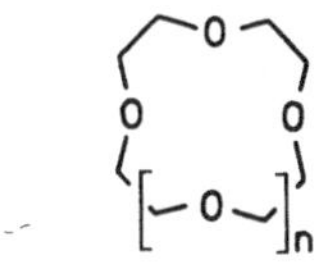

9a–c: n= 1–3

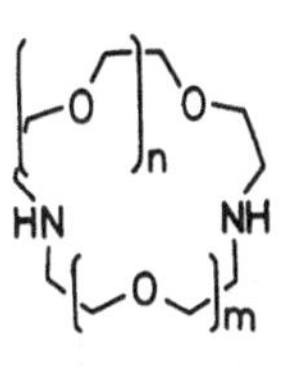

10a: n=m=1
10b: n=1, m=2
10c: n=2, m=3

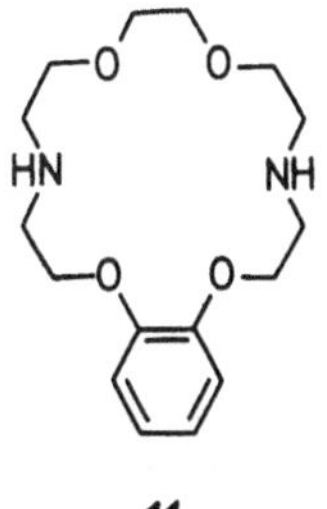

11

Fritz Vögtle, Heinz Sieger and Walter Manfred Müller

Meanwhile, it has been able, however, to obtain crystalline urea complexes of [18]crown-6 (*9c*) and also of diaza[2.2]coronand *10b*, some data are compiled in Table 1. Similarly, it has been possible to demonstrate that open-chain analogues of the cyclic crown ethers form stoichiometric complexes with urea and thiourea (cf. Sect. 3.1.1).

The 1:2 (host:guest) complex of *[18]crown-6* with *benzenesulfonamide*[21] (Fig. 1) is principally held together by NH ... O hydrogen bonds; the crown ether assumes almost the same conformation as in its uncomplexed form. Nevertheless, the elliptical shape of the crown shows, in comparison with the practically circle-like arrangement in other complexes, the ability of the [18]crown-6 framework to adapt to neutral guest molecules.

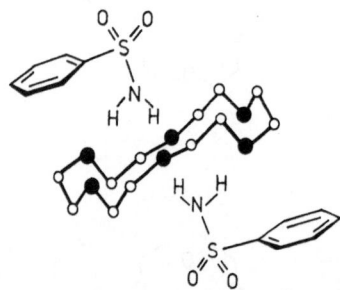

Fig. 1. Schematic representation of an ORTEP drawing of the 1:2 [18]crown-6 benzenesulfonamide complex[21]

2.2 Polar CH-Bond-Containing Neutral Guest Molecules

Crystalline complexes of crown compounds with neutral molecules, e.g. with CH-acidic compounds like *acetonitrile*[22, 23, 24], *dimethyl acetylenedicarboxylate*[25], *malonodinitrile*[26] etc. have been prepared (cf. Table 2).

Up to now, the comment of Cram and Liotta et al.[24a] has been valid more generally: *"The nature of the interactions between host and guest is not clearly understood. There is an obvious possibility that different substrates interact differently with the host, affording on different occasions a complex, a solvate, and so on. Intuitively, it appears that two possible factors favor the formation of a host-guest solid adduct. The large size of the 18-membered ring and its lack of rigidity might favor the interstitial trapping of other molecules to gain a more favorable crystal lattice. The second factor which probably influences the formation of such complexes is the multiplicity of electronegative heteroatoms distributed in the ring system which have the potential for interacting with the guest molecule in the lattice with further ordering. We therefore use the term 'complex' advisedly and are aware that probably only structural data derived from direct observations (e.g. X-ray) will resolve the nature of the complex in individual cases"*.

An X-ray analysis of the crystalline nitromethane complex was first tried by F. P. de Boer and P. P. North (cited in [22]) of the [18]*crown-6* acetonitrile complex prepared by Gokel, Cram, Liotta et al.[24a] in 1974. Due to the undefined, changing stoichiometry of the complex this analysis can apparently be carried out only with difficulty. It crystallizes as beautiful large, colorless, transparent crystals after dissolving [18]crown-6 (*9c*) in little acetonitrile and can serve to purify the ligand since the latter remains in a pure state upon subsequent distillation of the acetonitrile in vacuo [24].

Monocyclic aza coronands such as *10a–c* and *11* are able to form solid complexes with benzyl chloride in acetonitrile the complexation of which depends on the stoichiometric ratio of the starting components [26] (Table 2).

In 1975, using X-ray analysis, Goldberg [25] described the stereochemistry of the *dimethyl acetylenedicarboxylate* complex of *[18]crown-6* (Fig. 2). It is noteworthy that in this complex all of the six oxygen atoms of each crown ether molecule participate in the binding of two dimethyl acetylenedicarboxylate molecules, which are fixed on opposite sides of the crown ring "*by means of dipole-dipole interactions between the electronegative oxygen atoms of the macrocycle and the electropositive carbon atoms and methyl groups of the guest*". [25]

Recently, the stereochemistry of the *[18]crown-6* complex with *dimethyl sulfone* [20] (stoichiometry 1:2) has been evaluated by an X-ray analysis [27b]. It is necessary to coat the crystals in Araldite® to reduce the decomposition in air. Thus, two crystals were required for complete collection of intensities. It has been shown that the centrosymmetrical [18]crown-6 molecule exhibits approx. D_3d symmetry, the bond lengths and angles do not differ significantly from those of the uncomplexed molecule at room temperature [27a].

Two molecules of the dimethyl sulfone guest interact with the [18]crown-6 host via two or possibly three C(13)—H ... O contacts as shown in Fig. 3. Alternate oxygen atoms deviate from the most appropriate plane through all six oxygen atoms

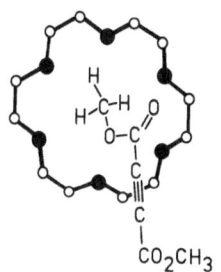

Fig. 2. X-ray structure of the complex [18]crown-6 dimethyl-acetylenedicarboxylate [25] (schematically)

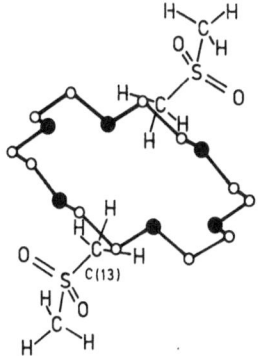

Fig. 3. X-ray structure of the [18]crown-6 dimethyl sulfone complex [27b] (schematically)

of the crown by 0.115, 0.098 and 0.187 Å resp. and C(13) is 1.632 Å from this plane with the S—C bond forming an angle of 18.9° with the normal to the plane. All the oxygen atoms of the crown molecule appear to be recipients of hydrogen bonds which are nearly coplanar with respect to the CH_2—O—CH_2 groups. This is consistent with a trigonal rather than a tetrahedral distribution of neighbours around the oxygen atoms. The interaction with all three methyl hydrogen atoms found in the complex contrasts with the bonding between the methyl groups of dimethyl acetylenedicarboxylate [25]. The volume occupied per complex molecule is not significantly different from that occupied by the components as packed in their pure crystals.

Similarly, the commercially available *dibenzopyridino[18]crown-6 (12a)* allows a broad entry into the field of neutral molecule complexation to be made [28a, b]. Crystalline, stoichiometric complexes of *12a* with acetonitrile as well as with a series of organic neutral compounds including other crown compounds have been isolated and characterized (Table 2).

A complex with a 1:1-stoichiometry has also been isolated with the basic 4-dimethyl-aminopyridino ligand *18* and CH_3CN as guest components [29] (cf. data in Table 2).

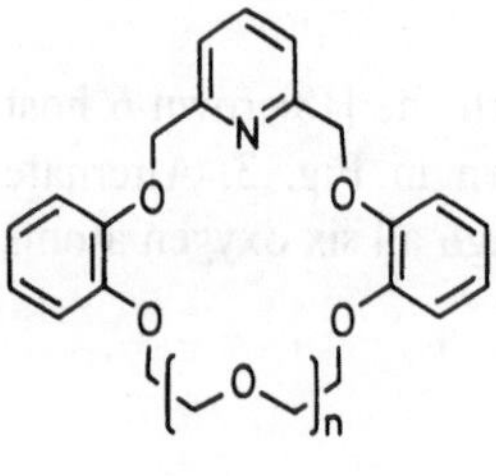

12a: n=1
12b: n=2

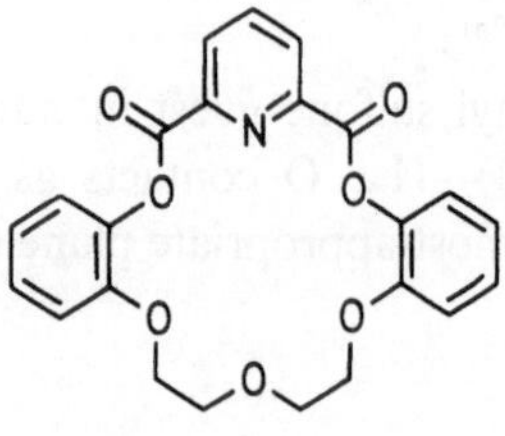

13

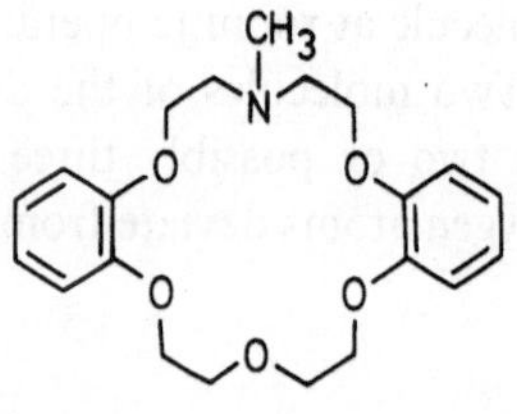

14

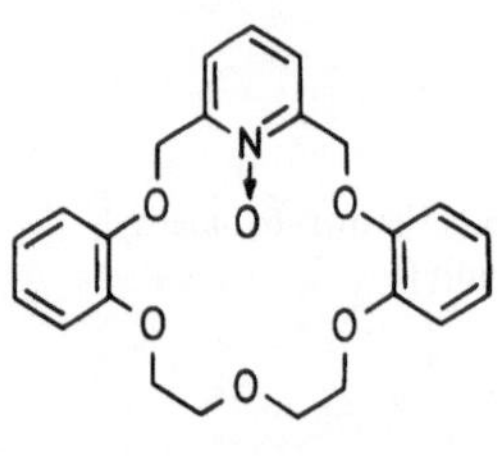

15

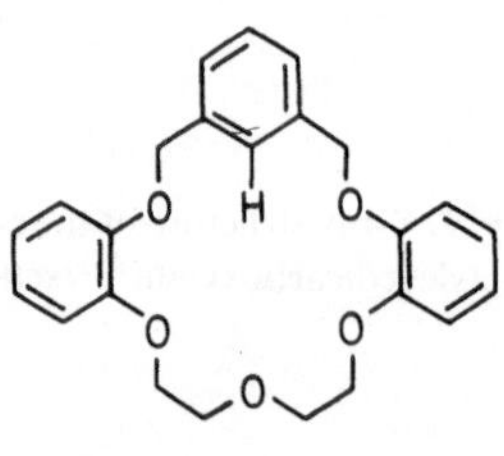

16

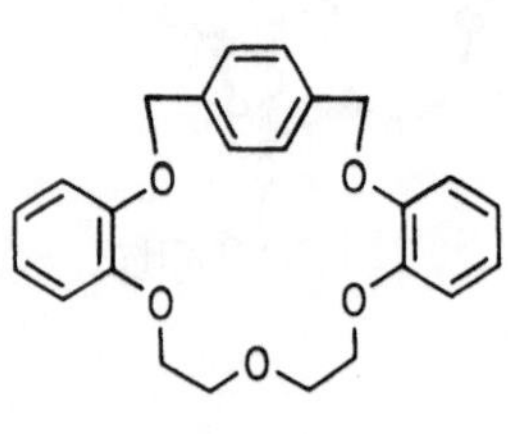

17

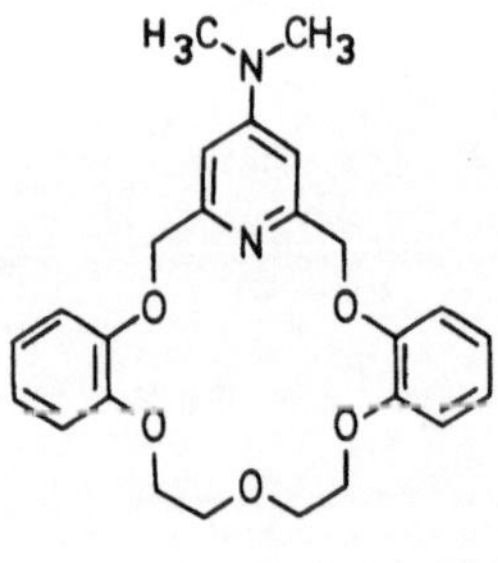

18

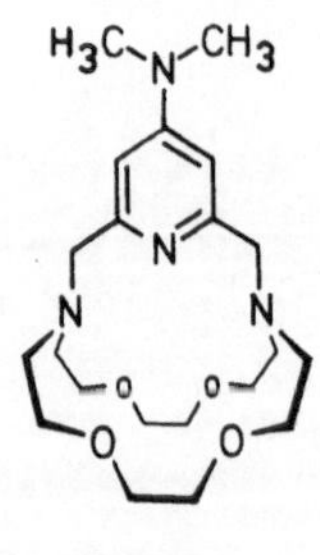

19

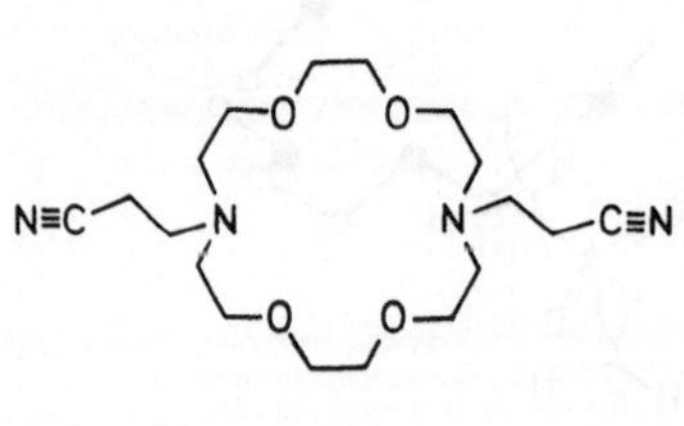

20

Another series of complexes of crown compounds with CH-acidic molecules, possessing a definite stoichiometry, have been described by Knöchel et al. [26, 30–32]. The neutral guest molecules employed are listed in Table 2.

CH-acidic molecules such as *acetoacetic ester* and *acetylacetone*, which are prone to enolization, do not seem, however, to form complexes with [18]crown-6. In the series of the dinitriles used, the CH-acidity strongly decreases from malonodinitrile to adipodinitrile, the melting points of the complexes formed fall accordingly. A hydrogen bond formation could be confirmed in the ^{1}H—NMR spectrum by a downfield shift of the signals of the participating protons, in accordance with the CH-acidity, and also by the temperature dependence of the resonances [26].

The crystal and molecular structure of the complex of malonodinitrile with *[18]crown-6* is shown in Fig. 4 [30]: both of the malonodinitrile molecules are fixed at opposing oxygen atoms through hydrogen bonds, as can be recognized in the two perspectives given.

The bond distances found clearly reveal that the complex, which is held together by hydrogen bonds and, if necessary, by additional dipole-dipole interactions, mostly resembles the *ion-dipole* complexes of [18]crown-6, e.g. with cations like Rb$^+$, Cs$^+$. All torsion angles about the C—C bonds are practically synclinal (*sc*), those about the C—O bonds are nearly antiperiplanar (*ap*). The conformation of the *[18]crown-6* in the *malonodinitrile* complex is thus comparable to that found in its K$^+$, Rb$^+$ and Cs$^+$ complexes.

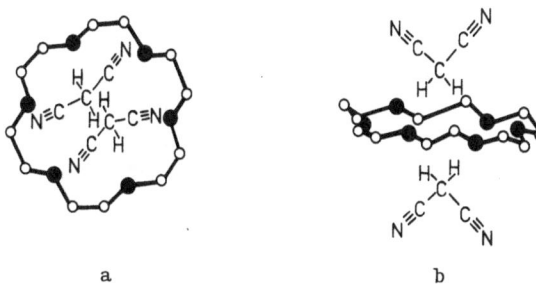

a   b

Fig. 4. Spatial arrangement in the complex malonodinitrile [18]crown-6 [30]: a) as seen perpendicular to the crown ether plane; b) as viewed in the plane of the crown ether

Microcalorimetric measurements [31] give an insight into the stability of the complexes. The conclusion can be drawn that the complex can even compete with protic solvents that form hydrogen bonds.

The crystal and molecular structure of the colorless 1:2-complex of the N,N'-disubstituted azacoronand *20* with malonodinitrile has lately been elucidated [32]. The spatial arrangement of the complex is shown in Fig. 5. One acidic H-atom of each malonodinitrile does not participate in hydrogen bond formation while the second H-atom, in accord with the bonding relationships found already for open-chain systems (cf. Sect. 3.1.1), forms a bifurcating bridge to a nitrogen and an oxygen atom. A further hydrogen bond between the 2-cyanoethyl group and a nitrogen atom of a guest molecule causes an additional increase in the complex stability.

Polar succinic acid derivatives such as succinodinitrile seem to bind —CH$_2$—CH- groups [31] more favorable than a —CH$_2$—CH$_2$-arrangement [32].

Another series of polar neutral molecules, e.g. *N-methylformamide, dimethylformamide, dimethylacetamide, acetoacetic ester, acetylacetone, α-picoline, collidine,*

Fritz Vögtle, Heinz Sieger and Walter Manfred Müller

Table 2. Neutral molecule complexes of coronands with guest molecules containing polar CH-bonds

Host	Guest	Stoichiometry Host:Guest	M.p. [°C]	Ref.
2a	malonodinitrile	1:1	38–40	26)
8	malonodinitrile	1:1	125–136	26)
9b	malonodinitrile	1:1	42–44	26)
9c	nitromethane	1:2 (1:1.8)	107–115	22, 26)
	acetonitrile	1:2 (1:1.6)	77	22, 24, 26)
	malonodinitrile	1:2	138	26)
	bromomalonodinitrile	1:2	58	26)
	succinodinitrile	1:2	88	26)
	glutarodinitrile	1:1	54	26)
	adipodinitrile	1:1	56	26)
	cyanoacetic acid	1:1:1 H_2O	77	26)
	maleic acid anhydride	1:2	83	26)
	dimethyl sulfoxide	1:2	128–137	20)
	dimethyl sulfone	1:2	94–95	20)
	dimethyl acetylenedicarboxylate	1:1	101	25)
10a	benzyl chloride	1:1.1	186–192 (dec.)	26)
10b	benzyl chloride	1:1.5	172–175	26)
10c	benzyl chloride	1:0.8	197–200	26)
11	benzyl chloride	1:0.9	160–172	26)
12a	[12]crown-4	1:1	170–174	28a)
	acetonitrile	1:1	131 dec. 139–141	28a)
	nitromethane	1:1	130 dec. 139–141	28a)
	dimethylformamide	1:1	117–125 dec. 139–141	28a)
	dimethylacetamide	1:1	65 des. 128–133	28a)
	dimethyl sulfoxide	1:1	125 dec. 139–141	28a)
	dimethyl acetylenedicarboxylate	1:1	85, 141–151	28a)
	diacetyl	1:1	125, from 133	28a)
	ethylene glycol	1:2	125, from 133	28a)
	mesitylene	3:1	70, from 173	28a)
12b	acetonitrile	3:2	79–80, 99–100	28a)
13	nitromethane	1:1	85–90, 179, 205–209	28a)
15	acetonitrile	3:2	140, 145–159	28a)
	nitromethane	1:1	140–152	28a)
	dimethyl sulfoxide	2:1	174, 210–211	28a)
16	nitromethane	3:2	150–160	28a)
17	acetonitrile	3:2	85, 112–114, 120–122	28a)
	nitromethane	3:2	100, 111–113, 120–122	28a)
18	nitromethane	1:1	136–138	29)
19	chloroform	1:1	218–220	29)
20	malonodinitrile	1:2	60	32a)

and even *ethylene glycol* and *[12]crown-4* give crystalline complexes with an intricate stoichiometry, however, or non-stoichiometric complexes with the pyridino crown *12a* (cf. Table 2).

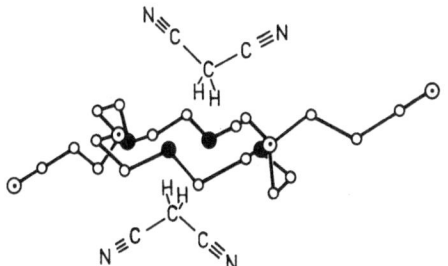

Fig. 5. Crystal structure of the malonodinitrile complex of azacoronand *20* [32a)]

As in the case of *12a* which represents a simple monocyclic receptor for neutral molecules with polarized CH_3 or CH_2 groups, the pyridino crown ester *13* [28b)] also forms a 1:1 complex with nitromethane (Table 2). No such complex has yet been isolated with the more flexible azacoronand *14* [28a)] that contains the same donor atoms but no pyridino nucleus. Several neutral molecule complexes of *2a, 9c, 15–19* are also given in Table 2.

Recently, crystalline, stoichiometric complexes of volatile, highly toxic chemicals containing acidic methyl or methylene groups as e.g. in dimethyl sulfate, N,N-dimethyl-nitrosamine, methyl tosylate, 2-chloroethyl tosylate, mesyl chloride, mesyl bromide, benzyl chloride, epichlorohydrin, acetic anhydride (cf. Table 3) were isolated [33)]. The complexes of these reagents with [18]crown-6 (*9c*) and dibenzopyridino-[18]crown-6 (*12a*) can be handled easily; the vapor pressure of these most probably "molecular-encapsulated" compounds is low as shown by Dräger tube tests with the dimethyl sulfate [18]crown-6 complex. Owing to their exact stoichiometric composition they may be used e.g. as more convenient alkylating and acylating agents instead of the free guest compounds. Whether a modified reactivity may be exploited for selective alkylations is a subject of further investigations.

Table 3. Crystalline stoichiometric complexes of toxic and volatile reagents with neutral host compounds of the crown type

Host	Guest	Stoichiometry Host:Guest	M.p. [°C]	Ref.
9c	dimethyl sulfate	1:1	88–94	33)
	mesyl chloride	1:2	87, 103, 106–110	33)
	mesyl bromide	1:2	79, 92, 103–107	33)
	methyl tosylate	1:2	62–64	33)
	2-chloroethyl tosylate	1:2	56–58	33)
12c	N,N-dimethylnitrosamine	2:1	125, 138–140	33)
	benzyl chloride	2:1	145–155	33)
	epichlorohydrin	2:1	125–132	33)
	acetic acid anhydride	2:1	128–132, 138–143	33)

2.3 Substituted Hydrazines as Uncharged Guest Molecules

With phenylhydrazine, 4-nitrophenylhydrazine and 2,4-dinitrophenylhydrazine [18]-crown-6 readily forms crystalline, stoichiometric complexes some of which are listed in Table 4 (cf. also Section 3.1.2). The 1:2 stoichiometry [34] of most of the complexes suggests that similarly as in the case of the [18]crown-6 benzenesulfonamide complex [21], a hydrazine molecule is bound to each face of the crown ether ring through hydrogen bonds.

The above interpretations are confirmed by the X-ray analysis of the [18]crown-6 2,4-dinitrophenylhydrazine complex [35]. The 2,4-dinitrophenylhydrazine molecules, as shown in Fig. 6, are almost perpendicular to the crown ether ring. Thereby, the hydrazine H-atoms form hydrogen bonds with the ether O-atoms of the ligand, and apparently one H-atom interacts with a pair of oxygen atoms through bifurcated hydrogen bonds. The second 2,4-dinitrophenylhydrazine molecule attacks in the same way from the back of the ligand ring. The crystals are extremely stable and remain intact for a very long time when exposed to X-rays, in contrast to the crystals of the simple phenylhydrazine complex. The acidification of the hydrazino group by both nitro groups thus seems to strengthen the hydrogen bond in the complex.

The phenylhydrazine crown ether complexes are suited for the stabilization of hydrazine reagents because of their crystallinity, stoichiometry and stability; they are also useful as *phase transfer reagents* for uncharged molecules in reactions with carbonyl compounds. Instead of aqueous inorganic acids or DMSO etc. [36], solvents such as toluene, nitromethane, acetonitrile, ethanol etc. may be employed for derivatizations under mild conditions. The reaction thus offers a new, convenient alternative to the preparation of phenylhydrazones from carbonyl compounds.

Other crown compounds such as *dibenzo[18]crown-6 (5a)*, *dibenzopyridino[18]-*

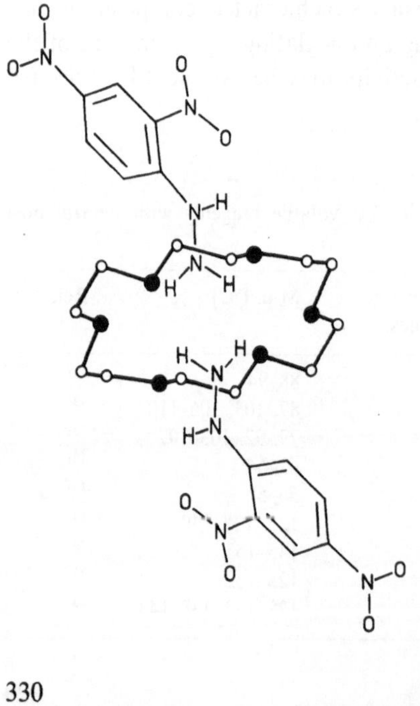

Fig. 6. X-ray structure of the [18]crown-6 2,4-dinitrophenylhydrazine 1:2 complex [35]. The two quest molecules are located at opposite sides of the crown ring

crown-6 (12a), *dicyclohexano[18]crown-6 (8)*, *[15]crown-5 (9b)*, *[12]crown-4 (9a)* and the *[2.2]diazacoronand (10b)* also raise the solubility of 2,4-dinitrophenyl-hydrazine in toluene and benzene, though up to now only with *9b* and *12a* (stoichiometric) complexes have been isolated [30].

The solubilisation of uncharged species by means of crown ligands opens novel possibilities for phase transfer reations and phase transfer catalysis which, in contrast to ion or salt phase transfer cannot be achieved with onium-type phase transfer reagents.

Table 4. Stoichiometric crystalline complexes of [18]crown-6 (*9c*) and substituted hydrazines

Guest	Stoichiometry Host:Guest	M.p. [°C]	Ref.
phenylhydrazine	1:2	90–91	34)
4-nitrophenylhydrazine	1:2	140–143	34)
2,4-dinitrophenylhydrazine	1:2	164–168	' 34)
4-toluenesulfonylhydrazine	1:2	109–111	34)
N,N'-diformylhydrazine	1:2	125–130	34)

2.4 Aromatic Amines and Phenols as Guest Molecules

Table 5 lists the crystalline complexes obtained so far from crown compounds and anilines as well as with phenols and aminophenols [37a].

Fig. 7 shows the result of an X-ray analysis of the 1:2 complex of 2,4-dinitraniline with [18]crown-6: in contrast to the hydrazine complex described above, the amino group forms hydrogen bonds to three adjacent ether oxygen atoms of the crown molecule, one additional hydrogen bridge directing to the nitro group of the 2,4-dinitraniline itself. The second 2,4-dinitraniline molecule is bound to the three other

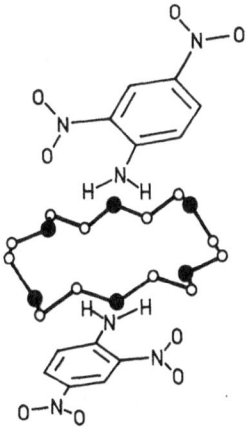

Fig. 7. X-ray structure of the 1:2 complex of 2,4-dinitraniline with [18]crown-6 [37b]

Table 5. Complexes of [18]crown-6 (*9c*) and aromatic amines, phenols and aminophenols

Guest	Stoichiometry Host:Guest	M.p. [°C]	Ref.
4-nitroaniline	1:2	89–97	[37a]
2,4-dinitroaniline	1:2	127–135	[37a]
2,6-diaminopyridine	1:2	81–88	[37a]
2-hydroxy-5-nitroaniline	1:2	105–113	[37a]
3-nitrophenol	1:2	49–56	[37a]
2,4-dinitrophenol	1:2	59–67	[37a]
2-hydroxypyridine	1:2	66–83	[37a]
2-nitroresorcinol	1:1	72–75	[37a]

crown oxygen atoms at the other side of the crown plane so that a centrosymmetric complex results [37b].

2.5 Water as Guest Molecule(s)

In the complexation by cyclic ligands [38] — such as natural ionophores — cations are wrapped in such a way that the polar regions (ether-, carbonyl-oxygen atoms) of the ligand are directed toward the interior where the cation is located (endo-hydrophilicity or endo-polarophilicity resp.). Several metal atoms may be taken up into a larger cavity. Thus, two potassium or sodium ions are coordinated in the potassium thiocyanate [39] or sodium p-nitrophenolate [40] complex of dibenzo[24]-crown-8. An X-ray structure of the same ligand binding barium *and* water molecules has been described [41] (such complexes which are composed of more than two components, e.g. those containing salts *and* H_2O or other neutral molecules in addition to the host molecule, are more closely described in Sect. 6.4).

Thanks to the determination of complex constants via equilibration experiments with aqueous solutions of tert-butyl-ammonium-hexafluorophosphate and $CDCl_3$ solutions of crown ethers, an essential role of water in the solvation could be found [42, 43]. It was shown that water can readily be "dissolved" in chloroform by the addition of crown ethers as e.g. *21a–f* and that this complexation ability rises with the ring size of the crown ether.

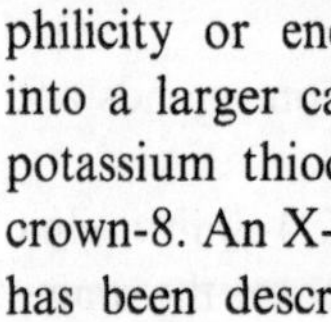

21a–f: n = 2–7

When the crown cavities in the case of *21a–f* are too large for the formation of stable 1:1-complexes with salts like tert-$BuNH_3^+$-hexafluorophosphate, dimers containing water molecules as indispensable components are formed [43].

The crystal structure of a binaphthol crown ether complex in which water is exclusively taken up as guest in the polyether ring, has recently been described [44]. Figure 8 shows the structure of this water complex.

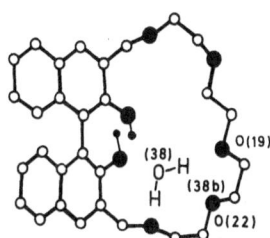

22

The water molecule is too small to fill the whole macrocyclic cavity and is, therefore, not coordinated by all the oxygen atoms. The skeleton of the crown ether, which is filled by the H_2O molecule, is largely fixed by hydrogen bonds. This is clearly revealed by the atomic distances and angles: *e.g.* O(38) ... O(19): 2.97 Å, H(38b) ... ... O(19) and O(38)—H(38b) ... O(19): 173°. Between H_2O and O(22) hardly any interaction seems to occur [O(38) ... O(22): 3.18 Å] [44 a].

Similar geometries have also been found in numerous complexed crown ether skeletons. The "empty part" of the ligand cavity shows a marked unchanged conformation (cf. also the molecular structures of the free [18]crown-6 (*9c*) [27a, 45] and of the tetramethyldibenzo[18]crown-6 ligand [46]), which is apparently stabilized by intramolecular C—H ... O dipole attraction, at least in the crystal).

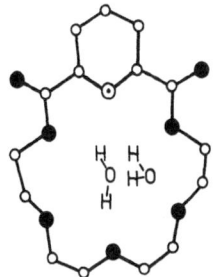

Fig. 8. X-ray structure of the binaphthol crown ether —H_2O complex (*22* · H_2O): the cavity is only partly filled by the small H_2O molecule. Coordination occurs through three H ... O hydrogen bonds and one interior-directed CH_2 group [44 a]

The pyridino crown ester *23* forms a 1:2-complex with water, as an X-ray study revealed [47]: As shown in figure 9 the two water molecules are hydrogen bonded to

Fig. 9. X-ray structure of the *23* · 2 H_2O complex [47]

each other and with two crown ether oxygen atoms, but not to the pyridine nitrogen.

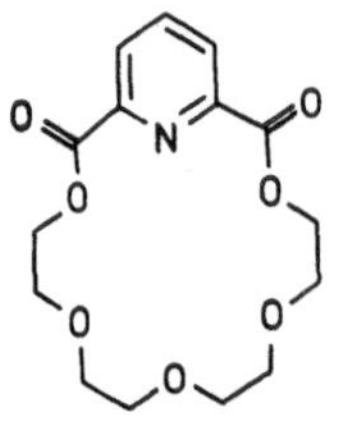

23

The triple crown ether 24 [48] derived from cyclotriveratrylene on account of its shell-type molecular structure could be able to enclose neutral molecules intramolecularly or into its crystal lattice. 24 crystallises as dihydrate, the structure of which is not known as yet.

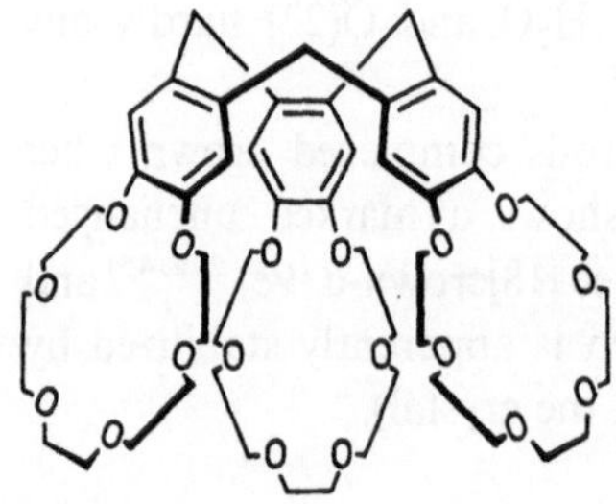

24

2.6 Inclusion of Alcohols

The pyridino crown 25 opens new possibilities for the complexation of small neutral molecules such as alcohols [49]. Unbranched alcohols from methanol up to n-butanol are stoichiometrically added in the crystalline state, whereas longer chained ones such as octanol as well as branched alcohols such as 2-propanol and 2-butanol do not form

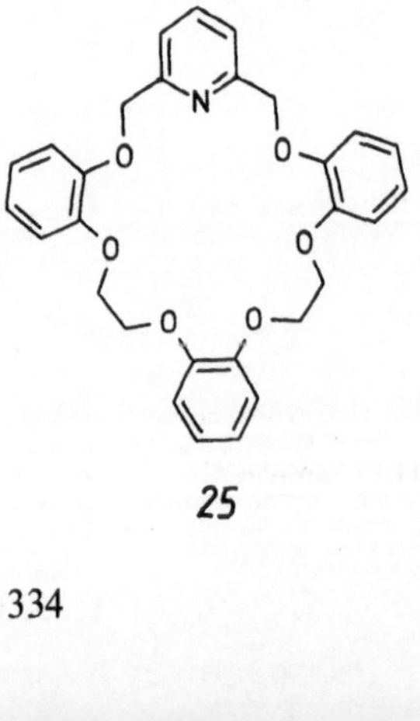

25

crystalline adducts. As the melting points of the complexes all lie around 50 °C, but that of the pure crown 25 at 102–114 °C, crystal lattice-inclusions similar to those of tri-o-thymotide [12] are most probable here. Tri-o-thymotide also crystallizes with a number of alcohols, but it does not clearly select between lower and higher molecular and unbranched and branched alcohols.

2.7 Molecular Halogen as Guest Molecules

Crystalline complexes between crown compounds and molecular bromine, especially of dibenzo[18]crown-6 (5a) and 4,4',5,5'-tetrabromodibenzo[18]crown-6 (26) have been obtained as yellow to brown crystalline powders [50]. The molecular proportion of bromine to dibenzo[18]crown-6 was found to be around 1:1.5, which was interpreted as resulting from mixtures of the 1:1- and 2:1-complexes. With other crown compounds stoichiometries lying between 0.9 and 1.85 have been obtained.

As the equilibrium constants of complexation in solution are low, the precipitation of the fairly stable crystalline powders seems to be due to their limited solubility.

26

The low values of the vapor pressure of bromine over such crystals and its independence of the bromine — to — ether ratio in the crystal seem to role out the possibility that a significant amount of free bromine may be trapped by occlusion. In chloroform solution there seems to be a competition between chloroform and bromine for the complexation with the complexing agent; hydrogen bonding between chloroform and ethers has been investigated by PMR spectroscopy.

Remarkably, polyfunctional dicyclohexano[18]crown-6 (8) and monofunctional tetrahydrofuran show a similar affinity toward bromine. This is in agreement with the basicity of the oxygens in the two ethers, but is different from their behavior in systems in which the complexing power of the crown ethers is enhanced on entropy grounds. Apparently, complexation of a bromine molecule with two ether oxygens belonging to the same crown molecule is configurationally unfavorable.

The formation constants of the bromine complexes of several polyethers of different sizes have been measured [51]: essentially, no selectivity of complex formation is found, i.e. K is 1.2 1/mol^{-1} for dioxane, 2.2 for [12]crown-4 (9a), 0.8 for [15]crown-5 (9b), 1.0 for [18]crown-6 (9c), and 1.2 for dicyclohexano[18]crown-6 (8). This result implies that the binding of the bromine to the polyether may not involve the cavity of the crowns.

The isolable complexes of dibenzo[18]crown-6 (5a) with molecular bromine have been shown to effect highly stereoselective brominations of cis- and trans-β-methylstyrene. The Br$_2$ complex exhibits the greatest stereoselectivity of the brominating agents utilized; for the bromination of E-β-methylstyrene, a stereospecific trans-addition is obtained regardless of the polarity of the solvent. The intermediate three-membered bromonium ion is supposed to be stabilized by crown ethers relative to the open carbonium ion in all solvents, the stabilization being greater for the trans-isomer.

The results also indicate that very little rate enhancement is achieved by the poly-ethers, i.e. k is 1260 1/mol^{-1} s^{-1} for dioxane, 1540 for [12]crown-4, 1330 for [18]crown-6, and 1070 for Br$_2$ alone. It is concluded that complexation of Br$_2$ with crown ethers tends to reduce the overall rate of the reaction; however, complexation of the bromonium ion by crown ethers tends to counteract this effect and an overall rate enhancement is found. The major effect of the presence of the crown ethers is, however, the very significant increase in stereoselectivity.

A nonstoichiometric complex between bromonitrile and [18]crown-6 has also been reported [26].

2.8 Metal(O) as Guest Particles

Ag$^+$ complexes of long-chain alkyl-substituted crown ether compounds like 27 have been reduced to the corresponding Ag(0) complexes, either photochemically or thermally [52]. A cyanine dye has been used as a sensitizer in the photochemical reduction. The observed absorption spectrum has been ascribed to silver atoms which are stabilized by the cyclic crown ligand. The absorptions are similar to those obtained via matrix isolation spectroscopy of silver atoms. Encapsulation by the crown compounds seems to hinder the aggregation of Ag(0). The crown allows a high local concentration of the acceptor Ag$^+$ to be obtained in the reduction process and thereby enhances the yield. The crown also stabilizes silver atoms and hinders the reversed transfer of electrons by means of the microscopic electrostatic barrier lipid/water interface.

27

Apart from the photochemical reduction, a quantitative conversion of the Ag$^+$-into the Ag(0) complex of 27 has also been achieved, using NaBH$_4$ as reducing agent.

The complex 27-Ag(0) has reducing properties: Acceptors such as copper(II) and Fe(CN)$_6^{3-}$ reoxidize it quantitatively to the Ag$^+$ complex. The redox process may also be conducted electrochemically. A very high complex formation constant of 10^{21} for Ag atoms with the macrocyclic ligand 27 is deduced from the standard potential of the redox pair Ag$^+$/Ag(0). This high constant is explained by the larger diameter of Ag(0) compared with Ag$^+$.

A [2.2.2]cryptand-Na$_2$-complex in powder and crystalline form with the ionic structure Na$^+$[2.2.2]Na$^-$ has been prepared [53]. The sodium is hence not bound molecularly but the Na$^+$ cation is enclosed by the cryptand whereas Na$^-$ is located outside the cryptand cavity. The complex is the first solid compound that contains alkali metal anions. The formation of the cryptand and the relative position of the Na$^+$- and Na$^-$ ions is quite similar to the geometry found in the Na$^+$[2.2.2]I$^-$ complex.

The hexagonal platelets of the complex that crystallize on careful cooling, have a metal-glooming gold colored surface. They melt at 83 °C with decomposition whereby sodium metal is liberated. A temperature dependence of the electric conductivity is found with pressed powder of Na$^+$[2.2.2]Na$^-$.

Apart from the sodium complex, other alkali metals have been split by cryptands into cation and anion, e.g. in the compounds K$^+$[2.2.2]K$^-$. K$^+$[2.2.2]Na$^-$, Rb$^+$[2.2.2]Rb$^-$ and Cs$^+$[2.2.2]Cs$^-$.

3 Complexes of Open-Chain Neutral Ligands

3.1 Podands as Host Molecules

The property of open-chain bioionophores like nigericin [54] to selectively complex alkali metal ions just as their cyclic counterparts, e.g. monactin and valinomycin [55], has stimulated the investigations of noncyclic oligoethers. The latter have the advantage that their synthesis is a simple one, not requiring high dilution or template synthesis, and is thus economic and allows the use of a diversity of ligand structures [6].

As has been shown in numerous studies, noncyclic neutral ligands readily complex alkali and alkaline earth metal ions [6a]. Thus, in addition to the complexation of salts, the stoichiometric complexation of uncharged molecules by such ligands should also be appealing in analogy to that observed with the monocyclic crown compounds [6a].

Though Pedersen could not isolate complexes of open-chain glyme compounds with neutral molecules, as he clearly emphasized [19], numerous neutral molecule complexes of acyclic ligands have been obtained meanwhile.

3.1.1 Thiourea and Urea as Guest Molecules

Podands like *28* and *29a* are able to form stoichiometric 1:1 complexes with neutral guest molecules such as urea and thiourea [56]. The complexes can be prepared in crystalline form, e.g. by simply mixing methanolic solutions of the substrate and a noncyclic neutral ligand like *28*.

The colorless complexes can be recrystallized unchanged from dry acetone. Elemental analyses confirm an exact 1:1-stoichiometry which is also obtained when the dissolved components are mixed in different stoichiometric ratios. This observation also holds for many other neutral molecule complexes of this type: The most stable complexes precipitate regardless of the stoichiometry of the educts.

It may be deduced from the stoichiometry that the ligand *28* or *29* is not trapped in thiourea host channels to form inclusion compounds, because for such channel-complexes stoichiometries involving five to six thiourea or urea molecules and one

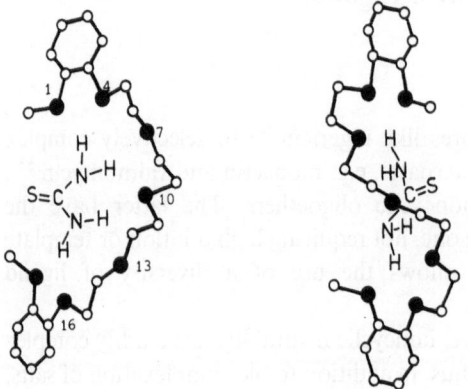

28

29a-c: n=1-3

30a: n=8
30b: n=9

guest molecule have always been found [1i, 8, 9]. Whilst quinoline, quinaldine, and other heteroaromtic compounds form 1:1-complexes [57] with urea and thiourea, in the case of **28** and **29** two aromatic terminal group units [58] instead participate in the complexation.

The X-ray structural analysis [59] of the crystallized product from **29c** and thiourea confirms a 1:1 crown ether-type complex (Fig. 10):

Fig. 10. X-ray structure of the **29c**-thiourea complex [59]

Noteworthy is that the amino groups of the thiourea are coordinated with all seven donor centers of the ligand with the participation of bifurcated hydrogen bonds to the ether oxygen atoms. Two bonds depart from the central O(10) atom. These bonds with N ... O distances in the range of 3.059 to 3.258 Å are longer than the comparable distances of approx. 2.90 Å between amino (or imino) groups and carbonyl oxygen atoms and are thus weaker.

All torsional angles CCOC and COCC along the polyether chain are antiperiplanar (*ap*) while the angles OCCO, on the other hand, are oriented synclinal (sc), as has already been observed in complexes of linear polyethers with metal cations [59].

An analogous 1:1-complex is formed between thiourea and the like-wise hept-adentate neutral ligand $O_5(quin)_2$ (**28**) [56, 58, 60]; the X-ray·structure is sketched in Fig. 11. The ligand **28** assumes a helical conformation in the complex. As in the corresponding RbI complex of the same ligand [61], a uniform bending of the torsional angle by a regular value is not observed [62], but an abrupt change in the geometry which causes the planes of both quinoline terminal groups to be oriented perpendicular to each other.

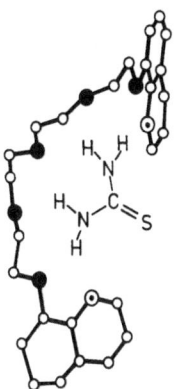

Fig. 11. X-ray structure of the *28*-thiourea complex [60]

Table 6. Urea and thiourea complexes of podands *28–38*

Host	Guest	Stoichiometry Host:Guest	M.p. [°C]	Ref.
28	thiourea	1:1	142	[56]
29a	urea	1:1	133–134	[56]
29c	thiourea	1:1	97–98	[56]
30a	urea	1:1	125 (dec.)	[63]
30a	thiourea	1:2	165–167	[56]
30b	urea	1:1	120 (dec.)	[63]
31	thiourea	1:3	153	[56]
32	thiourea	3:4	88–90	[56]
33	thiourea	1:2	194	[64]
34	thiourea	1:2	203–204	[64]
35	urea	1:1	157	[64]
35	thiourea	1:2	213–220	[64]
36	thiourea	1:1:2 H_2O	68–73	[66]
37	urea	1:1	43	[66]
38	thiourea	1:2	155–157	[20]

Parallel to the investigations about whether rigid donor end groups bridged only by alkyl chains as in *30* can form crystalline alkali/alkaline earth salt complexes, it has surprisingly been possible to also obtain stoichiometric urea complexes with a ligand-to-substrate ratio of 1:1 [63] (Table 6).

With thiourea, *30a* forms, however, a 2:1 crystalline complex. *31* and *32* also give thiourea complexes with 1:3- and 3:4-stoichiometries, respectively [56] (cf. Table 6).

31

32

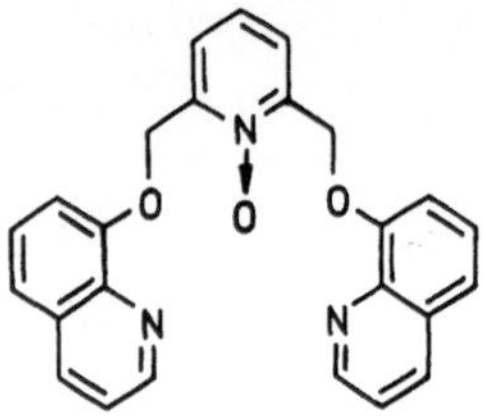

33 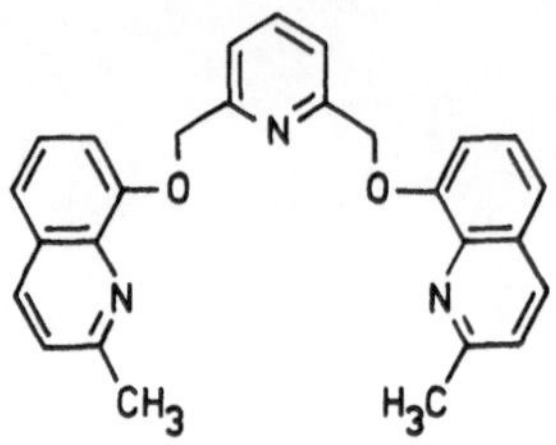   **34**

In order to gain more insight — using simple model compounds – into the possibility of complexation between neutral molecules, pyridine N-oxides have been used as host molecules. An analogous behavior between the pyridine N-oxide-containing ligand *33* and the similar pyridino compound *34* has been found [64], both forming crystalline complexes with thiourea. The 2:1 stoichiometry (thiourea:ligand) suggests that in both cases the neutral complexes are held together to a considerable extent through both quinoline and quinaldine nulcei resp. These complexes may be classified among the 1:1 complexes found by Wendt and Ried [57] in 1951 for various heterocycles as e.g. quinoline with thiourea, urea and acetamide.

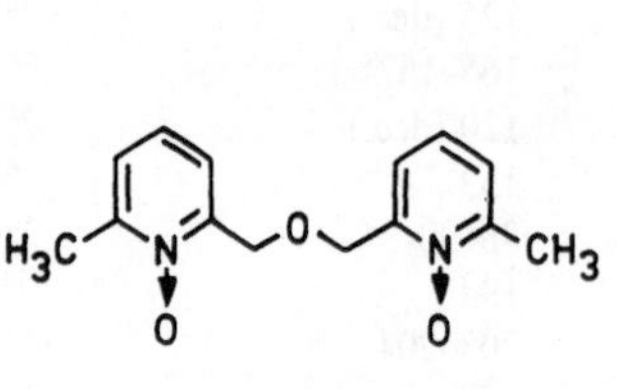

35 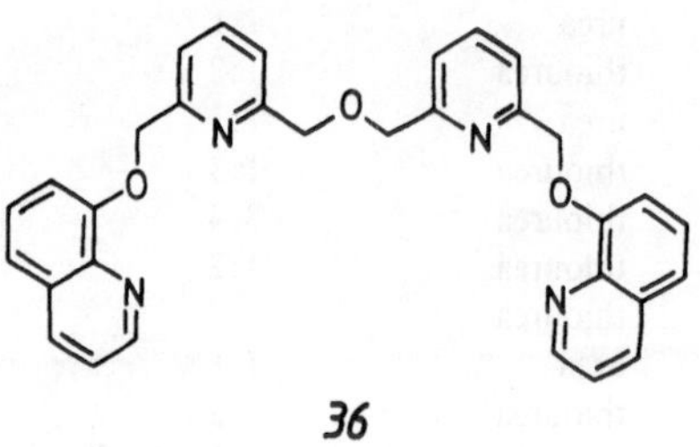   **36**

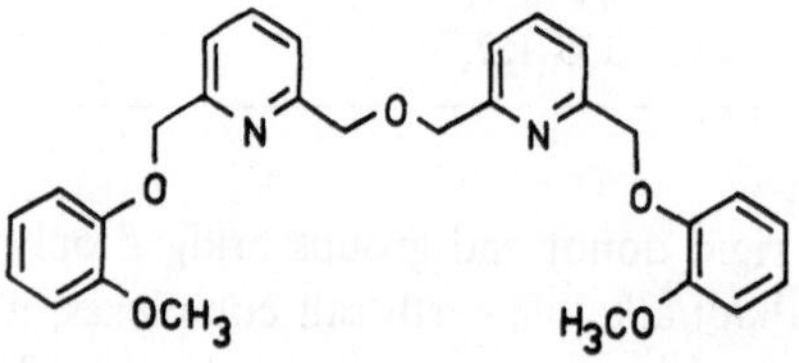

37 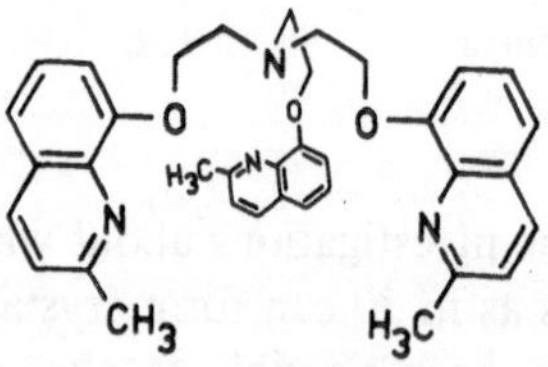   **38**

The ligand *35* containing two of the pyridine N-oxide units also gives stoichiometric complexes with urea and thiourea [64]. Apparently, the pyridine N-oxide moiety promotes the formation of the neutral complexes because no urea or thiourea complex has been isolated with the analogous hosts containing only pyridine rings.

The acyclic neutral ligands *36* and *37* arising from a combination of the end group concept [58b] with 2-pyridylmethylether units [65] similarly yield isolable crystalline complexes with other neutral molecules like urea and thiourea [66] (cf. Table 6). One may, perhaps, imagine a structure of the *36*-thiourea complex (1:2-stoichiometry) with a twisted ligand skeleton in which each half of the ligand takes up a thiourea molecule [67].

Finally, it has been possible recently to obtain a 2:1 guest-to-host complex between thiourea and a three-armed neutral ligand such as *38* [20] (cf. Table 6).

Thus, it has been shown that podands just as coronands are suited for the complexation of uncharged substrate molecules and therefore possess receptor model-character [68]. This is of some clinical interest in the case of urea as a guest molecule, owing to a possible selective dialysis by tailored urea receptors. The design of such improved urea receptors may hence also be oriented along the bonding types observed in the podand-urea complexes.

3.1.2 Polar CH-Bond-Containing Neutral Guest Molecules

In Section 2.2 neutral molecule complexes of *cyclic* crowns with CH-acidic compounds such as acetonitrile, malonodinitrile, dimethyl acetylenedicarboxylate etc. were reported. By a selective combination of appropriate substrates still more neutral molecule complexes can be obtained. Complexes of acyclic neutral ligands with CH-acidic compounds, on the other hand, have only recently been known.

Remarkably, the double acyclic pyridine N-oxide *35* readily forms crystalline complexes not only with *urea* and *thiourea* (cf. Section 3.1.1) but also with malonodinitrile [64].

The complex of tripodand *38* with 4-nitrophenylhydrazine, which bears an analogy to the neutral molecule complexes described in Section 2.3, with substituted hydrazines as guest molecules, is worth mentioning (stoichiometry 1:1, mp. 128–130 °C) [34].

3.1.3 Water as Guest Molecule(s)

In Section 2.5 the role of water in the crown ether complexation was already mentioned. Investigations with acyclic antibiotics reveal exactly how these pseudo-cycles wrap around the cavity in the course of the complexation, generally resulting in stronger fixation of the conformation of the mostly flexible, open-chain bio-ionophore. With monobasic acids the pseudocycles are stabilized by terminal group interaction [69] in the complex as in the free ligand [70]. The pseudocavity is usually filled with an additional water molecule [71].

Thus, the well-characterized carboxylic antibiotics, e.g. monensin [69a, b, 72] and nigericin [73], possess an acyclic primary structure which can, however, be fixed by hydrogen bonds in the pseudocyclic tertiary structures. X-ray analyses [69a] reveal that the *free acid monensin* (*39*) has such an arrangement with two intramolecular hydrogen bonds between the CO group and a water molecule binding both ends of the chain (Fig. 12). The water molecules, which are held by hydrogen bonds in the free ligands, seem to play an important role in the complexation of cations by influencing the hydration shell of the central ion [71b].

An interesting 1:2 complex of the synthetic, double pyridine N-oxide *35* [65] with water has also been obtained [64, 74]. Compared with the flexible open-chain podands *28* and *29* with donor end groups, which wrap around cations and neutral molecules [6a, 56, 58, 59, 60], the quite rigid monoether *35* coordinates in a different way. The X-ray analysis gave the following result (Fig. 13):

Remarkably, the ether oxygen of the host ligand does not participate in the interaction with the H_2O guest molecule. Two ligand molecules are pseudocyclized by a pair of H_2O molecules through hydrogen bonding at the N-oxide oxygen atoms,

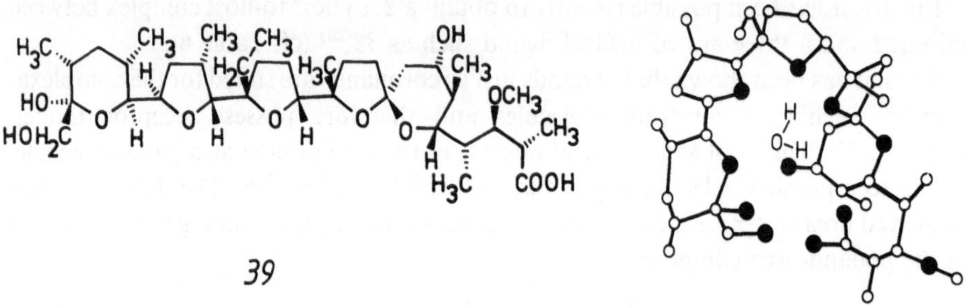

39

Fig. 12. Monensin (39) and structure of monensin monohydrate [69a]

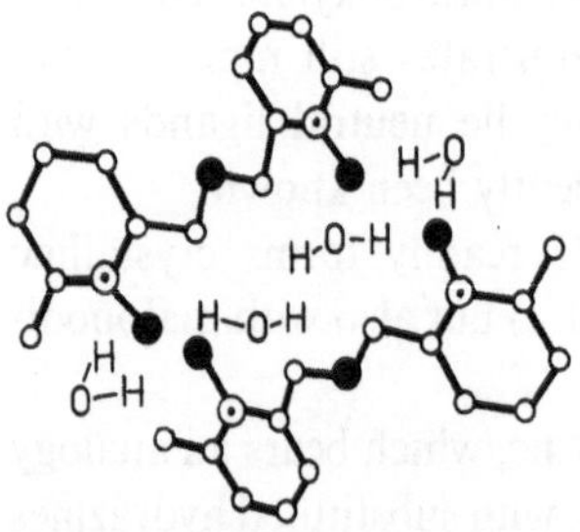

Fig. 13. X-ray structure of the H_2O complex of podand 35 [74]

whilst two other water molecules are similarly held together and bound to the pyridine N-oxide oxygen atoms through hydrogen bonds in the pseudocavity formed [75]. These dimeric units are further bound to each other forming an "endless hydrogen bond-chain". The torsional angles are all "normal", i.e. either 0° or 180° (±3°) so that an almost planar conformation of the ligand results. Only when the ligands were to assume a butterfly-like folded conformation or one with at least a synclinal torsional angle could all three oxygen atoms surround a guest molecule without steric hindrance. In the crystal lattice the molecules are arranged such that an additional π—π-interaction of the dimers appears feasible due to an overlap of the pyridine nuclei.

The liquid bis-phenol ligand 40 [28 a, 76 a] crystallizes in the presence of water whereby a monohydrate is formed. An X-ray analysis [76b] shows that the hexadentate ligand binds a water molecule intramolecularly forming hydrogen bridges to all donor centers (Fig. 14).

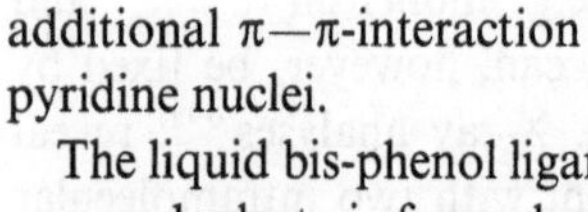

40

Fig. 14. X-ray structure of the 40-H_2O complex (schematically) [76b]

Tris(2-methyl-8-quinolyloxy)ethylamine *38* crystallizes as a dihydrate from ethyl acetate [77]. X-ray analysis [78] reveals that one chain of the tripode ligand is wraped around the solvent water molecule while the remaining two chains are left without obvious host-guest interactions. A second water molecule is hydrogen bonded to the first (Fig. 15). The geometry of the one wrapping loop of the ligand is close to that of a fragment of [18]crown-6 complexing a suitable metal ion [79]. The O atom of the water molecule W(1) and the heteroatoms N(16), O(27) and N(15) are coplanar within 0.12 Å.

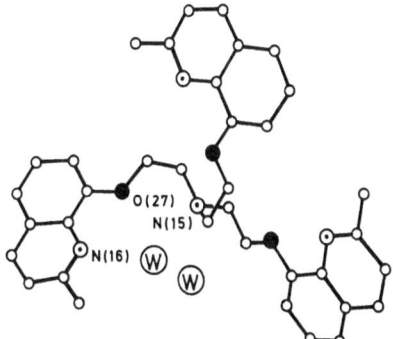

Fig. 15. X-ray structure of the *38* · 2 H$_2$O complex [78]

3.2 Phenanthroline, Bipyridine and Terpyridine as Hosts

The property of nitrogen heterocycles as organic reagents in analytical chemistry, e.g. bipyridine and 1,10-phenanthroline and their analogues, to complex transition metal cations is sufficiently known [80]. In 1938, Pfeiffer [81] already investigated the complexation ability of the unsubstituted o-phenanthroline. More recently, further progress in the field of alkali metal complexation with ligands of this type has been reported [82].

Starting from the urea/thiourea complexation of the podands, it has been investigated whether simple, inflexible pyridine nitrogen-containing donor molecules (with basic character) (e.g. o-phenanthrolines *41*, *42*, bipyridine *43* and terpyridine *44*) can, in analogy to the crown compounds, form complexes with urea and similar neutral molecules, thereby acting as simple open chain host molecules.

41 : R = H
42 : R = CH$_3$

43

44

The complexes of the heterocyclic host molecules *41–44* with urea, thiourea, N-phenylthiourea, p-tolylurea, catechol, resorcinol, and malonodinitrile are listed in Table 7 [83].

Table 7. Isolated neutral molecular complexes of o-phenanthrolines, bipyridine and terpyridine

Host	Guest	Stoichiometry Host:Guest	M.p. [°C]	Ref.
41	urea	1:1	204–206	[83]
	thiourea	1:1	212	[83]
	N-phenylthiourea	1:1	130–133	[83]
	o-tolylurea	1:1	110–113	[83]
	malonodinitrile	1:1	144–145	[83]
	catechol	1:1	116–118	[83]
42	urea	1:1	220–222	[83]
	N-phenylthiourea	1:1	137	[83]
	p-tolylurea	1:1	167–170	[83]
	catechol	1:1	134–138	[83]
	resorcin	1:1	209	[83]
43	urea	1:1	124–126	[83]
	N-phenylthiourea	1:1	87–89	[83]
44	urea	1:1	197	[83]

In the crystal of 2,9-dimethylphenanthroline (*42*) with the *1,3-dihydroxybenzene* (resorcinol) guest molecule an intramolecular rather than an *inter*molecular inclusion is formed: An X-ray analysis taken from material recrystallized from acetone/methanol shows (Fig. 16) that two host molecules form perpendicular parallel sheets which are held together by two weak hydrogen bonds between the phenolic OH groups and the nitrogen donor centers of the phenanthroline ligand [84].

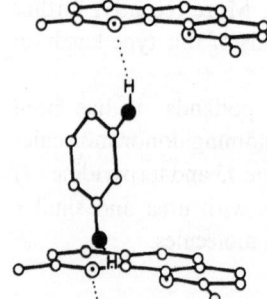

Fig. 16. Structure of the intermolecular complex of 2,9-dimethyl-phenanthroline and resorcinol [84]

3.3 Complexes of Hexahost and Octopus Molecules with other Uncharged Molecules

Many-armed neutral ligands — polypodands — as e.g. the six-armed benzene derivative *45* with numerous donor centers — "octopus compouns" [6a, 85] — show marked phase-transfer properties toward metal ions. The higher number of donor atoms is essential for an effective complexation, since the observed phase transfer is curtailed when the number of donor sites is diminished.

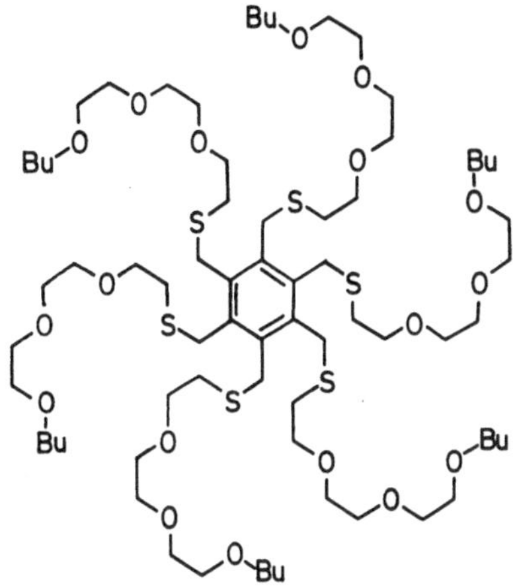

45

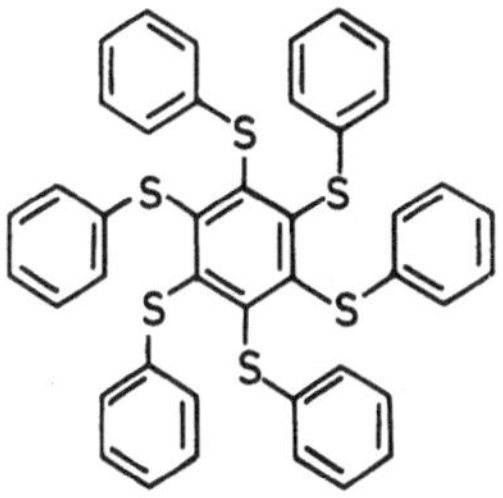

46

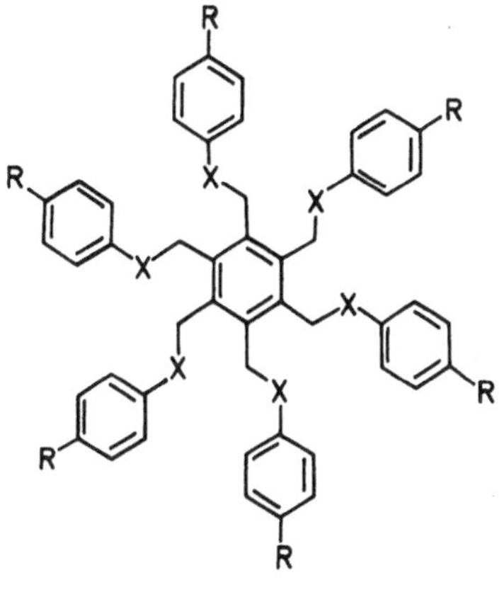

Nr	X	R
47	O	H
48	S	H
49	S	tBu
50	S	1-Adamantyl
51	S	OH
52	S	NH$_2$
53	Se	tBu
54	Se	H

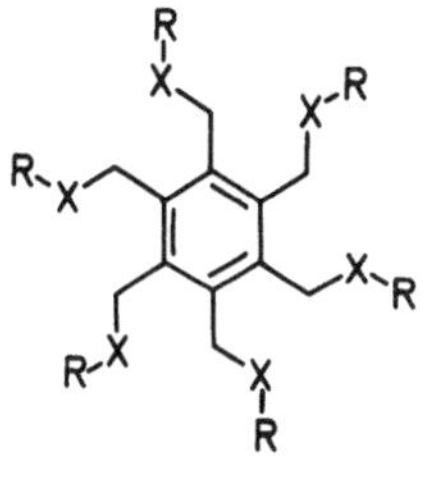

Nr	X	R
55	S	Benzyl
56	S	CH$_2$-CH$_2$-phenyl
57	O	Cyclopentyl
58	S	Cyclohexyl
59	S	2-Naphthyl
60	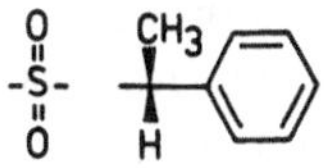 -S- (O,O)	CH$_3$ / H phenyl

Table 8. Selected inclusion compounds of hexahosts

Host	Guest	Stoichiometry Host:Guest	Ref.
47	toluene	1:2	87a, 88)
	1,4-dioxan	1:3	88)
	tetrahydrothiophen	1:1	88)
48	toluene	1:1	88)
	1,4-dioxan	1:2	88)
49	cyclohexane	1:2	88)
	cycloheptane	1:2	88)
	cyclooctane	1:2	88)
	toluene	1:2	88)
	iodobenzene	1:2	88)
	phenylacetylene	1:2	88)
	1-methylnaphthalene	1:2	88)
	2-methylnaphthalene	1:2	88)
	bromoform	1:2	88)
	squalene	2:1	88)
	hexamethyldisilane	2:1	88)
55	cyclohexane	1:1	88)
	toluene	1:1	87b, 88)
	1,4-dioxan	1:1	88)
	acetone	1:2	88)
56	1,4-dioxan	1:1	92)
	benzene	1:2	92)
	toluene	1:2	92)
	tetrahydropyran	2:3	92)
	fluorobenzene	1:2	92)
	chlorobenzene	1:2	92)
	bromobenzene	2:3	92)
	iodobenzene	1:1	92)
	1,1,1-trichloroethane	1:1	92)
58	benzene	2:3	88)
	toluene	2:3	88)
	methyl acetate	2:3	88)
	o-xylene	2:3	88)
	1,4-dioxan	2:3	88)

Starting from the hexameric units bound via hydrogen bridges in the clathrates of the dianin compounds, "hexahosts" [1b] such as 46–60 function as hosts for organic neutral molecules like acetone, dioxan, chloroform, carbon tetrachloride etc. as well as with a series of hydrocarbons to form a new type of stoichiometric inclusion compounds [86–95] (Table 8).

The decomposition of the complexes, which are stable at room temperature, is strongly enhanced by heating under reduced pressure. An X-ray analysis of the CCl_4 clathrate of the host compound 46 has been described (Fig. 17) [89].

As is typical of hexasubstituted benzenes [90], the SC_6H_5 side arms are displaced alternately upwards and downwards out of the plane of the central benzene ring, causing the formation of three-dimensional intermolecular cavities in the crystal packing. Into these cavities CCl_4 guest molecules are embedded. Both of the CCl_4 guests fit very well in the cage formed (17 Å long), with one C—Cl bond of each guest

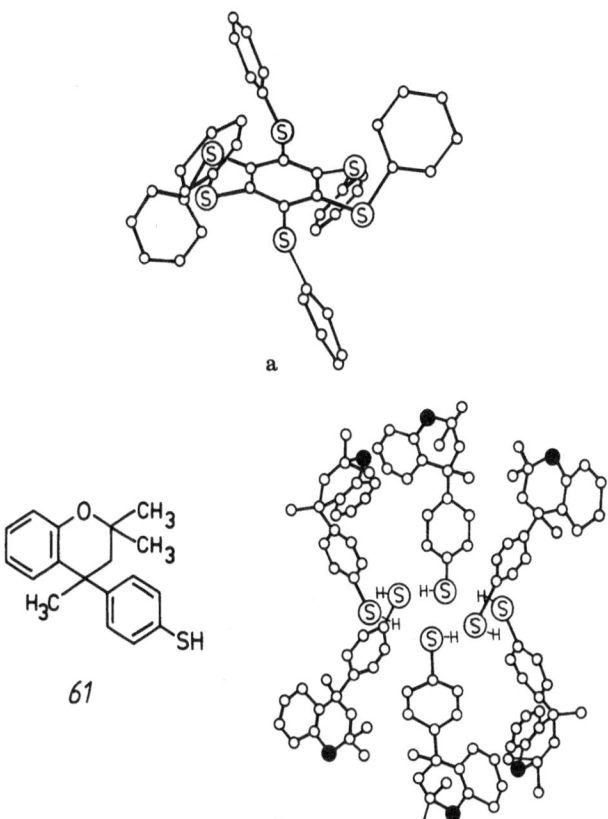

61

Fig. 17. a) Arrangement of the hexahost 46 in the CCl$_4$ clathrate [89]. b) For comparison, the intermolecular host cavity formed by six dianin molecules 61 [10d] is shown

molecule being colinear with the C-axis of the crystal (1:2 host-to-guest stoichiometry).

With the same ratio CCl$_3$CH$_3$ and CCl$_3$SCl can be embedded in the host lattice while CCl$_3$Br or CCl$_3$NO$_2$ give 1:1 complexes [89].

For comparison with the 46-CCl$_4$ adduct, an intermolecular host cavity, formed by six molecules of the dianin compound 61 and in which CCl$_4$ is likewise embedded [10d], is displayed in Fig. 17.

Thus, the hitherto mentioned hexahost neutral molecule complexes can be regarded as complexes of small guest molecules accommodated in the cages of clathrate inclusion compounds, in the sense of *inter*molecular cavity formation. This fact is underlined by the analogy of the crystal structures of the complexes. A remarkable aspect is the found guest selectivity of the hexahosts in the recrystallization from solvent mixturs (e.g. o- and p-xylene in equimolar amounts). Which of the two isomers is preferably enclosed depends primarily on the structure of the host. Thus, 55 and 59 favor the inclusion of p-xylene while 47 and 48 prefer to take up the o-isomer. For hexakis(phenylthiomethyl)benzene (48), a marked p/o-ratio of 90:10 is observed [88].

Investigations of the structural factors governing the host character show that no inclusion properties are found for the tris-substituted analogs 62–65 [92]. This amplifies the significance of a sixfold substitution in the hexahost analog and highlights the care required in the design of new trigonal host molecules [92].

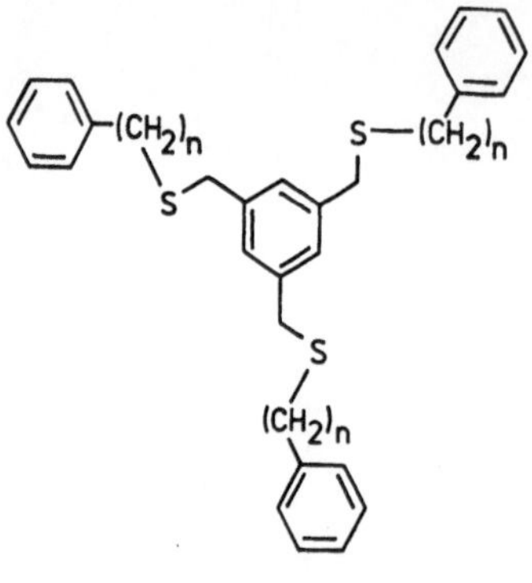

$$62: n = 0$$
$$63: n = 1$$
$$64: n = 2$$
$$65: n = 3$$

In contrast, the four-atom chain hexahost molecule *56* forms a 1:1 dioxane complex as well as 1:2 complexes with benzene and toluene, while its shorter chain counterpart *55* gives 1:1 complexes with these solvents. An even longer chain extension has given no evidence of inclusion behavior.

A chiral hexahost molecule, hexakis [4-(*R*-α-phenylethyl)phenylsulfonylmethyl]-benzene (*60*) forms an inclusion compound on crystallization from acetic acid [93], a host-to-guest ratio of 1:4 being determined by integration of the ^{1}H—NMR signals of a CDCl$_3$-solution of the complex.

Fig. 18 illustrates the molecular structure of the complex of host molecule *60* with

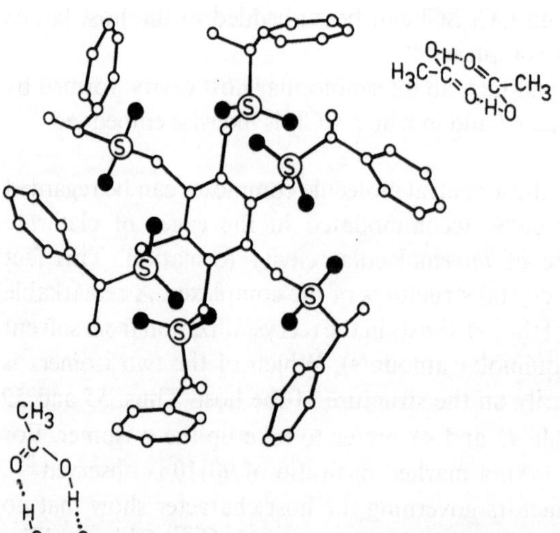

Fig. 18. X-ray structure of the *60* · 2 dimeric acetic acid complex [93]

acetic acid guest molecules. The legs of the hexahost are situated alternately above and below the mean plane of the central benzene ring as has been found for hexakis[4-(benzyl)phenylthiomethyl]benzene (55). Interestingly, the four acetic acid guest molecules appear as hydrogen-bonded dimeric pairs which are crystallographically independent. These discrete hydrogen-bonded units contrast with the infinite chains of hydrogen-bonded molecules found in the crystal structure of acetic acid and also in the 1:1-complex of deoxycholic acid and acetic acid [13b]. Each dimer consists of two planar acetic acid molecules whose mean planes intersect at an angle of approx. 18 degrees. The mean O ... O distance is 2.62 Å. The acetic acid complex of 60 allows the first direct observation of dimeric acetic acid.

The first example of a channel-type inclusion behavior of a hexahost molecule has been found in an X-ray study at −110 °C of the complex formed from hexakis(4-

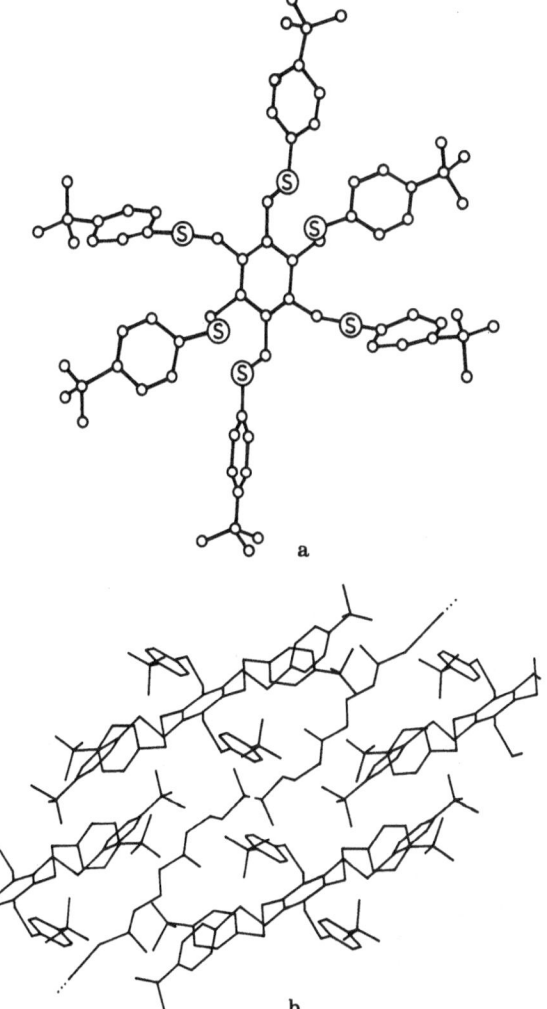

a

b

Fig. 19. a) Structure of the host molecule 49 in the crystal of the squalene complex. [94] b) Illustration of the host-to-guest packing in the complex of hexa host 49 with squalene (66)

349

tert-butylphenylthiomethyl)benzene (*49*) and squalene (*66*); the guest triterpene ($C_{30}H_{50}$) is observed in a novel conformation controlled by host-guest interactions [94].

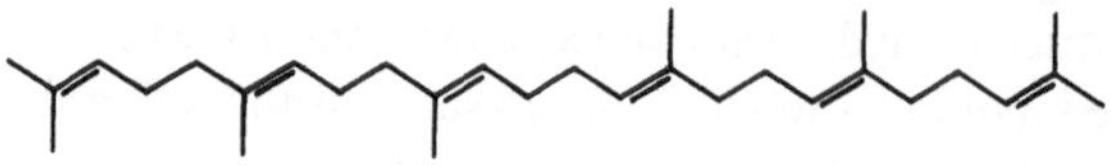

66

Figs. 17–19 show that the legs of hexahosts usually point alternatively above and below the plane of the central benzene ring. The squalene guests are accommodated in continuous channels running through the crystal. The fixation of pure conformations which are normally inaccessible to guest molecules (e.g. squalene) seems to be of theoretical interest because it suggests the possibility of a "conformational selection".

Similarly, the octopus molecules (polypodands) *67–70* [95] containing rigid donor end groups form crystalline complexes with neutral molecules like tetramethylurea, 2,6-dimethylpyridine, 2,4,6-trimethylpyridine, dimethylacetamide, hexamethyl phosphorustriamide, chloroacetonitrile, 2,2,2-trichloroethanol, and diglyme. Even with various crown ethers as guest molecules, e.g. [12]crown-4 (*9a*), [18]crown-6 (*9c*), crystalline complexes (for data see Table 9) may be obtained [95].

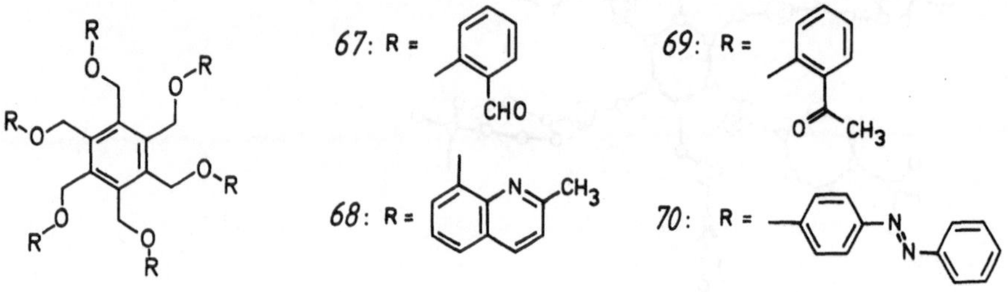

Whether the guest molecules in these complexes are embedded in an intramolecular pseudocavity of the octopus ligand or are again located in lattice cavities between the host molecules is still unknown at present. X-ray analyses of the new complexes will give further information about the nature of host-guest interaction.

A comparison of the *67*- and *69*-tetramethylurea complexes (host-to-guest = 1:2 and 1:1 resp.) shows that different stoichiometries appear, depending on the ligand structure.

Noteworthy is that not only crown ethers themselves (as quest molecules) are incorporated into the host lattice but also that a 1:1 stoichiometry is found for the complex of [18]crown-6 and *67* containing six salicylaldehyde units, while the smaller [12]crown-4 gives a 1:2 stoichiometry. The orange-colored hexaazo compound *70* with [12]crown-4 and tetramethylurea yields crystalline 1:2 complexes.

The above findings suggest that the ligands *67–69* can form different cavities — *intra*- and *inter*molecular ones — which can be filled by small guest molecules, the stoichiometries varying if necessary.

Table 9. Neutral molecule complexes of octopus molecules with rigid end groups and various neutral molecules

Host	Guest	Stoichiometry Host:Guest:H$_2$O	M.p. [°C]	Ref.
67			230–232	[95]
	[12]crown-4	1:2:1	135, (155), 220–226	[95]
	[15]crown-5	1:2:2	120, 215–220	[95]
	[18]crown-6	1:1	130, 185, 195–200	[95]
	pyridine	1:2	120, 245–253	[95]
	2,6-dimethylpyridine	1:2	220, 243–247	[95]
	2,4,6-trimethylpyridine	1:2	220, 247–252	[95]
	dimethylacetamide	1:1	190, 232, 253–254	[95]
	tetramethylurea	1:2	190, 215–220	[95]
	hexamethylphosphoric triamide	1:2:1	210, 230–232	[95]
	chloroacetonitrile	1:1	215, 248–253	[95]
68			247–250	[95]
	[12]crown-4	1:2	190, 245–248	[95]
	[15]crown-5	1:1	150, 243–246	[95]
	dimethylformamide	1:2	245–250	[95]
	dimethylacetamide	1:2	247–250	[95]
	hexamethylphosphoric triamide	1:2:2	147, (230), 240–244	[95]
	chloroacetonitrile	1:3	145, (264), 245–247	[95]
	2,2,2-trichloroethanol	1:3:3	120, 230–233	[95]
	diglyme	1:1:1	145, 160, (170), 235–241	[95]
69			186–190	[95]
	tetramethylurea	1:1	186, 194–197	[95]
70			247–250	[95]
	[12]crown-4	1:2	135, 243–249	[95]
	tetramethylurea	1:2	243–247	[95]

4 Synthetic Host Molecules with Large Endolipophilic Cavities

26 years ago, Stetter and Roos [96 a] described dimeric tri- and tetramethylene-bridged benzidine derivatives *72* and *73* which, after recrystallization, contained stoichiometric amounts of benzene (1:1) and dioxan resp. Even after drying at 80 °C for 5 h, *72* and *73* still contain 1 mol benzene which can only be removed by further drying at 150 °C for 5 h. Dioxan cannot be removed at all from *73*: After 30 h at 150 °C in vacuo, 0.75 mol of the solvent are found. In contrast, the tetra-N-tosylates do not show this behavior. The dioxan adducts of the tetraamines are stable to air, in contrast to the free amines *72* and *73*. In contrast to *72* and *73*, no addition compound is formed with *71* in the usual solvents. These observations led Stetter to the conclusion, in combination with molecular model considerations, that the host cavity should be able to enclose guest molecules of the dimensions of benzene and similar molecules (Fig. 20). Thus, he had *intra*molecular inclusion compounds available.

Similar to the situation in the crown ether series, in which the development starting with open-chain systems could have begun much earlier [6a], Stetter's complexes could have innovated the development in the field of cavity inclusion complexes

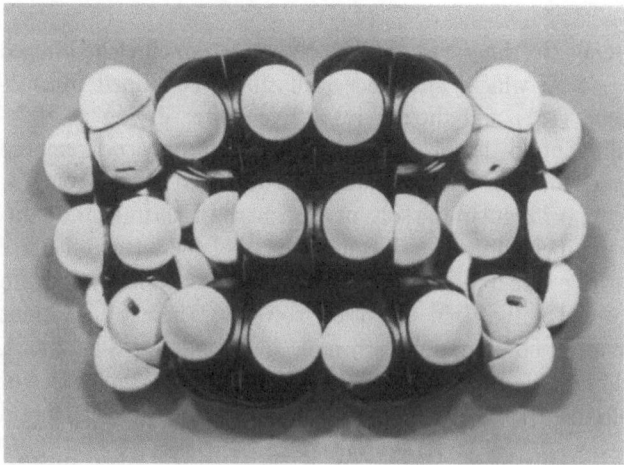

71: n = 2
72: n = 3
73: n = 4

already decades ago, perhaps all the more if at that time an X-ray analysis had unequivocally proven the intramolecular host-guest character of the complexes.

Some years after Stetter's observation, experiments to include guest molecules into lipophilic cavities of ring compounds of the phane type [97)] were performed with the binaphthyl instead of the biphenyl spacer with the result that the synthesized cavity molecule *74* formed analogous solvent complexes [96 b)].

74

Fig. 20. CPK-space filling models of the inclusion complex of the host molecule *73* with the guest molecule benzene

Whereas, after recrystallization from dioxan and methanol, *74* forms a 1:0.9 complex with dioxan, which does not loose the dioxan after drying at 100 °C in vacuo for 20 h, the tetratosylate of *74* binds two water molecules so strongly that it cannot be removed by drying.

Recently, however, the first example of a crystalline complex of a water-soluble paracyclophane with a hydrophobic substrate has been described[98]. This complex was isolated from an aqueous solution and characterized as an inclusion complex by X-ray analysis. The free 1,6,20,25-tetraaza[6.1.6.1]paracyclophane *75* is water-soluble below pH 2; its interactions with substrates having hydrophobic moieties were investigated in acidic aqueous solution:

The fluorescence intensity of 1-anilinonaphthalene-8-sulfonate (1,8-ANS) was markedly enhanced in the presence of *75*, suggesting that 1,8-ANS was transferred into a nonpolar environment and/or subjected to a conformational change by *75*. The Benesi-Hildebrand plot of the fluorescence intensity[99] indicates a 1:1 complex with the dissociation constant of 1.6×10^{-4} M, comparable with other complexes from the known water-soluble paracyclophanes. In the ^{1}H—NMR spectrum the signals of 2,7-dihydroxynaphthalene remarkably moved upfield in the presence of *75*; this is ascribed to a strong shielding effect of the aromatic rings of *75*. Acyclic reference compounds showed only a small effect in both the fluorescence and ^{1}H—NMR spectra. These spectral data suggest that *75* and the substrates are in an intimate contact that does not occur without the cyclic structure of *75*. Furthermore, *75* forms crystalline complexes from acidic solution with a variety of substrates having hydrophobic moieties, e.g. 1,3-dihydroxynaphthalene, 2,7-dihydroxynaphthalene, naphthalene, p-xylene, and durene. The durene complex has the stoichiometry $75 \cdot 4\,HCl \cdot durene \cdot 4\,H_2O$.

The X-ray analysis of this crystalline 1:1 complex is definite evidence of an intramolecular inclusion of a guest molecule into a synthetic host molecule. The shorter $(CH_2)_3$-bridge does not yield this type of complexation.

As shown in Fig. 21, the guest molecule durene is fully enclosed in the cavity of the host molecule; it is located exactly in the middle of the hole. The complex fits on a center of symmetry. The conformation of the host molecule is such that the four benzene rings are perpendicular to the mean plane of the macroring and the bridging chain moieties assume the *ap*-conformation except for the *sc*-conformation about the N(1)—C(2)- and N(20)—C(21)-bonds. As a result, a cavity with rectangularly

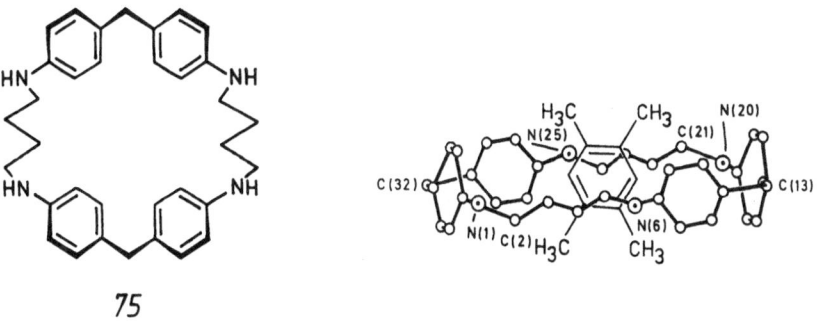

75

Fig. 21. X-ray structure of the $75 \cdot 4\,HCl \cdot durene \cdot 4\,H_2O$ complex (schematically)[98]

Fritz Vögtle, Heinz Sieger and Walter Manfred Müller

shaped open ends (approx. $3.5 \cdot 7.9$ Å = 350) and a depth of 6.5 Å is formed. Hydrophobic interactions as well as charge transfer interactions could play a role in the complex formation between $75 \cdot 4$ HCl and durene.

In contrast to Stetter's host molecule, a new feature is the exohydrophilic character of the protonated host in addition to the endolipophilic property of the cavity. These two properties, on the one side, render 75 soluble in hydrophilic solvents and, on the other side, allow lipophilic guest molecules to be incorporated into the cavity of 75 from aqueous solution. One can accept that this investigation, which is remarkable because of the simple structure of the host molecule, gives impetus so further syntheses of similarly simple and also of more sophisticated molecular cavities.

The following example shows that in spite of the above discussed encouraging studies, an intramolecular cavity must not necessarily enclose potential guest molecules: The complexation of cyclophane 76, which possesses a geometry with the two aromatic rings well separated by the rigid dioxaoctadiyne spacers [100], with aromatic guest molecules proceeds *not* via formation of an inclusion complex but is accompanied by a conformational flattening of 76 similar to a classical open-face stacking complex (cf. Fig. 22).

The hydrophobic nature of the interaction has been studied using the effect on the fluorescence intensity of 1-anilinonaphthalene-8-sulfonate (ANS). A two-fold enhancement was observed for complexed *vs.* uncomplexed ANS. From missing upfield shifts in the ^{1}H—NMR spectra it is concluded that structure 76A is more consistent with the interaction than 76B. Considering this result, the term "inclusion complex" should be used cautiously, since it implies the geometry of a complex that is not necessarily in accord with the anomalously large stability constants [100].

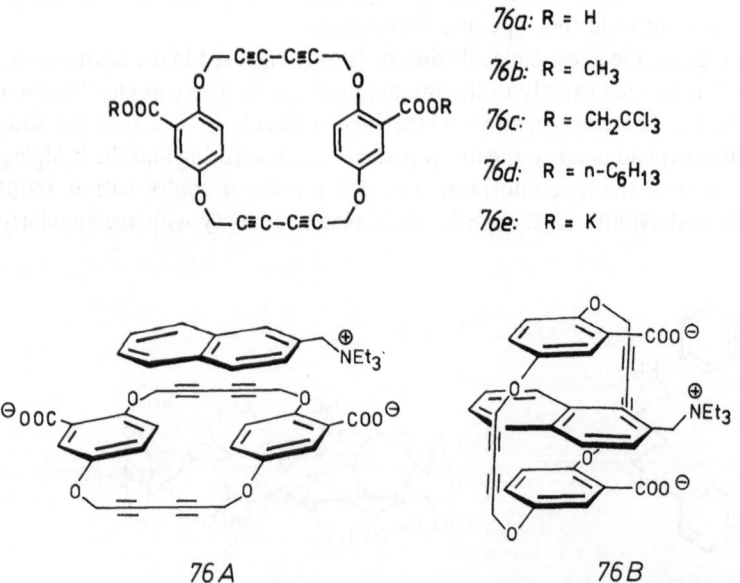

$76a$: R = H
$76b$: R = CH$_3$
$76c$: R = CH$_2$CCl$_3$
$76d$: R = n-C$_6$H$_{13}$
$76e$: R = K

76 A   76 B

Fig. 22. Possible complex geometries of 76 with 2-naphthylmethyl-triethylammonium chloride as guest molecule, structure 76 A being the more probable one [100]

The capped porphyrine "H$_2$-Cap" 77, first synthesized by J. E. Baldwin et al. [101], precipitates as the pentachloroform, monomethanol solvate in large crystals, when over the period of one year solvent diffusion of methanol at 5 °C into a chloroform solution of H$_2$-Cap is allowed to take place [102]. Solvent loss occurs rapidly when crystals are exposed to air at room temperature. During cooling to approx. −180 °C for X-ray data collection, the crystal undergoes a phase change, presumably a consequence of an increased ordering of the solvate molecules. As can be seen from the crystal packing diagram, the solvate molecules (chloroform and methanol) are not included inside the cavity of the molecule but are located between the capped porphyrine molecules. Hydrogen bonding of methanol and interactions of some of the chloroform solvate molecules with the H$_2$-Cap molecule dominate the crystal packing arrangement. In addition, numerous contacts of the van der Waals type exist, some of which are very short.

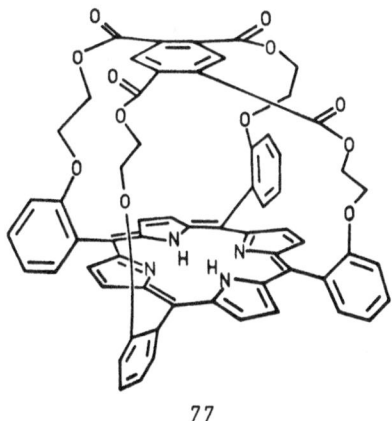

77

5 Complexes with Anions as Guest Particles

In 1968 Simmons [103] reported that protonated in-in isomers of diazabicycloalkanes ("catapinands") incorporate halogen ions in solution into their positive cavity ("catapinates"). The stability of the resulting bonding follows from the electrostatic attraction between the positive cavity and the anion as well as from hydrogen bonds of the kind $^+$N—H ... Cl$^-$ ... H—N$^+$ (Fig. 23) with a N(H) ... Cl distance of 3.10 ± 0.01 Å [104].

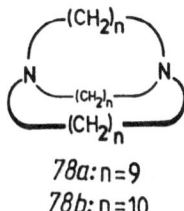

78a: n = 9
78b: n = 10

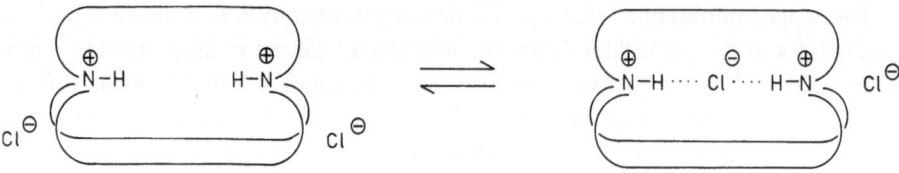

Fig. 23. Schematic representation of the embedding of the chloride ion in the protonated cavity of the catapinand 78 [103)]

The chloride ion is thereby better enclosed than the bromide ion (with a bridge consisting of 9 CH$_2$ groups). When the cavity size is diminished through shortening of the bridges, no enclosure is observed any more.

The heterocyclophane cations 79 and 80 display in aqueous solution a hydrophobic cavity into which hydrophobic guest substances can penetrate and bind. Thus, by means of fluorescence spectroscopy it has been possible to confirm the "complex formation" of 79 and 80 with sodium 1-anilinonaphthalene-8-sulfonate (1,8-ANS) [105)].

79

80

Starting from the metal ion- and neutral molecule-complexing properties of the cryptands, Lehn has been able to show by ^{13}C—NMR spectroscopy that spherical, protonated [106)] ligands such as 81–83 and 78b enclose halogen ions probably through hydrogen bondings in the positively charged cavity (Fig. 24a) [107)]. This has been confirmed by X-ray analysis (Fig. 24b).

81

82

83

The tetraprotonated ligands 81 ("soccer molecule") and 82 are selective for Cl$^-$ and Br$^-$ whereas 78b and 83 are less discriminating. The iodide ion with an anionic

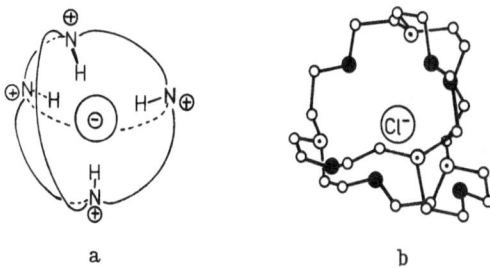

Fig. 24. a) Schematic representation of the enclosure of a chloride anion in a "soccer molecule" *81* [107]; **b)** crystal structure of the *81*H$_4^{4+}$ · Cl$^-$ complex

radius of 3.57 Å is too large for the cavity and is not enclosed. The chloride cryptand adduct has the highest stability; the latter is three orders of magnitude higher than that of bicyclic cryptates such as *78b* with 10 CH$_2$ bridge members [108].

Independent of the configuration and protonation equilibria at the nitrogen atoms, the quarternary cage molecules *84* and *85a,b* are able to coordinate inside their cavity chloride and bromide and even, in the case of *85a,b* the larger iodide ion [109].

84

85a: n=6
85b: n=8

The stability constants (log K$_s$) in water of the inclusion complexes of bromide and iodide with *84* and *85* as determined by halogenide electrodes lie above 2.4 and are thus remarkably high.

The macrotricyclic system *86* containing two TREN units is excellently suited in the hexaprotonated state for the binding of anions of the linear triatomic type XYZ (e.g. azide) as shown impressively by [13]C—NMR spectroscopy [110].

Crystalline salts (*86*—H$_6^{6+}$) 6 X$^-$ (X$^-$ = Cl$^-$, ClO$_4^-$) of *86* containing 6 equivalents of acid have been isolated from methanol/water solutions. The probable arrangement of the azide complex is depicted in Fig. 25.

The found selectivity sequence ClO$_4^-$, Cl$^-$, I$^-$ < CH$_3$CO$_2^-$, Br$^-$ < HCO$_2^-$ < NO$_3^-$, NO$_2^-$ ≪ N$_3^-$ is neither the lyotropic series nor the sequence of the hydration energies. It rather indicates a consequence of the topological discrimination based on the receptor properties of the hexaprotonated ligand *86*, i.e. on the size and form of the intramolecular cavity as well as on the arrangement of the binding sites. The elliptical cryptand *86*—H$_6^{6+}$ is thus a *molecular receptor* for linear triatomic species fitting into the cavity.

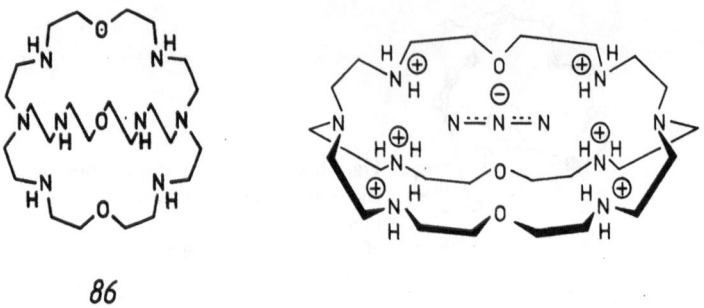

86

Fig. 25. Probable arrangement of the azide ion in the cavity interior of $86 \cdot 6\,H^+$ [110]

Further potential complexons for anions are represented by the macrocyclic polyguanidinium salts *87–89* [111a] and particularly by their acyclic counterparts *90–110* [111b]. These have been synthesized and their complexation with phosphate

87   **88**   **89**

and carboxylate anions relative to the polyammonium salts investigated by pH-metric titration methods. With phosphate ions, relatively stable complexes are obtained in water, the most stable ones being naturally formed with the species of highest charge density (e.g. for ligand 99^{3+}: $\log K_s\,P_2O_7^{4-} > \log K_s\,H_2P_2O_7^{3-} > \log K_s\,H_2P_2O_7^{2-}$). Similarly, the stability constant for a given anion rises with increasing charge of the ligand.

The remarkable chloride/carboxylate selectivity and the relative pH-independence of the polyguanidinium ligands show that such lipophilic anion complexes may serve as selective carriers for anion transport and anion-specific electrodes.

The existence of an imidazolate bridge in the Cu(II)-BIPM^{3+} compound of structure *111* prompted to the synthesis of a corresponding cyclic ligand *112* (Fig. 26), which encloses imidazolate in its interior via two central copper atoms [112].

Such complexes are known or suggested to form part of metallo proteins. An X-ray analysis of the imidazolate complex has already been reported. Thus, as expected, the

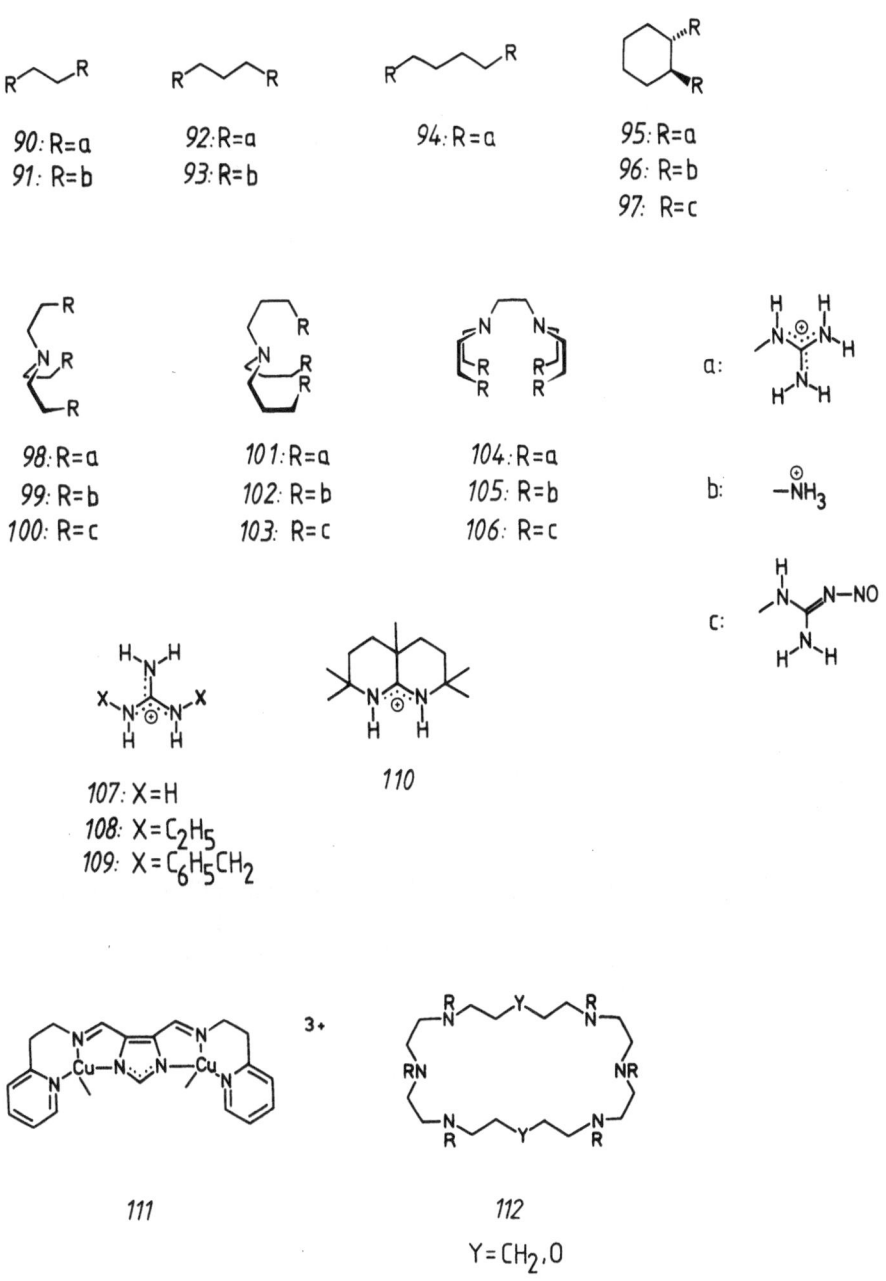

90: R=a
91: R=b

92: R=a
93: R=b

94: R=a

95: R=a
96: R=b
97: R=c

98: R=a
99: R=b
100: R=c

101: R=a
102: R=b
103: R=c

104: R=a
105: R=b
106: R=c

a:

b: $-NH_3$

c:

107: X=H
108: X=C₂H₅
109: X=C₆H₅CH₂

110

3+

111

112

Y=CH₂,O
R=H, Alkyl

Cu(II)$^{3+}$ ion is incorporated into the circular crown cavity in such a way that a penta-coordination with another coordination of a neutral imidazole arises (Fig. 26). The copper geometry is accordingly somewhat distorted trigonal-bipyramidally. The prevailing compound is of relevance with respect to the four Cu(II)-forms of bovine erythrocyte superoxide dismutase [113].

In the macrocyclic Schiff base *113*, the azide ion is likewise bound via metal ions (Fig. 27) [114].

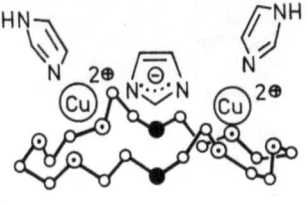

Fig. 26. Crystal structure of the imidazolate macrocyclic complex [112]

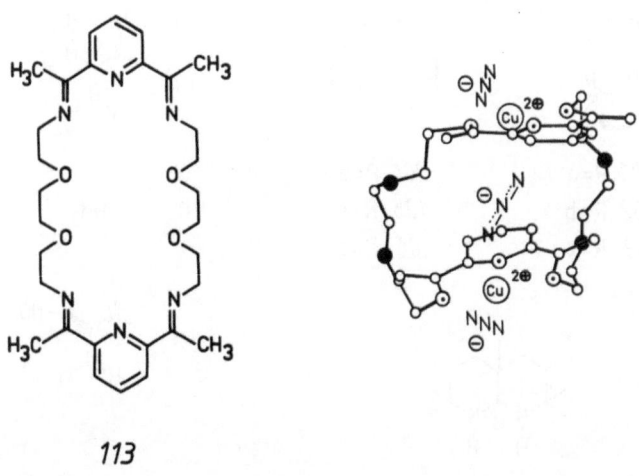

113

Fig. 27. Crystal structure of the azide complex of *113* [114]

With copper salts the macrocyclic ligand *114* forms a bimetallic complex which "provides" an additional cavity for the incorporation of a suitable substrate [115].

Thus, the addition of aqueous sodium azide solution to a methanolic solution of *114* containing $Cu(NO_3)_2$ leads to the formation of dark-green crystals of a tetraazido-di-Cu^{2+}-complex the structure of which is described in Fig. 28.

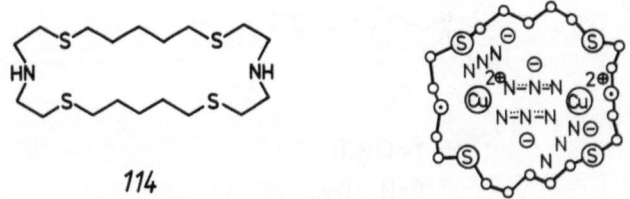

114

Fig. 28. Crystal structure of the tetraazido-di-Cu^{2+}-complex of *114* [115]

Through protonation (deuteration) of bicyclic diamines containing benzene substituents as stiffeners in the bridges, further potential catapinand systems (*115*, *116*) are accessible [116]. While neutral amines appear to be suitable for the incorporation of neutral particles, the protonated form should allow mono- or diamines to be enclosed.

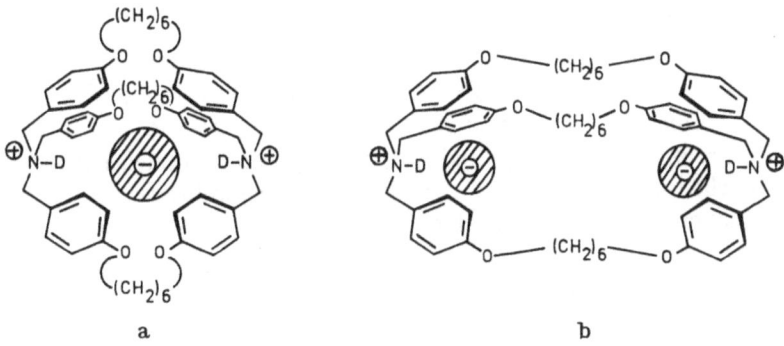

115   *in–in*-form

116

In analogy to the findings about catapinand and cryptand cavities described earlier, the inclusion of bromide and iodide ions in the interior of the protonated molecule *116* is suggested by NMR-spectroscopic investigations (Fig. 29).

Fig. 29. Probable arrangement of the mono- (**a**) and binuclear (**b**) anion inclusion in *116* · D_2^{2+} [116)]

The host molecules *117–122* with lipophilic cavities should, following protonation, also be suitable for the inclusion of anions [117)].

117   *118*

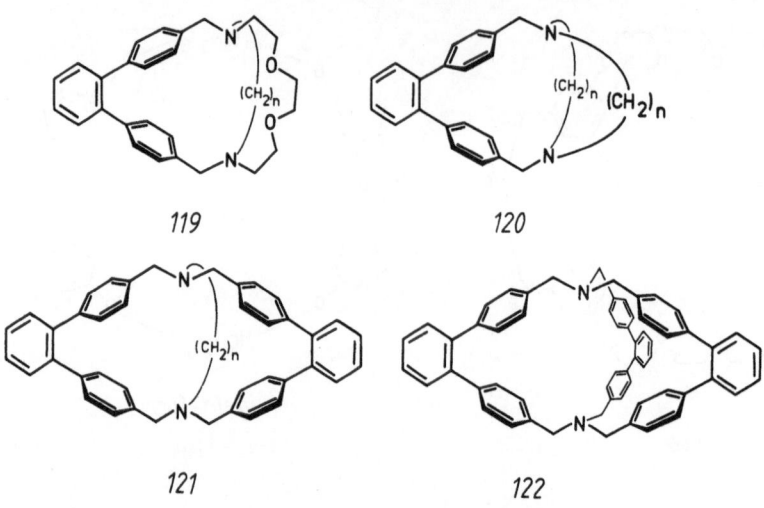

6 Complexes of more than two Components

6.1 Crown Compounds as Host Particles

A complex of dibenzo[18]crown-6, potassium iodide *and* thiourea first been obtained by Pedersen, who called it a "doubly wrapped salt" [19], has been prepared again [20] and investigated by the X-ray method [118]. The result is illustrated in Fig. 30: The seven-fold coordination of the potassium ion is very similar to that in the 1:1 complex of dibenzo[18]crown-6 (*5a*) and KI [119a]. The K^+ ion is located in the center of the macrocyclic ring with all the $K^+ \dots O$ distances being in the range 2.71–2.80 Å. The coordination sphere of the metal ion is completed by the iodide ion ($K^+ \dots I^-$ distance 3.57 Å).

Thiourea is not involved in the complexation of the cation nor does it have any contact to the polyether, but forms polymeric, hydrogen-bonded chains. One hydrogen

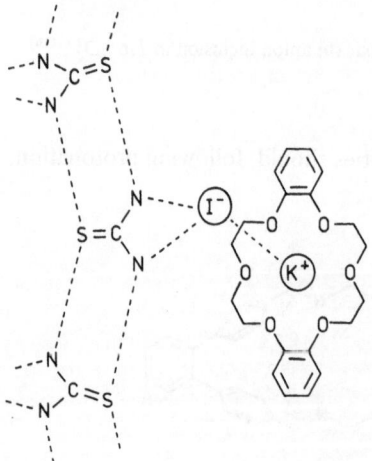

Fig. 30. X-ray structure of the dibenzo-[18] crown-6 · KI · thiourea complex (schematically) [118]

atom from each amino group is bound to the sulfur atom of the next molecule while the other hydrogen atom is in contact with the iodide ion. This arrangement resembles the situation met with some metal salt complexes of thiourea [38], the main difference being the lack of interactions between the sulfur and the metal ion.

It has also been possible to isolate a 1:1:1 [18]crown-6-thioacetamide KI complex [20].

"Ternary complexes" composed of a N-methyldihydropyridino crown dilactone, sodium perchlorate and acetonitrile or acetone have been isolated by R. M. Kellogg et al. [119b].

A three-component neutral molecular complex, starting from [18]crown-6, malono-dinitrile and water, has been obtained [31]. X-ray studies show the two guest components unsymmetrically hydrogen bonded to the crown, facing each other [31].

6.2 Cryptands as Host Compounds

The macrobicyclic heptaether *123* forms, besides a KCl · 5 H_2O complex in which the water molecules are disordered, a 1:1:1 complex with naphthalene-2,3-diol and water [120].

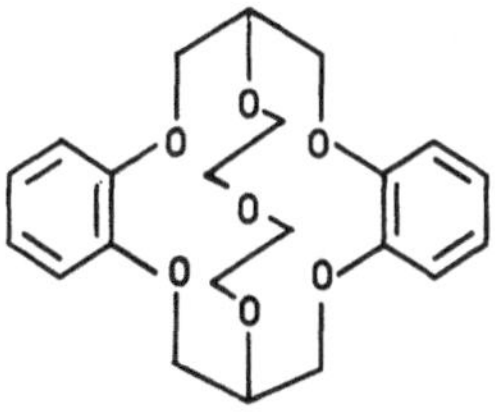

123

X-ray analysis indicates a monoclinic unit cell containing four molecules. Each unit cell consists of naphthalene-2,3-diol with the usual internal hydrogen bond and the second hydroxy hydrogen atom bonded to a water molecule [H ... O distance 1.79 Å] which, in turn, forms hydrogen bonds to four oxygen atoms of the bicyclic host compound. One of these bonds is bifurcated (H ... O distance 2.33 and 2.36 Å) while the other one has unequal H ... O distances (2.03 and 2.56 Å).

6.3 Podands as Host Compounds

Analogous to crown compounds with several acyclic neutral ligands, complexes containing stoichiometric amounts of thiourea *and* salts additionally could be isolated [20]. Data on the hitherto found stoichiometries of this type of three-component complexes are given in Table 10.

The elucidation of the bonding relationships in these symbiotic complexes by X-ray analysis [67] appears to be attractive because e.g. in the case of the acyclic ligand *28* both the structure of the pure thiourea complex [60] and that of the pure analogous salt complex (with RbI as complexed salt, however) [61,62] have been determined.

Table 10. Three-component complexes of coronands and open chain podands with neutral molecules and metal salts

Host	Guest	Salt	Stoichiometry Host:Guest:Salt	M.p. [°C]	Ref.
2a	thiourea	KSCN	2:1:1	159–162	[19]
5a	thiourea	NaSCN	1:1:1	230–231	[19]
5a	thiourea	KSCN	1:1:1	242–253	[19]
5a	thiourea	KSCN	1:6:1	164–183	[19]
5a	thiourea	KI	1:1:1	208–210	[19]
5a	thiourea	RbSCN	1:4:1	160–190	[19]
9c	thioacetamide	KI	1:1:1	125–130	[20]
28	thiourea	KSCN	1:1:1	115–119	[20]
38	thiourea	KSCN	1:1:1	136–140	[20]

Apart from the possibility that KSCN or thiourea may be embedded only in lattice spaces of the corresponding neutral complex, a comparison of the "super-complex" [121], the neutral molecule complex and the pure salt complex will give information about the preferred coordination sites and the resulting bondings of a neutral podand like 28 to thiourea and an additionally enclosed salt. Orientating X-ray investigations have confirmed that all three components are present in the unit cell of the 28-thiourea-KSCN complex [67].

Treatment of [12]crown-4 (9a) and LiSCN with the octopus compound 67 gives a crystalline complex (m.p. 230 °C, dec.) [95] that contains all three components and shows in its IR spectrum the SCN⁻ absorption at 2070 cm^{+1}.

6.4 Water-Containing Salt Complexes

Of the many known examples of this complex species only some newer ones will be discussed.

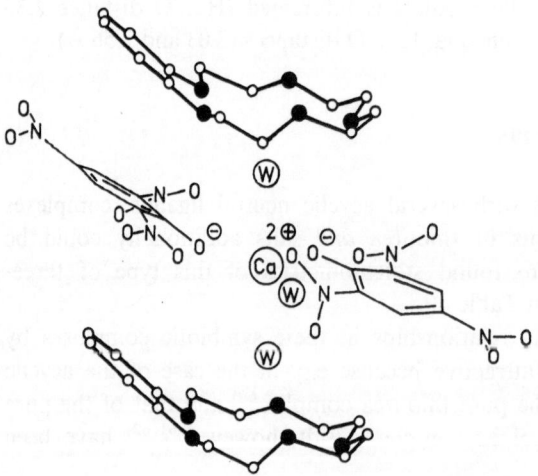

Fig. 31. X-ray structure of the Ca (picrate)$_2$benzo[15]crown-5 · 3 H$_2$O complex (schematically) [122]. W = Water

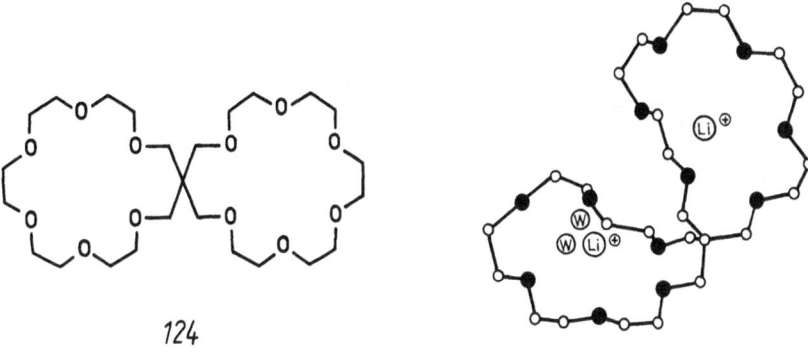

124

Fig. 32. X-ray structure of the LiI complex of the spiro crown *124* [124)]

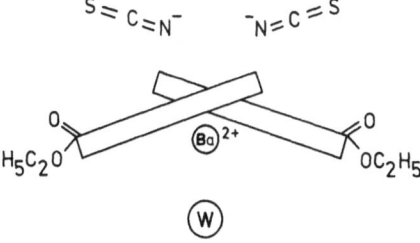

The calcium picrate complex of benzo[15]crown-5 (*2a*) has the stoichiometry Ca-(picrate)$_2$ · benzo[15]crown-5 · 3 H$_2$O [122)]. Whereas the picrate anions are bound directly — and partly via the NO$_2$ oxygen atoms — to the calcium, the water molecules form bridges to the benzo crown (Fig. 31). Here, we have not a calcium complex of the crown ether but the water molecules form hydrogen bridges to the five oxygen atoms of the crown and the oxygen atom of the water is bound to calcium.

In addition to two lithium atoms, a lithium complex of the spiro crown [123)] *124* contains four water molecules which are not located in the center of the crown rings, because of their small size [124)] (Fig. 32).

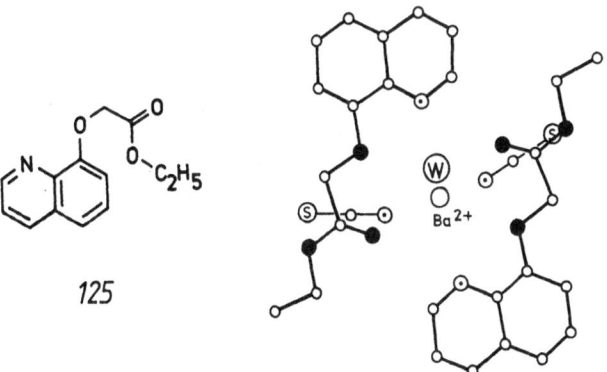

125

Fig. 33. Two views of the structure of the Ba(SCN)$_2$ complex of *125* [126)]

The Ba(SCN)$_2$ complex [125] of the open-chain ethyl 8-quinolyloxy-acetate *125* contains, in addition to the two ligand molecules and one Ba^{2+} cation, a water molecule which functions as a ligand at barium at the opposite side of the two coordinating SCN$^-$ anions [126]. Fig. 33 shows two different views of the complex, as X-ray analysis revealed. The two ligands are arranged approx. planar. The barium cation is coordinated by nine donor centers. The participation of the carbonyl oxygen atom of the ligand in the coordination is remarkable.

Similar to the calcium (picrate)$_2$-benzo[15]crown-5 · 3 H$_2$O complex, a 1:1:1 stoichiometry is found for the tetraethylene glycol · calcium dipicrate · 1 H$_2$O complex [127]. The calcium picrate here also forms a unity and water occupies open coordination positions at the calcium ion (Fig. 34).

126

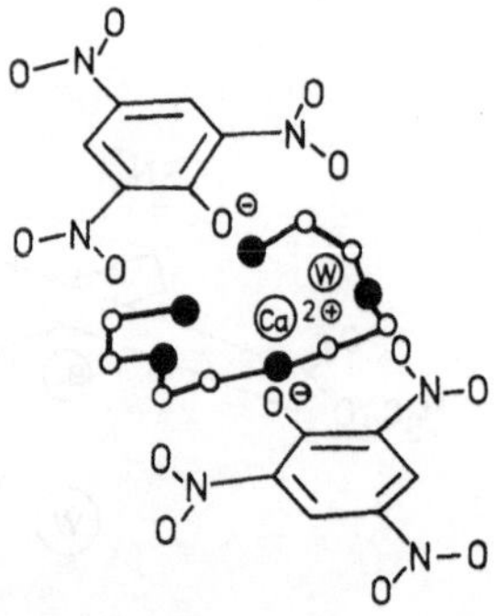

Fig. 34. X-ray diagram of the calcium picrate complex of *126* [127]

1,11-Bis(2-acetylaminophenoxy)-3,6,9-trioxaundecane (*127*) forms a KSCN complex that contains 1/2 mole of water [128]. Whereas the free ligand *127* is S-shaped and, upon K$^+$ complexation, changes the signs of some synclinal OCCO-torsion angles, CCOC-angles remain *ap*; the *127* · 2 KSCN · H$_2$O complex contains two different molecules, *127A* · KSCN · H$_2$O and *127B* · KSCN. The K$^+$-ions are circularly surrounded by the ligands and coordinated with only four of the five polyether oxygen atoms. K$^+$ ions further interact with H$_2$O, with the nitrogen of SCN$^-$ and with the acetylamino oxygen atoms of *127*. The latter interaction leads to polymeric structures which have previously not been observed for similar complexes [128] (Fig. 35).

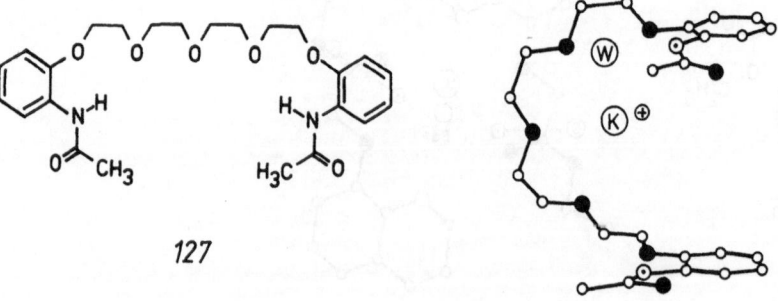

127

Fig. 35. X-ray structure of the *127* · KSCN · 1/2 H$_2$O complex [128]. W = Water

In analogy to the $3\,CoCl_2 \cdot 6\,H_2O$ complex of the octopus ligand *128* [95], the hexapodand structures *129* and *130* yield metal ion complexes containing several water molecules the binding geometry of which has not yet been elucidated [129].

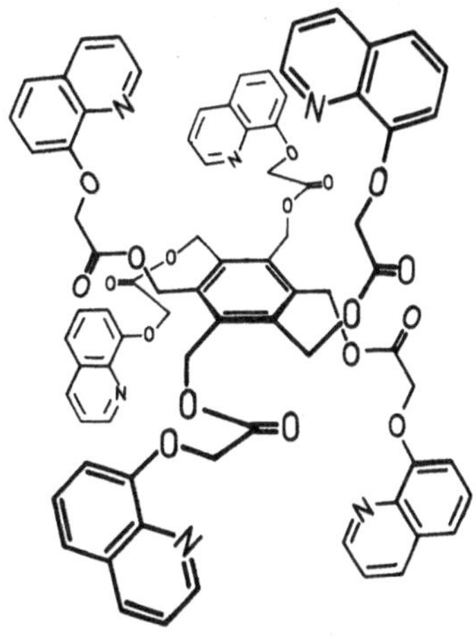

128

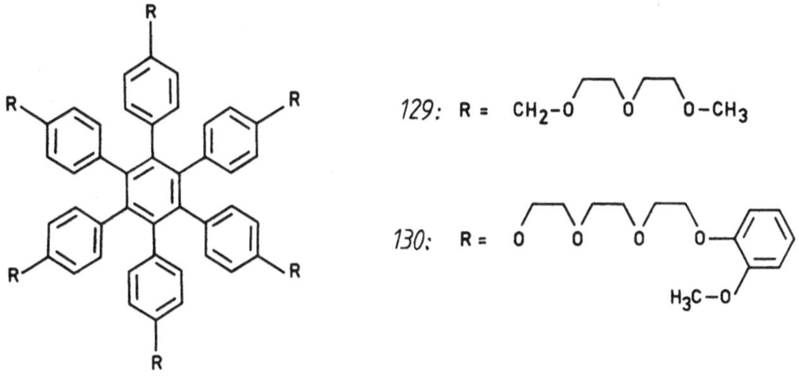

129: R = CH_2-O $\frown$ O $\frown$ $O-CH_3$

130: R = O $\frown$ O $\frown$ O $\frown$ $O-$ ... H_3C-O

6.5 Supercomplexes of γ-Cyclodextrin

In crown ether/cryptand metal ion complexes the ligand frame-work is known to be more rigidly fixed than in the free ligands.

Table 11. γ-Cyclodextrin supercomplexes

Guest	Stoichiometry γ-CD:Guest:H_2O	Darkening with decomposition at °C	Ref.
[12]crown-4 · LiSCN	1:1	273	130)
[2.1.1]cryptand · LiClO$_4$	1:1:1	283–288	130)
[2.2.1]cryptand · Ca(SCN)$_2$	2:1	276	130)
[2.2.2]cryptand · Ba(SCN)$_2$	2:1	281–292	130)

Nevertheless, as may be seen from Table 11, "supercomplexes" [121] are available through enclosure of rigid crown-ether metal ion complexes and cryptand metal ion complexes in the cyclodextrin cavity [130] whereby the cation is enclosed by the crown and the latter is again encapsulated by γ-cyclodextrin (131, "doll in the doll in the doll").

How far this will affect the selectivity of the stiffened crown ether cavity toward cations and other molecules remains to be investigated.

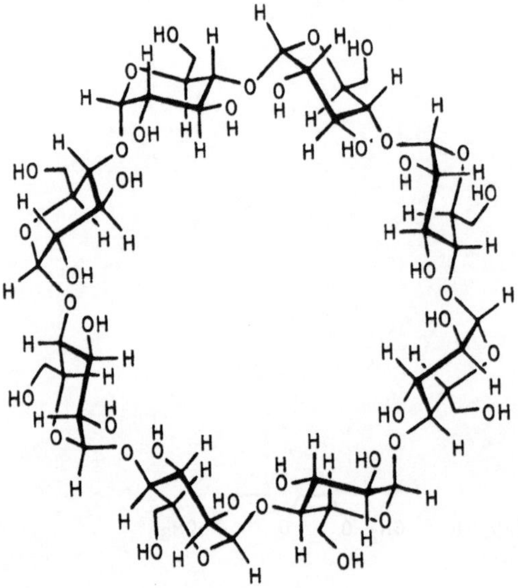

131

7 Outlook

The intramolecular inclusion of uncharged organic molecules into synthetic cavities seems to be generally possible today. The far aim of the "cavity chemistry" now can be considered to be closer, namely to include into larger cavities two different molecular species, e.g. substrate and reagent, and to specificly let them react with each other in the cavities. The formation of oligomers or unwanted by-products should then be prevented. For such "cavity catalysts" analogies exist, e.g. the formation of petrol from methanol within crystal lattice cavities of zeolithe catalysts (Mobil process).

The molecular encapsulation of reagents for the sake of changing reactivities/selectivities should also be an attractive aim of further studies.

The specific inclusion of water and similarly uncharged molecules for the dilution of materials or for the formation of dispersions could be possible.

Column materials consisting of polymers containing molecular cavities of different diameters with selective binding power may be valuable for affinity chromatography, and also for the removal of toxic substances or metabolites from physiological solutions.

"Cavity chemistry" as a branch of organic chemistry will not rest until concave cavities for a lot of important convex molecules have been synthesized and their host-guest relations and host-guest chemistry studied.

Acknowledgement. The authors thank Barbara Jendrny, Dipl.-Chem. Mechthild Wittek and cand. chem. Michael Palmer for their help in preparing the manuscript and Dr. Edwin Weber for his advice.

8 References

1. a) Saenger, W.: Umschau 74, 635 (1974); b) MacNicol, D. D., McKendrick, J. J., Wilson, D. R.: Chem. Soc. Rev. 7, 65 (1978); c) Mandelcorn, L. (ed.): Non-stoichiometric compounds. New York: Academic Press 1964; d) Kuhn, H.: Ber. Bunsenges. Phys. Chem. 80, 1209 (1976); e) Frömming, K.-H.: Pharm. in uns. Zeit 2, 109 (1973); f) Gawalek, G.: Einschluß-verbindungen, Additionsverbindungen, Clathrate. Berlin: VEB Deutscher Verlag der Wissen-schaften 1969; g) Hagan, M.: Clathrate inclusion compounds. New York: Reinhold Publ. Co. 1962; h) Cramer, F.: Angew. Chem. 64, 437 (1952); 68, 115 (1956); Cramer, F.: Einschluß-verbindungen. Berlin, Heidelberg, New York: Springer Verlag 1954; Cramer, F.: Methoden der Organischen Chemie. Houben-Weyl-Müller (eds.), vol. I/1, 4th edit., p. 445. Stuttgart: Georg Thieme Verlag 1958; i) Schlenk, Jr., W.: Chem. in uns. Zeit 3, 120 (1969); Schlenk, Jr., W.: Methoden der Organischen Chemie. Houben-Weyl-Müller (eds.), vol. I/1, 4th edit., p. 391. Stuttgart: Georg Thieme Verlag 1958; Schlenk, Jr., W.: Fortschr. Chem. Forsch. 2, 92 (1951); j) Pollmer, K.: Z. Chem. 19, 81 (1979); k) Davies, J. E. D.: J. Chem. Educ. 54, 536 (1977); l) Hein, F., Heyn, B.: Chemie der Komplexverbindungen. Leipzig: Hirzel Verlag 1978; m) Pfeiffer, P.: Organische Molekülverbindungen, 2nd edit. Stuttgart: Enke Verlag 1927
2. Kohler, F., Schneider, G.: Nachr. Chem. Techn. Lab. 25, 381 (1977); see also: Jones, D. S. in Coffey, S. (ed.): Rodd's chemistry of carbon compounds, vol. IV, 2nd ed., part L, p. 186 ff. Amsterdam, Oxford, New York: Elsevier Publ. Co. 1980
3. Foster, R. (ed.): Organic charge-transfer complexes, London: Academic Press 1969; Briegleb, G.: Elektronen-Donator-Acceptor-Komplexe. Berlin, Heidelberg, New York: Springer Verlag 1961; Foster, R. (ed.): Molecular complexes, vols. 1 and 2. London: P. Elek (Scientific Books) Ltd. 1973; Foster, R. (ed.): Molecular association, vol. 1. London: Academic Press 1975; Kover, K. A., Mayer, W.: Pharm. in uns. Zeit 8, 46 (1979)
4. a) Vögtle, F.: Chimia 33, 239 (1979); b) Vögtle, F.: Pure Appl. Chem. 52, 2405 (1980)
5. Review: Vögtle, F., Weber, E.: Kontakte (Merck) 1/77, 11; 2/77, 16; 3/77, 36; 2/78, 16; Vögtle, F., Weber, E., Elben, U.: Kontakte (Merck) 3/78, 32; 1/79, 3; 2/80, 36; 1/81, 24. Vögtle, F., Weber, E.: The chemistry of the ether linkage Part II. Patai, S. (ed.), chapter 2, p. 59 ff. New York, London: John Wiley & Sons 1980; Pedersen, C. J., Frensdorff, H. K.: Angew. Chem. 84, 16 (1972); Angew. Chem., Int. Ed. Engl. 11, 16 (1972); Lehn, J. M.: Struct. Bonding 16, 1 (1973); Lehn, J. M.: Pure Appl. Chem. 51, 979 (1979); Lehn, J. M.: Acc. Chem. Res. 11. 49 (1978); Prelog, V.: Pure Appl. Chem. 50, 893 (1978); Cram, D. J., Cram, J. M.: Science 183, 803 (1974); Cram, D. J., Cram, J. M.: Acc. Chem. Res. 11, 8 (1978); Stoddart, J. F.: Chem. Soc. Rev. 8, 85 (1979); Izatt, R. M., Christensen, J. J. (eds.): Synthetic multidentate macrocyclic compounds. New York: Academic Press 1978; Izatt, R. M., Christensen, J. J. (eds.): Progress in macrocyclic chemistry, vol. I. New York: John Wiley and Sons 1979; Bradshaw, J. S. et al.: Chem. Rev. 79, 37 (1979); Bradshaw, J. S., Stott, P. E.: Tetrahedron 36, 461 (1980)
6. a) Review: Weber, E., Vögtle, F.: Angew. Chem. 91, 813 (1979); Angew. Chem., Int. Ed. Engl. 18, 753 (1979); b) for classification and nomenclature see: Weber, E., Vögtle, F.: Inorg. Chim. Acta 45, L65 (1980)
7. The definitions complex, adduct, addition compound, molecular compound, molecular com-plex, associate, inclusion compound, clathrate, host-guest complex, key-lock complex etc. are inconsistently used in the literature. The new term "neutral molecule complex" denotes a complex between two or more uncharged molecules. In contrast, the term "neutral com-plexes" may designate complexes of e.g. crown ethers with metal salts where the total charge is also zero

8. a) Schlenk, Jr., W.: Liebigs Ann. Chem. *1973*, 1145; b) Otto, J.: Acta Crystallogr. *B28*, 543 (1972); c) Schiessler, R. W., Flitter, D.: J. Am. Chem. Soc. *74*, 1720 (1952)

9. Schlenk, Jr., W.: Liebigs Ann. Chem. *573*, 142 (1951)

10. a) Dianin, A. P.: J. Russ. Phys. Chem. Soc. *46*, 1310 (1914); Baker, W., McOmie, J. F. W.: Chem. Ind. (London) *1955*, 256; b) Collett, A., Jacques, J.: J. Chem. Soc. Chem. Commun. *1976*, 708; c) Gall, J. H. et al.: J. Chem. Soc. Perkin II, *1979*, 376; d) Hardy, A. D. U. et al.: J. Chem. Soc. Chem. Commun. *1977*, 292; e) Flippen, J. L., Karle, J., Karle, I. L.: J. Am. Chem. Soc. *92*, 3749 (1970)

11. a) Farina, M., Allegra, G., Natta, G.: J. Am. Chem. Soc. *86*, 516 (1964); b) Allegra, G. et al.: J. Chem. Soc. B *1967*, 1020; c) Allegra, G. et al.: J. Chem. Soc. B *1967*, 1028; d) Natta, G., Farina, M.: Struktur und Verhalten von Molekülen im Raum. Weinheim: Verlag Chemie 1976

12. Lawton, D., Powell, H. M.: J. Chem. Soc. *1958*, 2339; Downing, A. P., Ollis, W. D., Sutherland, I. O.: J. Chem. Soc. B *1970*, 24; Brunie, S., Tsoucaris, G.: Cryst. Struct. Commun. *3*, 481 (1974); Williams, D. J., Lawton, D.: Tetrahedron Lett. *1975*, 111; Baker, W., Gilbert, B., Ollis, W. D.: J. Chem. Soc. *1952*, 1443; Arad-Yellin, R. et al.: Tetrahedron Lett. *1980*, 387

13. De Sanctis, S. C. et al.: Acta Crystallogr. *B28*, 3656 (1972), and references cited therein; Craven, B. M., De Titta, G. T.: J. Chem. Soc. Chem. Commun. *1972*, 530

14. Child, Jr., W. C.: Quart. Rev. *18*, 321 (1964); Powell, H. M.: J. Chem. Soc. *1948*, 61; Mark, T. C. W. et al.: J. Chem. Soc. Perkin II, *1976*, 1169

15. Wong, R. Y. et al.: Acta Crystallogr. *B32*, 2396 (1976)

16. Allemand, A., Gerdil, R.: Acta Crystallogr. *A31*, s130 (1975)

17. Allcock, H. R. et al.: J. Am. Chem. Soc. *98*, 5120 (1976); Allcock, H. R., Stein, M. T.: J. Am. Chem. Soc. *96*, 49 (1974)

18. a) Saenger, W.: Angew. Chem. *92*, 343 (1980); Angew. Chem., Int. Ed. Engl. *19*, 344 (1980); b) Bender, M. L., Komiyama, M.: Cyclodextrin chemistry. Berlin, Heidelberg, New York: Springer-Verlag 1978; c) Szejtli, J.: Stärke *30*, 427 (1978); d) Bergeron, R. J.: J. Chem. Educ. *54*, 204 (1977); e) Saenger, W.: Environmental effects on molecular structure and properties. Pullman, B. (ed.), p. 265ff. Dordrecht, Holland; D. Reidel Publ. Co. 1976; f) Griffiths, D. W., Bender, M. L.: Adv. Catalysis *23*, 209 (1973); g) Cramer, F., Hettler, H.: Naturwiss. *54*, 625 (1967); h) Thoma, J. A., Stewart, L.: Starch: Chemistry and technology. Whistler, R. L., Paschall, E. F. (eds.), vol. 1, p. 209. New York: Academic Press 1965; i) French, D.: Adv. Carbohydr. Chem. *12*, 189 (1957)

19. Pedersen, C. J.: J. Org. Chem. *36*, 1690 (1971)

20. Vögtle, F., Müller, W. M.: unpublished results

21. Knöchel, A. et al.: J. Chem. Soc. Chem. Commun. *1978*, 595

22. McLachland, R. D.: Spectrochim. Acta Vol. *A30*, 2153 (1974)

23. Liotta, C. L., Harris, H. P.: J. Am. Chem. Soc. *96*, 2250 (1974); Liotta, C. L.: US. Patent 3,997, 562, 1976

24. a) Gokel, G. W. et al.: J. Org. Chem. *39*, 2445 (1974); b) Org. Synth. *57*, 30 (1977)

25. Goldberg, I.: Acta Crystallogr. *B31*, 754 (1975)

26. El Basyony, A. et al.: Z. Naturforsch. *31b*, 1192 (1976)

27. a) Maverick, E. et al.: Acta Crystallogr. *B36*, 615 (1980); b) Herbert, J. A., Truter, M. R.: Acta Crystallogr. *B37* (1981), in press.

28. a) Vögtle, F., Müller, W. M., Weber, E.: Chem. Ber. *113*, 1130 (1980); b) Weber, E., Vögtle, F.: Chem. Ber. *109*, 1803 (1976); c) Frensch, K., Oepen, G., Vögtle, F.: Liebigs Ann. Chem. *1979*, 858

29. Vögtle, F., Panagiotidis, I.: unpublished results

30. Kaufmann, R. et al.: Chem. Ber. *110*, 2249 (1977)

31. Knöchel, A.: private communication

32. a) von Deuten, K. et al.: J. Chem. Res. (S) *1979*, 358; (M) *1979*, 4035; b) Chang, C. K.: J. Am. Chem. Soc. *99*, 2819 (1977)

33. Vögtle, F., Müller, W. M.: Naturwissenschaften *67*, 255 (1980)

34. Vögtle, F., Müller, W. M.: Chem. Ber. *113*, 2081 (1980)

35. Saenger, W., Hilgenfeld, R.: Z. Naturforsch., in press

36. Shine, H. J.: J. Org. Chem. *24*, 252 (1959); Parrick, J., Rasburn, J. W.: Can. J. Chem. *43*, 3453

(1965); Hanna, J. G.: The chemistry of the carbonyl group. Patai, S. (ed.), vol. 1, pp. 375, 390. New York, London: Interscience Publishers 1966

37. a) Vögtle, F., Müller, W. M.: Chem. Ber. *114* (1981), in press; b) Weber, G., Sheldrick, G. M., Acta Crystallogr. 1981, in press. We thank Dr. G. Weber (Univ. Göttingen) for this communication before publication

38. Review: Truter, M. R.: Struct. Bonding *16*, 71 (1973)

39. Mercer, M., Truter, M. R.: J. Chem. Soc. Dalton *1973*, 2469

40. Hughes, D. L.: J. Chem. Soc. Dalton *1975*, 2374

41. Hughes, D. L., Wingfield, J. N.: J. Chem. Soc. Chem. Commun. *1977*, 80

42. Christensen, J. J., Eatough, D. J., Izatt, R. M.: Chem. Rev. *74*, 351 (1974)

43. de Jong, F., Reinhoudt, D. N., Smit, C. J.: Tetrahedron Lett. *1976*, 1371

44. a) Goldberg, I.: Acta Crystallogr. *B34*, 3387 (1978); b) Helgeson, R. C., Tarnowski, T. L., Cram, D. J.: J. Org. Chem. *44*, 2538 (1979)

45. Dunitz, J. D., Seiler, P.: Acta Crystallogr. *B30*, 2739 (1974)

46. Mallinson, P. R.: J. Chem. Soc. Perkin Trans. 2, *1975*, 266

47. Newkome, G. R., Fronczek, F. R., Kohli, D. K.: J. Am. Chem. Soc. *103* (1981), in press. Two further crystalline water complexes of coronands containing pyridino ketone units have lately been isolated and studied by X-ray analysis: Newkome, G. R., Vögtle, F. et al., in press.

48. Frensch, K., Vögtle, F.: Liebigs Ann. Chem. *1979*, 2121

49. Weber, E., Vögtle, F.: Angew. Chem. *92* (1980), 1067; Angew. Chem., Int. Ed. Engl. *19* (1980), 1030

50. Shchori, E., Jagur-Grodzinski, J.: Isr. J. Chem. *10*, 935 (1972)

51. Pannell, K. H., Mayer, A.: J. Chem. Soc. Chem. Commun. *1979*, 132

52. Humphry-Baker, R. et al.: Angew. Chem. *91*, 669 (1979); Angew. Chem., Int. Ed. Engl. *18*, 630 (1979)

53. Dye, J. L.: Angew. Chem. *91*, 613 (1979); Angew. Chem., Int. Ed. Engl. *18*, 587 (1979)

54. Harned, R. L. et al.: Proc. Indiana Acad. Sci. *59*, 38 (1959); Steinrauf, L. K., Pinkerton, M., Chamberlin, J. W.: Biochem. Biophys. Res. Commun. *33*, 29 (1968)

55. Brockmann, H., Schmidt-Kastner, G.: Chem. Ber. *88*, 57 (1955); Pinkerton, M., Steinrauf, L. K., Dawkins, P.: Biochem. Biophys. Res. Commun. *35*, 512 (1969)

56. Raßhofer, W., Vögtle, F.: Tetrahedron Lett. *1978*, 309

57. Wendt, B., Ried, W.: Angew. Chem. *63*, 218 (1951)

58. a) Cf. also: Weber, E., Vögtle, F.: Tetrahedron Lett. *1975*, 2415; b) Vögtle, F., Sieger, H.: Angew. Chem. *89*, 410 (1977); Angew. Chem., Int. Ed. Engl. *16*, 396 (1977)

59. Suh, I.-H., Saenger, W.: Angew. Chem. *90*, 565 (1978); Angew. Chem., Int. Ed. Engl. *17*, 534 (1978)

60. Weber, G., Saenger, W.: Acta Crystallogr. *B36*, 424 (1980)

61. Saenger, W. et al.: Metal-ligand interactions in organic chemistry and biochemistry. Pullman, B., Goldblum, W. (eds.), p. 363. Dordrecht, Holland: D. Reidel Publ. Co. 1977

62. Saenger, W., Brand, H.: Acta Crystallogr. *B35*, 838 (1979)

63. Heimann, U., Vögtle, F.: Chem. Ber. *112*, 3034 (1979)

64. Vögtle, F., Oepen, G., Raßhofer, W.: Liebigs Ann. Chem. *1979*, 1577

65. Newcomb, M., Gokel, G. W., Cram, D. J.: J. Am. Chem. Soc. *96*, 6810 (1974); Newcomb, M. et al.: J. Am. Chem. Soc. *99*, 6392 (1977)

66. Oepen, G., Vögtle, F.: Liebigs Ann. Chem. *1980*, 512

67. Saenger, W.: private communication

68. Hayward, R. C.: Nachr. Chem. Techn. Lab. *25*, 15 (1977)

69. a) Lutz, W. K., Winkler, F. K., Dunitz, J. D.: Helv. Chim. Acta *54*, 1103 (1971); b) Agtarap, A. et al.: J. Am. Chem. Soc. *89*, 5737 (1967); c) Pinkerton, M., Steinrauf, L. K.: J. Mol. Biol. *49*, 533 (1970); d) Lutz, W. K., Wipf, H.-K., Simon, W.: Helv. Chim. Acta *53*, 1741 (1970)

70. Lindenbaum, S., Sternson, L., Rippel, S.: J. Chem. Soc. Chem. Commun. *1977*, 268

71. a) Hodgson, K. O.: Intra-Sci. Chem. Rept. *8*, 27 (1974); b) Burgermeister, W., Winkler-Oswatitsch, R.: Top. Curr. Chem. *69*, 91 (1977)

72. Agtarap, A., Chamberlin, J. N.: Antimicrob. Agents Chemother. *1967*, 359; see also: Collum, D. B., McDonald, III., J. H., Still, W. C.: J. Am. Chem. Soc. *102*, 2117 (1980)

73. Kubota, T. et al.: J. Chem. Soc. Chem. Commun. *1968*, 1541; Shiro, M., Koyama, H.: J. Chem. Soc. B *1970*, 243

74. Weber, G., Saenger, W.: Acta Crystallogr. *B36*, 207 (1980)
75. Mootz, D., Wussow, H. G.: Angew. Chem. *92*, 559 (1980); Angew. Chem., Int. Ed. Engl. *19*, 552 (1980): In the pyridine 2.5 H_2O complex the water molecules form a hydrogen-bridged separate layer, some hydrogen bridges binding to the nitrogen atoms of the pyridine rings in another layer
76. a) Oepen, G., J. P. Dix, Vögtle, F.: Liebigs Ann. Chem. *1978*, 1592; b) Saenger, W. (MPI Göttingen): private communication
77. Vögtle, F. et al.: Angew. Chem. *89*, 564 (1977); Angew. Chem., Int. Ed. Engl. *16*, 548 (1977)
78. Weber, G., Sheldrick, G. M.: Acta Crystallogr. *B36*, 1978 (1980). Crystal structure of the RbI complex: Weber, G., Sheldrick, G. M.: Inorg. Chim. Acta *45*, L35 (1980)
79. Dunitz, D. J. et al.: Acta Crystallogr. *B30*, 2733 (1974)
80. a) Fries, J., Getrost, H.: Organische Reagentien für die Spurenanalyse (E. Merck, Darmstadt), 1975; b) Pfeiffer, P., Tappermann, F.: Z. Anorg. Allg. Chem. *215*, 273 (1933); c) Pfeiffer, P., Werdelmann, B.: Z. Anorg. Allg. Chem. *261*, 197 (1950)
81. Pfeiffer, P., Christeleit, W.: Z. Anorg. Allg. Chem. *239*, 133 (1938)
82. a) Buhleier, E., Vögtle, F.: Liebigs Ann. Chem. *1977*, 1080; b) Vögtle, F., Müller, W. M., Raßhofer, W.: Isr. J. Chem. *18*, 246 (1979); c) Gillard, R. D., Johns, K. W., Williams, P. A.: J. Chem. Soc. Chem. Commun. *1979*, 357; d) Summers, L. A.: Advances in heterocyclic chemistry. Vol. *22*. New York, San Francisco, London: Academic Press, 1978
83. Oepen, G., Vögtle, F.: Liebigs Ann. Chem. *1979*, 2114
84. We thank Prof. Dr. W. H. Watson, Texas Christian University, for this communication before publication
85. Vögtle, F., Weber, E.: Angew. Chem. *86*, 896 (1974); Angew. Chem., Int. Ed. Engl. *13*, 814 (1974)
86. MacNicol, D. D., Wilson, D. R.: J. Chem. Soc. Chem. Commun. *1976*, 494
87. a) Baker, H. J.: Rec. Trav. Chim. Pays-Bas *54*, 833 (1935); b) Baker, H. J.: Rec. Trav. Chim. Pays-Bas *54*, 905 (1935); c) Baker, H. J.: Rec. Trav. Chim. Pays-Bas *55*, 17 (1936)
88. Hardy, A. D. U., MacNicol, D. D., Wilson, D. R.: J. Chem. Soc. Perkin II *1979*, 1011
89. MacNicol, D. D., Hardy, A. D. U., Wilson, D. R.: Nature *266*, 611 (1977)
90. a) Cp. structure of hexakis (2-bromomethyl) benzene: Marsau, M. P.: Acta Crystallogr. *18*, 851 (1965); b) Baker, H. J.: Rec. Trav. Chim. Pays-Bas *54*, 745 (1935)
91. MacNicol, D. D., Wilson, D. R.: Chem. Ind. (London) *1977*, 84
92. MacNicol, D. D., Swanson, S.: J. Chem. Res. (S) *1979*, 406
93. Freer, A. et al.: Tetrahedron Lett. *1980*, 205
94. Freer, A. et al.: Tetrahedron Lett. *1980*, 1159
95. Weber, E., Müller, W. M., Vögtle, F.: Tetrahedron Lett. *1979*, 2335
96. a) Stetter, H., Roos, E.-E.: Chem. Ber. *88*, 1390 (1955); b) Faust, G., Pallas, M.: J. prakt. Chem. *11*, 146 (1960)
97. Cf. the toluene complex of a tetrahydroxy [1.1.1.1]metacyclophane host molecule: Andreetti, G. D., Ungaro, R., Pochini, A.: Chem. Commun. *1979*, 1005
98. Odashima, K. et al.: J. Am. Chem. Soc. *102*, 2504 (1980); Odashima, K. et al.: Tetrahedron Lett. *1980*, 4347; Soga, T., Odashima, K., Koga, K.: Tetrahedron Lett. *1980*, 4351
99. Benesi, H. A., Hildebrand, J. H.: J. Am. Chem. Soc. *71*, 2703 (1949)
100. Jarvi, E. T., Whitlock, Jr., H. W.: J. Am. Chem. Soc. *102*, 657 (1980)
101. Almog, J. et al.: J. Am. Chem. Soc. *97*, 226 (1975); Almog, J., Baldwin, J. E., Huff, J.: J. Am. Chem. Soc. *97*, 227 (1975)
102. Jameson, D. B., Evers, J. A.: J. Am. Chem. Soc. *102*, 2823 (1980)
103. Park, C. H., Simmons, H. E.: J. Am. Chem. Soc. *90*, 2431 (1968)
104. Bell, R. A. et al.: Science *190*, 151 (1975)
105. Tabushi, I., Sasaki, H., Kuroda, Y.: J. Am. Chem. Soc. *98*, 5727 (1976); Tabushi, I., Kuroda, Y., Kimura, Y.: Tetrahedron Lett. *1976*, 3327
106. Cf. kinetic studies for the protonation of cryptands: Cox, B. G., Knop, D., Schneider, H.: J. Am. Chem. Soc. *100*, 6002 (1978)
107. Graf, E., Lehn, J. M.: J. Am. Chem. Soc. *98*, 6403 (1976), ibid. *97*, 5022 (1975)
108. Cf. also cyclodextrin anion complexes: Wojcik, J. F., Rohrbach, R. P.: J. Phys. Chem. *79*, 2251 (1975); Rohrbach, R. P. et al.: J. Phys. Chem. *81*, 944 (1977)

109. Schmidtchen, F. P.: Angew. Chem. *89*, 751 (1977); Angew. Chem., Int. Ed. Engl. *16*, 720 (1977); Chem. Ber. *113*, 864 (1980)
110. Lehn, J. M., Sonveaux, E., Willard, A. K.: J. Am. Chem. Soc. *100*, 4914 (1978); see also Lehn, J. M. et al.: J. Am Chem. Soc. *103*, 1282 (1981)
111. a) Dietrich, B. et al.: J. Chem. Soc. Chem. Commun. *1978*, 934; b) Dietrich, B. et al.: Helv. Chim. Acta *62*, 2763 (1979)
112. Coughlin, P. K. et al.: J. Am. Chem. Soc. *101*, 265 (1979)
113. Fee, J. A., Briggs, R. G.: Biochem. Biophys. Acta *400*, 439 (1975)
114. Drew, M. G. B., McCann, M., Nelson, S. M.: J. Chem. Soc. Chem. Commun. *1979*, 481
115. Agnus, Y., Louis, R., Weiss, R.: J. Am. Chem. Soc. *101*, 3381 (1979)
116. Wester, N., Vögtle, F.: Chem. Ber. *113*, 1487 (1980)
117. Rossa, L., Vögtle, F.: Liebigs Ann. Chem. *1981*, 459
118. Hilgenfeld, R., Saenger, W.: 6th European Crystallographic Meeting, Barcelona, Spain 1980; Angew. Chem. *93*, 1981, in press
119. a) Myskiv, M. G.: Proceedings of the 12th International Congress on Crystallography, Warsaw 1978; b) van Bergen, T. J., Kellogg, R. M.: J. Am. Chem. Soc. *99*, 3882 (1977); Kellogg, R. M.: J. Chem. Soc. Chem. Commun. *1978*, 923
120. Herbert, J. A., Truter, M. R.: J. Chem. Soc. Perkin II *1980*, 1253
121. Lehn, J. M.: Pure Appl. Chem. *50*, 871 (1978)
122. Bhagwat, V. W., Manohar, H., Poonia, N. S.: Inorg. Nucl. Chem. Lett. *16*, 289 (1980)
123. Weber, E.: Angew. Chem. *91*, 230 (1979); Angew. Chem., Int. Ed. Engl. *18*, 219 (1979)
124. Czugler, M., Weber, E.: J. Chem. Soc. Chem. Commun. *1981*, 472
125. Raßhofer, W., Müller, W. M., Vögtle, F.: Chem. Ber. *112*, 2095 (1979)
126. Czugler, M.: private communication
127. Singh, T. P., Reinhoudt, N., Poonia, N. S.: Inorg. Nucl. Chem. Lett. *16*, 239 (1980)
128. Suh, I. et al.: Z. Naturforsch. *35b*, 352 (1980)
129. Vögtle, F., Herzhoff, M.: unpublished results
130. Vögtle, F., Müller, W. M.: Angew. Chem. *91*, 676 (1979); Angew. Chem., Int. Ed. Engl. *18*, 623 (1979)

The Calixarenes

C. David Gutsche

Department of Chemistry, Washington University, St. Louis, Mo., 63130, USA

Table of Contents

1 Introduction

The synthesis of compounds containing cavities of molecule-sized dimensions has captured the attention of numerous chemists in recent years, and this area of organic chemistry is acquiring the status of a recognizable and expanding subdiscipline. The principal reason for the burgeoning interest in these compounds is their imputed, and in some instances demonstrated, ability to form inclusion complexes, *i.e.* to participate in what has been variously described as "host-guest" chemistry [1] or "receptor-substrate" chemistry [2]. This review deals with certain [1ₙ]metacyclophanes possessing basket-like shapes, particular attention being given to those members which have been named "calixarenes" [3,4].

2 Nomenclature of the Calixarenes

The compounds discussed in this review are represented by the general structure *1*. In the IUPAC system of nomenclature [5], a specific member of this group (as represented by structure *2*) is named

> pentacyclo[19.3.1.1^{3,7}.1^{9,13}.1^{15,19}]octacosa-1(25),3,5,7(28), 9,11,13(27),15,17,19(26),21,23-dodecaene ,

and it is numbered as shown in Fig. 1. An alternative nomenclature for this type of ring structure was suggested by Cram and Steinberg [6,7] according to which *2* is named as [1.1.1.1]metacyclophane. Several research groups have reported syntheses of the tetrahydroxy derivatives of *2* (as represented by structure *3*) and have named them in various ways. Zinke and coworkers [8] called these compounds "cyclischen Mehrkernmethylene-phenolverbindungen", Hayes and Hunter [9] named them "cyclic tetranuclear novolaks", and Cornforth and coworkers [10] referred to them as "tetra-hydroxycyclotetra-*m*-benzylenes. For convenience of written and verbal discussion

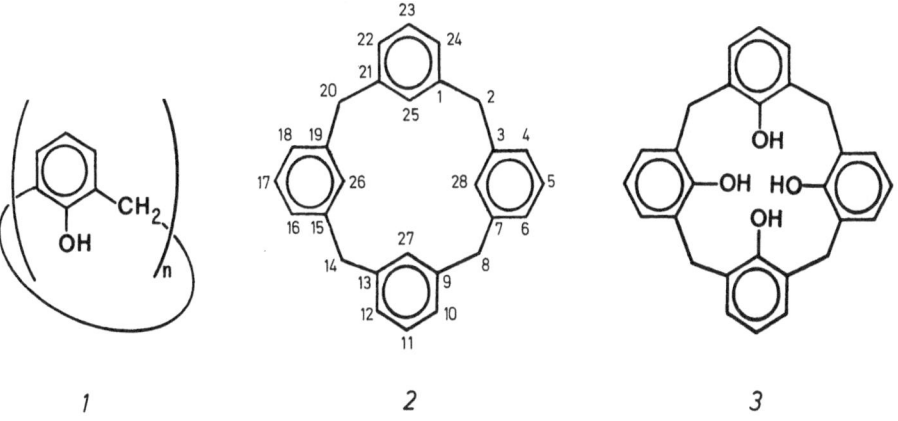

1   *2*   *3*

Fig. 1. Poly-aryl *m*-methylene-bridged macrocyclic compounds

we have chosen to call them "calixarenes" (Greek, *calix*, chalice; arene, indicating the incorporation of aromatic rings in the macrocyclic array), specifying the size of the macrocycle by a bracketed number inserted between *calix* and *arene* and specifying the nature and position of substitution on the aromatic rings by appropriate numbers and descriptors [11].

The structures and numbering for five types of calixarenes containing intraannular hydroxyl groups [12] which figure prominently in the following sections of this review are shown in Fig. 2. The cyclic tetramer composed of *p-tert*-butylphenol units and methylene units, for example, is named 5,11,17,23-tetra-*tert*-butyl-25,26,27,28-tetra-hydroxycalix[4]arene; in abbreviated fashion it will be referred to as *p-tert*-butyl-calix[4]arene.

3 Synthesis and Characterization of the Calixarenes

3.1 Arene-Aldehyde Condensations

3.1.1 Base-Catalyzed Condensation of *p*-Substituted Phenols and Formaldehyde

In 1872 Baeyer heated aqueous formaldehyde with phenol and observed a reaction which yielded a hard, resinous, noncrystalline product [13]. The chemical techniques at the time were not sufficiently advanced to allow characterization of such materials, however, and the structure remained unknown. Three decades later Baekeland devised a process for using this phenol-formaldehyde reaction to make a tough, resiliant resin (called a phenoplast) which he marketed under the name "Bakelite" [14] with tremendous commercial success. As a result, considerable industrial and academic attention was focused on phenol-formaldehyde processes, and a significant literature arose dealing with phenoplasts. Among these investigations were ones carried out by Zinke and coworkers in connection with the "curing" phase of the process [8,15-19]. In the investigation of this phenomenon they treated various *p*-substituted phenols with aqueous formaldehyde and sodium hydroxide, first at 50–55 °C, then at 110–120 °C for 2 hours and, finally, in a suspension of linseed oil at 200 °C for several hours. From *p*-methyl, *p-tert*-butyl, *p-tert*-amyl, *p*-(1,1,3,3-tetramethylbutyl), *p*-cyclohexyl, *p*-benzyl, and *p*-phenylphenol very high-melting, highly insoluble materials were obtained, all of which were postulated to be cyclic tetramers, *i.e.* calix[4]arenes of structure *4* in Fig. 2. The tacit assumption that a single product is formed in every instance was later shown to be incorrect by Cornforth and coworkers [10] who isolated higher- and lower-melting compounds from the condensations of formaldehyde with *p-tert*-butylphenol and *p*-(1,1,3,3-tetramethylbutyl)phenol (often referred to in the calixarene literature as *p*-octylphenol). Cornforth's conclusion that the materials were conformational isomers of the calix[4]arenes, however, was subsequently invalidated by Kämmerer and coworkers [20,21] and by Munch [22] whose temperature dependent ^{1}H NMR studies showed that rapid conformational interconversion occurs at room temperature. Finally, the recent work of Gutsche and coworkers [23] has revealed that mixtures comprising cyclic oligomers of *various ring size* are generally obtained in these condensation reactions. In the most thoroughly studied example [3,23-28] it has been shown that the condensation of *p-tert*-butyl-

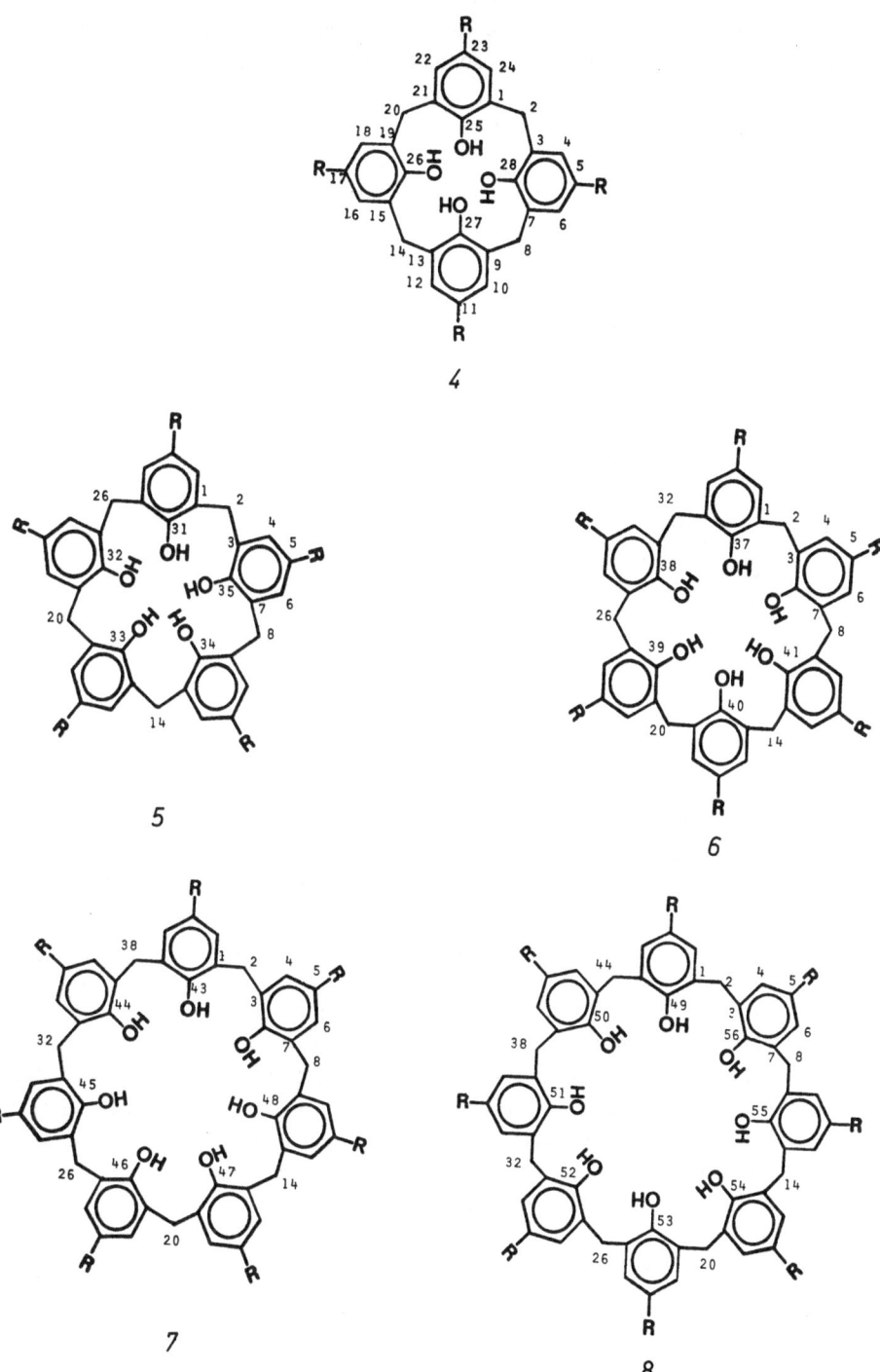

Fig. 2. Structures and numbering of intraannularly-hydroxylated calix[n]arenes

phenol and formaldehyde yields cyclic tetramer (*4*, R = *tert*-Butyl), cyclic hexamer (*6*, R = *tert*-Butyl), and cyclic octamer (*8*, R = *tert*-Butyl) as the major products as well as small amounts of cyclic pentamer [26] (*5*, R = *tert*-Butyl) and cyclic heptamer [27] (*7*, R = *tert*-Butyl) under some conditions. In addition, *p-tert*-butyldihomooxacalix[4]arene (*9*) is formed in ponderable quantity in certain cases [23,24,28]. The structures of the cyclic oligomers in the *p-tert*-butyl series have been well established. All are in complete agreement with chemical, spectral, and analytical data; and definitive structure proofs via x-ray crystallography have been provided for compounds *4* (R = *tert*-butyl) [29], *4* (R = 1,1,3,3-tetramethylbutyl) [30], *5* (R = *tert*-Butyl) [31], *5* (R=H) [32], *6* (R = *tert*-Butyl) [33], and *8* (R = *tert*-Butyl) [34].

9

Although the structures of the cyclic oligomers from the base-catalyzed condensation of *p-tert*-butylphenol and formaldehyde are now well understood, there is considerable confusion in the older literature concerning these compounds. As early as 1912 Raschig postulated the existence of cyclic compounds in Bakelite products [35], but Baekeland pointed out [36] that "we must not forget that one hypothesis is about as easy to propose as another as long as we are unable to use any of the methods for determining molecular size and molecular constitution". Not until three decades later did evidence begin to accumulate in support of cyclic oligomers as condensation products of phenols and formaldehyde. In 1941 Zinke and Ziegler [15] described a product which they obtained in good yield from a base-catalyzed condensation of *p-tert*-butylphenol and formaldehyde. It was stated to have a melting point above 340 °C and to form an acetate that had a molecular weight of 1725. No structure was suggested for this product, but in retrospect it seems quite certain that what these workers had isolated was *p-tert*-butylcalix[8]arene (*8*, R = *tert*-Butyl). Three years later Zinke and Ziegler again described a *p-tert*-butylphenol/formaldehyde condensation product [16], prepared under somewhat different and more carefully detailed conditions than those previously described, to which they assigned a cyclic *tetrameric* structure (*4*, R = *tert*-Butyl). However, they stated that molecular weight data could not be obtained, because neither the parent compound nor its acetate were sufficiently soluble. Thus, the cyclic tetrameric structure seems to have been based more on intuition than on solid data, cyclic tetrameric structures being "in the air" at the time. For example, Niederl and McCoy [37] claimed to have obtained a *p*-methylcalix[4]arene from an acid-catalyzed condensation of *p*-cresol with 2,6-*bis*-(hydroxymethyl)-4-methylphenol, repeating the work of Koebner [38] who had assigned a linear trimeric structure to this product. But, Koebner's contention was ultimately sustained [39,40], and it is quite certain that a cyclic tetramer is *not* formed under these conditions. Zinke and coworkers continued the investigation of the products of the base-catalyzed reactions of *p*-substituted phenols and formaldehyde and in 1952 published data on the products from the seven phenols cited above. They reported a molecular weight of 873 for the acetate of the product from *p*-octylphenol and formaldehyde, in agreement with a cyclic tetrameric structure. Thus, it was assumed that all of the other *p*-substituted phenol/formaldehyde products were also cyclic tetramers.

The first suggestion that the Zinke products were not pure entities came from Cornforth's experiments [10] in which he isolated mixtures from the condensations of *p*-*tert*-butylphenol and *p*-octylphenol with formaldehyde. The high-melting compounds were designated as HBC and HOC, respectively, and the low-melting compounds as LBC and LOC. Although the molecular weights of all of these compounds and/or their acetates seemed to be in agreement with a cyclic tetrameric structure and although preliminary x-ray crystallographic data seemed also to support this contention, more recent work [20-23,29,30,34] indicates that only the low-melting compounds (LBC and LOC) possess this structure. The high melting compounds (HBC and HOC) are now known to be the cyclic octamers [23,41]. Other workers, including the author of this review, also succumbed to the intuitively appealing and logical assumption that the products of the base-catalyzed condensation of *p*-substituted phenols and formaldehyde must possess cyclic tetrameric structures. Using a condensation procedure devised by chemists of the Petrolite Corporation, Gutsche and coworkers [42] reported the preparation of "cyclic tetramers" from *p*-methyl-, *p*-*tert*-butyl-, *p*-phenyl-, *p*-methoxy-, and *p*-carbomethoxyphenol with formaldehyde. Using a slightly modified version of the Petrolite procedure, Patrick and Egan [43] condensed the same five phenols and also imputed cyclic tetrameric structures to all of the products [44]. Subsequent experiments by Gutsche et al. [23,45,46], however, have shown that in none of these condensations is the cyclic tetramer a major product and that in most instances it is present in such low amounts as to be nonisolable. The Petrolite procedure [47], devised to simulate the factory production of phenol/formaldehyde resins for the manufacture of surfactant compounds, consists of refluxing a *p*-substituted phenol, paraformaldehyde, and a trace of 50% sodium hydroxide in xylene for several hours (the Patrick-Egan modification substitutes potassium *tert*-butoxide for sodium hydroxide and tetralin for xylene). The cooled reaction mixture deposits copious amounts of an insoluble product which, in the case of the *p*-*tert*-butylphenol reaction, is now known to be almost entirely cyclic octamer. Thus, from *p*-*tert*-butylphenol crystalline *p*-*tert*-butylcalix[8]arene (*8*, R = *tert*-Butyl) can be obtained in yields of 60–70%, making it a readily available cyclic oligomer.

In the process of unravelling the intricacies of the condensation of *p*-*tert*-butylphenol and formaldehyde [23], it was discovered that if a stoichiometric amount of base is used in the condensation instead of the catalytic amount employed in the original Petrolite procedure the major product is the cyclic hexamer, *p*-*tert*-butylcalix[6]arene (*6*, R = *tert*-Butyl). Yields as high as 70–75% of pure, crystalline material can be obtained, thus making this another abundantly available cyclic oligomer. Ironically, the cyclic tetramer is the even-numbered cyclic oligomer produced in lowest yield. Employing the Zinke procedure as modified by Cornforth by the substitution of Dowtherm (a eutectic of biphenyl and diphenyl ether) for linseed oil in the final step, one can obtain p-tert-butylcalix[4]arene in capriciously varying yields ranging from almost nothing to as high as 45%. Considerable effort has been expended in an attempt to understand the details of this reaction, but definitive results have yet to be obtained. One of the critical steps in the Zinke-Cornforth procedure is the last one in which the solid resinous material is powdered and heated (i.e. in linseed oil or Dowtherm). With regard to this step Zinke states [19] that "we believe we have isolated such cyclized compounds by heating resoles which had been condensed as far as the ether stage and which had not been washed free from alkali". Experiments in our laboratories have shown that acid washing the finely powdered resin fails to remove all of the base (sodium content of acid-washed resin is 1.3%). Only by dissolving the resin in an organic solvent, washing the solvent with acid followed by water, and evaporating the solvent can a base-free resin be obtained. The base-free material fails to yield cyclic oligomers when heated in Dowtherm, but upon the addition of a small amount of base (0.15 equivalent, based on the starting phenol, may be the optimum quantity [48]) cyclic tetramer is formed in 25–35% yield [48,49].

The odd-numbered calixarenes are more difficult to obtain in quantity than the even-numbered calixarenes. Employing the Patrick and Egan modification of the Petrolite procedure and changing the heating sequence (55 °C for 6 hours followed by 150 °C for 6 hours) Ninagawa and Matsuda [26] obtained a mixture from which they isolated 23% cyclic tetramer, 5% cyclic pentamer, and 11% cyclic octamer. Employing the Petrolite procedure but with dioxane as the solvent and a 30 hour heating period, Nakamoto and Ishida [27] obtained a mixture containing cyclic hexamer, heptamer, and octamer from which they isolated 6% of the heptamer.

Little is known about the overall mechanism of cyclic oligomer formation. although the mechanism of the initial stages of the sequence seems fairly clear. The first chemical event is the reaction of formaldehyde (formed in the Petrolite procedures by depolymerization of paraformaldehyde) with phenol to form 2-hydroxy-methyl- and 2,6-bis(hydroxymethyl)phenols in a base-catalyzed process, as shown in Fig. 3. Such compounds were characterized many years ago [50], obtained from the action of aqueous formaldehyde on phenol in basic solution at room temperature. Subsequent condensation between the hydroxymethylphenols and the starting phenol occurs to form linear dimers, trimers, tetramers, etc. via a pathway that might involve o-quinonemethide intermediates which react with phenolate ions in a Michael-like reaction, as portrayed in Fig. 4. The condensation of hydroxymethyl-

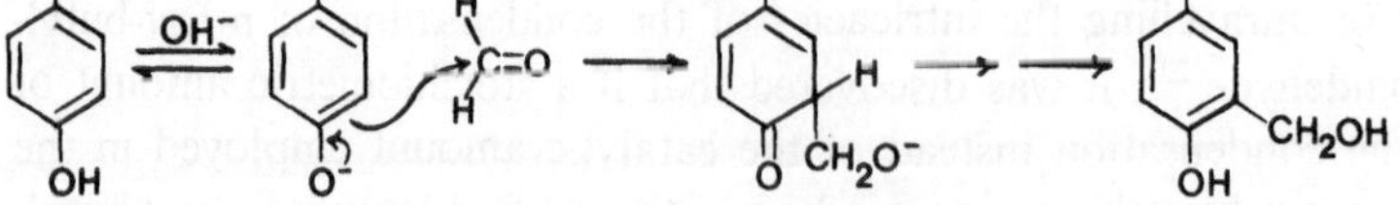

Fig. 3. Base-catalyzed hydroxymethylation of phenols

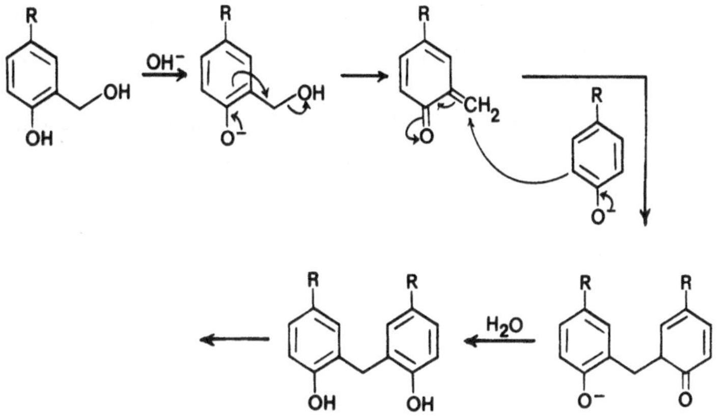

Fig. 4. Base-catalyzed formation of linear oligomers from phenols and formaldehyde

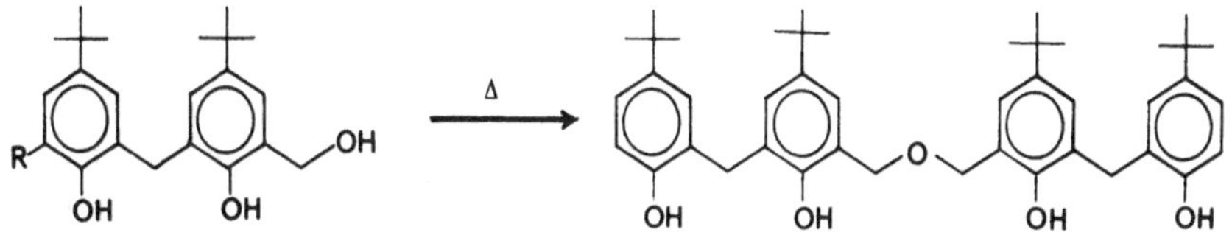

Fig. 5. Formation of dibenzyl ethers of 2-hydroxymethylphenols

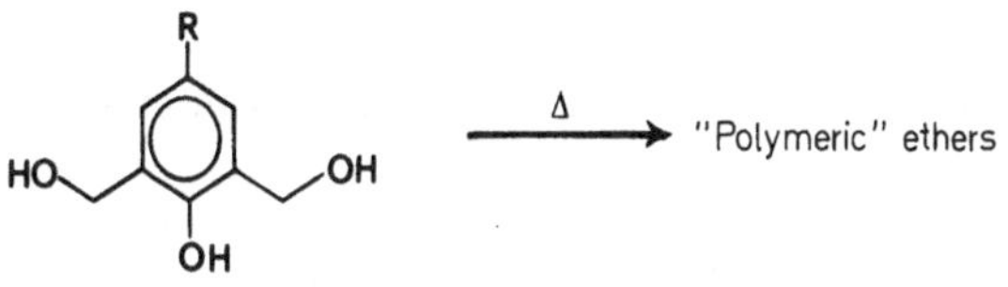

phenols to form oligomers also has been shown to occur under relatively mild conditions. The possibility of quinonemethide intermediates was suggested as far back as 1912 [51)] and has been promoted by Hutzsch [52)], v. Euler [53)], and others. However, because the formation of o-quinonemethides from compounds such as o-(methoxymethyl)phenol requires quite high temperatures (500–600 °C) [54)] doubt has been cast on the validity of this premise [55)]. On the other hand, it is known that oxy-Cope rearrangements occur far more readily with anions than with the corresponding neutral compounds [56)], so conversion of 2-hydroxymethylphenolates to o-quinonemethides may, in fact, occur under the basic conditions prevailing in the reactions under discussion. Intermolecular dehydration of 2-hydroxymethylphenols to form dibenzyl ethers also occurs under the conditions of the Zinke-Cornforth and Petrolite condensation reactions, as illustrated in Fig. 5. For example, 10a yields the ether 11 [57)], and 12 (R=CH$_3$ and R = tert-Butyl) yields polymeric ethers [58] [59] upon simple heating. Thus, the mixtures from which the calixarenes emerge contain linear oligomers of various lengths in which the o,o'-bridges are CH_2 as well as CH_2OCH_2 groups.

The events that occur in the terminal phases of the sequence leading to the cyclic oligomers remain a mystery. It is certain that the linear oligomers lose water and formaldehyde in the process of being converted to the cyclic oligomers [60], but the immediate precursors of the cyclic oligomers are not known. The preferential formation of the even-numbered cyclic oligomers in the *p-tert*-butyl series might suggest a common precursor such as *10a* or *10b* which, by dimerization, trimerization, and tetramerization could yield *4, 6*, and *8* (R = *tert*-butyl). The detection of cyclic pentamer and cyclic heptamer in the reaction mixtures, however, casts doubt on this pathway. Also uncertain is the nature of the driving forces that promote cyclization in high yields, particularly in the case of the cyclic hexamers and octamers; why are the larger sized cyclic oligomers formed with greater ease than the cyclic tetramer, which should be strongly favored on entropic grounds? Our current hypothesis invokes a combination of intramolecular hydrogen bonding and cation template phenomena. The cyclic tetramer is strongly intramolecularly hydrogen bonded, as indicated by the concentration-independent OH stretching absorptions at 3160 cm^{-1} in the infrared (see Sect. 4.4). Inspection of space-filling molecular models (CPK models) shows that the four OH groups of the cyclic tetramer are forced into close proximity (in the "cone" conformation; see Sect. 5.1). Surprising, however, is the fact that the more flexible cyclic hexamer and octamer show similar IR behavior, having OH stretching absorptions at 3150 cm^{-1} and 3230 cm^{-1}, respectively and indicating very strong intramolecular hydrogen bonding in these systems as well. Even more surprising is the fact that the linear tetramer, pentamer, and hexamer have OH stretching absorptions at 3200 cm^{-1}, again indicative of strong intramolecular hydrogen bonding [61]. Hydrogen bonding in the linear oligomers can either be intermolecular, giving rise to pseudocyclic arrays which we have designated as "hemicalixarenes" or intramolecular, giving rise to pseudocyclic arrays which we have designated as "pseudocalixarenes" [62], as illustrated in Fig. 6. Pseudocalixarene formation might constitute a major factor in determining the course of the cyclization process.

It is reported [23] that the yield of cyclic hexamer, which is the major product when a stoichiometric amount of base is used, is slightly better with RbOH than with CsOH, KOH, or NaOH; LiOH is an ineffective catalyst. This has been interpreted as suggesting that a template effect may play a part in the cyclization

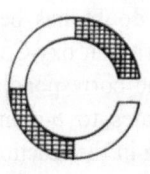 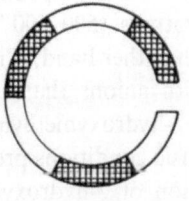 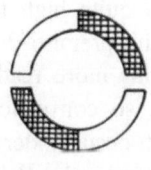 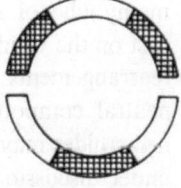

| Pseudocalix [4] arene | Pseudocalix [6] arene | Hemicalix [4] arene | Hemicalix [6] arene |

Fig. 6. Hemicalixarenes and pseudocalixarenes

process. The work of Izatt and coworkers [63] has shown that the calixarenes do, indeed, have ionophoric capacity (see Sect. 7.3) for NaOH, KOH, RbOH, and CsOH but nor for LiOH. The details of the role of cations in the cyclization process, however, remain to be clarified.

An observation that promises to have significant mechanistic implications concerns the interconvertibility of the cyclic oligomers. Contrary to an earlier report that they are non-interconvertible under the conditions of the reaction [23], it has recently been demonstrated [49] that when *p-tert*-butylcalix[8]arene and *p-tert*-butylcalix[6]arene are heated in boiling diphenyl ether in the presence of a small amount of potassium *tert*-butoxide a 20–35% yield of *p-tert*-butylcalix[4]arene is obtained.

3.1.2 Acid-Catalyzed Condensation of Resorcinols and Aldehydes

The acid-catalyzed reactions of *p*-substituted phenols and formaldehyde yield mixtures of linear oligomers which, under certain conditions [64], include compounds containing as many as 25 or more monomeric units. There is no evidence that any cyclic oligomers are present in these mixtures [65]. Resorcinol, on the other hand, has long been known to react with aldehydes (other than formaldehyde) to yield well defined materials. In 1883 [66] Michael isolated two crystalline compounds from the acid-catalyzed reaction between resorcinol and benzaldehyde, and he assigned a cyclic dimeric structure to one of them. In 1940 this assignment was reformulated by Niederl and Vogel [67] to a tetrameric structure, *i.e.* a calix[4]arene containing eight *extra*-annular hydroxyl groups [12], as illustrated in Fig. 7 for compound *13* (R' = H). Mass spectrometric support was supplied by Erdtman and coworkers [68] on the octamethyl ether (*13*, R = R' = CH$_3$), and x-ray crystallographic determinations [69, 70] conclusively established the structures.

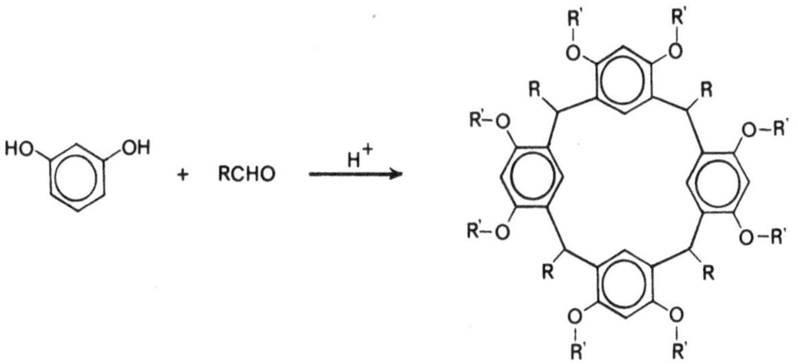

Fig. 7. One-step synthesis of octahydroxycalix[4]arenes

The mechanism of the reaction has been studied in some detail by Högberg [12, 71, 72]. In contrast to the base-catalyzed oligomerization, the acid catalyzed process involves electrophilic aromatic substitutions by cations, as outlined in Fig. 8. Although formaldehyde does not react with resorcinol to produce cyclic oligomers, other aldehydes such as acetaldehyde and benzaldehyde give excellent yields of

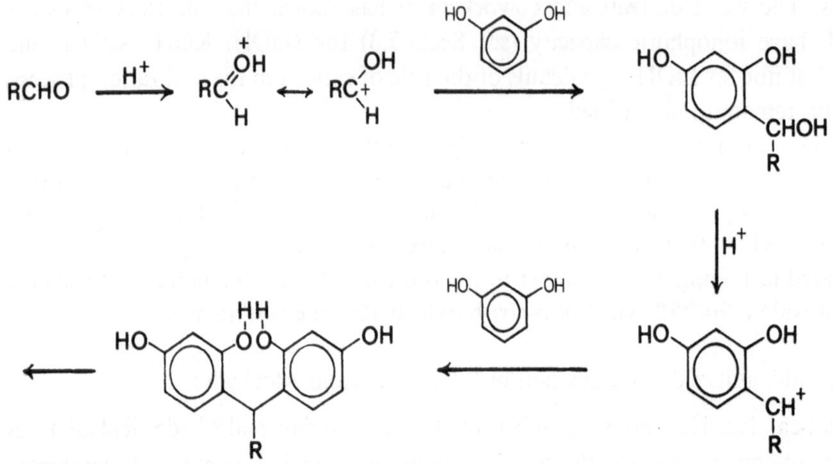

Fig. 8. Acid-catalyzed formation of resorcinol/aldehyde oligomers

cyclic tetramer *13*. The problem with formaldehyde arises from its great reactivity which leads to substitution at the 2-position as well as the 4- and 6-positions, resulting in the formation of cross-linked polymers. Other aldehydes, more bulky than formaldehyde, are less likely to react at the hindered 2-position. Of the four diastereomeric possibilities for *13*, assuming conformational mobility (see Sect. 5.1), only two comprise the bulk of the product, *viz. cis-cis-cis* and *cis-trans-cis*, presumably because of non-bonded interactions in the precursors leading to the final products. Högberg has demonstrated that the *cis-trans-cis* isomer is the kinetic product in the case of the resorcinol/benzaldehyde condensation but that it can be converted, *in situ*, to the *cis-cis-cis* product which is the more insoluble and which separates from solution. By taking advantage of this circumstance, a greater than 80% yield of the *cis-cis-cis* isomer can be obtained [72]. In the case of the product from 2-methylresorcinol and benzaldehyde [12] it is the *cis-trans-cis* isomer that is the more insoluble and, therefore, the one that can be obtained in high yield. The oligomerization in these cases is a reversible process [73], and insolubility is a driving force for the formation of a single isomer and, perhaps, for the cyclization itself. It is uncertain whether intramolecular hydrogen bonding in the acyclic precursor(s) plays a part comparable to that imputed to the base-catalyzed process involving *p*-substituted phenols and formaldehyde. The IR stretching frequencies reported [71] for the isomers of *13*, which are very broad (3700–2500 cm^{-1}), do not provide any clear-cut information on this point, and the IR data on the corresponding linear oligomers are not available.

3.1.3 Acid-Catalyzed Condensations of Alkylbenzenes and Formaldehyde

Calix[4]arenes *14a* and *14b* have been prepared from mesitylene and from 1,2,3,5-tetramethylbenzene by condensation with formaldehyde in the presence of acetic acid [74]. Calixarene *14a* has also been prepared by a Friedel-Crafts reaction with chloromethylmesitylene [75].

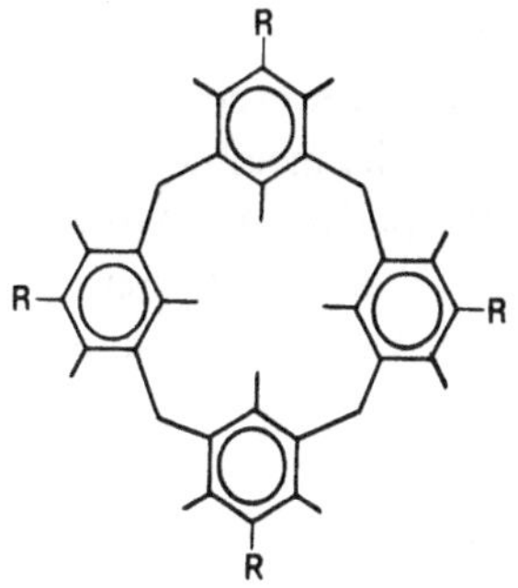

14a (R = H)

14b (R = CH₃)

3.1.4 Acid-Catalyzed Condensations of Heterocyclic Compounds and Aldehydes

Although not formally classified as calixarenes, compounds closely related in architecture to the calix[4]arenes have been prepared by the condensation of furans, thiophenes, and pyrroles with aldehydes and ketones. Furan undergoes acid-catalyzed condensation with aldehydes and ketones to give *15*[76-79], as shown in Fig. 9.

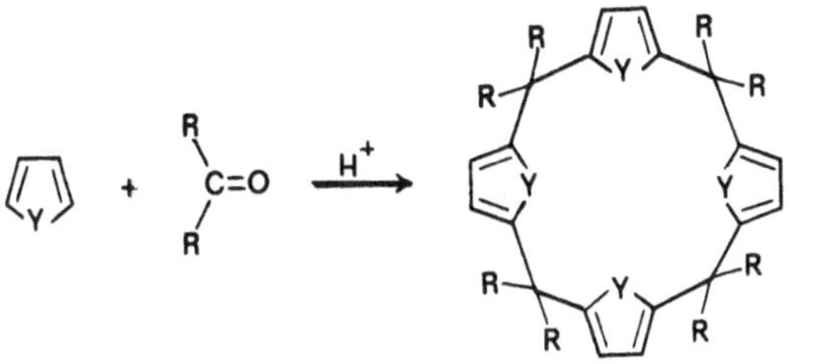

15 Y = O

16 Y = S

17 Y = N

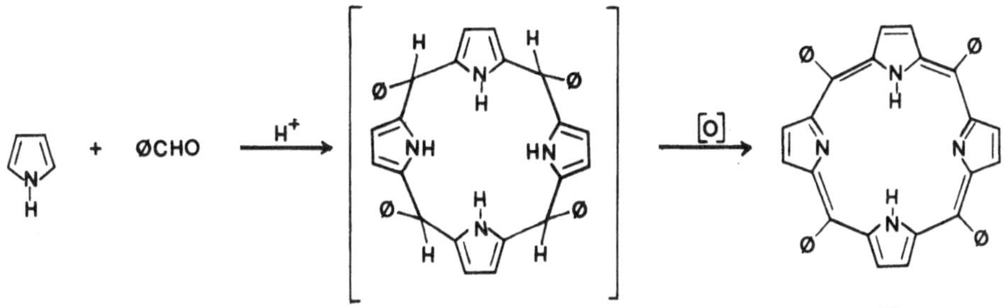

18

Fig. 9. Acid-catalyzed formation of cyclic tetramers from furan, thiophene, and pyrrole with aldehydes and ketones

The increase in yield of *15a* from *ca* 20% to over 40% in the presence of lithium perchlorate [77] has led to the suggestion of a template effect. However, recent experiments [78] indicate that the higher yields correlate not with the metal ion but with the acidity of the reaction medium. In similar fashion, thiophene and pyrrole undergo acid-catalyzed condensations with acetone to yield *16* [80] and *17* [81]. Benzaldehyde also condenses with pyrrole [82], but the initially formed compound loses six hydrogens to form the planar tetraphenylporphin *18*.

3.2 Related Condensations Involving Formaldehyde

Formaldehyde is an exceedingly reactive molecule and condenses with a wide variety of compounds, some of which have been illustrated by the examples in the previous parts of this Section. Other examples of formaldehyde condensations leading to macrocyclic compounds include veratrole (leading to cyclotriveratrylene [83]), glycoluril (leading to curcurbituril [84]), 2,5-dimethylthiophene (leading to a cyclic trimer [85]), 2-arylimidazole (leading to a cyclic trimer and a cyclic tetramer [86]), and N-methylindole (leading to a cyclic trimer [87]).

3.3 Stepwise Synthesis of Calixarenes

3.3.1 Hayes-Hunter-Kämmerer Synthesis

Noting Zinke's claim to have isolated cyclic tetramers from the condensation of *p*-substituted phenols and formaldehyde, Hayes and Hunter in 1956 sought to

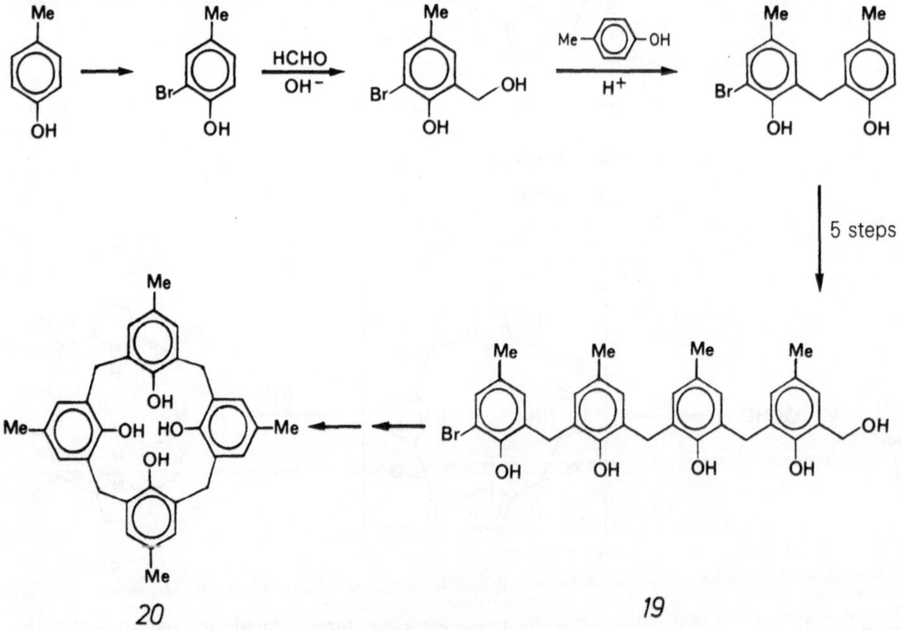

Scheme 1. 10-Step synthesis of *p*-methylcalix[4]arene (Hayes and Hunter method).

provide additional evidence for such structures by what they termed a "rational" synthesis [9]. Starting with *p*-cresol, they protected one of the *o*-positions by bromination and then sequentially added methylene groups (by base-induced hydroxymethylation) and aryl groups (by acid-catalyzed arylation). The *o*-bromo-*o'*-hydroxymethyl linear tetramer (*19*)' thus obtained was debrominated and then cyclized to *20*, as illustrated in Scheme 1. Although no comparison of the Hayes and Hunter product with the Zinke product was reported, this synthesis seems to have been generally accepted as implicit proof for the Zinke tetrameric structure [88]. More recently, Kämmerer and coworkers [20, 21, 89–94] have improved and expanded the Hayes and Hunter synthesis, demonstrating its potential by the preparation of a series of methyl- and *tert*-butyl-substituted calixarenes, including the cyclic tetramers, pentamers, hexamers, and heptamers shown in Fig. 10.

The Hayes and Hunter stepwise synthesis has certain drawbacks; it is long and

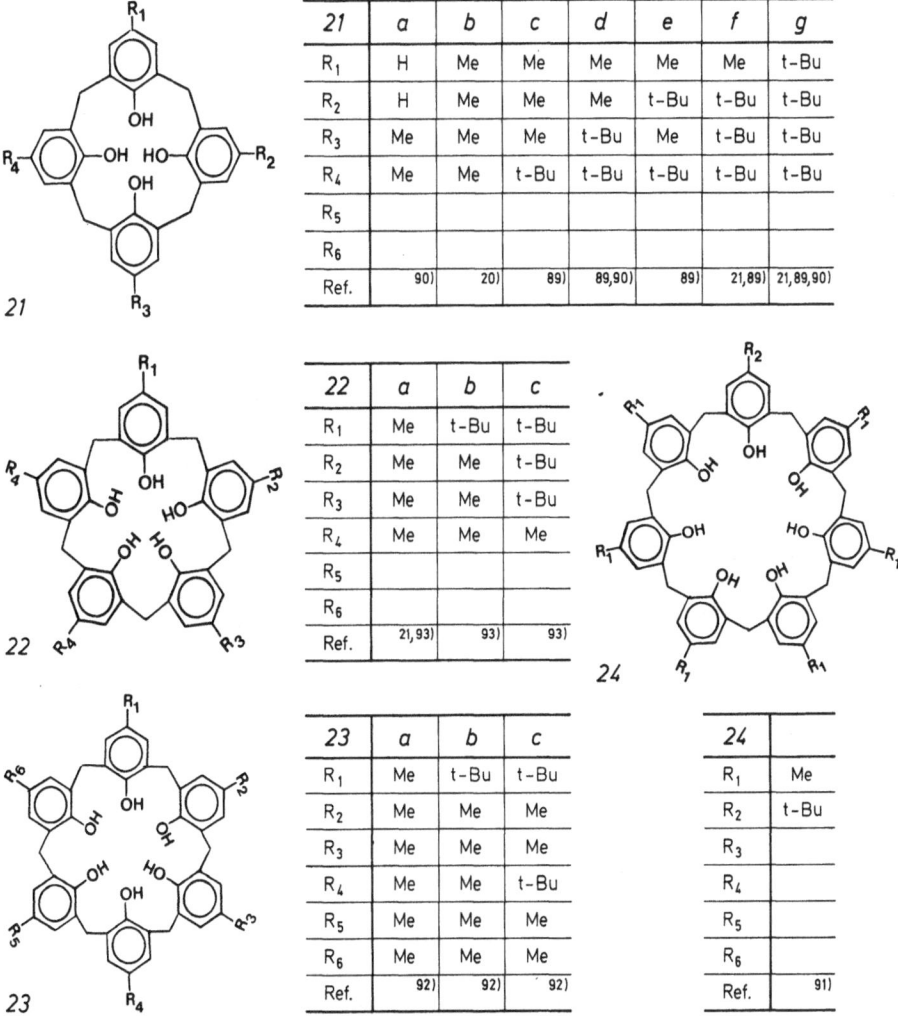

21	a	b	c	d	e	f	g
R_1	H	Me	Me	Me	Me	Me	t–Bu
R_2	H	Me	Me	Me	t–Bu	t–Bu	t–Bu
R_3	Me	Me	Me	t–Bu	Me	t–Bu	t–Bu
R_4	Me	Me	t–Bu	t–Bu	t–Bu	t–Bu	t–Bu
R_5							
R_6							
Ref.	90)	20)	89)	89,90)	89)	21,89)	21,89,90)

22	a	b	c
R_1	Me	t–Bu	t–Bu
R_2	Me	Me	t–Bu
R_3	Me	Me	t–Bu
R_4	Me	Me	Me
R_5			
R_6			
Ref.	21,93)	93)	93)

23	a	b	c
R_1	Me	t–Bu	t–Bu
R_2	Me	Me	Me
R_3	Me	Me	Me
R_4	Me	Me	t–Bu
R_5	Me	Me	Me
R_6	Me	Me	Me
Ref.	92)	92)	92)

24	
R_1	Me
R_2	t–Bu
R_3	
R_4	
R_5	
R_6	
Ref.	91)

Fig. 10. Calixarenes synthesized by Kämmerer et al. via the Hayes and Hunter stepwise method

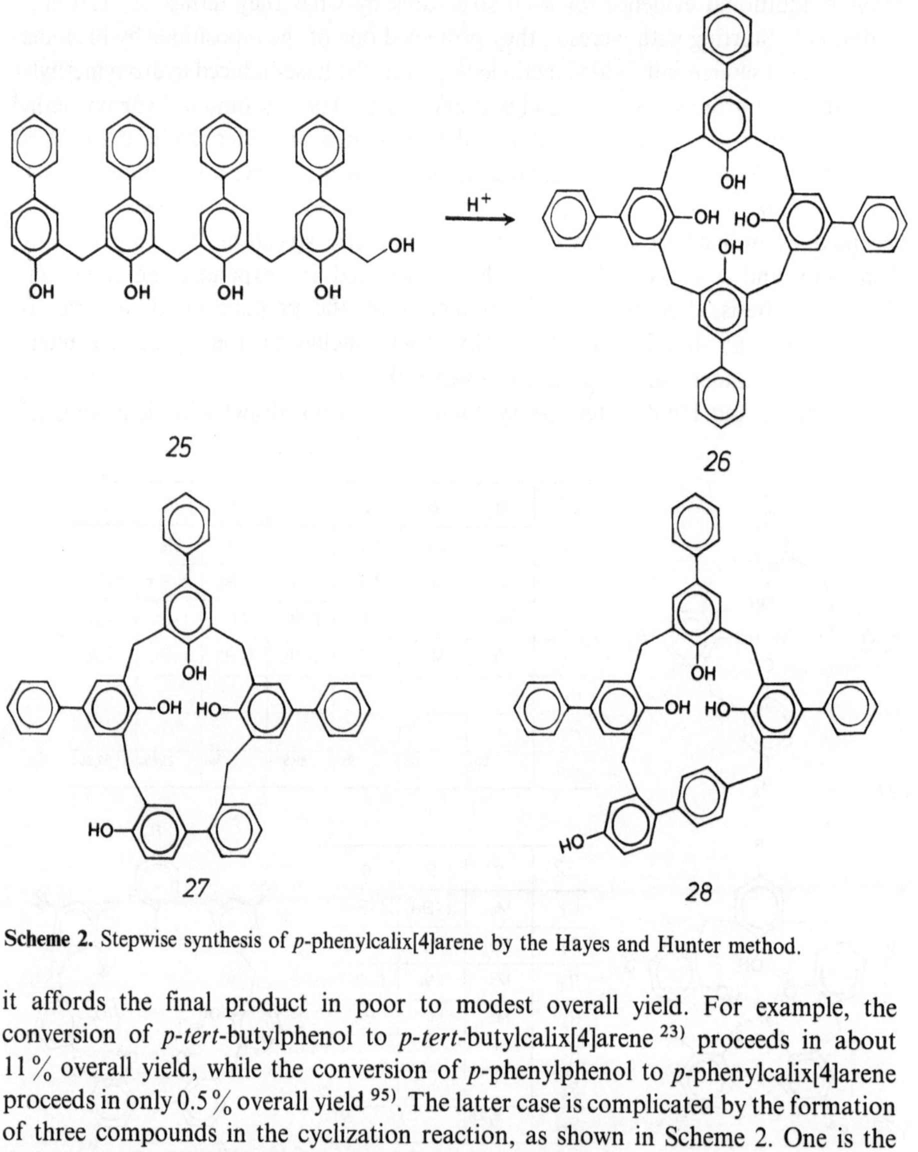

Scheme 2. Stepwise synthesis of p-phenylcalix[4]arene by the Hayes and Hunter method.

it affords the final product in poor to modest overall yield. For example, the conversion of p-tert-butylphenol to p-tert-butylcalix[4]arene [23] proceeds in about 11 % overall yield, while the conversion of p-phenylphenol to p-phenylcalix[4]arene proceeds in only 0.5 % overall yield [95]. The latter case is complicated by the formation of three compounds in the cyclization reaction, as shown in Scheme 2. One is the desired product (26), and the other two are postulated to be the isomeric compounds 27 and 28, formed as the result of cyclization into the other reactive sites of the terminal p-phenylphenyl moiety. Thus, p-phenylcalix[4]arene, a compound of considerable interest with respect to the formation of molecular complexes, is not readily accessible via this route.

3.3.2 Böhmer, Chhim, and Kämmerer Synthesis

Recognizing the deficiencies in the Hayes and Hunter synthesis, Böhmer, Chhim, and Kämmerer [96] have explored a more convergent approach which retains much of the flexibility of the sequential approach. It involves the condensation of a linear trimer (29) with a 2,6-bishalomethylphenol (30), as illustrated in Scheme 3. Although short, this method suffers from quite low yields in the cyclization step, ranging from

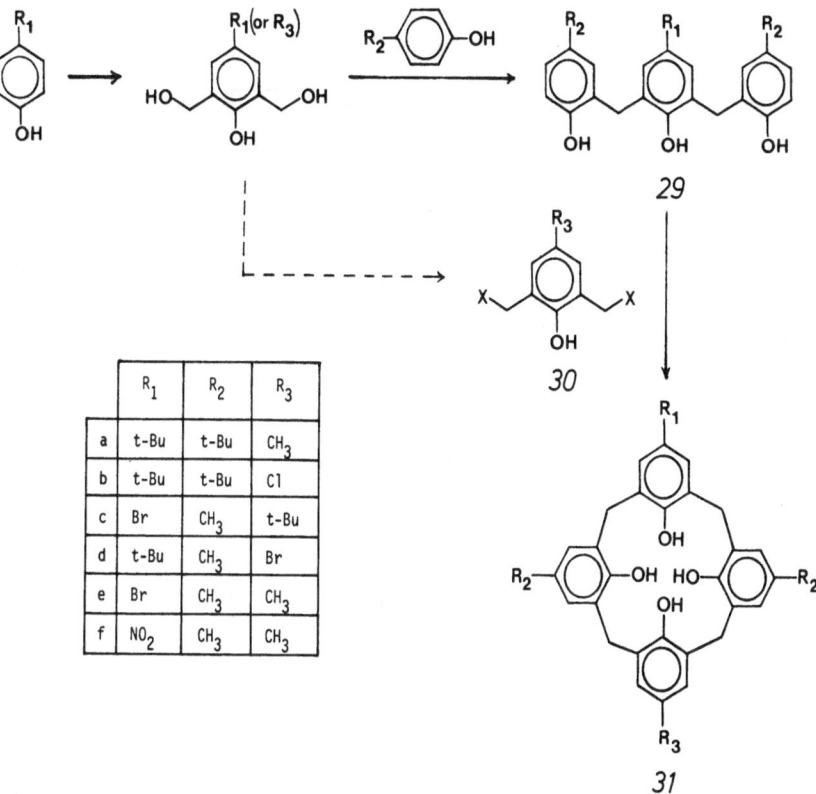

	R_1	R_2	R_3
a	t-Bu	t-Bu	CH₃
b	t-Bu	t-Bu	Cl
c	Br	CH₃	t-Bu
d	t-Bu	CH₃	Br
e	Br	CH₃	CH₃
f	NO₂	CH₃	CH₃

Scheme 3. Convergent, stepwise synthesis of calixarene[4]arenes (Böhmer, Chhim, and Kämmerer method).

10–20% in the best cases down to 2–7% in the more interesting cases in which a mixture of alkyl, bromo, and nitro functions are incorporated as R groups.

3.3.3 Moshfegh, Hakimelahi et al. Synthesis

As part of an extensive program dealing with analogues of phloroglucides, Moshfegh, Hakimelahi and coworkers [97–99] have investigated the synthesis of a variety of p-halocalix[4]arenes using a procedure similar to that of Böhmer et al. When the dimer 32 and the bis(hydroxymethyl) dimer 33 are used as starting materials calixarene 34 is stated to be produced in 65% yield (however see Ref. [124]), accompanied by 10% of the diooxa compound 35. In similar fashion the condensation of the dimer 32 (X = Cl) with 4-halo-2,6-bis(hydroxymethyl)phenol (36) is stated to

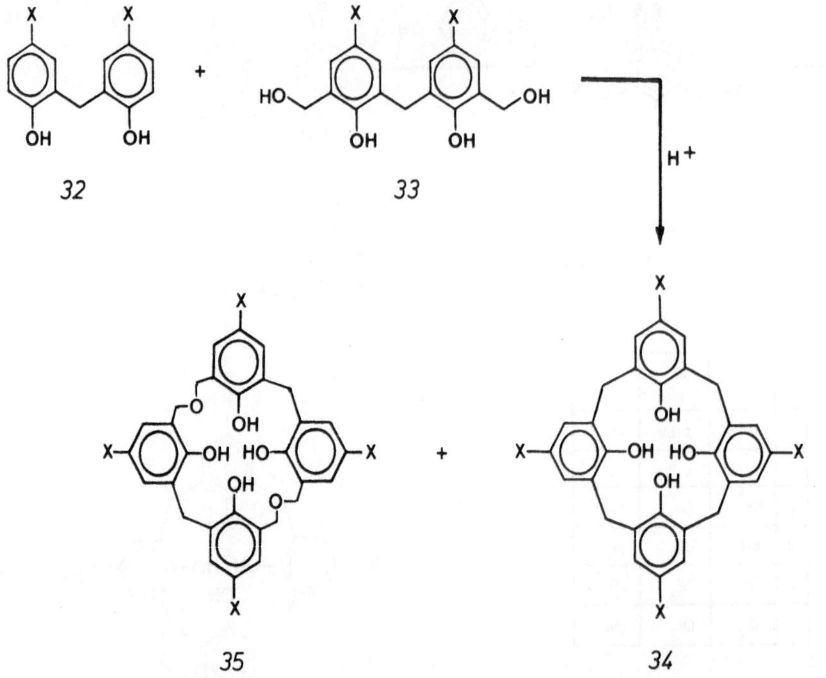

Scheme 4. Convergent, step-wise synthesis of calix[4]arenes (Moshfegh, Hakimelahi et al. method).

afford a 69–90% yield of *p*-halocalix[3]arenes (*37*). The striking contrast in yield between the Böhmer, Chhim, and Kämmerer method and the Moshfegh, Hakimelahi et al. method for the preparation of calix[4]arenes indicates that a careful investigation should be made to determine the true potentialities of this general approach. Also, since space filling models of the calix[3]arene ring system indicate considerably more non bonded strain than in the higher membered calixarenes, it is hoped that additional and more complete details in support of this structure will be forth-coming.

3.3.4 No and Gutsche Synthesis

Another convergent synthesis has been devised by No and Gutsche [100] which retains some of the functional group flexibility of the Hayes and Hunter method and which gives sufficiently high yields to make calixarenes available in quantities large enough for catalysis studies. In the 4-step sequence shown in Scheme 5 a *p*-substituted phenol (*e.g. p*-phenylphenol) is treated with formaldehyde under controlled conditions to produce the *bis*(hydroxymethyl) dimer *38*; the dimer is condensed with two equivalents of a *p*-substituted phenol (*e.g. p-tert*-butylphenol) to yield the linear tetramer *39* (*e.g.* R = *tert*-Butyl); and, the tetramer is monohydroxymethylated to yield *40* (*e.g.* R = *tert*-Butyl) and then cyclized to yield calix[4]arene (*e.g.* R = *tert*-Butyl). Although the overall yield is only ca 10%, the starting materials are cheap and the work-up and purification procedures are generally simple and straightforward. If improvements in the selective hydroxymethylation step can be realized the synthesis would excel the Zinke "one-flask" method not only with respect to flexibility but with respect to yield as well.

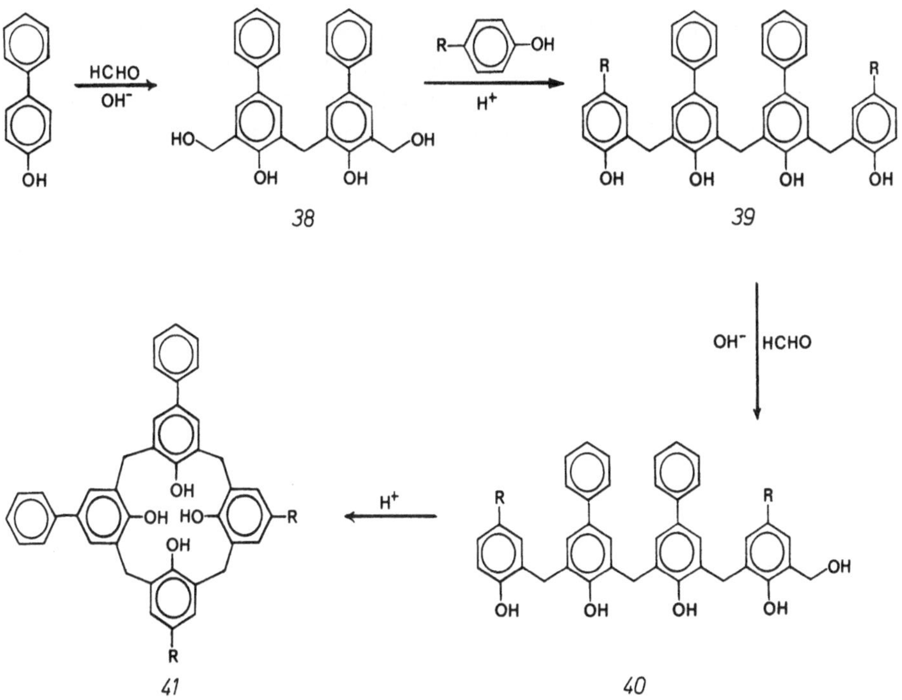

Scheme 5. Convergent, step-wise synthesis of calix[4]arenes (No and Gutsche method).

3.4 Oxacalixarenes

The Petrolite method for calixarene synthesis also leads to dihomooxacalix[4]arenes, as represented by the *p-tert*-butyl compound *9*. Proton-NMR spectra and TLC data indicate that the mixture obtained from *p-tert*-butylphenol and paraformaldehyde contains appreciable amounts of *9*, but it is difficult to obtain a pure sample of *9* from this mixture [23, 24, 28]. Compound *9* is much more readily accessible by the thermally-induced dehydration of the *bis*(hydroxymethyl)tetramer *46* which can be readily prepared by the convergent synthesis outlined in Scheme 5 followed by *bis* hydroxymethylation [62], as shown in Scheme 6. Other oxacalixarenes can also be obtained by thermally-induced dehydration. For example, 2,6-*bis*(hydroxymethyl)-phenols (*42*) yield hexahomotrioxacalix[3]arenes (*43*) [28, 62, 101]; *bis(hydroxymethyl)* dimers *44a* yield tetrahomodioxacalix[4]arenes *45a* [62, 97, 102]; and *bis*(hydroxymethyl) dimers *44b* yield octahomotetraoxacalix[4]arenes *45b* [103]; as illustrated in Scheme 6. The ease with which these dehydrations occur has been attributed [62] to intermolecular and intramolecular hydrogen bonding forces which establish the reacting systems in cyclic arrays prior to cyclization (see Fig. 6).

3.5 Calixarene Esters and Ethers

The phenolic OH groups of the calixarenes are readily converted to ester and ether moieties. The acetates of many of the calixarenes have been prepared, the earliest examples being those reported by Zinke and Ziegler [15, 16]. The acetates are frequently, but not invariably, lower melting than the parent calixarene and are more soluble in organic solvents. Advantage can be taken of the improved solubility in the purification of calixarene mixtures produced in the "one-flask" procedures. For example, the product obtained from *p*-phenylphenol is not amenable to purification by crystallization, fractional extraction, or column chromatography. Conversion of the crude mixture to the acetate gives a more tractible material from which pure compounds have been isolated [46] by flash chromatography [104]; the parent calixarenes can then be obtained by removal of the ester groups by means of

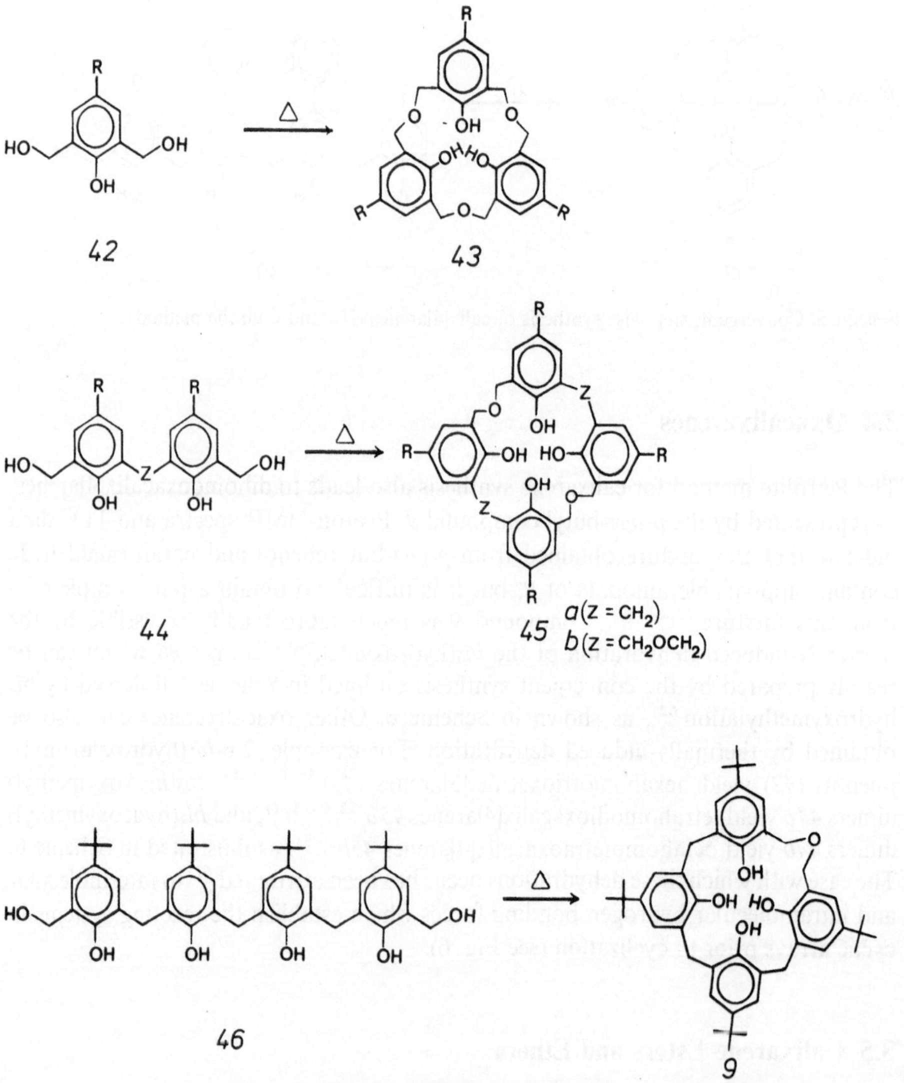

Scheme 6. Synthesis of oxacalixarenes by intra- and intermolecular dehydration.

ethylene diamine in DMF solution [105]. When an excess of acetylating agent is used the completely acetylated product is generally produced. However, instances of incompletely acetylated calixarenes have been reported. For example, two acetates have been isolated from p-tert-butylcalix[4]arene that are not conformational isomers (see Section 5.2) [23]; the one melting at 383–386° is assigned the tetraacetate structure [106, 107], and the one melting at 247–250 °C the triacetate structure [107]. Although the acetates are the most commonly prepared ester of calixarenes, others such as the benzoates and p-toluenesulfonates can be easily made [48, 65]. The mono- and di-camphorsulfonyl esters of p-tert-butylcalix[8]arene have been reported and the effects on the circular dichronic characteristics of the chiral camphorsulfonyl moiety noted [25].

A convenient method for preparing the completely alkylated ethers of the calixarenes involves treatment of the calixarene in THF-DMF solution with an alkyl halide in the presence of sodium hydride. Methyl, ethyl, allyl, and benzyl ethers have all been prepared in this fashion in excellent yields [107]. Various $(CH_2CH_2O)_nR$ ethers have been prepared via the action of the tosylate of the alkylating agent in the presence of potassium tert-butoxide [106, 108]. Under different conditions, partially alkylated calixarenes have been isolated. For example, treatment of p-tert-butylcalix[4]-arene with dimethyl sulfate in the presence of BaO—Ba(OH)$_2$ in DMF yields the trimethyl ether [107], and treatment with benzyl tosylate yields the dibenzyl ether [107]. Treatment with ethereal diazomethane yields a product possessing a ^{1}H nmr spectrum compatible with a dimethyl ether structure, but an x-ray crystallographic determination carried out by Professor G. G. Stanley of Washington University indicates the structure to be a monomethyl ether. To explain these data it is postulated that the product actually is a mixture in which the major component is the dimethyl ether but that the single crystal selected for x-ray analysis was the monomethyl ether.

2,4-Dinitrophenyl ethers of p-tert-butylcalix[8]arene have been prepared by heating a pyridine solution of the calixarene with 2,4-dinitrochlorobenzene [25]. Depending on the ratio of calixarene to arylating agent, the product contains one, two, or six 2,4-dinitrophenyl moieties. Although an excess of arylating agent fails to yield an octa-substituted compound, the remaining OH groups can be converted to acetoxy groups by treatment with acetyl chloride. It is of historical interest that these experiments gave one of the early clues that the materials previously thought to be cyclic tetramers are, in fact, cyclic octamers.

Still another calixarene derivative of interest and utility is the trimethylsilyl ether. The hexa-trimethylsilyl calix[6]arenes and octa-trimethylsilyl calix[8]arenes can be prepared [23] by using standard trimethylsilylating agents such as hexamethyldisilazene and chlorotrimethylsilane. The tetra-trimethylsilyl ethers of the calix[4]arenes, however, do not form under these conditions and require the use of the very reactive N,O-bis(trimethylsilyl)acetamide [109].

The formation of calixarene esters and ethers can be complicated by the problem of incomplete derivatization, as noted above, and also by the fact that conformational "fixing" generally occurs in the case of the calix[4]arenes. The consequences of the latter are discussed in Section 5.2.

4 Physical and Spectral Properties of Calixarenes

4.1 The Shape of the Calixarenes

Perceiving a similarity between the shape of a Greek vase known as a Calix Crater and the shape of the cyclic tetramer, as illustrated in Fig. 11., we assigned the name "calixarene" [3,4]. If the calixarenes assume the shape designated as the "cone" conformation (see Sect. 5.1) they are seen to have cavities whose dimensions increase as the number of arene moieties in the macrocyclic array increases, as illustrated in Fig. 12. Whether the shapes of these cavities are time-invariant depends on the flexibility of the calixarene (see Sect. 5.2); whether the "open" conformation exists depends on intramolecular hydrogen bond interactions. There is evidence (see Sect. 5.1), for example, that the calix[6]arenes and calix[8]arenes exist in solution in nonpolar solvents in transannularly "pinched" conformations, as illustrated in Fig. 13. x-Ray crystallographic determinations have established that in the solid state the calix[4]-arenes exist in the "cone" conformation [29,30], the calix[5]arenes in the "cone" conformation [31,32], and the derivatives of the calix[6]arenes and calix[8]arenes in the "alternate" rather than "cone" conformation [33,34].

Fig. 11. *p*-Phenylcalix[4]arene (left) and Calix Crater (right)

4.2 Melting Points of the Calixarenes

A characteristic feature of the calixarenes is their unusually high melting points, almost invariably higher than those of their acyclic counterparts. With the exception of the *p*-halocalix[4]arenes and the *p*-halocalix[3]arenes reported by Moshfegh, Hakimelahi and coworkers [97,99,124], all of the calixarenes prepared to date have melt-

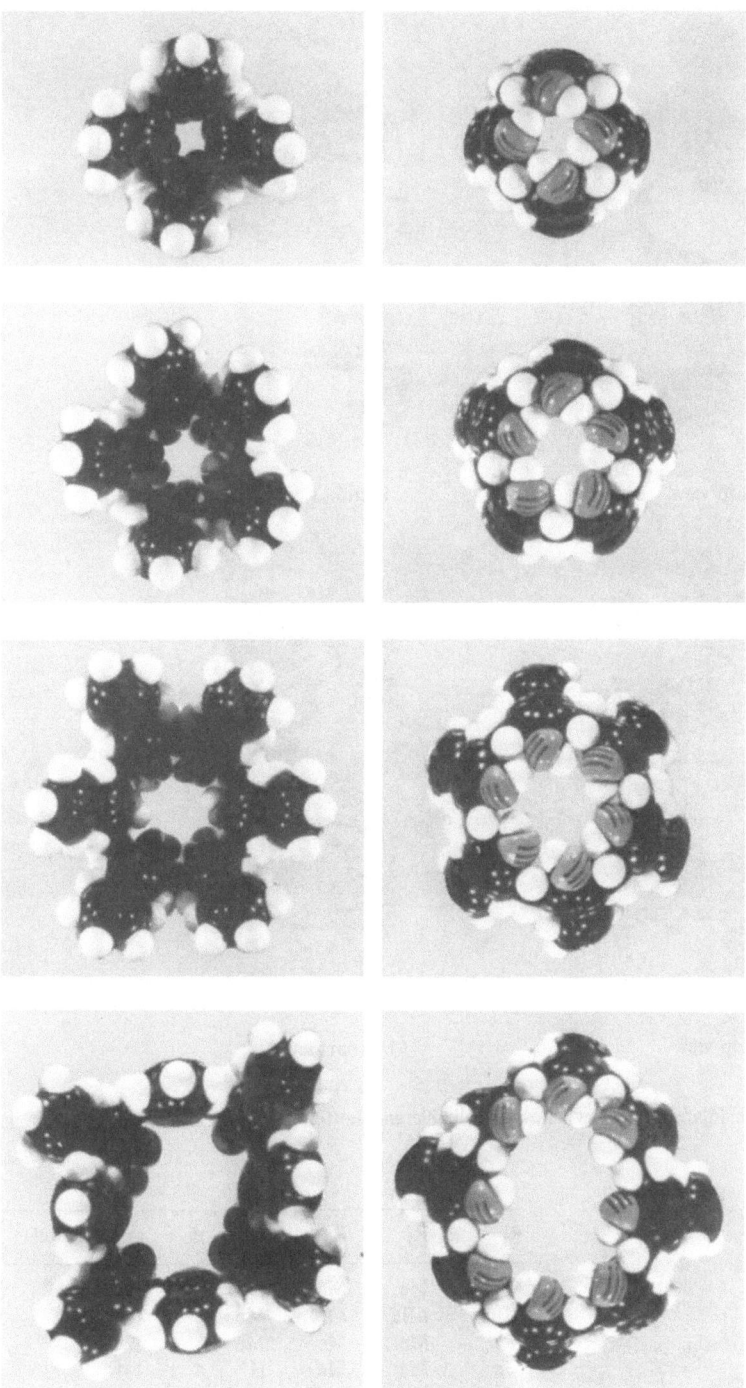

Fig. 12. Space filling molecular models of *p*-phenylcalixarenes. The *p*-phenyl rather than the *p-tert*-butyl substituent has been chosen to more clearly show the proportions of the cavities

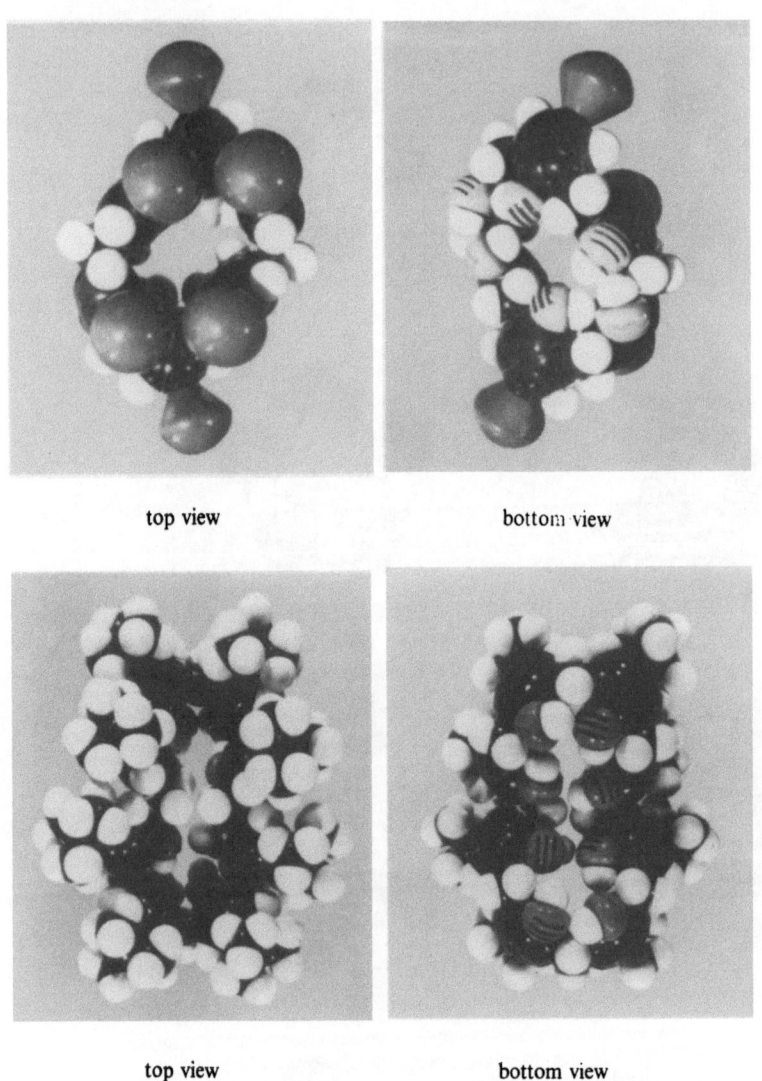

top view   bottom view

top view   bottom view

Fig. 13. ρ-R-Calix[6]arene (top row) and ρ-R-calix[8]arene (bottom row) in "pinched" conformations

R_1	R_2	R_3	R_4	n	m.p.	Ref.
Me	Me	Me	Me	2	192°	89)
t-Bu	t-Bu	t-Bu	Me	2	173°	89)
Me	Me	Me	Me	3	126°	93)
Me	Me	Me	H	4	141°	92)
Me	Me	t-Bu	H	4	201°	92)
Me	Me	t-Bu	H	5	210°	91)

Fig. 14. Melting points of linear oligomers of phenols and formaldehyde

	p-tert-Butyl-calix[4]arene		p-tert-Butyl-calix[6]arene		p-tert-Butyl-calix[8]arene		Calix[4]arene		Tetraacetyl-p-tert-butylcalix[4]arene	
	cold	hot	cold	hot	cold	hot	cold	hot	cold	hot
Cyclohexane	−	−	−	−	−	−	−	±	−	−
Heptane	−	±	−	−	−	−	−	±	−	−
Benzene	±	++	+	+	+	+	±	++	+	+
Toluene	±	++	+	++	+	+	±	++	+	+
Xylenes	−	++	+	++	±	+	+	+++	++	+++
CH₂Cl₂	±	±	±	±	±+	±	++	+++	++	+++
CHCl₃	−	±	−	++	+	+	++	+++	−	++
CCl₄	−	−	−	−	−	−	±	+	−	−
Ether	−	−	−	−	−	−	−	−	±	±+
Tetrahydrofuran	±	−	±+	+	±	+	++	+++	++	+++
Diglyme	−	+	−	−	−	−	±	++	++	+++
Dioxane	−	++	±+	±	−	+	±	++	++	+++
Diphenyl Ether	+	+	±	−	−	−	±	+	+	+++
MeOH, EtOH, PrOH and BuOH	−	−	−	−	−	−	−	−	±	+
Acetone	−	−	−	−	−	−	+	++	++	+++
Acetonitrile	−	−	−	−	−	−	−	+	++	+++
Pyridine	−	+	+	++	++	++	+	++	+	+
Triethylamine	±	±+	+	++	++	++	−	−	+	+
Nitromethane	−	±+	−	−	−	+	++	+	++	++
Nitrobenzene	±	+	+	±	+	±	±+	+	++	+
DMSO	−	−	−	−	−	−	−	++	±	+
DMF	−	++	−	±	−	+	++	++	++	++
Ethyl Acetate	−	+	−	−	−	−	−	−	+	+
Carbon Disulfide	±	±	+++	+++	+++	++	±	+	−	±+
Acetic Acid	−	±	−	±	+++	+++	−	+	++	++
Trifluoroacetic Acid	−	−	−	−	−	−	−	−		
Sulfuric Acid	−	−	−	−	−	−	−	−	+	

Fig. 15. Solubility characteristics of calixarenes. The symbols indicate the following: insoluble (−), very sparingly soluble (±), slightly soluble (+), moderately soluble (++), very soluble (+++)

ing points above 250 °C, in many instances very much higher. For example, *p-tert*-butylcalix[4]arene melts at 344–346 °C, *p-tert*-butylcalix[6]arene melts at 380–381 °C, and *p-tert*-butylcalix[8]arene melts at 411–412 °C; *p*-phenylcalix[4]arene melts at 407–409 °C, and *p*-phenylcalix[8]arene melts above 450 °C. Conversion to calixarene derivatives such as esters and ethers frequently lowers the melting points. For example, the tetramethyl ether of *p-tert*-butylcalix[4]arene melts at 226–228 °C, and the tetrabenzyl ether of *p-tert*-butylcalix[4]arene melts at 230–231 °C. However, the tetraacetate of *p-tert*-butylcalix[4]arene melts at 383–386 °C, and the tetra-trimethyl-silyl ether of *p-tert*-butylcalix[6]arene melts at 410–412 °C. In contrast to these high melting points, the acyclic oligomers melt considerably lower, as illustrated by the examples shown in Fig. 14.

4.3 Solubilities of the Calixarenes

A second characteristic feature of the calixarenes is their low solubility in organic solvents. This prevented Zinke from obtaining cryoscopic molecular weights for his cyclic oligomers, and it frequently poses a problem in purifying and characterizing these materials. However, many of the calixarenes have sufficient solubility in $CHCl_3$ to allow osmometric molecular weights to be obtained (concentrations as low as 0.2 g/L can be used [110a]) and in $CDCl_3$ and pyridine-d_5 to make FT-NMR measurements possible. The solubilities of the even-numbered *p-tert*-butylcalixarenes, the unsubstituted calix[4]arene, and the tetraacetate of *p-tert*-butylcalix[4]arene in a wide variety of solvents are recorded in Fig. 15. As would be anticipated, the nature of the *p*-substituent can have a considerable effect on the solubility characteristics of the calixarene. Among the calixarenes substituted with nonpolar groups, the *p*-allyl-calixarenes are the most soluble, and the *p*-phenyl and *p*-adamantylcalixarenes are the least soluble. A calix[4]arene substituted with *p*-2-hydroxyethyl groups shows considerable solubility in more polar solvents such as DMSO. Conversion of the calixarenes to the ethers and esters generally increases the solubility in nonpolar solvents. For example, the octamethyl ether of *p*-phenylcalix[8]arene is moderately soluble in $CHCl_3$, in striking contrast to the calixarene itself.

4.4 Infrared Spectra of the Calixarenes

The calixarenes show concentration-independent OH stretching bands in the 3200 cm^{-1} region of the infrared, indicative of very strong intramolecular hydrogen bonding, as illustrated in Fig. 16. This is not uniquely associated with the covalently cyclic nature of the calixarenes, however, for the linear oligomers also show OH stretching bands in the same region, as noted previously (see Sect. 3.1 and Fig. 6). The "fingerprint" regions of the IR spectra of the calixarenes have very similar appearances, but close inspection reveals differences that may be useful in establishing the size of the macrocyclic ring [111]. Thus, the cyclic tetramer is characterized by a moderately strong absorption at 830 cm^{-1}, the cyclic hexamer by absorptions at 750 and 800 cm^{-1}, and the cyclic octamer by the absence of any of these three absorptions. The alkyl ethers of these calixarenes show a unique absorption at 850 cm^{-1} for the cyclic tetramers, a unique absorption at 810 cm^{-1} for the cyclic hexamers, and the absence of these absorptions for the cyclic octamers.

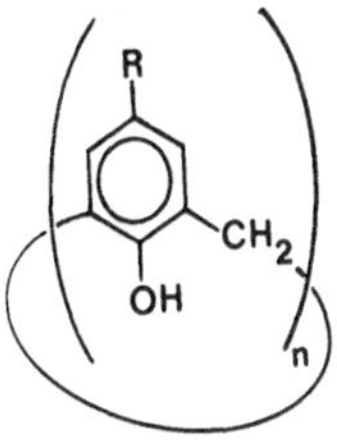

	n	ν_{OH}, cm^{-1}	1290	1180–1250 ($\nu_{fingerprint}$, cm^{-1})	990	910	870	830	800	780	750	730	700
	4	3130–3180	—	3 bands	—	w	w	m	—	m	—	w	w
	5	3280											
	6	3150–3160	m	3 bands	w	m	m	—	m	—	s	sh	—
	7	3155											
	8	3230	m	2 bands	w	w	w	—	—	w	—	—	w

Fig. 16. Infrared spectral characteristics of the calixarenes

4.5 Ultraviolet Spectra of the Calixarenes

Kämmerer and Happel [89] compared the UV spectra of several calix[4]arenes with the corresponding linear oligomers and found little difference; both classes of compounds have λ_{max} at ca 280 nm and 288 nm with approximately equal extinction coefficients. As the size of the macrocyclic ring increases, the wavelengths of the maxima remain invariant, although the molar extinction coefficients rise, as shown by the data in Fig. 17. The molar extinction coefficients per aryl ring remain approximately constant for the absorption at 280 nm and increase with ring size for the absorption at 288 nm (e.g. 2075–2225 for n = 4; 4000 for n = 8). The most significant change in the spectra as the ring size increases is in the ratio of the intensities of the 280 and 288 nm absorption bands. For the cyclic tetramer the intensity of the 280 nm band is greater than that of the 288 nm band (ratio ca 1.3), whereas for the cyclic octamer the reverse is true (ratio ca 0.75). For the cyclic pentamer, hexamer, and heptamer the bands are of approximately equal intensity.

R Groups		λ_{max}, mole^{-1} cm^{-1}		Solvent	Ref.
		280 ± 1 nm	288 ± 1 nm		
all Methyl	4	10,500	8,300	Dioxane	[89]
all tert-Butyl	4	9,800	7,700	CHCl$_3$	[110b]
Me and tert-Butyl	5	14,030	14,380	Dioxane	[93]
all tert-Butyl	6	15,500	17,040	CHCl$_3$	[110b]
Me and tert-Butyl	6	17,210	17,600	Dioxane	[92]
all tert-Butyl	7	18,200	20,900	CHCl$_3$	[110b]
Me and tert-Butyl	7	19,800	20,900	Dioxane	[91]
all tert-Butyl	8	23,100	32,000	CHCl$_3$	[110b]

Fig. 17. Ultraviolet absorption characteristics of the calixarenes

4.6 NMR Spectra of the Calixarenes

Among the spectral techniques, that of NMR provides the best indication for the macrocyclic structure of the calixarenes. For example, the ^{1}H NMR spectra of the

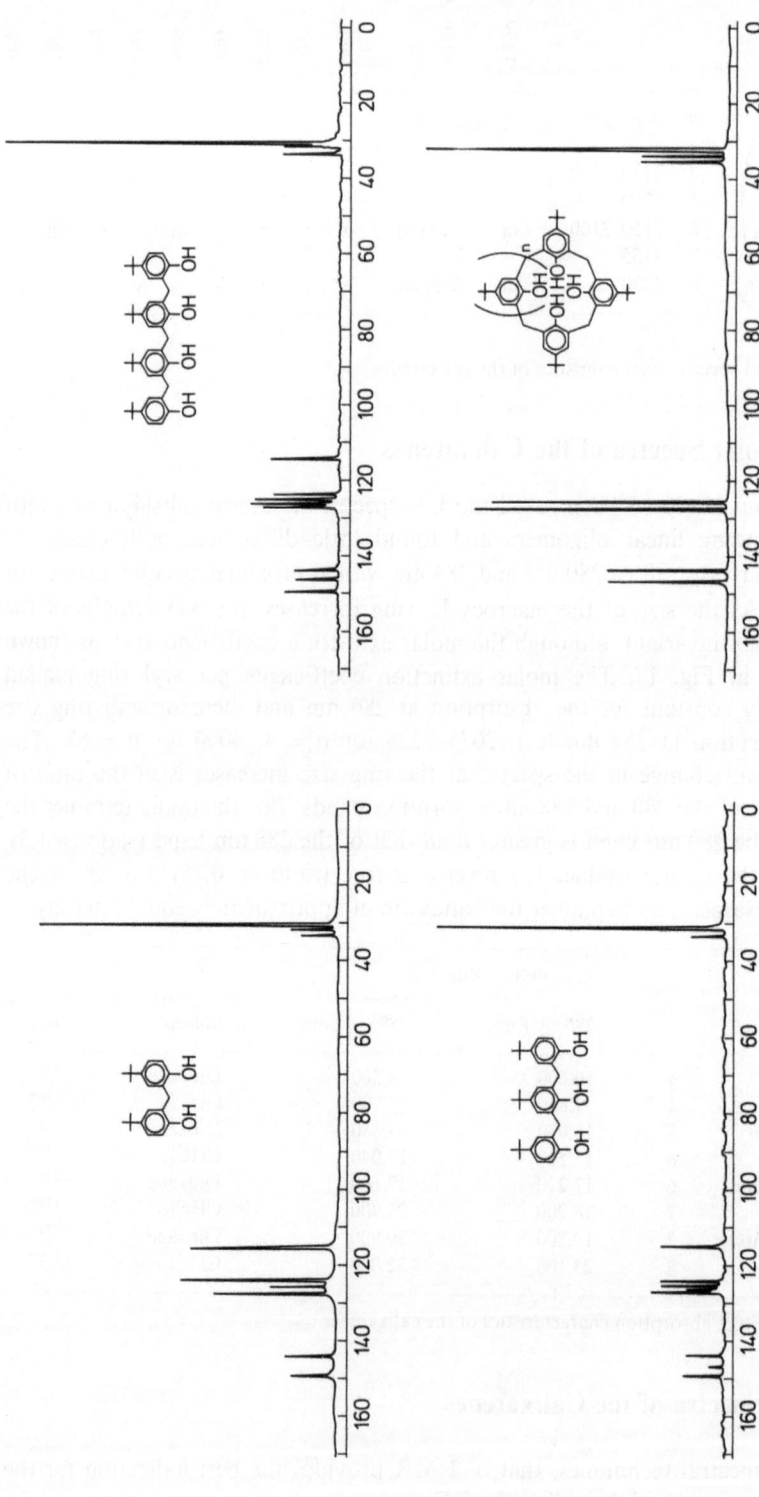

Fig. 18. ^{13}C NMR spectra of linear and cyclic oligomers obtained from the condensation of *p-tert*-butylphenol and formaldehyde

calixarenes in the *p-tert*-butyl series show a single resonance for the aromatic protons, a single resonance for the *tert*-butyl protons, and a temperature-dependent pattern for the methylene protons (discussed in more detail in Sect. 5.1). Shown in Fig. 18 are the ^{13}C NMR spectra of three linear oligomers in the *p-tert*-butyl series along with a cyclic oligomer, illustrating the increasing spectral complexity as the length of the linear oligomer increases but the dramatic decrease in complexity upon cyclization. Due to the symmetry of the cyclic oligomers only four aromatic resonances, one methylene resonance, and two *tert*-butyl resonances are observed.

4.7 Mass Spectra of the Calixarenes

Mass spectra provide a means for determining the molecular weights of the calixarenes via the parent ion signal, but such information must be used with caution. For example, the mass spectrum of *p-tert*-butylcalix[8]arene shows a moderately strong signal at m/e 648, seeming to correspond to a cyclic tetramer. For several years we accepted this as evidence that the major product prepared by the Petrolite procedure was, indeed, the cyclic tetramer. The persistent appearance of small signals at m/e values higher than 648 was disconcerting, however, and when a fully trimethylsilylated sample of the supposed cyclic tetramer was subjected to mass spectral analysis it showed a strong signal at m/e 1872, corresponding to the cyclic octamer. This material showed another quite strong signal at m/e 936, corresponding either to a tetramer resulting from the cracking of the octamer or to the dication of the octamer, thereby providing a possible explanation for the m/e 648 signal from the calixarene itself. In similar vein, mass spectral molecular weights have been reported for calixarenes obtained from *p*-cresol and *p*-methoxy-phenol [43] that seem to indicate a cyclic tetramer but that probably arise from compounds of considerably higher molecular weight.

Kämmerer et al. [20, 21, 89, 91−93] have obtained mass spectral data for almost all of the *p*-alkylcalixarenes that they have prepared, including the calix[7]arene *24* which has a parent ion at m/e 883. Kämmerer has stated [94] that the cyclic oligomers show distinct and characteristic differences in the mass spectrum as compared with the analogous linear oligomers, viz. the cyclic oligomers preferentially lose methyl or *tert*-butyl groups and conserve their ring structure, whereas the linear oligomers preferentially cleave into their phenolic units. It has been noted, however, that the calix[5]arene *22a* shows its strongest signal at m/e 480, corresponding to the apparent extrusion of one of the aryl moieties [21], and the cyclic tetramers themselves also show signals corresponding to the extrusion of one, two, and three aryl moieties.

5 Conformational Properties of the Calixarenes

5.1 Conformationally Mobile Calixarenes

The possibility of conformational isomerism in the calix[4]arenes, adumbrated by Megson [112] and Ott and Zinke [113], was made explicit by Cornforth et al. [10] who

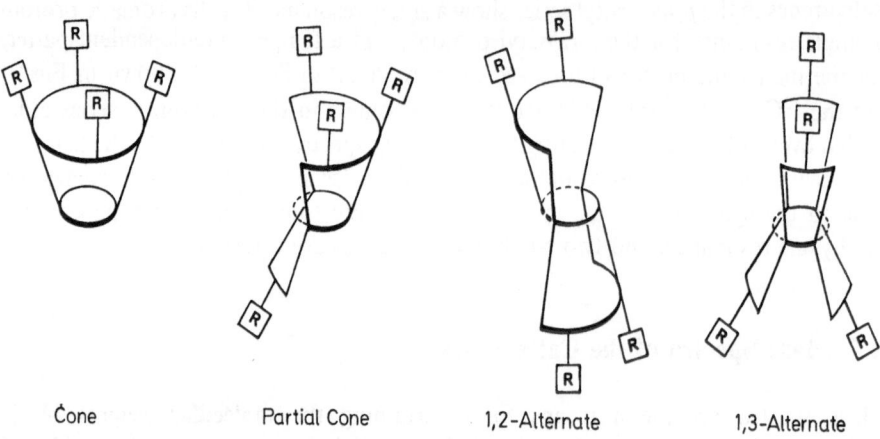

| Cone | Partial Cone | 1,2-Alternate | 1,3-Alternate |

Fig. 19. Conformations of the calix[4]arenes

pointed out that four discrete forms can exist which we refer to [107] as "cone", "partial cone", "1,2-alternate", and "1,3-alternate" conformations, illustrated in Fig. 19. That these are readily interconvertible was first shown by Kämmerer et al. [20, 21] who carried out dynamic ^{1}H NMR studies of *p*-alkylcalix[4]arenes prepared by the stepwise procedure (see Scheme 1). Observing the resonances arising from the ArCH$_2$Ar methylene hydrogens of the calix, they noted that above room temperature the pattern is a sharp singlet while below room temperature it is a pair of doublets. This can be interpreted in terms of a "cone" conformation that is interconverting rapidly on the NMR time scale at the higher temperature and slowly at the lower temperature. From the coalescence temperature of ca 45 °C it can be calculated that the rate of interconversion is ca 100 sec^{-1}. A similar study was reported in 1977 by Munch [22] using a calixarene prepared by the Petrolite method [47] from *p*-(1,1,3,3-tetramethylbutyl)phenol. Virtually identical results were obtained, and this appeared to provide definite proof (in conjunction with the mass spectral data cited in Sect. 4.7) for the cyclic tetrameric structure of the Munch compound. It seemed inconceivable that the quite rigid calix[4]arene and the much larger, and presumably much more flexible, calix[8]arene might manifest the same dynamic ^{1}H NMR behavior. Yet, this is precisely what occurs. In 1981 Gutsche and Bauer [114] showed that authentic samples of *p-tert*-butylcalix[4]arene and *p-tert*-butylcalix[8]arene do, indeed, show virtually identical dynamic ^{1}H NMR behavior in CDCl$_3$ and C$_6$D$_5$Br, as shown in Fig. 20. In pyridine-d$_5$, however, a difference is observed; whereas the calix[4]arene resolves into a pair of doublets at ca 15 °C, the calix[8]arene shows only a singlet at temperatures as low as —90 °C. The similarity of the dynamic ^{1}H NMR spectra in CDCl$_3$ or C$_6$D$_5$Br (nonpolar solvents) is ascribed to intramolecular hydrogen bonding in the cyclic octamer which creates a "pinched" conformation (see Fig. 13) that has the superficial aspect of a pair of tetramers "stuck" together, making the cyclic octamer behave, quite fortuitously, as though it were a cyclic tetramer. The type of hydrogen bonding present in the cyclodextrins has been referred to as "circular" by Saenger [115] who has suggested that circularity confers a special stability on such systems. The cyclic octamer, by "pinching"

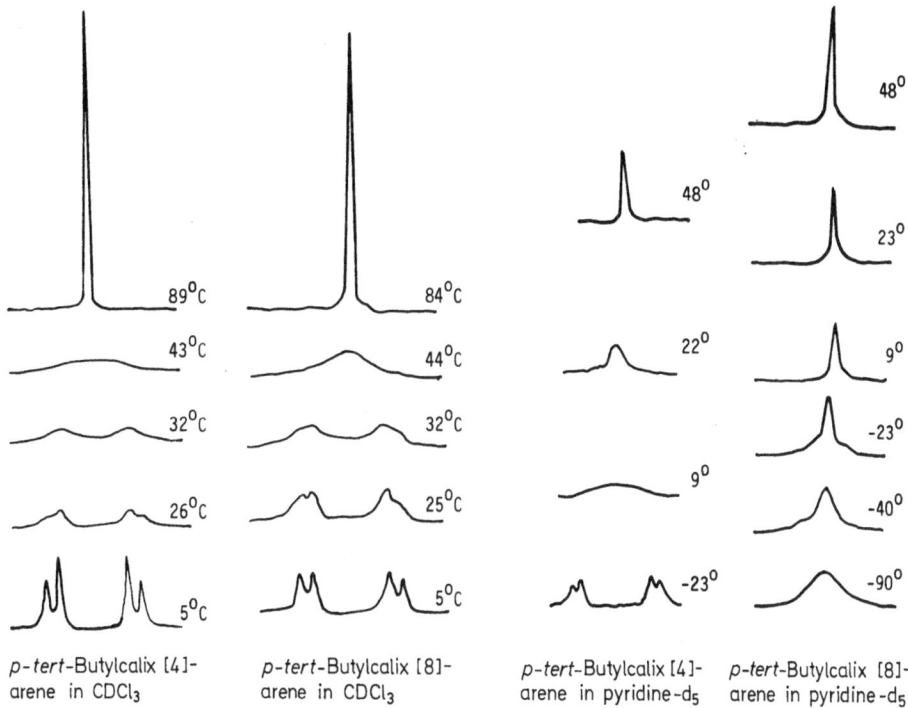

Fig. 20. ^{1}H NMR spectra of *p-tert*-butylcalix[4]arene and *p-tert*-butylcalix[8]arene at various temperatures in CDCl$_3$ and pyridine-d$_5$

transannularly, can assume a conformation containing two such circularly hydrogen bonded arrays, each containing four OH groups [114]. A necessary corollary of this postulate, however, is that a pseudorotation is operative which averages the CH$_2$ *groups* at a rate that is fast on the NMR time scale at the temperatures at which the dynamic NMR studies are conducted, *i.e.*

In a fashion similar to the calix[8]arenes, the calix[6]arenes are postulated on the basis of their dynamic ^{1}H NMR behavior to exist in "pinched" conformations (see Fig. 13) in nonpolar solvents. Transannular "pinching" establishes two circular arrays each containing three intramolecularly hydrogen bonded OH groups. The result is to force the system into a conformation in which the aryl groups in the 1- and 4-positions project outward like a pair of wings perpendicular to the aryl groups in the 2-, 3-, 5-, and 6-positions which are colinear, forming a channel into which other molecules might comfortably fit.

Inversion Rate a Function of:	p-Substituents	Ring Size	Solvent	ΔG*, kcal/mole	$T_{°C}$	Ref.
Ring Size	four t-Bu	4	CDCl₃	15.3	52	114)
	five Me	5	CDCl₃	12.1		93)
	six t-Bu	6	CDCl₃	13.0	8	116)
	one t-Bu; six Me	7	CDCl₃	12.25		91)
	eight t-Bu	8	CDCl₃	15.3	53	114)
Solvent	four t-Bu	4	C₆D₅Br	15.2	44	114)
	four t-Bu	4	C₅D₅N	13.4	15	114)
	four H	4	CDCl₃	14.6	36	116)
	four H	4	(CD₃)₂CO	13.3	−5	116)
	four H	4	CD₃CN	13.3	0	116)
	four H	4	(CD₃)₂SO		20	116)
	four C₆H₅	4	CDCl₃	15.4	44	116)
	four C₆H₅	4	C₆D₅Br	14.8	36	116)
	four C₆H₅	4	(CD₃)₂CO	13.9	8	116)
	four C₆H₅	4	C₅D₅N	12.4	−2	116)
	eight t-Bu	8	C₆D₅Br	15.2	43	114)
	eight t-Bu	8	C₅D₅N	9	< −90	114)
Substituents	four H	4	CDCl₃	14.6	36	116)
	three Me; one t-Bu	4	CDCl₃	15.9		21)
	one Me; three t-Bu	4	CDCl₃	15.4		93)
	four i-Pr	4	CDCl₃	14.8	33	116)
	four t-Bu	4	CDCl₃	15.2	44	114)
	four t-C₅H₁₁	4	CDCl₃	14.5	27	116)
	four CH₂CH=CH₂	4	CDCl₃	15.0	37	116)
	four t-C₈H₁₇	4	CDCl₃	14.7	30	116)
	four C₆H₅	4	CDCl₃	15.4	44	116)
	two t-Bu; two Cl	4	CDCl₃	15.0	38	116)
	eight t-Bu	8	CDCl₃	15.3	53	114)
	eight t-C₈H₁₇	8	CDCl₃	15.1	53	116)

Fig. 21. Free energies of activation for the conformational inversion of calixarenes

A number of calixarenes with various p-substituents and ring sizes have now been studied by means of dynamic ¹H NMR, and the free energies of activation for the conformational inversion have been calculated, as shown in Fig. 21.

5.2 Conformationally Immobile Calixarenes

Cram has defined a "cavitand" as a *synthetic* compound containing an "enforced cavity" [117] large enough to engulf ions or other molecules [118]. The calix[4]arenes are synthetic compounds, and they have the capacity to entrap other molecules. They do not have *enforced* cavities, however, because they are conformationally mobile. To make them cavitands it is necessary to "fix" them either in a "cone" or "partial cone" conformation. One way for accomplishing this is to replace the hydrogens of the OH functions with larger groups. When transformation between conformations occurs an aryl group rotates around the C-2/C-6 axis in a direction that brings the OH groups through the center of the macrocyclic ring. Space filling models indicate that while the OH groups provide little hindrance to this process, larger groups can make the transformation virtually impossible. Ungaro and co-workers [106] and Gutsche and coworkers [107] have studied the process of conformational fixing by this means in some detail.

Each of the conformations for the calix[4]arenes (see Fig. 19) has a distinctive pattern of resonances in the ^{1}H NMR [107], allowing easy structure assignment of the conformationally "fixed" derivatives of the calix[4]arenes. Acetylation of *47* yields a tetraacetate (*50f*) which, by means of ^{1}H NMR, can be shown to be fixed in the "1,3-alternate" conformation [107]. This result was not surprising, because it had already been established that the octahydroxycalix[4]arenes of structure *13* are fixed in a "1,3-alternate-like" conformation [69, 70, 72]. It appears, however, that the "1,3-alternate" conformation is rather rare among the derivatives of the tetrahydroxy-calix[4]arenes (*e.g.* *47–49*) and that the "partial cone" and "cone" conformations are generally favored. For example, *48* yields a tetraacetate (*51f*) that is fixed in a "partial cone" conformation, a result that has been verified by x-ray crystallography [119]. Methylation and ethylation of *47* and *48* and allylation of *47* yield the corresponding ethers *50a*, *50b*, *50c*, *51a* and *51b* in the "partial cone" conformation [107]. Somewhat surprisingly, the tetramethyl ethers *50a* and *51a* are more conformationally mobile than is predicted by inspection of CPK models. At room temperature the ^{1}H NMR spectra of these compounds show only a broad resonance for the methylene hydrogens, quite analogous to the parent calixarenes. These sharpen to simple patterns at elevated temperatures and to more complex patterns at lower temperatures, commensurate with a "partial cone" conformation, and this provides a good example of the fact that solid, space-filling models tend to overestimate the barriers to conformational changes.

"Cone" conformations are established when *48* is converted to its tetra-allyl ether *51c*, benzyl ether *51d*, or trimethylsilyl ether *51e*, when *47* is converted to its

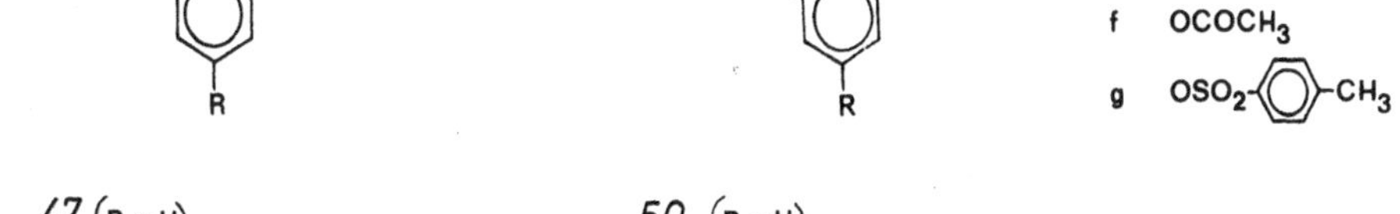

Y	
a	OCH$_3$
b	OC$_2$H$_5$
c	OCH$_2$CH=CH$_2$
d	OCH$_2$C$_6$H$_5$
e	OSi(CH$_3$)$_3$
f	OCOCH$_3$
g	OSO$_2$-⟨C$_6$H$_4$⟩-CH$_3$

47 (R = H)

48 (R = t-C$_4$H$_9$)

49 (R = CH$_2$CH=CH$_2$)

50 (R = H)

51 (R = t-C$_4$H$_9$)

52 (R = CH$_2$CH=CH$_2$)

Fig. 22. Calixarenes, calixarene esters, and calixarene ethers

tetrabenzyl ether *50d* or tetra-tosylate *50g*, and when *49* is converted to its tri-methylsilyl ether *52e* or tetra-tosylate *52g*. In only a few instances have mixtures of conformational isomers been obtained upon derivatization as, for example, in the acetylation of *41* (R = *tert*-Bu) which yields a "1,3-alternate" and two "partial cone" conformers [107, 120]. The "cone" conformation of *41* is established as the only one produced when conversion to the trimethylsilyl derivative occurs.

By appropriate choice of derivatizing agent calix[4]arenes can be fixed either in the "cone" or "partial cone" conformation. Thus, acetylation appears to favor the latter in most cases, and benzylation and trimethylsilylation favor the former. This represents a particularly useful facet of calixarene behavior, for not only does it bring the calixarenes into the family of cavitands but it provides a means for contouring the cavity by design. The explanation for the different conformational outcomes upon derivatization is not known, and the roles played by the derivatizing agent and the reaction conditions remain to be determined. It is clear, however, that the derivatizing agent not only can play a part in determining the conformation of the product but its structure as well, as discussed in Section 3.5. For the partially alkylated calixarenes a "flattened partial cone" conformation is assigned to the trimethyl ether and a "flattened 1,3-alternate" conformation to the monomethyl and dibenzyl ethers. The trimethyl and monomethyl ethers are conformationally less mobile than the tetramethyl ethers, a result that is ascribed to intramolecular hydrogen bonding arising from the free hydroxyl group(s) in the molecules.

Although the calix[8]arenes resemble the calix[4]arenes with respect to conforma-tional mobility, their ester and ether derivatives show no conformational fixation, even at very low temperatures. This difference in behavior arises from the differences in the sizes of the annuli in the two series, the calix[4]arenes having a very small annulus through which groups only as small as OH and OMe can pass but the calix[8]arenes having a much larger one through which groups even as large as tosyl can pass. The calix[6]arenes occupy an intermediate position. Their derivatives are more conformationally mobile than those of the calix[4]arenes but can be frozen out (on the NMR time scale) at low temperatures. Dradi, Pochini and Ungaro [121]

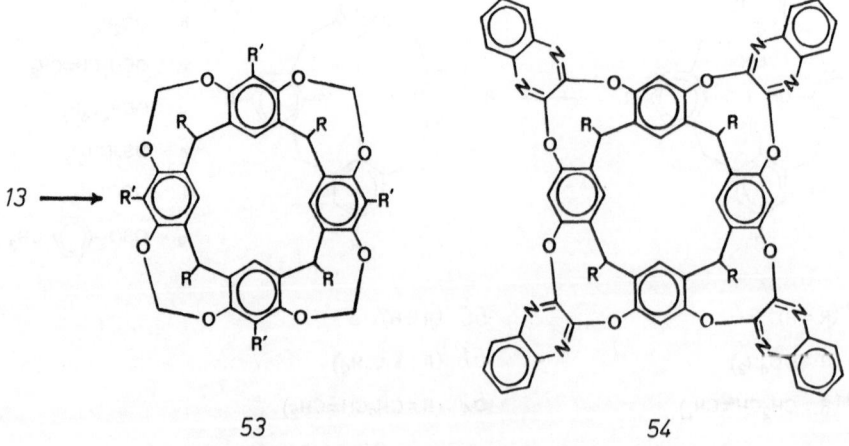

53   54

Fig. 23. Synthesis of a rigid, fully bridged calixarene

have studied the dynamic ^{1}H NMR behavior of the hexa-(2-methoxyethyl)ether of p-tert-butylcalix[6]arene and conclude that the preferred conformation(s) is either a "1,3,5-alternate" (i.e. three "up" and three "down") and/or a "1,4-alternate" (i.e. a "winged" conformation similar to that suggested for the free calix[6]arene (see Sect. 5.1). An x-ray crystallographic determination of the compound [33] pictures it in the solid state as an "up, down, out, down, up, out" conformation.

Another way to reduce the conformational mobility of the calixarenes is to bridge two or more regions of the molecule with an appropriate number of atoms. This approach has been successfully applied by Cram [118] to the octahydroxycalix[4]-arene 13 which has been converted to the conformationally rigid compounds 53 (R=H) and 54 by treatment with ClCH$_2$Br and 2,3-dichloro-1,4-diazanaphthalene, respectively. Working in the tetrahydroxycalix[4]arene series, Gutsche and Bailey [122] have synthesized the dimer unit 55 and have incorporated it, using the stepwise synthesis depicted in Scheme 5, into a dihomooxacalix[4]arene (57). Although the single bridges in 56 and 57 do not completely curtail conformational flexing, as do the four bridges in 53 and 54, complete inversion is no longer possible, and this is manifested in higher coalescence temperatures in the dynamic ^{1}H NMR spectra.

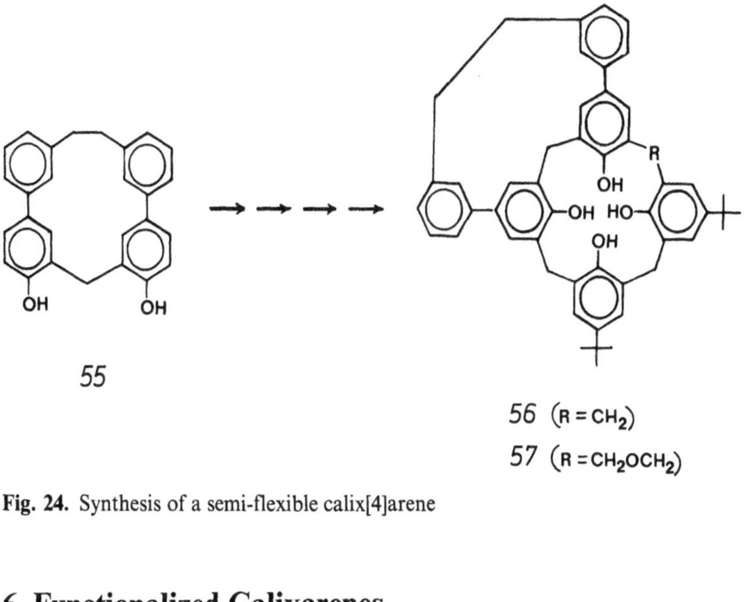

55

56 (R = CH$_2$)
57 (R = CH$_2$OCH$_2$)

Fig. 24. Synthesis of a semi-flexible calix[4]arene

6 Functionalized Calixarenes

One of the principal reasons for the interest in the calixarenes is their potential for serving as enzyme mimics. If they are to operate in this capacity, however, it is necessary that they carry functional groups of various types. The p-tert-butylcalix-arenes have proved to be excellent precursors for functional group introduction, for the p-tert-butyl group is easily removed by trans-alkylation [123]. As outlined in Scheme 7, de-tert-butylation of p-tert-butylcalix[4]arene (58) yields compound 59 which is amenable to functionalization in the p-position. For example, bromination of the tetramethyl ether of calix[4]arene (60) proceeds smoothly, affording the tetra-bromo compound 61 [124] which, by lithiation followed by carbonation yields the tetramethyl ether of *p*-carboxycalix[4]arene (*65a*) or by cyanation yields the tetra-methyl ether of *p*-cyanocalix[4]arene (*65b*) [46]. In similar fashion, Friedel-Crafts acylation of *60* yields *p*-acetylcalix[4]arene, amenable to transformation to *65a* by a haloform oxidation [125].

The bromination and acetylation reactions show that the calixarenes are capable of undergoing conventional electrophilic substitution reactions without rupture of the macrocyclic ring. Zinke et al. [18] reported the nitration of *p-tert*-butylcalixarene (ring size uncertain) and obtained a material that failed to melt up to 400 °C, exploded when heated more strongly, and had a nitrogen analysis in agreement with a tetranitro-calixarene. Attempts to obtain a nitration product from calix[4]arene *59*,

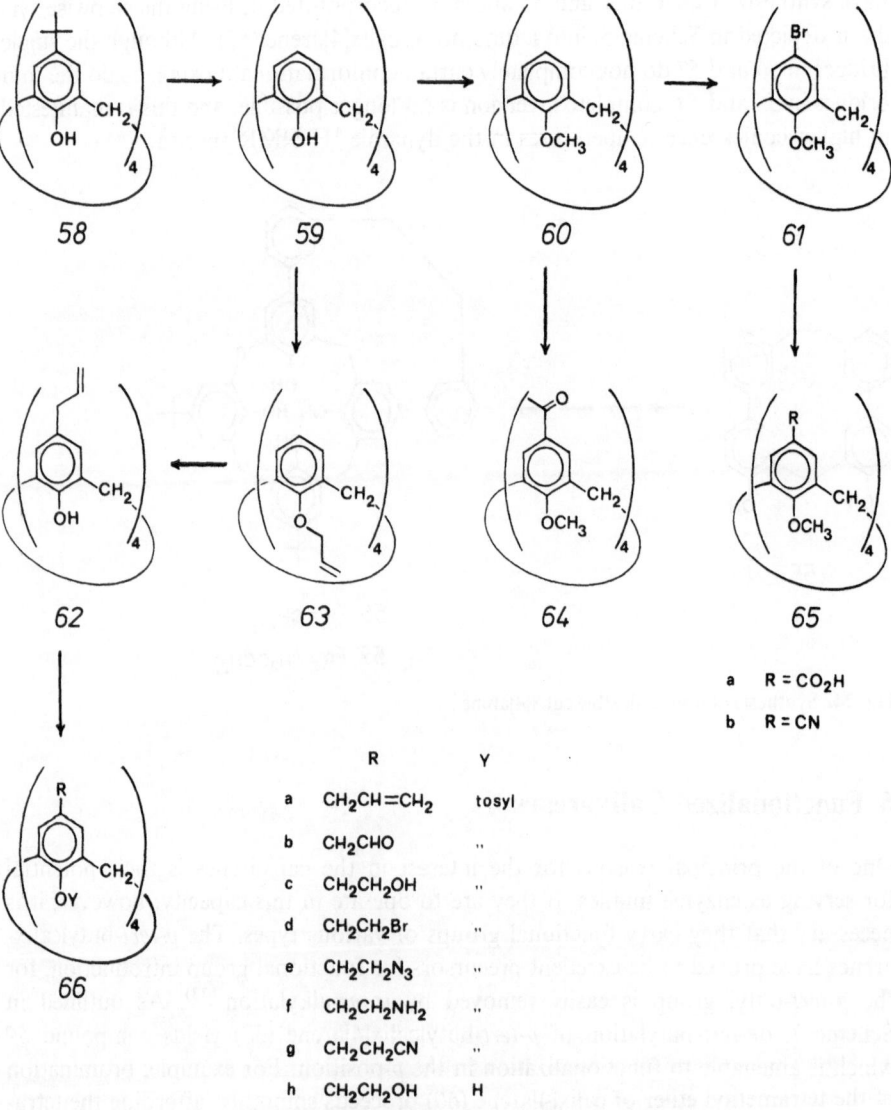

Scheme 7. Synthesis of functionalized calix[4]arenes.

however, have not been successful, a result that is in agreement with Högberg's finding that the octahydroxycalix[4]arenes *13* undergo nitrodealkylation to yield lower molecular weight nitration products [73, 12].

As an alternative to electrophilic substitution as a means for introducing functional groups into the calixarenes, a reaction sequence has been developed that involves the conversion of calix[4]arene (*59*) to the tetraallyl ether *63*. When *63* is heated in diethylaniline it undergoes a four-fold *p*-Claisen rearrangement to afford *p*-allyl-calix[4]arene (*62*) in excellent yield [126]. From the tetra-tosyl ester of *62* (i.e. compound *66a*) a variety of functionalized calixarenes have been obtained, including the aldehyde *66b*, alcohol *66c*, bromide *66d*, azide *66e*, amine *66f*, and nitrile *66g*. Removal of the tosyl group occurs under mildly basic conditions to yield, for example, *p*-(2-hydroxyethyl)calix[4]arene (*66h*).

Reaction sequences similar to those portrayed in Scheme 7 should also be applicable to the other calixarenes obtained by de-*tert*-butylation of *67* (n = 5, 6, 7, 8), as depicted in Scheme 8. Some of these transformations have already been demonstrated [125], while others remain to be investigated. De-*tert*-butylation of *p*-*tert*-butylcalix[6]-arene (*67*, n = 6) and *p*-*tert*-butylcalix[8]arene (*67*, n = 8) proceed fairly smoothly to afford the corresponding calix[6]arene (*68*, n = 6) and calix[8]arene (*68*, n = 8). Conversion of *68* (n = 6) to the hexa-allyl ether (*69b*, n = 6) followed by Claisen rearrangement produces a modest yield of *p*-allylcalix[6]arene (*70*, n = 6). Similar reactions have been shown to take place in the calix[8]arene series as well [125], although the yields are much lower and the products have not yet been fully characterized.

The cavitand *53* (R' = Br) provides another precursor for the preparation of a functionalized calixarene. Bromination of the octahydroxycalixarene *13* prior to the

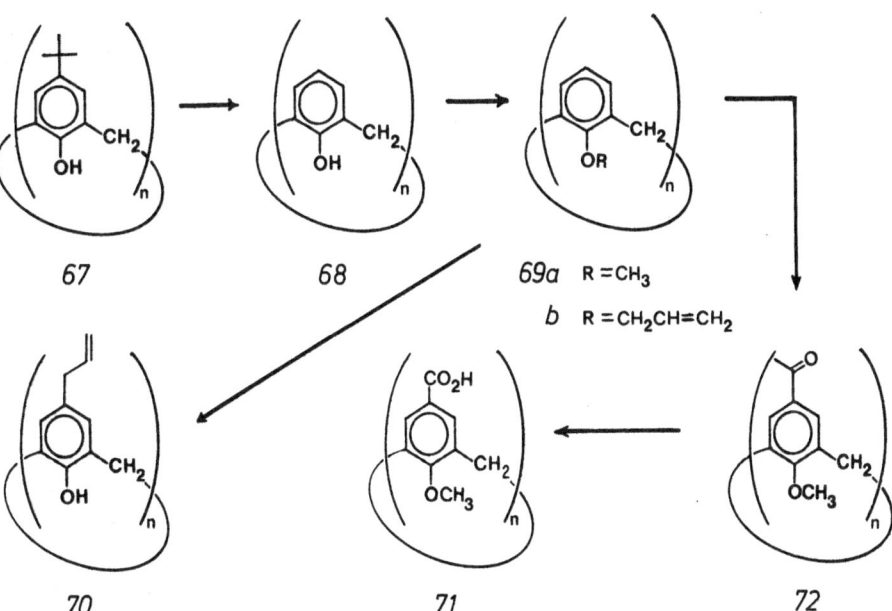

Scheme 8. Synthesis of functionalized calix[n]arenes.

introduction of the methylene bridges between the oxygens affords *53* (R′ = Br), and lithiation followed by treatment with carbon dioxide then yields the tetracarboxy compound *73* [118].

53 ⟶

73

7 Complex Formation Involving Calixarenes

7.1 Solid State Complexes with Neutral Molecules

It has long been known that many of the calixarenes retain the solvent from which they are crystallized. For example, *p-tert*-butylcalix[4]arene forms solid state complexes with chloroform, benzene, toluene, xylene, and anisole [32]; *p-tert*-butylcalix[5]arene forms complexes with isopropyl alcohol [26] and acetone [32]; *p-tert*-butylcalix[6]arene forms a complex containing chloroform and methanol [23]; *p-tert*-butylcalix[8]arene forms a complex with chloroform [23]; *p-tert*-butyldihomooxacalix[4]arene forms a complex with methylene chloride [23]. The tenacity with which the guest compound is held varies widely within this group, however. Whereas the cyclic octamer loses chloroform upon standing a few minutes at room temperature and atmospheric pressure, the cyclic hexamer retains some chloroform even after heating for six days at 257 °C at 1 mm pressure. Also, the location of the guest molecule varies, depending on the guest and host, as demonstrated by the x-ray crystallographic work of Andreetti et al. An x-ray crystallographic determination of the *p-tert*-butylcalix[4]-arene-toluene complex shows the calixarene to be in the "cone" conformation and the toluene to be located in the center of the calix [29], *i.e.* an "*endo*-calix" complex as illustrated in Fig. 25. Benzene, *p*-xylene, and anisole are stated to form similar types of complexes [32]. *p*-(1,1,3,3-Tetramethylbutyl)calix[4]arene (*4*, R = 1,1,3,3-tetramethylbutyl), on the other hand, forms *exo*-calix complexes (channel complexes [127]) with aromatic molecules [128], and calix[5]arene (*5*, R = H) forms a 1:2 complex with acetone in which one acetone molecule is in the calix and the second acetone molecule is external to the calix [32], as illustrated in Fig. 25. An x-ray crystallographic examination of solvent-free *p*-(1,1,3,3-tetramethylbutyl)calix[4]arene [30] shows that the calixarene is in the "cone" conformation and that two of the 1,1,3,3-tetramethylbutyl groups are oriented inwards to cover the calix, *i.e.* intramolecular complexation. Calix[4]arene, the parent compound lacking *p*-alkyl substituents, also fails to form a tight *endo*-calix complex with aromatic compounds, leading to the postulate [30] that the *tert*-butyl group plays a special role in promoting complexation because

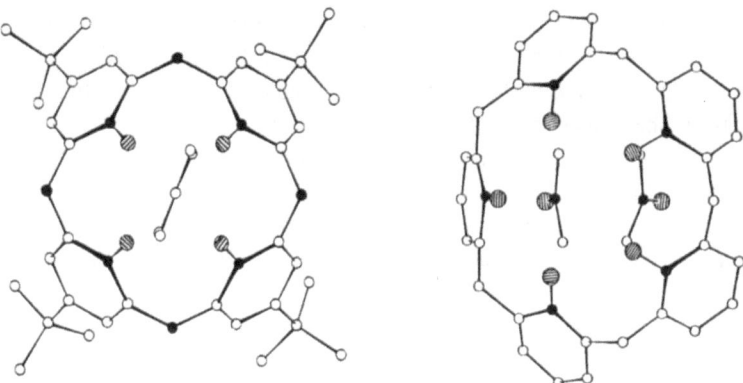

Fig. 25. x-Ray crystallographic structures of 1:1 complex of *p-tert*-butylcalix[4]arene and toluene (left) and 1:2 complex of calix[5]arene and acetone (right)

(a) it is not flexible enough to bend inward and fill the calix itself and (b) it provides CH$_3$/π interactions [129] between its methyl groups and the aromatic ring of the guest molecule. As has been noted in other studies of clathrate formation [130, 131], subtle changes in the structure of the host and guest can significantly alter the stability and the structure of the host-guest complex.

Selective complexation has been demonstrated [132] by crystallizing *p-tert*-butyl-calix[4]arene from 50:50 mixtures of two guest molecules such as benzene and *p*-xylene. It was found that anisole and *p*-xylene are complexed in preference to most other simple aromatic hydrocarbons. *p-tert*-Butylcalix[4]arene has been shown to have some selectivity for mixtures of aromatic hydrocarbons when used as the immobile phase in a chromatographic column [133].

7.2 Solution Complexes with Neutral Molecules

The isolation and characterization of solid state complexes does not necessarily indicate that similar complexes exist in solution. In fact, there are no published data in support of solution complexes of calixarenes. Experiments in our laboratory, using UV/Vis and NMR probes, have failed so far to give any convincing evidence for the formation of solution complexes by various *p*-alkylated-calix[4]arenes, calix[6]arenes, and calix[8]arenes, using chloroform, benzene, toluene, bromobenzene, and a variety of other putative guests. However, certain amines do appear to form such complexes. For example, when *p*-allylcalix[4]arene and *tert*-butylamine are mixed in an acetonitrile solution, the ^{1}H NMR resonance position of the *tert*-butylamine hydrogens shifts downfield by 0.3 ppm, the resonances of the aromatic hydrogens of the calixarene shift by 0.4 ppm, the rate of conformational inversion of the calixarene is reduced, the relaxation time (T_1) of the *tert*-butylamine hydrogens is shortened, and nuclear overhauser effects on the protons of the allyl group of the calixarene are observed [116]. These phenomena are interpreted as arising from an initial acid-base reaction between the calixarene and the amine to form the calixarene anion (*74*) and the *tert*-butylammonium cation (*75*) followed by association of the ions. The structure of the ion-pair seems best interpreted as an *endo*-calix complex (*76*) as shown in Fig. 26. In support of the idea that a proton transfer constitutes the first event is the fact that weaker amines, such as aniline do not form complexes and the fact that the ^{1}H NMR shifts in the calixarene and amine are comparable to those observed when the calixarene is treated with a strong base and the amine with a strong acid. In support of the idea that the result is an *endo*-calix complex are the ^{1}H NMR data concerning the conformational inversion rate, the relaxation rate, and the nuclear overhauser effects and the fact that some selectivity with respect to the amine component is manifested. For example, neopentyl amine, which has approximately the same basicity as *tert*-butylamine, forms a considerably less stable complex. This is attributed to the "angular" structure of neopentylamine. Assuming

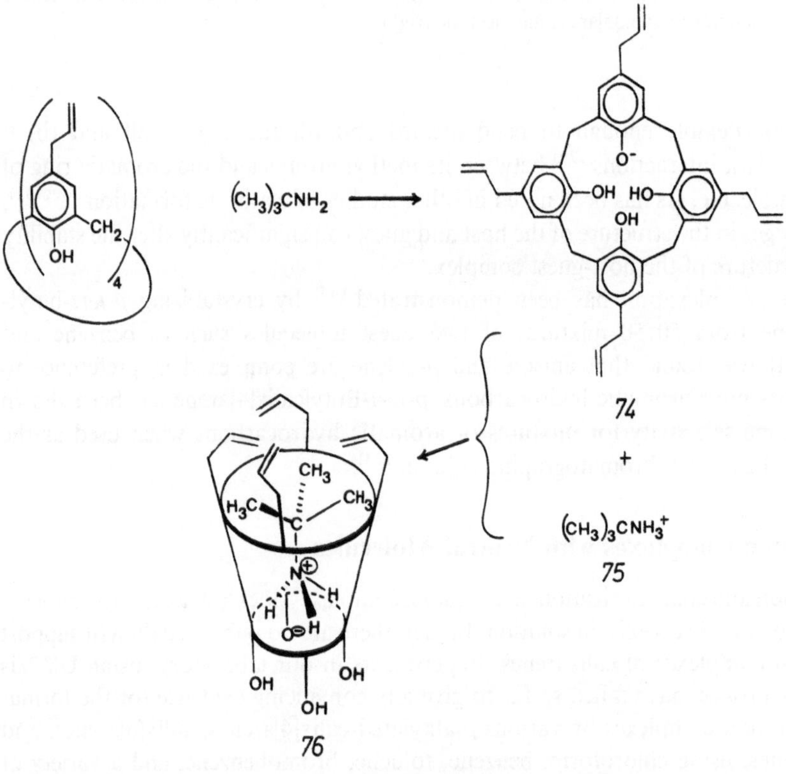

Fig. 26. Complex formation between *p*-allylcalix[4]arene and *tert*-butylamine

that the preferred mode of association between the calixarene and the amine brings the three ammonium hydrogens as close to the calixarene oxygens as possible, the carbon framework of *tert*-butylamine is seen to fit comfortably in the calix, whereas that of neopentylamine presses into the side of the calix. Thus, good evidence is at hand in support of the formation, in solution, of an *endo*-calix complex; but, whether this should be regarded as complexation with a neutral molecule perhaps is debatable.

7.3 Solution Complexes with Cations

The ability of the p-tert-butylcalixarenes to transport metal ions through hydrophobic liquid membranes has been studied by Izatt and coworkers [63]. Although the calixarenes are ineffective toward cations in neutral solution, they show considerable transport ability in basic solution, as illustrated by the data in Fig. 27. This contrasts with 18-crown-6 compounds which are far more effective for KNO_3 than KOH. The p-tert-butylcalix[4]arene, p-tert-calix[6]arene, and p-tert-butylcalix[8]-arene all show the same relative cation transport abilities, viz. $Cs^+ > Rb^+ > K^+ > Na^+ \gg Li^+$. The cyclic tetramer shows the greatest selectivity for Cs^+ compared with the other cations, while the cyclic octamer shows the greatest absolute transport ability for Cs^+. Control experiments with p-tert-butylphenol, which shows little or no transport ability, indicate that the macrocyclic ring plays a critical role. It seems doubtful that the calixarenes are playing a crown ether-type role by surrounding the cations with a planar array of oxygens. The annulus of the calix[4]arene is too small to do this for an ion as large as Cs^+, and the annulus of the calix[8]arene is too large to do this effectively. The results with the calixarene-amine complexation, discussed in Section 7.2, suggest that the calixarene-metal cation complexes may behave in similar fashion, i.e. that they are endo-calix complexes. As pointed

	p-tert-butylcalixarene			18-crown-6	p-tert-butylphenol
Source Phase	*4*	*6*	*8*		
LiOH	—	10 ± 1	2.0 ± 0.2	0.9	0.9
NaOH	1.5 ± 0.4	13 ± 2	9 ± 2		
KOH	0.4 ± 0.1	22 ± 3	10.0 ± 0.4		
RbOH	5.6 ± 0.7	71 ± 8	340 ± 20		
CsOH	260 ± 90	810 ± 80	1200 ± 90		

Fig. 27. Cation transport from basic solution by calixarenes, 18-crown-6, and p-tert-butylphenol; data given as flux in moles/second-meter$^2 \times 10^8$

out by Izatt et al. [63], cesium ion loses its hydration sphere more easily than the other monovalent cations, and this may be the factor that determines the selectivity favoring cesium. The calix[8]arene in a transannularly "puckered" conformation (see Fig. 13) should have the capacity for carrying two Cs^+ ions per molecule and might be expected to be twice as effective as the calix[4]arene; in fact, it is about four times as effective. The study by Izatt and coworkers indicates that the calixarenes may possess useful features as ion carriers because of their low water solubility, their ability to form *neutral* complexes with cations through loss of a proton, and their potentiality for allowing the coupling of cation transport with the reverse flux of protons.

It has been reported that a complex between the hexa-(methoxyethyl) ether of p-tert-butylcalix[6]arene and guanidinium tetraphenyl borate can be detected by ^1H NMR measurements which show large shifts in the resonances of the hydrogens in the methoxyethyl groups and a reduced conformational inversion rate [121]. The complex formed in this instance, however, is different from those formed between the calixarenes and amines or alkaline hydroxides inasmuch as it carries a positive charge and, apparently, involves the methoxyethyl side chains rather than the calix as the site of complexation.

8 Physiological Properties of Calixarenes and Calixarene Derivatives

Phenolic compounds have long been recognized as eliciting various kinds of physiological responses. For example, the urushiols, which are long chain alkyl-substituted catechols, are the active vessicant principle of poison ivy. Somewhat comparable responses, viz. contact dermatitis, have been noted with p-tert-butyl-phenol/formaldehyde resins, particularly the linear tetramer [134]. The p-tert-butyl-calix[4]arene and p-tert-butylcalix[8]arene have been subjected to the Ames' test for mutagenicity [135] and found to give negative readings. Whether this arises from an inherent lack of physiological effect or simply from the great insolubility of these materials is uncertain. Halo-substituted phenols have elicited recurring interest as bacteriostatic agents, and some of this attention has been focused on phenol/formaldehyde oligomers. An early example comes from the Monsanto group [136] who prepared and tested the bacteriostatic properties of linear dimers and trimers from p-halophenols and formaldehyde. A recent example is the work of Moshfegh, Hakimelahi et al. [97-99,137] who have prepared a long series of linear as well as cyclic oligomers from p-halophenols and formaldehyde (see Scheme 4) and have tested their in vitro activity against various pathogenic organisms. Significant activity was noted for linear tetramers and cyclic tetramers, and it is suggested [97] that this correlates with the chelating ability (e.g. toward Fe^{++}) of these compounds.

An extensive medical and biochemical literature has arisen during the last several decades concerning the oxyalkyl derivatives of simple phenols as well as phenol/formaldehyde condensation products. Of particular note in the context of this review is the work of Cornforth and coworkers [10,108] on the polyoxyethyl derivatives of the p-tert-butyl and p-(1,1,3,3-tetramethylbutyl)calixarenes. In the earlier work [10] oxyethylation was effected simply by treating the calixarenes with ethylene oxide to give products designated as "macrocyclons". In the later work [108] a more carefully controlled set of experiments was carried out using the high melting compound from the condensation of p-(1,1,3,3-tetramethylbutyl)phenol and formaldehyde. Tests of these compounds for tuberculostatic activity led to the conclusion that the lipophilic-hydrophilic balance of the molecule may be the most critical factor but that resistance to chemical breakdown in vivo is also important if activity is to be shown against the relatively slow course of experimental tuberculosis. Macrocyclon and other oxyalkylated phenols have been tested against a variety of other organisms [138] and as carcinostatic agents [139].

9 Conclusion

The calixarenes, along with the cyclodextrins, crown ethers, and other macrocyclic compounds, provide an entry into a field of research that has been referred to in an earlier volume of this series as "cavity chemistry" [140]. Much of the work in the field of calixarenes has focused on the synthesis of these compounds. Methods are now available for constructing the basic ring system either by single step processes, which are severely limited with respect to substituent group possibilities, or by multi-step processes, which permit greater substituent group flexibility. Methods have been developed for the introduction of functional groups onto the calixarene framework, either at the p-positions of the aromatic rings or the oxygens of the phenolic groups. Thus, a variety of cavity-containing calixarenes with various ring sizes, substituents, and conformations are now available for study. The structures of several solid state complexes formed between small molecules and calixarenes have been verified by x-ray crystallography, and the formation of complexes in solution between calixarenes and amines has been demonstrated. It remains yet to demonstrate the catalytic and enzyme mimic properties of the calixarenes, and work is in progress with compounds 66c and 66h (as hydrolysis catalysts) [141], compound 66f (as metal chelator and oxygen carrier) [48], and various other functionalized calixarenes.

10 Acknowledgements

I am indebted to my coworkers, whose names appear in the references of this review article, for their splendid work in calixarene chemistry. Their conscientiousness and ingenuity are primarily responsible for the progress that has been made in our laboratories. Particular thanks are given to Drs. J. H. Munch and F. J. Ludwig of the Petrolite Corporation, whose long and continuing interest and collaboration in this work is greatly appreciated. Finally, I am indebted to Washington University, the National Science Foundation, and the National Institutes of Health for financial support and to the John Simon Guggenheim Memorial Foundation for a fellowship, during the tenure of which the invitation to write thise review was received and a portion of the manuscript completed.

11 References

1. Cram, D. J., Cram, J. M.: Science 183, 803 (1974)
2. Lehn, J. M.: Pure Applied Chem. 50, 871 (1978)
3. Gutsche, C. D., Muthukrishnan, R.: J. Org. Chem. 43, 4905 (1978)
4. Gutsche, C. D.: Accts. Chem. Res. 16, 161 (1983)
5. Patterson, A. M., Capell, L. T., Walker, D. F.: The Ring Index, 2nd. ed., Amer. Chem. Soc., Washington, D.C., 1960, Ring Index No. 6485
6. Cram, D. J., Steinberg, H.: J. Am. Chem. Soc. 73, 5691 (1951)
7. IPUAC Tentative Rules for Nomenclature of Organic Chemistry, Section. E. Fundamental Stereochemistry; cf. J. Org. Chem. 35, 2849 (1970)
8. Zinke, A., Kretz, R., Leggewie, E., Hössinger, K.: Monatsh. Chem. 83, 1213 (1952)

9. Hayes, B. T., Hunter, R. F.: Chem. Ind., 193 (1956): J. Applied Chem. *8*, 743 (1958)

10. Cornforth, J. W., D'Arcy Hart, P., Nicholls, G. A., Rees, R. J. W., Stock, J. A.: Brit. J. Pharmacol. *10*, 73 (1955)

11. We are indebted to Dr. Kurt L. Loening of Chemical Abstract Services for his guidance in questions concerning the numbering of these compounds, which follows the IUPAC guidelines as noted for compound *2* in Fig. 1

12. Högberg, A. G. S.: PhD Dissertation, Royal Institute of Technology, Stockhom, 1977. We are indebted to Dr. Högberg for making available to us this excellent thesis which suggests the "intraannular" and "extraannular" terminology

13. Beyer, A.: Ber. dtsch. chem. Ges. *5*, 25, 280, 1094 (1872)

14. Numerous accounts of the history of phenol-formaldehyde resins are available, including Baekeland, L. H.: Ind. Eng. Chem. (Industry) *5*, 506 (1913); Carswell, T. S.: Phenoplasts, Interscience Publishers, Inc., New York, 1947; Gillis, J., Oesper, R. E.: J. Chem. Ed. *41*, 224 (1964)

15. Zinke, A., Ziegler, E.: Ber. dtsch. chem. Ges. *74*, 1729 (1941)

16. Zinke, A., Ziegler, E.: ibid. *77B*, 264 (1944)

17. Zinke, A., Zigeuner, G., Hössinger, K., Hoffmann, G.; Monatsh. Chem. *79*, 438 (1948)

18. Zinke, A., Ott, R., Garrana, F. H.: Monatsh. Chem. *89*, 135 (1958)

19. Zinke, A.: J. Appl. Chem. *1*, 257 (1951)

20. Kämmerer, H., Happel, G., Caesar, F.: Makromol. Chem. *162*, 179 (1972)

21. Happel, G., Mathiasch, B., Kämmerer, H.: Makromol. Chem. *176*, 3317 (1975)

22. Munch, J. H.: Makromol. Chem. *178*, 69 (1977)

23. Gutsche, C. D., Dhawan, B., No, K. H., Muthukrishnan, R.: J. Am. Chem. Soc. 103, 3782 (1981)

24. Gutsche, C. D., Muthukrishnan, R., No, K. H.: Tetrahedron Lett. 2213 (1979)

25. Muthukrishnan, R., Gutsche, C. D.: J. Org. Chem. *44*, 3962 (1979)

26. Ninagawa, A., Matsuda, H.: Makromol. Chem. Rapid Comm. *3*, 65 (1982)

27. Nakamoto, Y., Ishida, S.:: ibid. *3*, 705 (1982)

28. Mukoyama, Y., Tanno, T.: Org. Coating & Plastics Chem. *40*, 894 (1979)

29. Andreetti, G. D., Ungaro, R., Pochini, A.: J. Chem. Soc. Chem. Comm. 1005 (1979)

30. Andreetti, G. D., Pochini, A., Ungaro, R.: J. Chem. Soc., Perkin II, 1773 (1983)

31. Ninagawa, A.: private communication

32. Coruzzi, M., Andreetti, G. D., Bocchi, V., Pochini, A., Ungaro, R.: J. Chem. Soc., Perkin II, 1133 (1982)

33. Andreetti, G. D., Ungaro, R., Pochini, A.: Abstr. 2nd Internat. Symp. on Clathrate Compounds and Molecular Inclusion Phenomena, Parma, 1982, p. 88

34. Andreetti, G. D., Ungaro, R., Pochini, A.: J. Chem. Soc. Chem. Comm. 533 (1981)

35. Raschig, F.: Z. für angew. Chem. *25*, 1939 (1912)

36. Baekeland, L. H.: J. Ind. Eng. Chem. (Industry) *5*, 506 (1913)

37. Niederl, J. B., McCoy, J. S.: J. Am. Chem. Soc. *65*, 629 (1943)

38. Koebner, M.: Z. Angew. Chem. *46*, 251 (1933)

39. Finn, S. R., Lewis, G. J.: J. Chem. Soc. Ind. *69*, 132 (1950)

40. Foster, H. M., Hein, D. W.: J. Org. Chem. *26*, 2539 (1961)

41. The *p-tert*-butylcalix[8]arene and *p*-(1,1,3,3-tetramethylbutyl)calix[8]arene have been correlated by removing the alkyl groups from each of these compounds and obtaining the same calix[8]-arene (*8*, R = H)[32]

42. Gutsche, C. D., Kung, T. C., Hsu, M.-L.: Abstr. 11th Midwest Regional Meet. Amer. Chem. Soc., Carbondale, IL, 1975, no. 517

43. Patrick, T. B., Egan, P. A.: J. Org. Chem. *42*, 382 (1977)

44. Patrick and Egan [43] reported molecular weights for their products using either mass spectroscopy or the Rast method (camphor as solvent). However, unless the *absence* of m/e signals higher than that of the anticipated parent ion (i.e. cyclic tetramer) is demonstrated, the mass spectral data can be misleading. The Rast method, requiring the compound to be at least moderately soluble in camphor, would appear to be inapplicable in the case of the cyclic oligomers, all of which are quite insoluble in most organic materials

45. Chen, S. I.; unpublished observations

46. Pagoria, P. F.: unpublished observations

47. Buriks, R. S., Fauke, A. R., Munch, J. H.: U.S. Patent 4,259,464, filed 1976, issued 1981

48. Levine, J. A.: unpublished observations
49. Stewart, D.: unpublished observations
50. Ullman, F., Brittner, K.: Ber. dtsch. chem. Ges. *42*, 2539 (1909)
51. Wohl, A., Mylo, B.: ibid. *45*, 2046 (1912)
52. Hultzsch, K.: ibid. *75B*, 106 (1942)
53. v. Euler, H., Adler, E., Cedwall, J. O., Törngren, O.: Ark. f. Kemi Mineral. Geol. *15A*, No. 11, 1 (1941)
54. Gardner, P. D., Sarrafizadeh, H., Brandon, R. L.: J. Am. Chem. Soc. *81*, 5515 (1959)
55. Wegler, R., Herlinger, H.: Methoden der Organischen Chemie (Houben-Weyl), XIV/2; Makromolecular Stoffe, Thieme Verlag, Stuttgart, 1963, p. 257
56. Evans, D. A., Golob, A. M.: J. Am. Chem. Soc. *97*, 4765 (1975); Thies, R. W., Seitz, E. P.: J. Org. Chem. *43*, 1050 (1978)
57. Adler, E.: Arkiv. f. Kemi Mineral. Geol. *14B*, No. 23, 1 (1941)
58. v. Euler, H., Adler, E., Bergstrom, B.: Arkiv. f. Kemi Geol., *14B*, No. 25, 1 (1941); Hultzsch, K.: Kunststoffe *52*, 19 (1962)
59. Zinke, A., Ziegler, E., Hontschik, I.: Monatsh. Chem. *78*, 317 (1948); Ziegler, E., Hontschik, I.: ibid. *78*, 325 (1948)
60. In addition to water and formaldehyde, other extrusion products, such as benzaldehydes, have also been shown to be formed upon heating mixtures above 150 °C [59]
61. Cairns, T., Eglinton, G.: Nature *196*, 535 (1962)
62. Dhawan, B., Gutsche, C. D.: J. Org. Chem. *48*, 1536 (1983)
63. Izatt, R. M., Lamb, J. D., Hawkins, R. T., Brown, P. R., Izatt, S. R., Christensen, J. J.: J. Am. Chem. Soc. *105*, 1782 (1983)
64. Burke, W. J., Craven, W. E., Rosenthal, A., Ruetman, S. H., Stephens, C. W., Weatherbee, C.: J. Polymer Sci. *20*, 75 (1956)
65. Dhawan, B.: unpublished observations
66. Michael, A.: Amer. Chem. J. *5*, 338 (1883)
67. Niederl, J. B., Vogel, H. J.: J. Am. Chem. Soc. *62*, 2512 (1940)
68. Erdtman, H., Haglid, F., Ryhage, R.: Acta Chem. Scand. *18*, 1249 (1964)
69. Erdtman, H., Högberg, S., Abrahamsson, S., Nilsson, B.: Tetrahedron Lett. 1679 (1968); Nilsson, B.: Acta. Chem. Scand. *22*, 732 (1968)
70. Palmer, K. J., Wong, R. Y., Jurd, L., Stevens, K.: Acta Crystallogr. *B32*, 847 (1976)
71. Högberg, A. G. S.: J. Org. Chem. *45*, 4498 (1980)
72. Högberg, A. G. S.: J. Am. Chem. Soc. *102*, 6046 (1980)
73. Additional evidence for the acid-lability of the cyclic tetramers (*13*) is seen in their tendency to undergo nitrodealkylation. For example, treatment of *13* (R = Phenyl, R′ = Methyl) with concentrated nitric acid yields 4,6-dinitroresorcinol dimethyl ether and 3,5-dinitrobenzoic acid [12]
74. Ballard, J. L., Kay, W. B., Kropa, E. L.: J. Paint Technol. *38*, 251 (1966)
75. Bottino, F., Montaudo, G., Maravigna, P.: Ann. Chim. (Rome) *57*, 972 (1967)
76. Brown, W. H., Hutchinson, B. J.: Can. J. Chem. *56*, 617 (1978) and preceding papers back to Brown, W. H., Sawatzky, H.: ibid. *34*, 1147 (1956)
77. Chastrette, M., Chastrette, F.: J. Chem. Soc. Chem. Comm. 534 (1973)
78. Healy, M. de S., Rest, A. J.: J. Chem. Soc. Chem. Comm. 140 (1981)
79. Högberg, A. G. S., Weber, M.: Acta. Chem. Scand. B *37*, 55 (1983)
80. Ahmed, M., Meth-Cohn, O.: Tetrahedron Lett. 1493 (1969); J. Chem. Soc. C 2104 (1971)
81. Rothemund, P., Gage, C. L.: J. Am. Chem. Soc. *77*, 3340 (1955)
82. Collman, J. P., Gagne, R., Reed, C., Halbert, T. R., Lang, G., Robinson, W. T.: J. Am. Chem. Soc. *97*, 1427 (1975)
83. Robinson, G. M.: J. Chem. Soc. *107*, 267 (1915); Collet, A., Gabard, J.: J. Org. Chem. *45*, 5400 (1980) and references therein
84. Freeman, W. A., Mock, W. L., Shih, N.-Y.: J. Am. Chem. Soc. *103*, 7367 (1981)
85. Meth-Cohn, O.: Tetrahedron Lett. 91 (1973)
86. Sawa, N., Nomoto, T., Aida, K., Suzuki, T.: J. Synth. Org. Chem. *33*, 1007 (1975)
87. Bergman, J., Högberg, S., Lindström, J.-O.: Tetrahedron *26*, 3347 (1970); Raverty, W. D., Thomson, R. H., King, T. J.: J. Chem. Soc., Perkin I, 1204 (1977)

88. Hunter, R. F., Turner, C.: Chem. Ind., 72 (1957) also reported the synthesis of a cyclic octamer containing six *m*-bridges and two *p*-bridges, the remaining four *m*-positions being "blocked" with methyl groups
89. Kämmerer, H., Happel, G.: Makromol. Chem. *179*, 1199 (1978)
90. Kämmerer, H., Happel, G., Böhmer, V., Rathay, D.: Monatsh. Chem. *109*, 767 (1978)
91. Kämmerer, H., Happel, G.: Makromol. Chem. *181*, 2049 (1980)
92. Kämmerer, H., Happel, G.: ibid. *112*, 759 (1981)
93. Kämmerer, H., Happel, G., Mathiasch, B.: ibid. *182*, 1685 (1981)
94. Kämmerer, H., Happel, G.: in: Weyerhaeuser Science Symposium on Phenolic Resins, Tacoma, Washington, 1979, p. 143
95. Gutsche, C. D.; No, K. H.: J. Org. Chem. 47, *2708* (1982)
96. Böhmer, V., Chhim, P., Kämmerer, H.: Makromol. Chem. *180*, 2503 (1979)
97. Moshfegh, A. A., Badri, R., Hojjatie, M., Kaviani, M., Naderi, B., Nazmi, A. H., Ramezanian, M., Roospeikar, B., Hakimelahi, G. H.: Helv. Chim. Acta. *65*, 1221 (1982)
98. Moshfegh, A. A., Mazandarani, B., Nahid, A., Hakimelahi, G. H.: Helv. Chim. Acta *65*, 1229 (1982)
99. Moshfegh, A. A., Baladi, E., Radnia, L., Afsanch, S. L. R., Hosseini, A. S., Tofigh, S., Hakimelahi, G. H.: Helv. Chim. Acta 65, 1264 (1982)
100. No, K. H., Gutsche, C. D.: J. Org. Chem. 47, 2713 (1982)
101. Hultzsch, K.: Kunststoffe *52*, 19 (1962)
102. von Euler, H., Adler, E., Bergstrom, B.: Ark. Kemi. Mineral Geol. *14B*, No. 30, 1 (1941)
103. Kämmerer, H., Dahm, M.: KunstPlast (Solothurn Switz.) *6*, 20 (1959)
104. Still, W. C., Kahn, M., Mitra, A.: J. Org. Chem. *43*, 2923 (1978)
105. Kricheldorf, H. R., Kaschig, J.: Liebigs Ann. Chem. 882 (1976)
106. Bocchi, V., Foina, D., Pochini, A., Ungaro, R.: Tetrahedron *38*, 373 (1982)
107. Gutsche, C. D., Dhawan, B., Levine, J. A., No, K. H., Bauer, L. J.: Tetrahedron *39*, 409 (1983)
108. Cornforth, J. W., Morgan, E. D., Potts, K. T., Rees, R. J. W.: Tetrahedron *29*, 1659 (1973)
109. Klebe, J. F., Finkbeiner, H., White, D. M.: J. Am. Chem. Soc. *88*, 3390 (1966)
110. (a) Gutsche, A. E.: unpublished observations; (b) Ludwig, F. J.: unpublished observations
111. We are indebted to Mr. Christopher Cramer for devising a computer program for comparing the "fingerprint" region data for a variety of calixarenes
112. Megson, N. R. L.: Oesterr. Chem. Z. *54*, 317 (1953)
113. Ott, R., Zinke, A.: Oesterr. Chem. Z. *55*, 156 (1954)
114. Gutsche, C. D., Bauer, L. J.: Tetrahedron Lett. 4763 (1981)
115. Saenger, W., Betzel, C., Brown, G. M.: Angew. Chem. Int. Ed. Engl. *22*, 883 (1983); Saenger, W., Betzel, C., Hingerty, B., Brown, G. M.: Nature 296, 581 (1982); Saenger, W.: ibid, 279, 343 (1979)
116. Bauer, L. J.: unpublished observations
117. Helgeson, R. C., Mazaleyrat, J.-P., Cram, D. J.: J. Am. Chem. Soc. *103*, 3929 (1981)
118. Moran, J. R., Karbach, S., Cram, D. J.: J. Am. Chem. Soc. *104*, 5826 (1982); Cram, D. J.: Science *219*, 1177 (1983)
119. Rizzoli, C., Andreetti, G. D., Ungaro, R., Pochini, A.: J. Molec. Structure *82*, 133 (1982)
120. The formation of two "partial cone" compounds is the result of the reduced symmetry of *41* (R = *tert*-Bu) which gives rise to six conformers; viz. one "cone", two "partial cone", two "1,2-alternate", and one "1,3-alternate"
121. Dradi, E., Pochini, A., Ungaro, R.: Abstr. 2nd Internat. Symposium on Clathrate Compounds and Molecular Inclusion Phenomena, Parma/Italy, 1982, p. 84
122. Bailey, D. W.: unpublished observations
123. See Tashiro, M.: Synthesis, 921 (1979) for general references and see Böhmer, V., Rathay, D., Kämmerer, H.: Org. Prep. Proc. Int. *10*, 113 (1978) for a closely analogous example
124. Compound *61*, which melts at 248–250 °C, can be demethylated with BBr$_3$ to *p*-bromocalix[4]-arene, which melts above 430 °C. The structure of *p*-bromocalix[4]arene has been established by the criteria discussed in this article except x-ray crystallography. *p*-Chlorocalix[4]arene, claimed to be the product of the synthesis illustrated in Scheme 4 [97], is reported to melt at 239–242 °C. The enormous difference in melting points between the *p*-bromo and *p*-chloro compounds casts considerable doubt on the structure of the *p*-chloro compound.
125. Lin, L. G.: unpublished observations

126. Gutsche, C. D., Levine, J. A.: J. Am. Chem. Soc. *104*, 2652 (1982)
127. Hagan, M.: Clathrate Inclusion Compounds, Reinhold, New York, 1962
128. Andreetti, G. D.: Internat. Symp. on Clathrate Compounds and Molecular Inclusion Phenomena, Jackranka/Poland, 1980; quoted in ref. [32]
129. Uzawa, J., Zushi, S., Kodama, Y., Fukuda, Y., Nishihata, K., Umemura, A., Nishio, M., Hirota, M.: Bull. Chem. Soc. Japan *53*, 3623 (1980)
130. MacNicol, D. D., McKendrick, J. J., Wilson, D. R.: Chem. Soc. Rev. *7*, 65 (1978)
131. Hilgenfeld, R., Saenger, W.: Angew. Chem. Suppl. 1690 (1982)
132. Andreetti, G. D., Mangia, A., Pochini, A., Ungaro, R.: Abstr. 2nd Internat. Symposium on Clathrate Compounds and Molecular Inclusion Phenomena, Parma/Italy, 1982, p. 42
133. Smolková-Keulemansová, E., Feltl, L.: ibid., p. 45
134. Schubert, H., Agatha, G.: Dermatosen im Beruf und Umwelt *27*, 49 (1979)
135. We are indebted to Professor Barry Commoner for making available the testing facilities of the Center for the Study of Biological Systems, Washington University, St Louis, Mo.
136. Beaver, D. J., Shumard, R. S., Stoffel, P. J.: J. Am. Chem. Soc. 75, 5579 (1953)
137. Hakimelahi, G. H., Moshfegh, A. A.: Helv. Chim. Acta *64*, 599 (1981)
138. A representative but not inclusive list of references includes: (a) D'Arcy Hart, P., Payne, S. N.: Brit. J. Pharmacol. *43*, 190 (1971); (b) D'Arcy Hart, P., Gordon, A. H.: Nature *222*, 672 (1969); (c) Niffenegger, J., Youmans, G. P.: Brit. J. Exptl. Pathol. *41*, 403 (1960); (d) Fulton, J. D.: Nature 187, 1129 (1960); (e) Depamphilis, M. L.: J. Bact., *105*, 1184 (1971); (f) Allwood, M. C.: Microbios *7*, 209 (1973)
139. A representative but non-inclusive list of references includes: (a) Carter, R. L., Birbeck, M. S. C., Stock, J. A.: Int. J. Cancer *7*, 32 (1971); (b) Miller, G. W., Janicki, B. W.: Cancer Chem. *52*, 243 (1968); (c) Franchi, G., Morasca, L., Reyers-Delgi-Innocenti, I., Garattini, S.: Europ. J. Cancer *7*, 535 (1971); (d) Rosso, R., Donelli, M. G., Franchi, G., Garattini, S.: Cancer Chemotherapy Rpt. *54*, 79 (1970)
140. Vögtle, F., Sieger, H., Müller, W. M.: Host Guest Complex Chemistry I, in Topics in Current Chemistry, Springer-Verlag, Berlin, 1981, p. 107
141. Sujeeth, P. K.: unpublished observations